T0399763

Sub-Micron Semiconductor Devices

Sub-Micron Semiconductor Devices

Design and Applications

Edited by

Ashish Raman, Deep Shekhar, and Naveen Kumar

CRC Press
Taylor & Francis Group
Boca Raton London New York

CRC Press is an imprint of the
Taylor & Francis Group, an **informa** business

First edition published 2022
by CRC Press
6000 Broken Sound Parkway NW, Suite 300, Boca Raton, FL 33487-2742

and by CRC Press
2 Park Square, Milton Park, Abingdon, Oxon, OX14 4RN

Library of Congress Cataloging-in-Publication Data

Names: Raman, Ashish, editor. | Shekhar, Deep, editor. | Kumar, Naveen, 1993- editor.
Title: Sub-micron semiconductor devices : design and applications / edited by Ashish Raman, Deep Shekhar, Naveen Kumar.
Description: First edition | Boca Raton : CRC Press, 2022. | Includes bibliographical references and index.
Identifiers: LCCN 2021054051 (print) | LCCN 2021054052 (ebook) | ISBN 9780367648091 (hardback) | ISBN 9780367648107 (paperback) | ISBN 9781003126393 (ebook)
Subjects: LCSH: Semiconductors. | Nanoelectronics. | Microelectronics.
Classification: LCC TK7871.85 .S83 2022 (print) | LCC TK7871.85 (ebook) | DDC 621.3815/2--dc23/eng/20220124
LC record available at https://lccn.loc.gov/2021054051
LC ebook record available at https://lccn.loc.gov/2021054052

ISBN: 978-0-367-64809-1 (hbk)
ISBN: 978-0-367-64810-7 (pbk)
ISBN: 978-1-003-12639-3 (ebk)

DOI: 10.1201/9781003126393

Typeset in Times LT Std
by KnowledgeWorks Global Ltd.

Contents

Preface

Modern society is very dependent on electronic gadgets, and there is a thirst-quenching need for smaller and more advanced electronics. As conventional devices try to bridge the problems that arise due to smaller dimensions, other contenders are trying to keep alive the journey of electronic devices toward the molecular regime. Thus, it is necessary to keep track of ongoing research on nano-marvels suitable for various applications. *Sub-Micron Semiconductor Devices* conceptualizes the novel devices required for boosting today's sophisticated electronic systems.

Miniaturization is the key behind the next-generation devices, which can help humanity advance to a more prosperous future. For a few decades researchers have tried to scale down semiconductor devices; however, this needs extremely advanced fabrication facilities, which saturate the pace toward sub-micron technologies. Thus, new optimization techniques, novel materials, advanced devices, and unorthodox device architectures are explored that can help achieve the feat of cost-effective, efficient, and modern electronics. Negative capacitance, feedback field-effect transistors (FETs), label-free biosensors, recessed-gate high electron mobility transfer (HEMT), terahertz semiconductor devices, multi-quantum well solar cells, organic light-emitting diodes (LEDs), and trench/U-shaped/nanotube-based architectures are a few of the examples that have the potential to replace conventional devices. Contemporary applications of semiconductor devices require dedicated focus on optimizing the device with the help of existing fabrication technologies.

To follow Moore's law, semiconductor devices are scaled down without compromising performance. Semiconductor devices are supposed to have reduced dimensions and work at lower operating biases, but the problem arises during the manufacturing of the devices. Thus, it is necessary to opt for a solution that can help to continue the path of performance improvement. Steady performance enhancement using optimization techniques can support the time required for advancements in fabrication technologies. This publication elucidates the novel semiconductor devices, issues with conventional devices, and optimization techniques and solutions for performance enhancement. With the presence of such a vast amount of data regarding semiconductor devices, it is hard for researchers to go through most of the recent advancements and clearly understand them. The motive behind this book is to comprehensibly present the material related to the recent advancements in the field of semiconductor devices and allow the reader to interpret the possible concepts behind the content. The study of novel semiconductor devices may help to unravel the mystery behind the problems during the fabrication of molecular devices.

The recent advancements in junctionless and dopingless devices show enhanced device performance with higher ON current, better subthreshold slope, and lower leakage current. The inclusion of gate-all-around architecture provides higher electrostatic control to spike the controlled device operation. Similarly, the inclusion of dopingless techniques can improve the performance of semiconductor devices used for various other applications. *Sub-Micron Semiconductor Devices* covers the important parts that get left out about novel techniques and device architectures.

This book describes detailed demonstrations about sub-micron semiconductor devices, design issues, and mechanisms for device design and their applications. Chapter 1, "Fundamental Phenomena in Nanoscale Semiconductor Devices," introduces the reader to nanoscale, nanodevices, various short channel effects, quantum confinement effects, quantum confined structures, etc. These effects need to be understood before going deep into the designing of nanodevices. Chapter 2, "Recent Advances in Growth and Stability of Phosphorene: Prospects for High-Performance Devices," deals with the structure and properties of phosphorene, challenges related to air instability and mass production of phosphorene, deposition of phosphorene by various strategies, applications of stable phosphorene, etc. The information in this chapter will accelerate further development of high-performance phosphorene-based electronics and optoelectronics devices. Chapter 3, "Study of Transition Metal Dichalcogenides in Junctionless Transistors and Effect of Variation in Dielectric

Oxide," proposes the junctionless transistor based on different transition metal dichalcogenide materials. Transition metal dichalcogenide materials, device properties (bandgap, potential, electric field), quantum effects (density of states, transmission probability), I_{DS}-V_{GS}, and linearity properties are demonstrated here. Chapter 4, "GNRFET-Based Ternary Repeaters: Prospects and Potential Implementation for Efficient GNR Interconnects," presents prospects of implementing GNRFET-stemmed ternary buffers as repeaters and investigates their potential implementation on high-performance on-chip graphene nanoribbon (GNR) interconnecting wire networks. Chapter 5, "An Effective Study on Particulate Matter (PM) Removal Using Graphene Filter," demonstrates how the graphene filter is used and which particulate matters can be removed. The commercial filters are not appropriate for electrostatic precipitation because they have electrical conductivity.

Further, Chapter 6, "Recent Trends in Fabrication of Graphene-Based Devices for Detection of Heavy Metal Ions in Water," discusses the working principles of conventional techniques for heavy metal ion detection and its limitations. Various synthesis techniques of graphene are described, with special focus on chemical exfoliation of graphite and fabrication of functionalized graphene nano-composites. In Chapter 7, "Vertical Tunnel FET Having Dual MOSCAP Geometry," the geometry of vertical tunnel FET (V-TFET) with dual metal oxide semiconductor (MOS) capacitors is demonstrated, designed, analyzed, and simulated using the Synopsys TCAD tool. The geometry and concepts of conventional TFETs are also presented. Chapter 8, "Leakage Current and Capacitance Reduction in CMOS Technology," presents various ways to minimize the leakage current and parasitic capacitances. Leakage current and capacitances affect the working of complementary MOS (CMOS) integrated circuits. In Chapter 9, "Design of Gate-All-Around TFET with Gate-on-Source for Enhanced Analog Performance," gate-all-around (GAA)-based extended-gate nanowire TFET is designed and termed as nanowire TFET gate-on-source. A comparative study among nanowire TFET with gate-on-drain, nanowire TFET, and nanowire TFET with gate-on-source are also demonstrated based on electrical and analog performances. Chapter 10, "Solving Schrodinger's Equation for Low-Dimensional Nanostructures for Understanding Quantum Confinement Effects," provides, learners with the fundamental understanding of quantum confinement effects. The wave functions for quantum wells, quantum wires, and quantum dots are analyzed analytically using Cartesian coordinates, cylindrical coordinates, and spherical coordinates, respectively.

Chapter 11, "Simulation of Reconfigurable FET circuits Using Sentaurus TCAD Tool," introduces the basic tool flow of the Sentaurus TCAD tool and explores the device characteristics' simulations and circuit-level simulations using the single-gate reconfigurable FET (SG-RFET) device. In Chapter 12, "NEGF Method for Design and Simulation Analysis of Nanoscale MOS Devices," provides a comprehensive understanding of the accurate carrier transport model that can be employed with advanced nanoscale devices where the classical theory fails. It is also capable of setting a strong background for readers and researchers who are actively working in the quantum and ballistic transport regime. Chapter 13, "Performance Investigation of a Novel Si/Ge Heterojunction Asymmetric Double-Gate DLTFET for Low-Power Analog/RF and IoT Applications," starts by discussing the needs and fundamentals of TFET and dopingless TFET (DLTFET). Further, it is followed by a description of published reports in the literature about how to overcome the limitations of conventional DLTFETs. After that, the device heterojunction asymmetric double-gate DLTFET (HJ-ADG-DLTFET) is proposed and discussed in the context of device design, simulation setup, and fabrication feasibility of the device, and a comparative analysis is reported. Chapter 14, "Synthesis of Graphene Nanocomposites Toward the Enhancement of Energy Storage Performance for Supercapacitors," summarizes the modern fundamental synthesis methods for graphene nanosheet fabrication and its excellent energy research properties. Chapter 15, "Design and Analysis of Dopingless Charge Plasma-Based Ring Architecture of Tunnel Field-Effect Transistor for Low-Power Application," proposes a dopingless charge plasma-based ring architecture of TFET (CP-Ring-TFET). Further, brief works regarding TFET followed by performance analysis of CP-Ring-TFET based on various device parameters are analyzed. In Chapter 16, "Hybrid Intelligent Technique-Based Doping Profile Optimization in a Double-Gate Hetero-Dielectric TFET," discusses maximizing the current ratio in DLTFET. To achieve the maximum current hybrid, an optimization technique is used that combines the evolutionary algorithm and swarm intelligence-based techniques.

Chapter 17, "Graphene Nanoribbon Devices: Advances in Fabrication and Applications," presents the recent manufacturing processes aimed at overcoming issues such as difficulties in producing defect-free, atomically narrow, reproducible, and transferrable GNRs and enabling high-density GNR-based electronic devices. It also presents the latest applications along with performance characteristics. Chapter 18, "Design and Analysis of Various Neural Preamplifier Circuits," implements the voltage doubling technique and flip voltage follower (FVF) in the standard circuit designs. The circuits have been designed with the perspective of being used as preamplifiers to record the incoming neural signals. In Chapter 19, "Design and Analysis of Transition Metal Dichalcogenide-Based Feedback Transistor," a transition metal dichalcogenide material-based feedback FET is proposed and analyzed. Molybdenum ditelluride ($MoTe_2$) is used as channel material as it has the lowest bandgap among the transition metal dichalcogenide materials, which further aids in high-speed applications. Chapter 20, "Reduced Graphene-Metal Phthalocyanine-Based Nanohybrids for Gas-Sensing Applications," emphasizes rGOs and their nanohybrids for gas-sensing applications. Different synthesis methods to prepare rGOs and their nanohybrids, their gas-sensing applications, and sensing mechanisms have been discussed. This will help researchers to understand the potential of these novel materials. Further, Chapter 21, "Phosphorene Multigate Field-Effect Transistors for High-Frequency Applications," demonstrates the characteristics of the phosphorene layer and phosphorene FET fabrication procedures. It also explores the two-dimensional (2D) phosphorene-based multigate MOSFETs for future analog/radiofrequency (RF) applications. In Chapter 22, "Analytical Modeling of Reconfigurable Transistors," explores the analytical modeling of a reconfigurable FET (RFET) and a reconfigurable TFET (RTFET) as well as the single-gated RFET and double-gated RTFET. Further, a novel technique to solve the 2D Poisson's equation considering exponentially varying charge density in the channel of the device is proposed. Chapter 23, "Flexi-Grid Technology: A Necessity for Spectral Resource Utilization," concludes this book by dealing with flexi-grid technology, i.e., superchannel transmission for the purpose of efficient bandwidth utilization to satiate extreme bandwidth requirement by the exponentially grown data traffic. This flexi-grid technology is the backbone of the next generation to support terabit transmission capacity along with providing excellent spectral efficiency.

This book is a unique combination of sub-micron semiconductor devices and their applications, covering recent development in the field of electronics such as semiconductor devices, materials, nanoelectronics, spintronics, and other research areas for efficient design of post CMOS-based devices, circuits, and systems.

Editors

Ashish Raman received a B.E. degree in electronics and communication engineering in 2003 and a M.Tech degree in microelectronics and VLSI design from the Shri Govindram Seksaria Institute of Technology and Science, Indore, India, in 2005; and PhD degree from the Dr. B. R. Ambedkar National Institute of Technology Jalandhar, India, where he works as an assistant professor. Dr. Raman's main areas of interest are electronics devices and circuits, device modeling, device design and simulation, MEMS, high-power devices, etc. He works as a principal investigator and member of various projects funded by the Science and Engineering Research Board (SERB), Ministry of Electronics, IT (MeitY), FIST, ISRO, and many more. He has contributed research articles/papers in SCI, Scopus, and other reputed journals like *IEEE Transactions of Electron Devices*, *IEEE Transactions on Nanotechnology*, Elsevier and Springer journals, and at international conferences. Dr. Raman is a member of the IEEE Electron Devices Society, the IEEE Solid-State Circuits Society, and the Institution of Engineers Society, India.

Deep Shekhar completed his B.Tech in electronics and communication engineering from Uttar Pradesh Technical University and his M.Tech in VLSI from the National Institute of Technology, Jalandhar, India, in 2012 and 2015, respectively. He has been a member of the faculty in the Department of Electronics and Communication Engineering at the National Institute of Technology, Jalandhar, since 2016. Mr. Shekhar is also pursuing a PhD from the Dr. B. R. Ambedkar National Institute of Technology, Jalandhar, India. His expertise is in solid-state devices, analog complementary metal oxide semiconductor (CMOS) integrated circuits, nanoscale device design and simulation, etc.

Naveen Kumar was born in India in 1993. He received a B.E. degree from the Rustamji Institute of Technology, Gwalior, India; M.Tech. degree in VLSI design from the Centre for Development of Advanced Computing, Mohali, Punjab; and PhD from the Dr. B. R. Ambedkar National Institute of Technology, Jalandhar, India. Dr. Kumar is currently working as a research associate at the University of Glasgow, Scotland. He has authored/co-authored more than 35 research articles/papers in reputed international journals and conference proceedings. His main areas of research interest include semiconductor device physics, MEMS/NEMS, and spintronics.

Contributors

Arash Ahmadivand
Metamaterial Technologies, Inc.
Pleasanton, USA

Aryan
J.C. Bose University of Science and
    Technology, YMCA
Faridabad, India

Krishna Lal Baishnab
Department of ECE
National Institute of Technology Silchar
Silchar, India

Rikmantra Basu
Department of Electronics and Communication
    Engineering
National Institute of Technology Delhi
Delhi, India

Katyayani Bhardwaj
Abdul Kalam Technical University
Lucknow, India

Arkaprava Bhattacharyya
Device Modeling Lab
SASTRA Deemed University
Thanjavur, India

Tarun Kanti Bhattacharyya
Department of Electronics and Electrical
    Communication Engineering
IIT Kharagpur
Kharagpur, India

Brinda Bhowmick
Department of ECE
National Institute of Technology Silchar
Silchar, India

Sagarika Choudhury
National Institute of Technology Silchar
Silchar, India

Swagata Devi
Department of ECE
National Institute of Technology Silchar
Silchar, India

K. J. Dhanaraj
Assistant Professor
Department of Electronics
    and Communication
NIT Calicut
Kerala, India

Vivek Garg
Department of Electronics
    and Communication Engineering
National Institute of Technology Surat
Surat, India

Dipak Kumar Goswami
Department of Physics
IIT Kharagpur
Kharagpur, India

Koushik Guha
Department of ECE
National Institute of Technology Silchar
Silchar, India

Ashok Kumar Gupta
Dr. B R Ambedkar National Institute of
    Technology
Jalandhar, India

Maneesha Gupta
Netaji Subhas University of Technology
Delhi, India

Nezhueyotl Izquierdo
Department of Electrical and Computer
    Engineering
University of Minnesota
Minneapolis, Minnesota

Andres Jaramillo-Botero
Division of Chemistry and Chemical
    Engineering
California Institute of Technology
Pasadena, California;
Omicas Program
Pontificia Universidad Javeriana-Cali
Santiago de Cali, Columbia

Remya Jayachandran
Department of Electronics
and Communication
NIE Mysuru
Karnataka, India

Rajib Kar
Department of Electronics
and Communication Engineering
National Institute of Technology
Durgapur
Durgapur, India

Baljit Kaur
Department of Electronics
and Communication Engineering
National Institute of Technology Delhi
Delhi, India

Kavita Khare
Electronics and Communication
Engineering
Maulana Azad National Institute
of Technology
Bhopal, India

Afreen Khursheed
Electronics and Communication Engineering
Department
Indian Institute of Information Technology
Bhopal, India

Suja K. J.
Assistant Professor
Department of Electronics
and Communication Engineering
National Institute of Technology Calicut
Kozhikode, India

Rama S. Komaragiri
Professor and Head
Department of Electronics
and Communication Engineering
School of Engineering and Applied Science
Bennett University
Greater Noida, India

Amit Kumar
Bhaskaracharya College of Applied Sciences
University of Delhi
New Delhi, India

Amitesh Kumar
Department of Electrical Engineering
National Institute of Technology Patna
Patna, India

Naveen Kumar
Dr. B R Ambedkar National Institute of
Technology
Jalandhar, India

Prateek Kumar
Faculty of Technology
University of Delhi
New Delhi, India

Aman Mahajan
Department of Physics
Guru Nanak Dev University
Amritsar, India

Santanab Majumder
School of Nanoscience and Technology
IIT Kharagpur
Kharagpur, India

Durbadal Mandal
Department of Electronics and Communication
Engineering
National Institute of Technology Durgapur
Durgapur, India

Juan M. Marmolejo-Tejada
Omicas Program
Pontificia Universidad Javeriana-Cali
Santiago de Cali, Colombia

Monojit Mondal
School of Nanoscience and Technology
IIT Kharagpur
Kharagpur, India

Sushil Kumar Pandey
Department of Electronics and Communication
Engineering
National Institute of Technology Karnataka
Mangalore, India

Adhithan Pon
Device Modeling Lab
SASTRA Deemed University
Thanjavur, India

Puspa Devi Pukhrambam
Department of ECE
NIT Silchar
Silchar, India

Ranjith R.
Assistant Professor
Department of Electronics
    and Communication Engineering
School of Engineering and Technology
CHRIST (Deemed to be University)
Bangalore, India

Ashish Raman
Dr. B R Ambedkar National Institute of
    Technology
Jalandhar, India

R. Ramesh
Device Modeling Lab
SASTRA Deemed University
Thanjavur, India

Zeinab Ramezani
Department of Electrical and Computer
    Engineering
College of Engineering
University of Miami, USA

Avik Sett
Department of Electronics and Electrical
    Communication Engineering
IIT Kharagpur
Kharagpur, India

Divya Sharma
Department of Electronics
    & Communication Engineering
Thapar Institute of Engineering and Technology
Patiala, India

Suruchi Sharma
Department of Electronics
    and Communication Engineering
National Institute of Technology Delhi
Delhi, India

Deep Shekhar
Dr. B R Ambedkar National Institute
    of Technology
Jalandhar, India

Jeetendra Singh
NIT Sikkim
Sikkim, India

Kunwar Singh
Netaji Subhas University
    of Technology
Delhi, India

Laxmi Singh
Department of Electronics
    & Communication
Rabindranath Tagore University (AISECT)
Bhopal, India

Navaneet Kumar Singh
Department of Electronics
    and Communication Engineering
National Institute of Technology
    Durgapur
Durgapur, India

Shivam Singh
Department of Electronics
    & Communication Engineering
Motilal Nehru National Institute
    of Technology Allahabad
Prayagraj, India

Manreet Kaur Sohal
Department of Physics
Guru Nanak Dev University
Amritsar, India

Ajay Somkuwar
Department of Electronics
    & Communication
MANIT Bhopal
Bhopal, India

Sofyan A. Taya
Department of Physics
Islamic University of Gaza
Gaza Strip, Palestine

Anurag Upadhyay
Department of Applied Science
    & Humanities
Rajkiya Engineering College
Azamgarh, India

Jaime Velasco-Medina
Bionanoelectronics Group
School of Electrical and Electronics
 Engineering
Universidad del Valle
Cali, Colombia

Chhaya Verma
NIT Sikkim
Sikkim, India

Vandana Devi Wangkheirakpam
Department of ECE
NIT Silchar
Silchar, India

Ravindra Kumar Yadav
Abdul Kalam Technical University
Lucknow, India

1 Fundamental Phenomena in Nanoscale Semiconductor Devices

Zeinab Ramezani and Arash Ahmadivand

CONTENTS

1.1 INTRODUCTION

Nanoscience is a fascinating and promising context used to keep continuing technological tendency toward device miniaturization, as well as the characterization, manipulation, and control of well-engineered architectures to operate at scales down to nanometer dimensions. It seeks to explore, describe, and manipulate the individual features to create new capabilities with immense applications in a wide range of science and engineering contexts. This topic offers great opportunities and continues to attract copious interest due to its potential advantages in diverse industries and applied sectors. Therefore, there will be a rapid evolution that will be expanded quickly in the coming years. According to Moore's law (1965), the number of devices placed on one chip is almost doubled every 2 years [1]. Hence, one of the indices of the semiconductor industry is that the circuit dimensions scale is continued in each new generation of technology [2, 3]. This enables important advantages including, but not limited to, lower power consumption and increased speed, as well as improved functionality of the tailored devices and circuits [4].

To further improve the speed and responsivity of nanodevices on the chip area, the transistors, as a common electrical device, can be manufactured in much smaller sizes. Consequently,

DOI: 10.1201/9781003126393-1

it would be possible to design tightly integrated circuits (ICs) with superior performance. By increasing the density of transistors according to Moore's law and the physical reduction of the device dimensions, subsequently, the speed of circuits and their density can be improved [5]. According to the International Terrestrial Reference System (ITRS) survey, the silicon (Si) transistors have been fabricated recently in which the channel length was reduced to below 7 nm [6]. However, by reducing the critical element dimensions of devices down to atomic sizes, quantum tunneling and some other quantum effects appear and conventional understanding of device operations collapses. Researchers are therefore pursuing alternative mechanisms to address these limitations in the fundamental aspects of physics [7–10].

In practice, shrinking the size of semiconductor platforms leads to substantial variations in the principal operation of the entire system. This stems from the quantum effect, that is, physics governed the electron's motion and its interaction in atoms and deals with some physical phenomena at nanoscale. The description of quantum effects in quantum confined structures, i.e., quantum dots (QDs), quantum wells, and nanowires, is important to perceive their spectacular optical and electronic properties.

Theoretically, the quantum confinement effects overcome electrical and optical properties of any system. When the quantum confinement is robust, the discrete nature of the state space becomes dominant. Considering rapid progress in nanotechnology and nanoscience, the precise analysis of the concept of discrete states of energy became a crucial requirement. Moreover, the charge transport at specific interfaces can be affected by defects on the subsurface. In such limits, Fermi level (FL) pinning, as the strongest component at defect locations, must be considered in the development of semiconductor nanodevices.

1.2 NANODEVICES

Nanotechnology is the development of materials and structures in atomic and molecular levels by considering matter [11], which implies that it utilizes very tiny particles of the materials to make the new ones on a large scale [12]. An important concern in nanoscale research is how to translate and implement changes in a scientific paradigm to new technological processes. All nanostructures in size ranging between 1 and 100 nm are typically found as the tailored precursors in nanostructured materials and corresponding nanodevices such as transistors, diodes, amplifiers, rectifiers, etc. For example, transistors that possess channel lengths smaller than 100 nm have been extensively employed in the production of microprocessors and memory modules. The development of such apparatuses can be used in diverse areas such as chemicals and advanced materials, information and communication technologies, energy, and pharmacological applications. In this section, we are mainly going to focus on various types of nanotransistors including metal oxide semiconductor (MOS) transistors, bipolar junction transistors (BJTs), single-electron transistors, carbon nanotube (CNT) transistors, Memristor, phase-change memory (PCM), and sensors, which have been employed in computation and information storage and transmission purposes.

1.2.1 MOS TRANSISTORS

In 1948, J. Bardeen and W. Brattain began a new revolution in electronics by inventing a transistor [13]. Since the availability of functional materials, i.e., single crystal silicon, the excellent potential of miniaturized transistors for high-speed operations has been widely studied and exploited practically. Among them, for example, are BJTs, which are well known for being the first generation of solid-state transistors that served in several analog and digital circuits, and sensor revolution [14, 15]. A BJT is constructed by connecting 2-PN junction diodes, which have a back-to-back arrangement with three terminals. The central and critical specification of the BJT is the current gain [16]. In nanoscale limit, this tool can be exploited only during the self-adjusting of the bipolar procedure. The transit frequencies of the BJT IC technology have been seen ranging between 40 and 120 GHz for transistors fabricated with pure Si and Si-germanium (SiGe) switching elements owing to the self-adjustment or self-assembly of the doping relative to each other. The BJT basic construction consists of three terminals labeled the base, emitter, and collector with 2-PN junctions. From the fabrication viewpoint, in collector and base, instead of using the implantation or diffusion process, very thin epitaxial films are used at different doping levels, and the diffusion process from polysilicon layer into the crystal is utilized just for the emitter [17]. Although BJT devices have been used in high frequencies, this technology requires a technological innovation and, alternatively, the heterojunction bipolar transistors have been proposed as alternative platforms for this purpose.

The other technology is the MOS transistor, which is a result of tremendous progressive downsizing of electronic structures. The MOS transistors are considered as the heart of the semiconductor industry because they can be simply assembled in ICs. However, scaling down of these structures has important drawbacks known as short channel effects (SCEs), which will be explained in the following section.

Generally, there are two rules for further shrinking of transistors. The first rule involves full scaling based on a constant field, and the second rule is to reduce the supply voltage. Shrinkage based on a constant electric field (E) implies that all dimensions of the device reduce by a specific factor of α, and the supply bias and other voltages are too small to keep the electric field constant. Indeed, this is not a preferred solution because the supply biases cannot be reduced arbitrarily, and multiple power circuits are too costly to combine. The second solution is the contraction based on a constant voltage. This method addresses the problem of changing the supply voltage and is a more suitable method for scaling. However, this technique also suffers from important drawbacks, such as increasing the electric field in small dimensions, which results in negative physical effects. Negative physical effects include velocity saturation, reduced mobility, increased leakage current, and reduced breakdown voltage.

On the other hand, miniaturization of a transistor not only includes the reduction of both length and width of the gate but also the other parts of the transistor, such as the oxide thickness (t_{ox}) related to the gate, the depth of the junctions, and the density of the doping must be varied. To have a successful scaling, the thinner gate oxide and higher doping level are essential for a high drive current and less SCEs. Also, the high level of doping may cause destructive effects, such as punching through, which is accountable for the increment of threshold voltage (V_{th}) that complicates turning the device "on" and decreases the channel (Ch) mobility. Similar to the supply voltage, the V_{th} must be reduced, and this defect will be eliminated by reducing the oxide thickness. However, by reducing the t_{ox} below 10 nm, the total gate capacitance will be smaller and eventually the transconductance of the device decreases. Further measurement of transitions, like either further diminution of the t_{ox} or the decrement of whole doping depths, are confined for mass production purposes, both technologically and physically. These results include the importance of quantum effects on the nanoscale transistor operation in both "on" and "off" states, which constrain the MOS field-effect transistor (MOSFET). The constraints mentioned earlier include (1) quantum tunneling of carriers (through t_{ox}, source [S] to drain [D], and D to the body); (2) control of the doping density to have a good I_{ON}/I_{OFF} ratio; and (3) control of the voltage-related effect, such as subthreshold swing and threshold voltage, known as SCEs [5].

1.2.2 INVESTIGATION OF SHORT CHANNEL EFFECTS

Today, MOSFET is the most common microchip transistor. The efficiency of MOSFETs can be achieved by diminishing the size of the device and entering the nanometer geometries. Despite its advantages, this process creates critical problems known as SCEs, which result in poor performance and increased power loss in these devices. New challenges appear as the dimensions of transistors and Ch lengths become smaller. Technically, a MOSFET is supposed to be short,

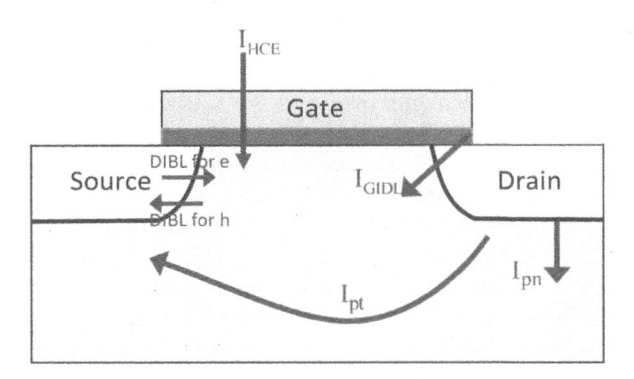

FIGURE 1.1 Leakage current with bulk technology in MOSFET.

whereas the length of the Ch is in equal order as the width of the D and S depletion layers. By decreasing the Ch length, the number of components in each chip causes the increment of the device speed. However, the SCEs arise as well, which can deteriorate the performance of the device and should be resolved.

The nanometer scale in a device causes SCEs, including drain-induced barrier lowering (DIBL) and gate-induced drain leakage (GIDL), which reduce the device performance, V_{th} roll-off, degradation of subthreshold slope (SS), hot-carrier effect (HCE; that possesses an effect on the reliability of device), and leakage power consumption, which are fully explained with its degradation mechanism in the following section. In the "off" state, the current passing through the device is called off current (leakage current), and has various factors involved in its production. Importantly, the leakage current in short-Ch transistors includes I_{GIDL}, I_{DIBL}, PN junction current (I_{pn}), hot electron effect current (leakage Gate current) (I_{HCE}), and punching through current (I_{pt}), as shown in Figure 1.1.

1.2.2.1 Charge Sharing Effect and DIBL

In long Ch devices, the depletion of the source (S) and drain (D) regions are neglected because these regions form a small part of the effective length of the Ch. Nevertheless, in nanoscale and short Ch devices, these areas are important parts of the effective Ch length (Figure 1.2). In this regime, some of the ionization atoms encompass the Ch carriers with their field lines. This effect is known as charge sharing. Technically, S and D penetration through the Ch and charge sharing give rise to a smaller barrier height, called DIBL. A potential barrier exists between the S and the Ch region in the weak inversion. The barrier height (H_b) can be determined by the balance between the drift-diffusion currents of these two regions. In a high D voltage (V_D) regime, for the influence of E_D, the H_b for Ch carriers decreases at the edge of the S and the DIBL effect occurs.

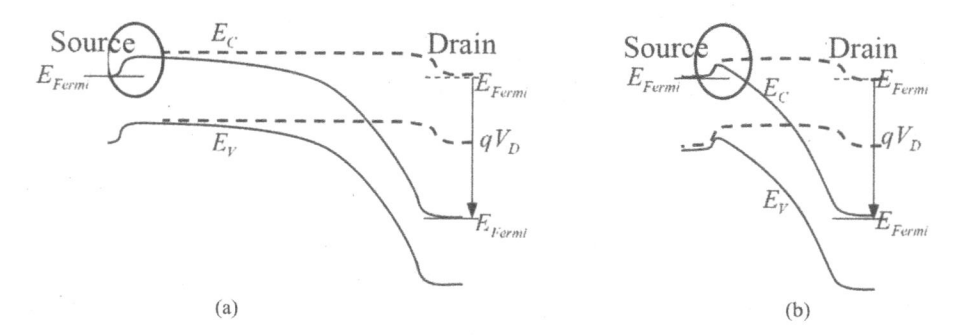

FIGURE 1.2 Energy band diagram. (a) Long Ch transistor. (b) Short Ch transistor.

Therefore, this effect causes a decrease in the S or potential barrier after applying the high V_D. This will increase the I_{OFF} because of an increment in the number of carriers, which can be injected from the S to the Ch. Thus, the drain current can be controlled by both gate and D voltages. This reduces the gate control and finally the leakage current increases [18]. Consequently, the V_{th} reduces as well. In nanometer scales, the S and D electrodes are close to all midpoints of the Ch, so they can have a gatelike effect. Here, the height of the thermal potential barrier is controlled by D; hence, the effect of V_G on the height of the potential barrier is reduced. It should be noted that DIBL is an effect that appears in high V_D applications.

1.2.2.2 Gate-Induced Drain Leakage (GIDL)

In higher V_{DS}, by enhancement of the electric field to a critical value (i.e., 3×10^5 in Si or 3×10^7 V/cm in SiO$_2$ [19]), the mobile carriers acquire adequate kinetic energy in a strong electric field. Thereby, their interactions with the Ch atoms generate the electron-hole (e/H) pairs in the semiconductor, which has been acknowledged as impact ionization [20–22]. This phenomenon is the physical mechanism for substrate current (I_{SUB}) generation and has a direct ratio with the maximum electric field (E_{Max}) and the carrier concentration. As the Ch length decreases, the electric field near the drain turns into a specific parameter. The created hole and electron carriers in n-Ch and p-Ch FET, respectively, flow through the substrate and result in the I_{SUB}. It can cause critical problems including latch-up, V_{th} shift, and degradation of the transconductance in short Ch devices. All these parameters contribute to the corresponding output conductance in the saturation regime and also breakdown (Br) characteristics. It is defined using the drain current, I_D, and the lateral electric field peak in the Ch, E_m. Eventually, I_{SUB} can be expressed as follows [18]:

$$I_{SUB} = \frac{I_D A_i E_m l}{B_i} \exp\left(-\frac{B_i}{E_m}\right) \tag{1.1}$$

where A_i and B_i are two constant values in the impact ionization rate and l is the characteristic length associated with the adjustable parameter.

Moreover, E_m can be described as a function of drain voltage (V_D) as below:

$$E_m = \sqrt{\left(\frac{V_D - V_{DSAT}}{l} + E^2_{SAT}\right)} \approx \frac{V_D - V_{DSAT}}{l} \tag{1.2}$$

The field of the Ch where the carriers reach their saturation velocity is defined by E_{SAT}, V_{DSAT} is the corresponding V_D, and the term l is the length of the effective ionization expressed by [23]:

$$l = 0.22 T_{ox}^{0.33} X_j^{0.5} \tag{1.3}$$

where X_j is the depth of the S/D junction.

Forasmuch as in P-MOSFET the S of electrons produced by impact-ionization caused I_{SUB}, the gate current can be expressed as [24, 25]:

$$I_{GATE} \approx 0.5 \frac{I_{SUB} T_{ox}}{B_r} \left(\frac{\lambda E_m}{\phi_b}\right) p\left(E_{ox}\right) \exp\left(-\frac{\phi_b}{\lambda E_m}\right) \tag{1.4}$$

It is well known that I_{GIDL} is related to the tunneling that occurs in the deep depleted D under G region. To describe the behavior of I_{GIDL}, multiple mechanisms have been presented, such as the band-to-band (B-B) tunneling model as an indirect band [26], known as the basic leakage mechanism [27]. The difference between the voltages of drain and gate (V_{DG}) directly influences I_{GIDL}, which leads to the formation of a strong depletion regime under the G/D overlap region. Higher V_{DG} causes enhancements in the vertical field over the D depletion layer, and the depletion width

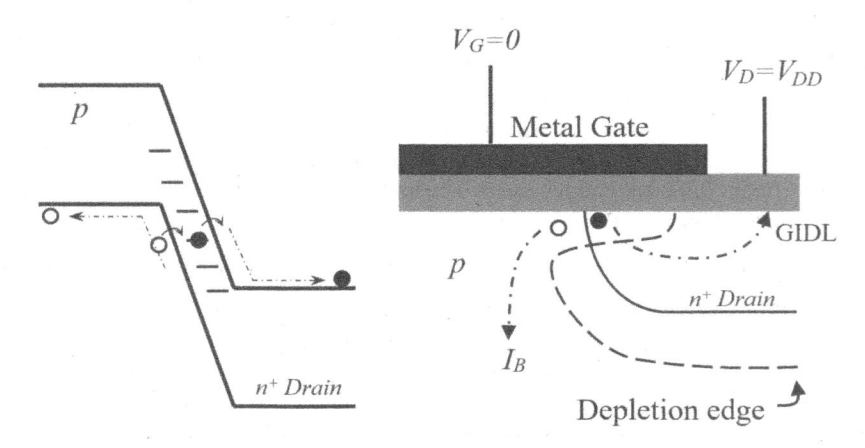

FIGURE 1.3   Schematic of GIDL tunneling current process in n-type structure.

at the *G/D* overlap area decreases, which results in GIDL [28]. In this situation, the energy bands will have overlap and minority carriers' transit by B-B tunneling from the valence band (VB) to the conduction band (CB) of the material. This process is depicted in Figure 1.3. In the largest electric field, the highest rate of the B-B tunneling occurs. Importantly, *D* collects the emitted electrons over the surface of the deep depletion layer and moves them toward the substrate [29]. At the semiconductor-oxide interface, the electric field has the doping concentration dependency in V_{DG} and diffusion region.

The simple model of B-B tunneling current has previously been presented in Ref. [30].

$$I_{GIDL} = AE_a \exp\left(\frac{B}{E_s}\right) \tag{1.5}$$

where *A* and *B* are constants, given by [31]:

$$A = \frac{q^2 . m_r^{\frac{1}{2}}}{18.\Pi.h^2.E_{gap}^{\frac{3}{2}}} \tag{1.6}$$

$$B = \frac{\prod m_r^{\frac{1}{2}} E_{gap}^{\frac{3}{2}}}{2\sqrt{2}qh} = 21.3 \text{ MV/cm} \tag{1.7}$$

with $m_r = 0.2\ m_o$ (effective mass of electron). E_{gap} is the direct energy gap of material [32].

The surface electric field in the *G/D* overlap region at the tunneling point is E_s and can be defined as follows [33]:

$$E_s \approx \frac{V_{GD} - 1.2}{3T_{ox}} \tag{1.8}$$

As a crucial parameter, temperature influences I_{GIDL} as:

$$E_{gap} = 1.12 - 2.4(T - 100)10^4 \tag{1.9}$$

where *T* represents the temperature in Kelvin.

1.2.2.3 Hot Carrier and Punching through Currents

Beyond the phenomena explained previously, additional transport processes occur at high electric fields, which depend on large V_D. Here, carriers known as hot carriers attain sufficient energy and the temperature of the carriers increases, which will be higher than the lattice temperature. In principle, hot carrier refers to the generated e/H pairs by impact ionization that gain above average energy in the D region of a MOSFET. They generate charge defects between the gate oxide layer and semiconductor interface, which degrade reliability and the operation of short Ch transistors. This involves the generation of undesired currents.

In short Ch transistors, due to the penetration of the depletion region of S/D into the Ch, the Ch length becomes less than the actual length, and punching through as a leakage current (I_{pt}) happens. This phenomenon is explained as a reverse bias, which is applied to D and causes an extended depletion region as seen in Figure 1.4. The two depletion regions of the D and S are intersectional with each other. This enables the formation of a depletion region, and flow of leakage current and finally Br voltage between S and D of the short-Ch MOSFET. If designed carefully, I_{pt} is modulated by V_G and can be utilized to increase the transconductance [23]. To avoid punch through, the doping of the Ch should be increased. However, this would lessen the mobility of charge carriers. A general solution to this challenge is the use of the low-doped drain, where the space charge will extend more toward the drain rather than the Ch. Recently, extensive research have been carried out to decrease off currents (I_{OFF}) [24–28].

1.2.2.4 Subthreshold Slope (SS) and Threshold Voltage (V_{th})

For scaled devices, the subthreshold is determined as an important parameter due to its importance on switching "off" the devices, especially in both low power and voltage applications. In MOSFET, under weak inversion, the Ch surface potential is a constant value, and diffusion of minority carriers determines the current flow because of a lateral concentration gradient. Diffused carriers from S to D make it difficult or impossible to turn off the device below the threshold. Therefore, an important parameter must be defined as the subthreshold swing in this region. At room temperature, SS confines to 60 mV per decade (mV/dec) and represents the reduction in the amount of V_{GS} for a decrement in I_D by an order of magnitude. A smaller SS implies that a lower V_{GS} is required to turn off the device and causes a lower distortion and nonlinear gain. In the subthreshold (or weak inversion) regime, the I_D depends on the V_{GS} exponentially [18]:

$$I_{d,weak} \approx \exp\left(\frac{V_{GS}}{n.V_T}\right) \tag{1.10}$$

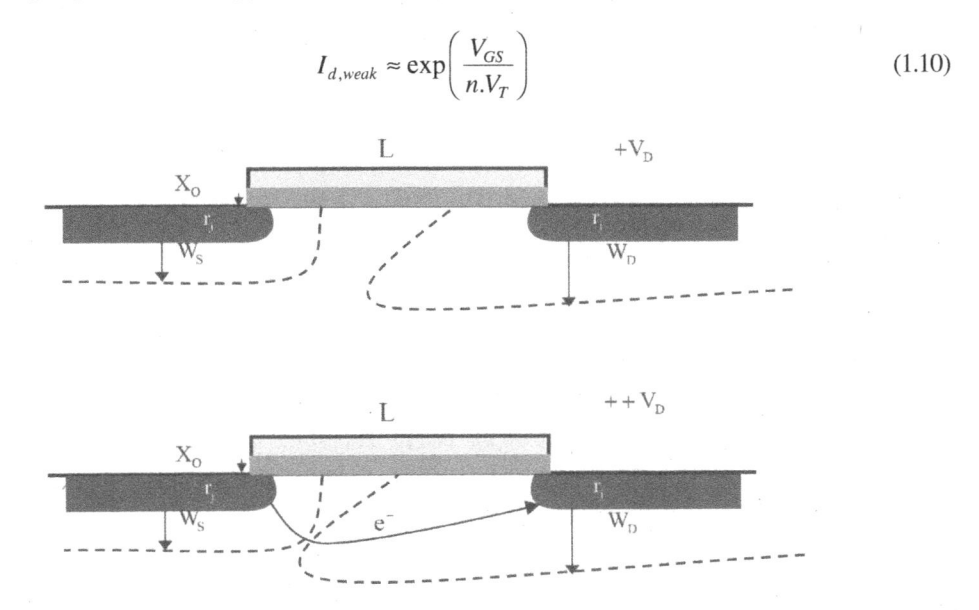

FIGURE 1.4 Schematic illustration of the S and D depletion region edges in punching through.

where V_T is the temperature voltage driven by $V_T = K.T/q$ (here K is the Boltzmann constant, T is the absolute temperature, and q is the electron charge). It is assumed that the Ch is doped uniformly, and the SS factor of a device (n) can be calculated using capacitance of the depletion layer (C_d) and capacitance of the gate oxide (C_i), respectively:

$$n = 1 + \frac{C_d}{C_i} \tag{1.11}$$

And SS can be expressed as:

$$S = \left[\frac{d\left(\log_{10} I_{ds}\right)}{dV_{GS}} \right]^{-1} = \frac{KT}{q}\left(1 + \frac{C_d}{C_i}\right) \tag{1.12}$$

Equation 1.12 explains that how the flow of I_D can be stopped when V_{GS} is reduced below V_{th}.

For the low power consumption, there is a need to have a low operating voltage. Thus, the threshold voltage (V_{th}) is a critical factor for switching devices, which defines the transport properties. To have a reliable device, a stable value of V_{th} is required. As mentioned earlier, in the process of ionization and the production of electron and hole pairs at high voltages, important defects occur in the interface between the gate oxide and the semiconductor. It should be underlined that due to the formation of a large E on the D side of the MOSFET, some traps can be generated in the gate oxide layer. Therefore, the V_{th} of devices will be defined by these defects, as well as the difference in the work function between the semiconductor and the top electrode. In practice, the observed changes in V_{th} shift can be attributed to the density of defects in the oxide layer.

$$\Delta V_T = -\frac{1}{\varepsilon_{SiO_2}} \int_0^{t_{ox}} x\rho_{ox}(x)\,dx \tag{1.13}$$

where x is a variable and specified to be 0 at the interface between metal and oxide.

In modern designs, while the dimensions are becoming smaller, the effects related to the small dimensions should be avoided. Many approaches have been proposed to deal with the SCEs [29–31]. The majority of these methods focused on engineering of Ch doping and gate metal, which significantly reduces such destructive effects [32, 33]. In addition, multi-gate structures contain double-gate (DG) transistors; FET with triple-gate (TG), such as the FinFET [31, 34]; quantum wire [35], and π-Ch [36]. Quadruple-gate structures consist of a device with gate covered all around [37] and DELTA transistor [38], and vertical pillar MOSFETs [39] enable many benefits, including increasing current, reducing the negative effects of short Chs, and increasing the mobility of carriers.

1.2.3 Single-Electron Transistors

Due to the advanced miniaturization techniques utilized in complementary metal oxide semiconductor (CMOS) devices, the existing technologies are encountered with fundamental and crucial limitations. In this technology, the leakage current flows quantum mechanically via tunneling of the electrons as a result of thinning the oxide layers. Therefore, scientists have been looking for an alternative hybrid CMOS/nanoelectronics technology to improve the downsizing of such platforms, and eventually reach the dimensions of single atoms and molecules. In 1985 [40], a single-electron tunneling (SET) transistor was proposed by three terminals, which operated based on the quantum effect and were practically implemented in 1987 [41]. The quantum properties of electrons and atoms play a main role in making transistors denser. Later, the metal SET was developed using the Coulomb blocked effect to further reduce the size of these devices below 100 nm.

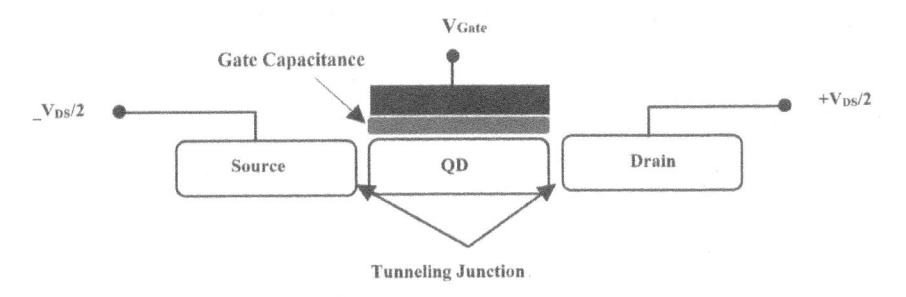

FIGURE 1.5 Schematic illustration of a double barrier structure of an SET.

Compared with conventional FETs, two tunnel junctions are used for SET's Ch. These are separated by QD or metallic layers. The electrons must pass through two tunneling junctions and form an isolated conductive electrode called an "island." This island is often attributed to a QD due to the confinement of electrons within each of the three directions, as portrayed in Figure 1.5. It is clear that an electron box contains two distinct junctions for entry and exit of the single electrons [42]. Electrons cross over the charge of the island and discharge, and the system's relative energies contain zero or one extra electron with a strong dependency on V_G. By applying gate bias, the system becomes energetic to vary the number of electrons on the island. At low source-drain voltages (V_{SD}), if these two configurations of charges have equal energies, only a current will make it through the SET. In recent years, several excellent reviews on the SET transistor have surfaced in the literature [43–46].

1.2.4 Carbon Nanotube Transistors

Carbon-based nanostructured materials have exquisite properties including high stability, good conductivity, low cost, extensive potential windows, and easy surface performance. Of particular interest is the use of one-dimensional (1D), cylindrically shaped allotropes of carbon known as CNTs. They are used as 1D excitonic configurations in photonic, electronic, and optoelectronic applications. In other words, CNTs can be demonstrated as graphene sheets that are rolled into cylinders. Monolayer graphene and CNTs are the most well-known carbon-based nanostructures to have been widely exploited for several electroanalytical applications [47–51]. The metallic or semiconducting characteristics of CNTs in electronic structures remarkably depend on their interconnection geometry between the tubes [52]. Tightly integrated devices in a semiconducting CNT with high amounts of charge carrier mobilities enable the development of CNT-based devices for applications in microelectronics.

There are two types of CNTs: multiwalled (MWCNTs) and single-walled (SWCNTs) [53]. In particular, SWCNTs are appropriate for the SET's fabrication due to their small diameters [54], which were defined in the order of 1–2 nm. In particular and as a practical example, the CNTFET devices are promising nanoscale instruments used to implement very dense and low-threshold circuits with high performance. This refers to a FET device that uses an array of CNTs or a stand-alone CNT for material of the Ch, where they are arranged in cylindrical fashion compared with the conventional silicon-based system. Hence, the core of a CNTFET is a CNT [55].

There has been a special interest in CNTs because they have 1D geometries and an absence of dispersion to a large extent. Also, its intriguing surface conductance occurs essentially because of the saturated and stable chemical bonds. Beyond that, there is a Schottky barrier in nanotube contact within the metal. With these unique features, CNTFET becomes a device of special interest for modern technologies.

1.2.5 Memory Resistor (Memristor)

In recent years, there has been a growing interest in the use of passive memories, i.e., flash memory [56]. In electronic designs, however, the focus should be on finding novel compact devices that are functioning based on Moore's Law with equal capability compared with traditional transistors. Initially, Leon Chua conceived a serious need for another fundamental circuit component instead of the resistor, capacitor, and inductor [57]. He considered the relationship between four basic circuit variables, i.e., voltage, current, flux, and charge, and found a missing component that should have a relationship between the flux and charge. This component was not fulfilled by available circuit elements. He explored and proposed the fourth fundamental component of electronics and called it "Memory Resistor" (Figure 1.6).

The discovery of a memory resistor (memristor) generated new enthusiasm and created a revolution in circuit design to neuromorphic and analog applications. Memristor possesses analog properties and a nonlinear nature, which is capable of remembering its history between two opposite states with minimum and maximum resistance without requiring external stimuli. This tool saves its values until the next programming voltage is applied and decreases the overall power consumption. These features and other advantages, including negligible leakage current, resealable scalability, nonvolatility, and compatibility with CMOS technology compared with conventional and standard memory technologies, make this instrument a potential choice for diverse applications in electronics [58–61]. Simply, from the structural perspective, this device mainly consists of two layers of titanium dioxide (TiO_2) placed between two platinum electrodes. The basis of its performance is the displacement of the boundary between two different parts of TiO_2: one part is basically pure, and the adjacent part has impurities. Unlike most samples of impurities in semiconductors, which contain external compounds, in this device, the impurity is defined by the lack of oxygen atoms. In the crystal structure of TiO_2, these empty places are flowing.

1.2.6 Phase-Change Memory (PCM)

By scaling down silicon devices under 10 nm, the leakage of tiny capacitors increases, and there is a demand to have a novel memory. PCM is a nanometer scale technology that is currently available as a nonvolatile electrical data storage memory [62]. PCMs can be reversibly switched between different phases of the phase-change compound that exist in amorphous (high electrical resistivity) and crystalline phases (low electrical resistivity) by external stimuli, i.e., bias and thermal energy.

PCMs have been broadly considered since the 1960s. They use the behavior of PCMs, which switches between the crystalline phases amorphous of various electrical resistivity reversibly [62]. In the 1990s, these components were used in the development of optical memory devices, in which the PCM can be heated by a laser source [62, 63]. The results of the progress and success in the implementation of optical storage by PCM led to further interest in the first decade of the 21st century [64–66].

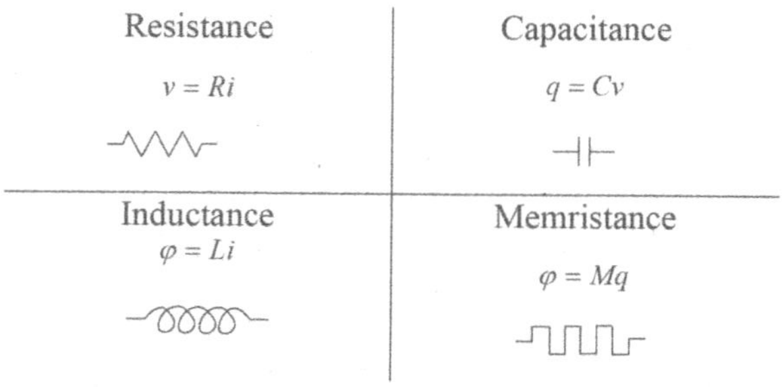

Resistance	Capacitance
$v = Ri$	$q = Cv$
—⌵⌵⌵—	—⊣⊢—
Inductance	**Memristance**
$\varphi = Li$	$\varphi = Mq$
—⌒⌒⌒⌒—	—⊓⊔⊓⌐

FIGURE 1.6 Illustration of the four fundamental circuit elements and their relationships.

1.2.7 NANOSENSORS

One of the primary applications of nanotechnology is in sensing and detection of environmental perturbations. Nanosensors can be categorized according to their structural properties, energy source, and applications [67]. These devices can be tailored as chemical, biological, optical, electrical, thermal, magnetic, or mechanical sensors, which take advantage of unique features of nanoparticles and nanomaterials for the detection, monitoring, and measurement applications [68–71]. Particularly, nanosensors are integrated devices (approximately 10–100 nm in size), and along with sensing, they are capable of simple functions, such as computation and local actuation. Considering the nanomaterials that have been employed for sensing applications, for example, a CNT has a very large surface area and electron conductivity, and nanowires have the capability of high detection sensitivity of thin films, polymers, and biomaterials. The metal and metal oxide nanoparticles are nanostructured materials that have been utilized in the architecture of the nanosensor [72–74]. In addition, sensors can be classified as chemical nanosensors, deployable nanosensors, electrometers, and biosensors based on their applications [67].

Among electronic sensing devices, the FET has many advantages, including but not limited to low manufacturing costs, mass production capability, and ultrasensitivity detection [75]. Ion-sensitive FETs (ISFETs), organic FETs, silicon nanowires, graphene FETs, and compound-semiconductor FETs are well-known tools that can be developed based on FETs with an immense potential in biosensing applications [76]. In the past decade, sensors based on silicon nanowire FETs have transpired as the main class of label-free sensors due to their high precision in detection and fabrication compatibility in silicon processing technology [49, 50, 77, 78].

1.3 QUANTUM CONFINEMENT EFFECTS IN NANOSTRUCTURE

With the rapid progress in shrinking the size of semiconductor devices toward molecular levels, the quantum nature of atoms and electrons becomes important in specifying how the devices are manufactured. Generally, such behavior can be classified into traditional and quantum reliability effects. Quantization effects emerge because of the movement confinement of electrons as a straightforward effect of the materials size reduction down to nanoscale. In semiconductors, by reducing particles dimensions below the bulk semiconductor – in particular, Bohr exciton radius – quantization effects turn to an important phenomenon. Depending on the size of the structure, the discrete set of electron energy levels is known as the simple potential well in quantum mechanics. Hence, the geometric changes make the electrons act within limited boundaries and respond by tuning their energy via the variation in the particle size. This phenomenon is defined as quantum size effect. This varies the density of states (DOS) or band structure in the nanomaterial. Indeed, such variations can cause transformation of material types, including non-metals to metals, or insulators to semiconductors, or even non-magnetic to magnetic systems. The effect of quantum size can be realized as the physical limitation of the electrons in the nanoparticles because of their small size, which causes dissociation in the energy states. Thus, both states of energy and electron distribution positions in the energy state variations lead to the rise of several phenomena, like increment of the bandgap in semiconductors, behavior of ferromagnetic, and surface plasmon phenomenon in metals. The characteristics influenced by the effect of quantum size exhibit a discontinuous behavior because of electrons' discontinuous nature in the energy level. For semiconductors in the IV, III–V [79], and II–VI group, quantum confinement appears in the length scale between 1 and 25 nm. Similarly, semiconductor nanomaterials are known to have interesting features when their size and shape reduce less than 100 to 10 nm. In general, quantum effects have three manifestations in nanodevices: quantum-mechanical size quantization, quantum interference, and tunneling [5]. Some of the fundamental physical effects that are observed in nanoelectronics research are discussed in Refs. [5, 80].

1.3.1   CLASSIFICATION OF QUANTUM CONFINED STRUCTURES

Quantum confined structures can be divided into 2D, 1D, and 0D potential wells, according to the dimensions number, in which the confined particle can move freely [81]. Two-dimensional structures are thin films in the order of a few nanometers thickness that are usually deposited on a bulk material (quantum well/superlattices). A quantum well is a special type of heterostructure in which a thin film is surrounded by two-barrier layers. Electrons and holes both experience lower energies in the well layer, which is called the potential well. Electrons and holes are the waves confined within this extremely thin layer, which is typically around 100 Å in thickness. In such structures, the allowed states correspond to standing waves to the perpendicular direction of the layers. The system is quantized due to the standing waves, which are merely particular waves, known as quantum wells [82]. Quantum wells are semiconductor structures with a thin layer, in which many quantum mechanical effects can be controlled. Most of their characteristic features are derived from the quantum confinement of electron and hole carriers in thin layers of a semiconductor material and sandwiched between the other barrier layers of the semiconductor. A particle in a box is a proper model to figure out the basic properties of a quantum well.

On the other hand, 1D structures include cylinder-like objects, such as tubes and wires, with lengths and diameters typically at microscale and nanoscale, respectively. In semiconductors, quantum sub-bands are formed by the electron energy's quantization in quantum wires with cross-sectional lengths in the order of 10 nm (the order of an electron wavelength). Thus, the electron transfer characteristics are expected to be significantly different in quantum wires compared with bulk semiconductors. Therefore, it is expected that new 1D electronic devices can be devised through the use of the quantum transport specifications [83].

Ultimately, typical examples of 0D nanoparticles are clusters, colloids, nanocrystals, and fullerenes. They are called QD/nanocrystals composed of ten to a few thousand atoms. It is essential to consider the phenomenon of QDs as an interesting class of materials to understand more about quantum confinement. QDs are a subatomic group of nanomaterial families that include insulators, metals, semiconductors, and organic materials. The quantum confinement happens only in semiconductor QDs as a result of their tunable bandgap nature. The specific features of the tunable bandgap are made only from the II–VI, III–V, and IV–VI materials.

QDs are very small semiconductor crystals or molecules that can be considered as e/H pairs (extinction) within a material confined in all three dimensions. Due to the Coulomb force exchange interaction and different polarity charges, a noteworthy connection exists between electrons and holes. This e/H pair looks like a bosonic quasiparticle, which is known as an exciton. Excitons have an average of physical separation between these two carriers (electron and hole) that is attributed to the exciton Bohr radius. This physical distance differs from one material to another one. In the case of semiconductors, in particular QDs, the particle size is smaller compared with the Bohr radius and the electron stimulated by an outer source of energy that tends to create a weaker bond with its own hole.

The exciton can be discussed as a hydrogen-like system, and the exciton radius can be compared with an approximation of the Bohr radius utilized to compute the spatial separation of the e/H pair easily, as follows:

$$r = \frac{\varepsilon h^2}{\pi m_r e^2} \tag{1.14}$$

where r is a 3D sphere comprising the excitation radius, m_r is the excitation reduced mass, ε is the dielectric constant of the related material, h is Planck's constant, and e is the elementary electron charge. Recently, there has been a growing interest in QD-based nanomaterials in many optoelectronic applications, including transistors, QD light-emitting diodes (Q-LEDs), laser diodes, solar cells, quantum computing, etc.

1.3.2 ELECTRON STATES AND QUANTUM CONFINEMENT

In general, for a bulk semiconductor, CB and VB are empty and filled states form the separated continuums, respectively. According to the quantum mechanical principles in semiconductors, the hole (electron) is created (occupied) as a result of the excitation of an electron from VB to CB. The finite energy level of the CB and VB is separated by the energy bandgap (E_g). Due to the thermal excitation or absorption of photons, an electron achieves sufficient energy to overcome the E_g and goes from VB to CB, and creates a hole in the VB. After energy is released, the hole will recombine by the charge carrier in a semiconductor device. However, for QDs, the energy levels are inside the empty/filled states and stay discrete. The E_g levels are in higher states between the highest and lowest occupied molecular orbitals (HOMO and LUMO) than that of the bulk materials, as illustrated in Figure 1.7.

Due to the charge carrier confinement, the electronic bands are disarticulated into the discrete energy levels, which leads to charge quantum confinement. By decreasing the semiconductor particle size (radius), E_g increases in magnitude. Several parameters are supposed to be constant for bulk semiconductors; however, they strongly depend on the dimensions of the material for quantum-confined systems. Because of such quantum confinement, variations in the E_g, band edge positions, and DOS have a dramatic effect on the probability of transitions [84, 85].

In the quantum-confined situation, E_g is defined as a correction form of the bulk bandgap ($E_{g,bulk}$) by solving Schrödinger's equation for the first state of the excitation of an e/H pair bounded in radius R for the kinetic energy and an effective average approach along with the dielectric screening for the potential energy [86]:

$$E_g = E_{g,bulk} + \frac{\hbar^2 \pi^2}{2R^2}\left(\frac{1}{m_e^*} + \frac{1}{m_h^*}\right) - \frac{1.8e^2}{\varepsilon R} + \frac{e^2}{R}\sum_{n=1}^{\infty}\alpha_n\left(\frac{r_e + r_h}{R}\right)^{2n} \qquad (1.15)$$

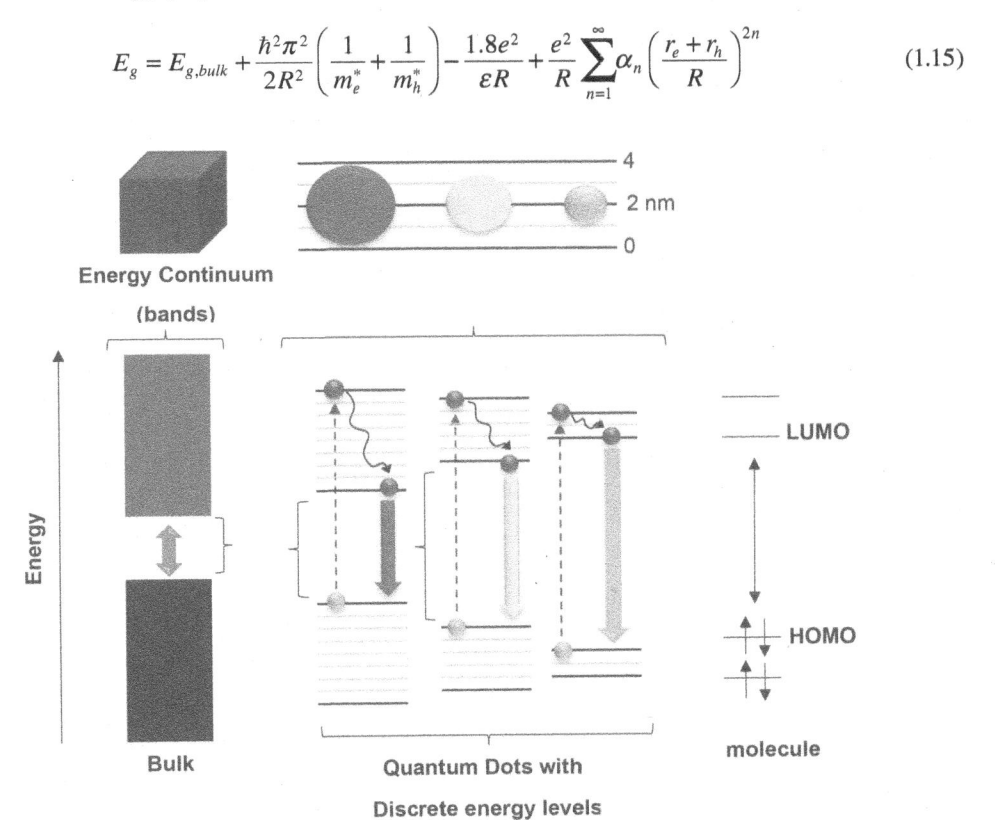

FIGURE 1.7 Energy-level diagram illustrating extinction of an electron from VB to the CB in molecule, bulk semiconductor, and also quantum nanostructure.

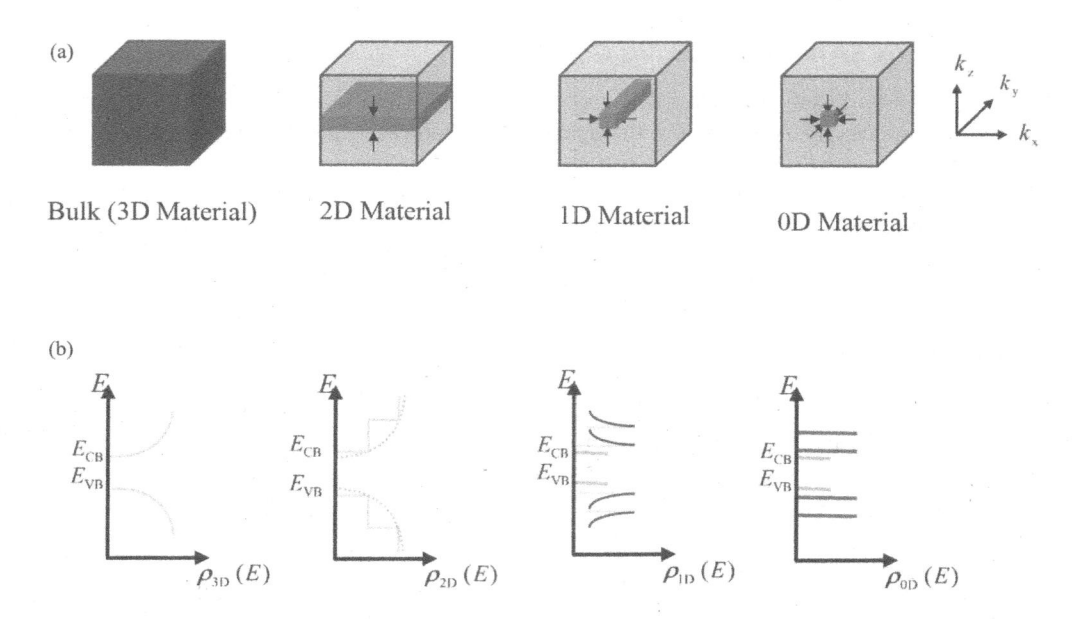

FIGURE 1.8 Schematic illustration of (a) 3D, 2D, 1D, and 0D materials and (b) their DOS.

where m_e^*/m_h^* is the electron/hole effective mass. In the second term, the enhanced kinetic energy from the e/H pair localization is described inside a sphere with a radius of R and scales as R^{-2}. The third term defines the Coulomb attraction in an environment scaled as R^{-1}. Also, the polarization term, given as an average polarization, holds an identical scaling as the screened Coulomb attraction and depends on the bandgap variations of the kinetic energy confinement.

Quantum confinement is defined as a spatial confinement of e/H pairs within a material in one or more dimensions. It causes discretization of electronic energy levels because of the electronic wave function confinement of the particles to the physical dimensions. This confinement can be performed in 0D, 1D, 2D, and 3D as shown in Figure 1.8(a). Quantum wells, as 1D systems, include free carriers in a plane. The movement of charge carriers (e/H) are limited in a plane, which means they have free movement in the 2D regime. Their properties might be influenced by the surface and interface effects or reflect electron confinement in the perpendicular direction to the film. Also, the effect of electron confinement potentially emerges in the transverse direction if the electrons acquire free movement along the structure. In a quantum wire as 2D systems, for example, carriers move freely downward, and in QDs as a 3D system, carriers are limited in all directions. This causes variations in the energy level from a continuous to discrete energy spectrum like an atomic one.

1.4 DENSITY OF STATES

DOSs have been utilized to quantify the number of electrons in a material. This aspect plays a significant role in every single quantum process, in which a final quantum particle state belongs to a continuous spectrum, as for the case of electrons in the CB and holes in the VB in solids. In this case, the probability of a quantum process is proportional to the density of the final states. This can be attributed to the theory of the quantum mechanical perturbation consequence. Basically, the state space is discrete because of the confined size of the domains and particle wave properties. According to the Louis de Broglie theory, the light and particles with a mass similar to electrons possess both wave and particle properties. According to this theory, the waves have a specific wavelength known as the de Broglie wavelength, which can be calculated from the ratio of

Planck's constant (h) and the momentum (p) of the particle. If a particle is significantly larger than its own de Broglie wavelength or interacts with other objects on a scale that is much larger than its de Broglie wavelength, its wavelike properties will not be remarkable. For everyday objects with normal speeds, λ_{dB} is too small to see any observable quantum effects. Thereby, if the size of the domain is significantly larger than de Broglie wavelength (λ_{dB}) of particles, discreteness will usually be neglected. This causes a continuous DOS function [87].

Theoretically, for particles in order of λ_{dB}, there is at least one of the domain sizes. For these confined domains, confined continuum approximation exhibits the state space more appropriately, in which the non-zero value of the ground states of the momentum components should be considered, while their discrete nature is neglected. Herein, for the eigenvalue's asymptotic behavior, Weyl's conjecture utilizes confined continuum approximation and gives a more accurate enumeration of states. Thus, it offers a more precise DOS function. Both functions are based on continuous approximation as they use the assumption of an infinitesimal energy interval. Nevertheless, when quantum confinement is strong, a quantum-mechanically minimum permissive energy interval is bounded, and the state space discrete nature becomes pronounced. When a low energy level of the system near to the ground state is occupied by a large portion of particles, the ground state energy prevails and the deviation from the continuous approximation tends to be important. Therefore, its occupation probabilities are higher at strong confinement conditions or low temperatures. In this limit, considering eigenvalues of discrete energy can be a more accurate DOS function. This concept can serve as a useful tool for a more accurate understanding of nanoscale materials and enables a more profound understanding of the physical behavior of matter.

1.4.1 Density of States in Quantum Confined Structures

The quantum well compared with the bulk semiconductors has a much higher DOS beside the edges of the CB and VB, and, subsequently, a higher carrier's concentration can take place in the band-edge emission. The more confinement dimensions create more discrete energy levels as a result of movement confinement of carriers in each dimension. The electron DOSs of bulk, 2D, 1D, and 0D semiconductor structures are depicted in Figure 1.8(b). As can be seen in the graph and by considering the theoretical calculations, the DOS becomes discontinuous by reducing the dimension of the constituent [88]. Conversely, 0D structures have a well-specified and quantized energy level.

The effect of quantum confinement on the nanoscale size can be computed through a simple model of effective mass approximation for electrons, which are confined in a QD, quantum wire, and quantum well in Equations 1.16–1.18, respectively [89]. This method anticipates the energy levels that are confined by solving Schrodinger's equation and taking into account an infinite confining potential of the barriers.

$$E_{n,m,l} = \frac{\pi^2 \hbar^2}{2m^*} \left(\frac{n^2}{L_z^2} + \frac{m^2}{L_y^2} + \frac{l^2}{L_x^2} \right), \psi = \phi(z)\phi(y)\phi(x) \tag{1.16}$$

$$E_{n,m}(k_x) = \frac{\pi^2 \hbar^2}{2m^*} \left(\frac{n^2}{L_z^2} + \frac{m^2}{L_y^2} \right) + \frac{\hbar^2 k_x^2}{2m^*}, \psi = \phi(z)\phi(y)\exp(ik_x x) \tag{1.17}$$

$$E_n(k_x k_y) = \frac{\pi^2 \hbar^2 n^2}{2m^* L_z^2} + \frac{\pi^2}{2m^*}(k_x^2 + k_y^2), \psi = \phi(z)\exp(ik_x x + ik_y y) \tag{1.18}$$

where n, m, $1 - 1, 2\dots$ are the numbers of quantum confinement; L_x, L_y, and Lz are dimensions of confinement; and $\exp(ik_x + ik_y y)$ expresses the wave function and explains the electronic movement along the x- and y-direction, which is the same as the wave functions of free electron.

Moreover, to estimate the DOS in different structures, the effective number of carriers at the bottom band, i.e., CB, can be formulated by integrating over the Fermi-Dirac distribution function $f_F(E)$ multiplied by the density of state $\rho(E)$:

$$N_c = \int_{E_c}^{E_\infty} \rho(E) f_F(E) dE \qquad (1.19)$$

The general form of the DOS can be calculated by:

$$\rho(E) = dN(E)/dE \qquad (1.20)$$

This equation can be derived from the effective mass approximation in quantum confined structures, and finally the DOS for bulk can be described as:

$$\rho_{3D}(E) = \frac{1}{2\pi^2} \left(\frac{2m^*}{\hbar^2} \right)^{3/2} E^{1/2} \qquad (1.21)$$

It is noteworthy that this is just for the DOS within a band and assumed that energy is given as a parabolic function of the motion of the crystal momentum (k).

Using Equation 1.21, the N_c is calculated as:

$$N_c^{3D} = \frac{1}{\sqrt{2}} \left(\frac{kTm^*}{\pi\hbar^2} \right)^{3/2} \qquad (1.22)$$

DOS for a 1D confined system is affected by the reduced number of charge carriers' freedoms (Equations 1.23 and 1.24). The number of states per unit space, in all cases, is determined by the space confined via k in the various dimensions multiplied by the spin-degeneracy factor, which is separated by the occupied area by each state.

$$N_{2D} = \frac{2\pi k^2}{4\pi^2} \qquad (1.23)$$

For 2D material:

$$N_{3D} = \frac{2.4\pi k^3}{3(2\pi)^3} \qquad (1.24)$$

For 3D material:

Utilizing the expression for the charge carriers numbers that are bounded into 2D and using an approach similar to the 3D case, Equation 1.25 can be obtained, which is independent of the energy [86].

$$\rho_{2D}(E) = \frac{m^*}{\pi\hbar^2} \qquad (1.25)$$

Using Equation 1.25, the N_c is calculated as:

$$N_c^{2D} = \frac{m^* kT}{\pi\hbar^2} \qquad (1.26)$$

The DOSs for the 1D and 0D materials are given by:

$$\rho_{1D}(E) = \frac{m^*}{\pi\hbar\left(m^*/2E\right)^{-1/2}}$$
(1.27)

$$\rho_{0D}(E) = 2\delta E$$
(1.28)

Equation 1.28 only uses the spin-degeneracy factor, which is multiplied by the energy levels' delta function in the range of energy. Using Equation 1.27, the N_c is calculated as:

$$N_c^{1D} = \sqrt{\frac{m^* kT}{2\pi\hbar^2}}$$
(1.29)

Further, the delta function in $\rho_{0D}(E) = 2\delta E$ reduces N_c^{0D} to number 2 at the CB edge if it is occupied.

1.4.2 Surface States and Fermi Level Pinned Surfaces

The electronic states that emerge at the surface of materials are surface states. Regarding the sharp transition from a solid material, which ends with a surface, surface states are formed and only found at the atomic layers close to the surface. They may occur between VB and CB by typical energies inside the gap. They are generally partly filled; thus, the chemical potential is located through the surface band. Thereby, bending of the energy bands happens, and the FL becomes pinned, which is of utmost importance for semiconductor heterostructures. To gain energies and wave functions, Schrödinger's equation should be solved straightforwardly in a realistic potential.

In semiconductor surfaces, the chemical absorption of species is indicated. There are two contributions to the energy absorption: charge transfer and creation of bands [90]. It is well expressed that the electric field of the surface has an effect on the quantum surface state's energy; however, the electron potential and the nuclei are canceled. Thus, the adsorption energy related to pinned FL at the surface depends on the Fermi energy, which is related to bulk doping. Adsorption energy analysis has successfully been explained by utilizing an absolute energy scale, in which valence/conduction and quantum states are well determined values in the bulk and separate atom/molecule, respectively. In most of the atomic configurations, FL is pinned at the surface of the semiconductor, which involves a suitable charge change on the surface state. This inadvertently leads to an electric field associating with the alignment of the band at the surface in the bulk. Nevertheless, the surface state pinning FL energy is defined by the position of the FL in the bulk [90]. In the absolute scale and during the adsorption process, the quantum state's energy can be altered at surfaces through FL pinned by surface donor and acceptor.

Therefore, different doping levels with the FL pinning in the bulk lead to the emergence of the electric fields in different directions. The molecule/atom adsorption gives rise to the variations in the energy of quantum states. Figure 1.9 exhibits the energy of the molecule or evolution of atomic quantum states from the far distance to the adsorbed position. The demonstrated surface has a quantum state, which pins FL at the surface. In the absolute scale, the pinning state holds less energy for p-type than n-type semiconductors. Naturally, the surface state energy generated by the related molecule/atom does not require energies similar to the pinning state energy. However, it might vary equally for both p-type and n-type bulks by the same difference (ΔE). It should be underlined that the surface state energy difference (ΔE) is only influenced by the inverse of the direction of the field. In the absolute scale, the molecular/atomic state's energy in the initial, far distance configuration is identical. In the final position for the adsorbed species at the surface, these energy states are different. This is determined by the bulk and ΔE Fermi energy on the virtue of pinning.

In addition, for n-type and p-type materials, the adsorption caused by the molecular/atomic quantum state's energy variations are different. Because the difference between the pinning state

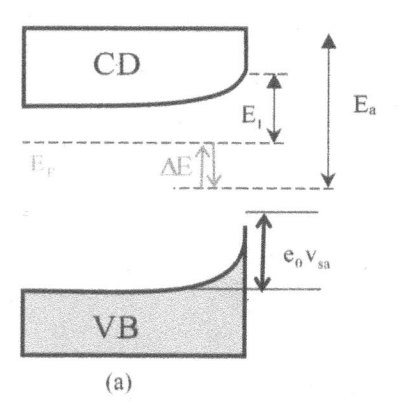
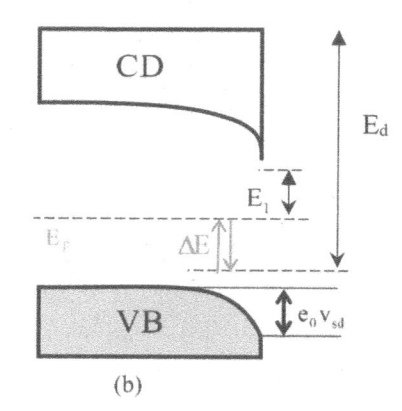

FIGURE 1.9 Energy of the quantum states in the absolute scale depict the variation during the process of adsorption at surfaces with FL free at the surface having a (a) surface acceptor and (b) surface donor.

energy and the surface bands is different from the difference of the FL and the bulk bands, the bands are bent upward/downward (acceptor/donor) as a result of the surface state charge associated with the electrostatic effect. It is worthwhile mentioning that the energy change of the negatively and positively charged states (electron and nuclei) are the same in magnitude but hold opposite signs to the energy change. Consequently, these two contributions are canceled. Hence, in the process, the adsorption energy at the surface with FL pinned should be equal to the vanishing surface-bulk charge transfer. In addition to the previous explanation, the adsorption is mainly accompanied by the electron transition to/from the bulk, which affects the surface charge state that can be defined by the average potential slope. More precisely, the electron is transferred to/from the defect state of the energy fairly close to the Fermi energy. As the surface state pins FL, the energy change can be quantified as the ΔE of the pinning state and the state generated by molecule/atom adsorption. In the case of the two different states, the contribution is finite, and the contribution is zero in the same state. Note that in different states, for p-type and n-type semiconductors, the contribution is the same. Thus, the adsorption energy should be identical for the FL pinned at the both types (n and p) of semiconductor surfaces, which are independent of the doping in the bulk [90].

To conclude, nanotechnology offers intriguing and exquisite opportunities and continues to attract a great deal of attention due to its immense potential on an inconceivably wide range of industries, sectors, and sciences. Consequently, this technology is improving rapidly and will expand faster in the future. The new features of nanotechnology will evolve and be utilized in the development of new and upcoming generations of integrated and on-chip instruments. Furthermore, it is also important to understand the fundamental physics of nanodevices and address the uncertainties and the possible challenges in the implementation of high-responsive, fast, and efficient nanoscale tools.

REFERENCES

1. Moore, G.E., *No exponential is forever: but" Forever" can be delayed! [semiconductor industry]*. IEEE International Solid-State Circuits Conference, 2003. Digest of Technical Papers. ISSCC, 2003. IEEE.
2. Tsividis, Y., and C. McAndrew, Operation and Modeling of the MOS Transistor. 2011: Oxford University Press.
3. Yoshimi, M., et al., *Two-dimensional simulation and measurement of high-performance MOSFETs made on a very thin SOI film*. IEEE Transactions on Electron Devices, 1989. **36**(3): pp. 493–503.
4. Braccioli, M., et al., *Simulation of self-heating effects in different SOI MOS architectures*. Solid-State Electronics, 2009. **53**(4): pp. 445–451.
5. Vasileska, D., et al., Quantum and Coulomb Effects in Nano Devices, in Nano-Electronic Devices. 2011: Springer. pp. 97–181.

6. Wilson, L., *International technology roadmap for semiconductors (ITRS)*. Semiconductor Industry Association, 2013. 1–26.

7. Saleem, M., *The failure of classical physics and the advent of quantum mechanics*. Quantum Mechanics. 2015: IOP Publishing.

8. Horodecki, M., and J. Oppenheim, *Fundamental limitations for quantum and nanoscale thermodynamics*. Nature Communications, 2013. **4**(1): pp. 1–6.

9. Credi, A., *Quantum dot–molecule hybrids: a paradigm for light-responsive nanodevices*. New Journal of Chemistry, 2012. **36**(10): pp. 1925–1930.

10. Khan, Q., et al., *Overcoming the electroluminescence efficiency limitations in quantum-dot light-emitting diodes*. Advanced Optical Materials, 2019. **7**(20): p. 1900695.

11. Roco, M., R. Williams, and P. Alivisatos, *Nanotechnology Research Directions: IWGN Research Report*. Committee on Technology, Interagency Working Group on Nanoscience, Engineering and Technology (IWGN), 1999: National Science and Technology Council.

12. Mann, S., *Nanotechnology and Construction. Nanoforum Report (2006)*, 2008.

13. Bardeen, J., and W.H. Brattain, *The transistor, a semi-conductor triode*. Physical Review, 1948. **74**(2): p. 230.

14. Agnihotri, P., P. Dhakras, and J.U. Lee, *Bipolar junction transistors in two-dimensional WSe2 with large current and photocurrent gains*. Nano Letters, 2016. **16**(7): pp. 4355–4360.

15. Feller, B.E., et al., Biosensors Including Surface Resonance Spectroscopy and Semiconductor Devices. 2017: Google Patents.

16. Sze, S.M., and K.K. Ng, Physics of Semiconductor Devices. 2006: John Wiley & Sons.

17. Fahrner, W., Nanotechnology and Nanoelectronics. 2005: Springer.

18. Ramezani, Z., and A.A. Orouji, *Amended electric field distribution: a reliable technique for electrical performance improvement in nano scale SOI MOSFETs*. Journal of Electronic Materials, 2017. **46**(4): pp. 2269–2281.

19. Aminbeidokhti, A., A.A. Orouji, and M. Rahimian, *High-voltage and RF performance of SOI MESFET using controlled electric field distribution*. IEEE Transactions on Electron Devices, 2012. **59**(10): pp. 2842–2845.

20. Ramezani, Z., A.A. Orouji, and H. Agharezaei, *A novel symmetrical 4H–SiC MESFET: an effective way to improve the breakdown voltage*. Journal of Computational Electronics, 2016. **15**(1): pp. 163–171.

21. Ramezani, Z., A.A. Orouji, and M. Rahimian, *High-performance SOI MESFET with modified depletion region using a triple recessed gate for RF applications*. Materials Science in Semiconductor Processing, 2015. **30**: pp. 75–84.

22. Orouji, A.A., Z. Ramezani, and S.M. Sheikholeslami, *A novel SOI-MESFET structure with double protruded region for RF and high voltage applications*. Materials Science in Semiconductor Processing, 2015. **30**: pp. 545–553.

23. Lohstroh, J., et al., *Punch-through currents in P+NP+ and N+PN+ sandwich structures—I: Introduction and basic calculations*. Solid-State Electronics, 1981. **24**(9): pp. 805–814.

24. Ramezani, Z., et al., *A nano junctionless double-gate MOSFET by using the charge plasma concept to improve short-channel effects and frequency characteristics*. Journal of Electronic Materials, 2019. **48**(11): pp. 7487–7494.

25. Afzali, S.S., A.A. Orouji, and Z. Ramezani, *A nano scale triple-gate transistors to suppress the aggregated body holes*. Silicon, 2019. **11**(4): pp. 2177–2184.

26. Ramezani, Z., and A.A. Orouji, *A novel double gate MOSFET by symmetrical insulator packets with improved short channel effects*. International Journal of Electronics, 2018. **105**(3): pp. 361–374.

27. Bousari, N.B., M.K. Anvarifard, and S. Haji-Nasiri, *Improving the electrical characteristics of nanoscale triple-gate junctionless FinFET using gate oxide engineering*. AEU-International Journal of Electronics and Communications, 2019. **108**: pp. 226–234.

28. Anvarifard, M.K., et al., *A nanoscale-modified band energy junctionless transistor with considerable progress on the electrical and frequency issue*. Materials Science in Semiconductor Processing, 2020. **107**: pp. 104849.

29. Ramezani, Z., and A.A. Orouji, *A new DG nanoscale TFET based on MOSFETs by using source gate electrode: 2D simulation and an analytical potential model*. Journal of the Korean Physical Society, 2017. **71**(4): pp. 215–221.

30. Ramezani, Z., and A.A. Orouji, *Dual metal gate tunneling field effect transistors based on MOSFETs: A 2-D analytical approach*. Superlattices and Microstructures, 2018. **113**: pp. 41–56.

31. Karimi, F., et al., *Electrothermal analysis of novel NPP FinFET with electrically doped drain: a dual material gate device for reliable nanoscale applications*. Applied Physics A, 2020. **126**(8): pp. 1–11.

32. Ramezani, Z., and A.A. Orouji, *Investigation of vertical graded channel doping in nanoscale fully-depleted SOI-MOSFET.* Superlattices and Microstructures, 2016. **98**: pp. 359–370.

33. Ramezani, Z., and A.A. Orouji, *An Asymmetric nanoscale SOI MOSFET by means of a PN structure as virtual hole's well at the source side.* Silicon, 2019. **11**(2): pp. 761–773.

34. Huang, X., et al. *Sub 50-nm finfet: PMOS.* International Electron Devices Meeting 1999. Technical Digest (Cat. No. 99CH36318). 1999: IEEE.

35. Colinge, J.-P., et al., *A silicon-on-insulator quantum wire.* Solid-State Electronics, 1996. **39**(1): pp. 49–51.

36. Jiao, Z., and C.A.T. Salama. *A fully depleted Δ-channel SOI nMOSFET.* Proceedings of the Electrochemical Society, 2001. **2001–2003**: pp. 403–408.

37. Colinge, J.-P., et al. *Silicon-on-insulator "gate-all-around device."* International Technical Digest on Electron Devices. 1990: IEEE.

38. Hisamoto, D., et al. *A fully depleted lean-channel transistor (DELTA)-a novel vertical ultra thin SOI MOSFET.* International Technical Digest on Electron Devices Meeting. 1989: IEEE.

39. Auth, C.P., and J.D. Plummer, *A simple model for threshold voltage of surrounding-gate MOSFET's.* IEEE Transactions on Electron Devices, 1998. **45**(11): pp. 2381–2383.

40. Averin, D., and K. Likharev, *Coulomb blockade of single-electron tunneling, and coherent oscillations in small tunnel junctions.* Journal of Low Temperature Physics, 1986. **62**(3–4): pp. 345–373.

41. Fulton, T.A., and G.J. Dolan, *Observation of single-electron charging effects in small tunnel junctions.* Physical Review Letters, 1987. **59**(1): p. 109.

42. Takahashi, Y., et al. *Silicon single-electron devices and their applications.* Proceedings 30th IEEE International Symposium on Multiple-Valued Logic (ISMVL 2000). 2000: IEEE.

43. Liu, K., et al., *Simple fabrication scheme for sub-10 nm electrode gaps using electron-beam lithography.* Applied Physics Letters, 2002. **80**(5): pp. 865–867.

44. Takahashi, Y., et al., *Silicon single-electron devices.* Journal of Physics: Condensed Matter, 2002. **14**(39): p. R995.

45. Hu, S.-F., et al., *Proximity effect of electron beam lithography for single-electron transistor fabrication.* Applied Physics Letters, 2004. **85**(17): pp. 3893–3895.

46. Patel, R., Y. Agrawal, and R. Parekh, *Single-electron transistor: review in perspective of theory, modelling, design and fabrication.* Microsystem Technologies, 2020: pp. 1–13.

47. Fattah, A., et al., *Efficiency improvement of graphene/silicon Schottky junction solar cell using diffraction gratings.* Optical and Quantum Electronics, 2020. **52**(9): pp. 1–18.

48. Ahmadivand, A., B. Gerislioglu, and Z. Ramezani, *Gated graphene island-enabled tunable charge transfer plasmon terahertz metamodulator.* Nanoscale, 2019. **11**(17): pp. 8091–8095.

49. Anvarifard, M.K., Z. Ramezani, and I.S. Amiri, *Proposal of an embedded nanogap biosensor by a graphene nanoribbon field-effect transistor for biological samples detection.* Physica Status Solidi (A), 2020. **217**(2): p. 1900879.

50. Anvarifard, M.K., Z. Ramezani, and I.S. Amiri, *Label-free detection of DNA by a dielectric modulated armchair-graphene nanoribbon FET based biosensor in a dual-nanogap setup.* Materials Science and Engineering: C, 2020. **117**: p. 111293.

51. Ghodrati, M., A. Mir, and A. Naderi, *Proposal of a doping-less tunneling carbon nanotube field-effect transistor.* Materials Science and Engineering: B, **265**: p. 115016.

52. McEuen, P.L., M.S. Fuhrer, and H. Park, *Single-walled carbon nanotube electronics.* IEEE Transactions on Nanotechnology, 2002. **1**(1): pp. 78–85.

53. Zhang, J., et al., *Room-temperature carbon nanotube single-electron transistors with mechanical buckling-defined quantum dots.* Advanced Electronic Materials, 2018. **4**(5): p. 1700628.

54. Zhang, J., et al., *Nanogap-engineerable electromechanical system for ultralow power memory.* Advanced Science, 2018. **5**(2): p. 1700588.

55. Alvi, P., et al., *Carbon nanotubes field effect transistors: a review.* Indian Journal of Pure and Applied Physics, 2005. **43**: pp. 899–904.

56. Wang, R., et al., *Recent advances of volatile memristors: devices, mechanisms, and applications.* Advanced Intelligent Systems, 2020. **2**(9): p. 2000055.

57. Chua, L., *Memristor-the missing circuit element.* IEEE Transactions on Circuit Theory, 1971. **18**(5): pp. 507–519.

58. Rozenberg, M., I. Inoue, and M. Sanchez, *Nonvolatile memory with multilevel switching: a basic model.* Physical Review Letters, 2004. **92**(17): p. 178302.

59. Borghetti, J., et al., *"Memristive" switches enable "stateful" logic operations via material implication.* Nature, 2010. **464**(7290): pp. 873–876.

60. Jo, S.H., et al., *Nanoscale memristor device as synapse in neuromorphic systems.* Nano Letters, 2010. **10**(4): pp. 1297–1301.

61. Linares-Barranco, B., and T. Serrano-Gotarredona, *Memristance can explain spike-time-dependent-plasticity in neural synapses.* Nature Proceedings, 2009. pp. 1–1.

62. Le Gallo, M., and A. Sebastian, *An overview of phase-change memory device physics.* Journal of Physics D: Applied Physics, 2020. **53**(21): p. 213002.

63. Wuttig, M., and N. Yamada, *Phase-change materials for rewriteable data storage.* Nature Materials, 2007. **6**(11): pp. 824–832.

64. Zhou, P., et al., *A durable and energy efficient main memory using phase change memory technology.* ACM SIGARCH Computer Architecture News, 2009. **37**(3): pp. 14–23.

65. Clarke, P., *Exclusive: micron drops phase-change memory–for now.* Electronics360. 2014: globalspec.com.

66. Le Gallo, M., and A. Sebastian, *Phase-change memory.* Memristive Devices for Brain-Inspired Computing. 2020: Elsevier. pp. 63–96.

67. Abdel-Karim, R., Y. Reda, and A. Abdel-Fattah, *Nanostructured materials-based nanosensors.* Journal of The Electrochemical Society, 2020. **167**(3): p. 037554.

68. Anvarifard, M.K., et al., *Profound analysis on sensing performance of nanogap SiGe source DM-TFET biosensor.* Journal of Materials Science: Materials in Electronics, 2020. pp. 1–14.

69. Hierold, C., et al., *Nano electromechanical sensors based on carbon nanotubes.* Sensors and Actuators A: Physical, 2007. **136**(1): pp. 51–61.

70. Li, C., E.T. Thostenson, and T.-W. Chou, *Sensors and actuators based on carbon nanotubes and their composites: a review.* Composites Science and Technology, 2008. **68**(6): pp. 1227–1249.

71. Akyildiz, I.F., and J.M. Jornet, *Electromagnetic wireless nanosensor networks.* Nano Communication Networks, 2010. **1**(1): pp. 3–19.

72. Maduraiveeran, G., and W. Jin, *Nanomaterials based electrochemical sensor and biosensor platforms for environmental applications.* Trends in Environmental Analytical Chemistry, 2017. **13**: pp. 10–23.

73. Franke, M.E., T.J. Koplin, and U. Simon, *Metal and metal oxide nanoparticles in chemiresistors: does the nanoscale matter?.* Small, 2006. **2**(1): pp. 36–50.

74. Azharuddin, M., et al., *A repertoire of biomedical applications of noble metal nanoparticles.* Chemical Communications, 2019. **55**(49): pp. 6964–6996.

75. Pulikkathodi, A.K., et al., *A comprehensive model for whole cell sensing and transmembrane potential measurement using FET biosensors.* ECS Journal of Solid State Science and Technology, 2018. **7**(7): p. Q3001.

76. Syu, Y.-C., W.-E. Hsu, and C.-T. Lin, *Field-effect transistor biosensing: devices and clinical applications.* ECS Journal of Solid State Science and Technology, 2018. **7**(7): p. Q3196.

77. Zafar, S., et al., *A comparison between bipolar transistor and nanowire field effect transistor biosensors.* Applied Physics Letters, 2015. **106**(6): p. 063701.

78. Anvarifard, M.K., Z. Ramezani, and I.S. Amiri, *High ability of a reliable novel TFET based device in detection of biomolecule specifies-a comprehensive analysis on sensing performance.* IEEE Sensors Journal, 2020. **21**: p. 5.

79. Cipriano, L.A., et al., *Quantum confinement in group III–V semiconductor 2D nanostructures.* Nanoscale, 2020. **12**(33): pp. 17494–17501.

80. Vasileska, D., and S.M. Goodnick, *Computational electronics.* Synthesis Lectures on Computational Electromagnetics, 2005. **1**(1): pp. 1–216.

81. Ramalingam, G., et al., *Quantum confinement effect of 2D nanomaterials.* Quantum Dots-Fundamental and Applications. 2020: IntechOpen.

82. Scully, M.O., and M.S. Zubairy, Quantum Optics. 1999: American Association of Physics Teachers.

83. Sone, J.I., *Electron transport in quantum wires and its device applications.* Semiconductor Science and Technology, 1992. **7**(3B): p. B210.

84. Brus, L.E., *Electron-electron and electron-hole interactions in small semiconductor crystallites: the size dependence of the lowest excited electronic state.* The Journal of Chemical Physics, 1984. **80**: pp. 4403–4409.

85. Dabbousi, B., et al., *Electroluminescence from CdSe quantum-dot/polymer composites.* Applied Physics Letters, 1995. **66**(11): pp. 1316–1318.

86. Edvinsson, T., *Optical quantum confinement and photocatalytic properties in two-, one-, and zero-dimensional nanostructures.* Royal Society Open Science, 2018. **5**(9): pp. 180387.

87. Aydin, A., and A. Sisman, *Discrete density of states.* Physics Letters A, 2016. **380**(13): pp. 1236–1240.

88. Dresselhaus, M.S., et al., *New directions for low-dimensional thermoelectric materials.* Advanced Materials, 2007. **19**(8): pp. 1043–1053.

89. Singh, V., et al., *Solar radiation and light materials interaction.* Energy Saving Coating Materials. 2020: Elsevier. pp. 1–32.

90. Krukowski, S., P. Kempisty, and P. Strąk, *Fermi level influence on the adsorption at semiconductor surfaces—ab initio simulations.* Journal of Applied Physics, 2013. **114**(6): p. 063507.

2 Recent Advancements in Growth and Stability of Phosphorene

Prospects for High-Performance Devices

Sushil Kumar Pandey, Vivek Garg,
Nezhueyotl Izquierdo, and Amitesh Kumar

CONTENTS

2.1 INTRODUCTION

Presently, two-dimensional (2D) materials have attracted a lot of attention in the scientific community due to their superior electronic and optical properties.[1,2] The 2D materials have a wide range of applications in the field of electronics and optoelectronic devices such as field-effect transistors (FETs), photodetectors, optical modulators, light-emitting diodes, lasers, and solar cells.[1–6] The 2D materials have the potential to shift the present research from the nanometer scale to the 2D regime due to their essential ultrathin 2D nature.[1–6] The remarkable ability of these materials

DOI: 10.1201/9781003126393-2

is the tuning nature of energy bandgap with the thickness of their layers.[1-8] Presently, the explored 2D materials either have a very low energy bandgap (e.g., from 0 to 0.3 eV) for graphene or metallic dichalcogenides or a moderate bandgap from 1 to 2 eV for dichalcogenides with a semiconducting nature.[1-8] The black phosphorus (BP) material can act as a bridge between zero bandgap graphene and large bandgap transition metal dichalcogenides (TMDCs) because it has an energy bandgap variation between ~0.3 eV (bulk) and 2 eV (monolayer).[7,8] The BP has only a direct bandgap nature for all its thicknesses, whereas TMDCs show direct bandgap only in its monolayer form and indirect bandgap for higher thicknesses, indicating the significance of BP for optoelectronic applications.[7-19] Additionally, the mobility of charge carriers in BP is much higher than TMDCs and other 2D materials. A monolayer, or a few layers, of BP is called phosphorene. The phosphorene has a similar mobility as graphene, but the phosphorene has a significant bandgap, which is not present in graphene. The BP material has anisotropic electrical and optical properties, which makes it very useful for next-generation electronic and optoelectric devices.[7-19] This material has strong resonant absorption in the infrared region and ultrafast carrier dynamics, resulting in its potential application in ultrafast laser photonics as a saturable absorber.[7,8]

It is very important to note that in the entire phosphorus family, BP in the only stable form. BP can be mechanically exfoliated in the same way as other 2D materials. Generally, red phosphorus is used to produce BP by keeping red phosphorus in high pressure and high temperature. [7,8] In BP, different layers are held together by interlayer weak van der Waals forces, allowing the production of a single atomic layer, called monolayer, by a simple mechanical exfoliation process. The physical properties of a monolayer or a few layers, known as phosphorene, are significantly different from bulk BP. A significant improvement is possible if phosphorene is used to prepare novel nanodevices, e.g., transistors, batteries, and sensors, nanomechanical resonators, photodetectors, and photovoltaics.[7-21]

The biggest challenge in realizing phosphorene-based devices is attributable to their sturdy reaction with oxygen and water, making them unstable in the atmosphere.[22-25] The degradation of phosphorene can be protected by passivating or encapsulating it by different stable materials.[22-25] It is observed in the literature survey that more than 80–85% of research papers published until now are based on theoretical work. Many things still need to be explored in this area, especially device fabrication. For phosphorene device fabrication, such as the production of defect-free layers, uniform and cheap phosphorene material is required. These things are explored in this chapter.

In this chapter, initially, the structure and properties of phosphorene are discussed in detail. Then, challenges related to air instability and mass production of phosphorene are discussed thoroughly. The deposition of phosphorene by various strategies is discussed under two principal headings: the top-down and bottom-up methods. The different passivation strategies to stabilize phosphorene in air and at high temperature are elaborated in a case study. After passivation, different applications of stable phosphorene are discussed in detail. Finally, in the conclusion, the status and viewpoint about the growth and stability of phosphorene are discussed.

2.2 STRUCTURE AND PROPERTIES OF PHOSPHORENE

It has been set up that phosphorene has numerous superb properties due to its novel structure. The physical, chemical, electronic, and optical properties of phosphorene rely incredibly on its structure and morphology. In the accompanying section, a short depiction of phosphorene is given by featuring the present status of the investigation into its key properties. A far-reaching portrayal of phosphorene structure and properties has been explored in recent reviews by the researchers in this material research field.

2.2.1 STRUCTURE

Phosphorene acquires a puckered structure with diminished symmetry, in contrast with graphene, which offers ascension to two anisotropic in-plane directions.[26-30] Because of the puckered configuration of phosphorene, anisotropy is seen in its electronic, optical, thermoelectric, and mechanical properties, and its Poisson's ratio. Additionally, the high in-plane anisotropic conductivity of phosphorene has been demonstrated in the past.[26-30] The puckered structure with quasi-2D characteristics brings about an exceptionally nonlinear and anisotropic Young's modulus and extreme strain.[26-30] The biaxial or uniaxial strain applied on BP can tune these anisotropic properties, which could establish the way for novel electronic, optoelectronic, and electromechanical devices utilizing BP. In BP, the atomic layers are stacked together due to the presence of van der Waals forces. Curiously, monolayer BP is covalently bonded with sp^3-hybridized phosphorus atoms. In this monolayer, every phosphorus atom has a lone pair of electrons, which is covalently bonded to three neighboring phosphorus atoms, resulting in a quadrangular pyramid-shaped structure.[30] The monolayer of BP is also called 2D-phosphane or stratophosphane as the monolayer comprises tervalent phosphorus atoms in concurrence with the International Union of Pure and Applied Chemistry (IUPAC) nomenclature of the phosphane group.[30] There are three stacking structures AC, AA, and AB for bilayer phosphorene.[30] It is found that the AC stacking comprises a maximum energy barrier during the sliding processes, which results in harder exfoliation for this stacking.[30]

2.2.2 Properties

BP is a direct bandgap semiconductor material, which shows layer-dependent bandgap variation from ~2 eV to 0.3 eV for monolayer and bulk, respectively.[7-19] Owing to tunable direct bandgap, p-type BP has significant preferences for its application in optoelectronic devices.[7-19] The optical absorption spectra and optical conductivity of BP account for the change in the layer thickness, light polarization, and doping at different frequencies. The BP materials have demonstrated interesting electronic properties, as their hole mobility of $100–1000\ cm^2\ V^{-1}\ s^{-1}$ at 300 K.[7,8,30] The BP has higher photoresponsivity than graphene, and its mobility is much higher than MoS_2.[7,8,30] The high mobility and photoresponsivity of BP show its potential application for high-performance solar cell and a broadband photodetector.[7,8,30]

Theoretical investigations have demonstrated that the in-plane ideal strain and Young's modulus in the perpendicular direction of pucker are 0.48 and 41.3 GPa for monolayer BP, respectively, while in the parallel direction, these values are 0.11 and 106.4 GPa, respectively.[30,31] BP can withstand a tensile strain of up to 32% and 30% for bulk BP and monolayer, respectively; the unmatched flexibility of phosphorene can likewise be used for high magnitude strain engineering.[30,31] First principal calculation demonstrated the presence of a negative Poisson's ratio in the out-of-plane direction under a uniaxial deformation toward the path corresponding to the pucker in the monolayer BP.[30,31] Quantum oscillations are obtained in BP as evidenced by Hall mobility in the 2D electron gas of BP. The proof of a normally occurring p-type conductivity observed as Seebeck coefficient is $+335 \pm 10\ \mu V\ K^{-1}$ at room temperature for measurement at a temperature of 300–385 K.[32] Apart from its capacity to tune the energy bandgap, strain additionally works as a significant function in tuning the effective masses, and along these lines influencing the binding strength and exciton anisotropy. The different properties of BP rely on the thickness, stacking order, applied strain force, and applied electric field, allowing its applications for electronics, optoelectronics, and the realization of devices for different applications such as electronics, optoelectronics, sensing, saturable absorber, and energy storage devices. In the past, BP and its composite showed saturable absorption, which is a nonlinear optical property.[30] The saturable absorber can be used for a solid-state laser for the generation of ultrashort pulses. The BP-based saturable absorbers with 272 fs, 786 fs, and 940 fs mode-locked pulses less than 1.5 μm have been generated with the use of Er-doped fiber lasers, which is demonstrated maximum average powers of 0.5 mW, 1 mW, and 1.5 mW, respectively.[30] The absorption bleaching in BP, in which countless photogenerated carriers cause band filling, instigates from Pauli blocking processes. The saturation intensity obtained for BP is $1.53\ MW\ cm^{-2}$, which is in the range of reported value for semiconductor saturable absorber mirrors and graphene.[33] The BP has a similar modulation depth as the carbon-nanotube, which has resonant absorption in the band of telecommunication.

The upsides of BP as a saturable absorber lie in its high absorption at infrared wavelength region and ultrafast carrier dynamics, suggesting its application for ultrafast photonics.[30,33]

2.3 PRESENT CHALLENGES OF PHOSPHORENE

The extraordinary properties of phosphorene, such as excellent electronic transport, high-performance tunable band structure, etc., offer a competitive edge over other 2D materials for applications in the field of energy storage, optoelectronic devices, and FETs. However, in spite of such a wide range of potential applications, the following challenges limited its practical implementation for the device applications: air instability and mass production.

2.3.1 AIR INSTABILITY

The degradation of phosphorene properties under ambient conditions has limited its practical implementation for device applications. The main cause of phosphorene degradation is its highly reactive behavior with oxygen under ambient conditions.[22–25] Among the 2D layered materials, phosphorene suffers the most severe degradation. Phosphorene nanoflakes degrade in a few hours after exfoliation, which makes them very difficult to handle in the air. Additionally, the defects caused by the edge and surface degradation during the preparation process degrade a few layers of phosphorene within 2 hours. Moreover, multilayer phosphorene is relatively more stable, which can be sustained up to a few days. Therefore, stability in phosphorene can be achieved by utilizing the passivation techniques by investigating the chemistry behind the process of stabilization is needed for phosphorene-based device applications.

2.3.2 MASS PRODUCTION

For commercial applications, large-scale production of the phosphorene material is required. However, the phosphorene material family is still facing major challenges in the upscaling of the production of the layer.[7,8] Generally, mechanical exfoliation, liquid exfoliation, and vapor phase growth are commonly adopted strategies for phosphorene production.[7–19] The mechanical exfoliation technique is used for the discovery of graphene, and it is still a promising technique for 2D materials preparation. It can provide the best quality 2D materials, making it an ideal tool for fundamental physics studies. However, it is not a suitable method for the large-scale production of 2D materials. Additionally, it is difficult to control the number of layers and their size, orientation, and phase. On the other hand, liquid phase exfoliation and pulsed laser deposition methods are suitable for the low-cost production of 2D materials with the scale-up capabilities. However, the size and the quality of the solution-prepared 2D materials are of concern for some applications.[7–19] Fabrication of the bulk phosphorene is a huge challenge to researchers because of the need for high pressure and temperature, which are not achievable in the ordinary crystal growth approach. Additionally, the growth of phosphorene using the bottom-up approach is also not suitable for achieving higher temperature and pressure. Therefore, there is a need for a novel growth process by which the production of good quality phosphorene with large area growth can be achieved.[7,8] The new technique for the growth of the phosphorene should be simple and include maximum production, scalable, cost-effective, and suitable for industrial applications. In addition to mass production, grown phosphorene should be stable under the ambient conditions for the application of device realization.

2.4 GROWTH TECHNIQUES FOR PHOSPHORENE

Generally, atomically thin phosphorene is prepared using two fundamental methodologies, specifically "top-down techniques" (e.g., mechanical exfoliation and liquid exfoliation) and "bottom-up techniques" (e.g., chemical vapor deposition [CVD] and wet-chemical deposition).[7,8,30] There are weak interplanar van der Waals forces and strong intraplanar covalent bonding in all 2D-layered materials. A very notable case of this materials family is graphite, which is a layered planar structure comprising feebly stacked graphene layers. In the top-down method, the layered bulk materials are exfoliated into mono- or few-layered nanosheets by the application of driving forces to break the interplanar weak van der Waals interaction. Like other 2D materials, in BP the individual atomic layers are stacked together by weak van der Waals forces, which have good conductivity. BP has an interlayer distance in the range of 3.20–3.73 Å relying on the stacking pattern.[7,8,30] These values of spacing between layers make this material exceptionally suitable for the exfoliation process to produce atomically thin layers of phosphorene from bulk BP crystal. Additionally, there are bottom-up techniques, including CVD and wet-chemical strategies, which use a direct synthesis approach to grow 2D materials from certain precursors via chemical reactions.

2.4.1 TOP-DOWN METHODS

In 2014, two individual groups reported the deposition of phosphorene using the mechanical exfoliation method by sticky tape.[7,8,30] This triumphant show of phosphorene deposition has created another research field. The FETs fabricated using ultrathin phosphorene showed an ON/OFF ratio of up to 10^4 and a mobility of hole of 286 cm^2 V^{-1} s^{-1} at room temperature.[34] Since the properties of phosphorene material are very sensitive to its thickness, a new growth process should be developed to obtain the controlled thickness of grown material. Mechanical exfoliation is a significant technique because it produces high-purity single-crystal ultrathin 2D nanosheets.[7,8,30] This deposition technique is used for the cleavage of various 2D materials. Also, this technique can produce a comparatively large area of phosphorene nanosheets. Nonetheless, a few impediments of this strategy, including the absence of efficient control of shape, size, thickness, and low production yield, limit its application. This mechanical cleavage technique has another recognizable issue related to chemical traces left from the adhesive tapes after the peeling. These issues have prompted ongoing endeavors focused on elective ways to deal with the conventional mechanical cleavage limitations. Gomez and associates have changed the mechanical exfoliation technique to improve the deposition of ultrathin phosphorene flakes.[35,36] They reported the reduction in contamination and improvement in yield by utilizing an intermediary viscoelastic surface.[35,36] In their process, the all-dry transfer method is adopted for efficient transfer of phosphorene. Liquid exfoliation is another important method for the preparation of monolayered phosphorene or a few layers of phosphorene. In this technique, a few important strategies, for example, ion exchange, oxidation, or surface passivation by different solvents, are used. In the past, liquid exfoliation produced ultrathin nanosheets of different 2D nanomaterials, which have demonstrated outstanding performances of devices fabricated using these exfoliated materials. Additionally, microwave-assisted exfoliation and lithiation methods are also used for exfoliation of high-quality phosphorene.[7,8,30]

2.4.2 BOTTOM-UP METHODS

In different bottom-up methods, the CVD technique is a prominent one to deposit 2D materials at the suitable temperature on given substrates.[7,8] It ought to be noticed that this technique has the potential to produce a large-area high-quality ultrathin film of 2D materials such as graphene and TMDs with tunable thickness and size. Presently, the researchers are putting more effort into producing large-area ultrathin phosphorene. The production techniques are designed for considering the deposition of large-area single crystals of phosphorene to study the anisotropic properties of this material. Recently, a very efficient CVD process is demonstrated by Smith and his group to produce phosphorene film of size more than 3 μm^2.[37] Additionally, wet-chemical synthesis is used as a bottom-up method for deposition of different 2D materials.[7,8] Unfortunately, there is no report to deposit ultrathin phosphorene film using this synthesis technique as per the author's knowledge. Therefore, more efforts are needed to explore the bottom-up method for producing large-area phosphorene for its practical applications.

2.5 INSTABILITY OF PHOSPHORENE AND PASSIVATION STRATEGIES: A CASE STUDY

It has been observed that the newly deposited phosphorene material is very prone to the oxidation process, which started promptly upon the uncovered surface under suitable conditions for the process.[22–25] The formation of POx occurs at the surface or edge of phosphorene in the presence of moist air and/or light. The degradation of phosphorene under ambient conditions limits the performance of phosphorene-based devices. Therefore, a detailed study to stabilize the phosphorene is very critical for its application for devices. Here, the authors are presenting a case study in detail to passivate phosphorene by different capping layers such as Al_2O_3, Si_3N_4, and Al_2O_3/Si_3N_4 stack to stabilize it in atmospheric conditions and at high temperature.[38] The detailed study consists of the sample preparation process and characterization to know the properties of passivated phosphorene followed by device fabrication using passivated phosphorene.

2.5.1 SAMPLE PREPARATION PROCESS

Nilges' method was employed to grow bulk BP crystals in which red phosphorus, Sn, and SnI_4 were used as precursors.[39] In this method, the fused silica ampoule consisting of loaded precursors is evacuated and sealed. Then this sealed ampoule is loaded into a furnace and heated at appropriate temperature. After the reaction and cooling, the ampoule is broken to collect BP crystals. Then BP (phosphorene) nanosheets are deposited on to AZO/Si by the mechanical exfoliation process. Finally, the passivation of phosphorene/AZO/Si samples is performed by the Al_2O_3, Si_3N_4, and Al_2O_3/Si_3N_4 stack.

The direct write lithography process, etching, metallization, etc., were employed to fabricate electrodes of phosphorene/AZO p-n heterojunction diode.

Bruker nanoscope AFM was employed to measure the thickness and surface imaging of phosphorene. The Raman spectra of different samples were investigated employing confocal Raman spectroscopy, which has a 50× objective lens to focus the laser beam and to collect the Raman signal.

2.5.2 CHARACTERIZATION OF PASSIVATED PHOSPHORENE AND DEVICE FABRICATION

Figure 2.1 demonstrates the height profile and AFM image of Si_3N_4 passivated phosphorene flakes. The thicknesses of phosphorene flakes are in the range of 8–16 nm as confirmed by height profiles in AFM measurements.

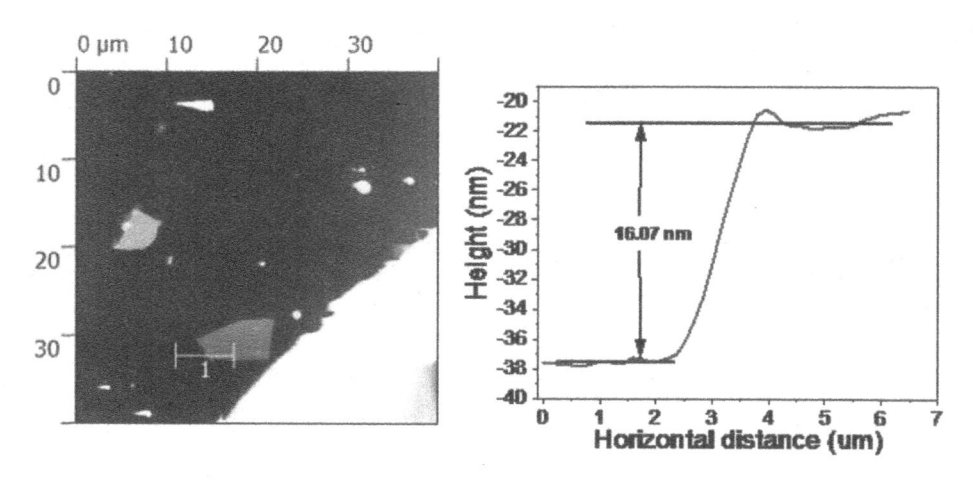

FIGURE 2.1 AFM image and the corresponding AFM height profile of typical phosphorene nanosheet. (Adapted from S. K. Pandey et al., RSC Adv. **7**, 46201–46207, 2017.)

The out-of-plane (A_g^1) and in-plane ($A2_g$ and B_{2g}) vibrational modes were detected in the Raman measurement. We found that the probe power during the measurement was an important parameter. It was observed that powers more than 23 mW/cm^2 damaged the phosphorene, resulting in peak position shifts, even with the passivation layers. For that reason, power densities of 2–5 mW/cm^2 were employed. Figure 2.2(a–c) demonstrates the effect of annealing temperature on the different vibrational modes for Si_3N_4, Al_2O_3, and Al_2O_3/Si_3N_4 passivation, respectively.

For Raman spectroscopy measurement, the same sample in the as-deposited state (unannealed) as well as after each annealing process at temperatures from 100°C to 600°C was used. The clear phonon peaks of in-plane ($A2_g$ and B_{2g}) and out-of-plane (A_g^1) vibrational modes up to 550°C annealing were observed for the sample with Si_3N_4 passivation. However, samples passivated by Al_2O_3 and Al_2O_3/Si_3N_4 hybrid layers were found to have phonon peaks up to 450°C and 500°C annealing, respectively. Thus, a nitride-only (Si_3N_4) passivation enables the highest temperature anneals. Due to stress present in phosphorene, the shifts in the Raman peak positions were observed. Blue shifts indicate decreasing tensile stress or increasing compressive stress. The phosphorene lattice adopts the puckered honeycomb crystal structure. It is found that the biaxial strain can exist in both the armchair and zigzag directions.[38]

Figure 2.3(a–c) demonstrates the graphs related to Raman peak positions. The dashed line in each figure shows the peak positions for a similar flake without a passivation layer, which helps to evaluate the shift. The unannealed outcome shows the effect of the passivation layer on the generation of stress in phosphorene. The similar effects of annealing were observed on out-of-plane modes and in-plane modes. Although the effects are less pronounced for out-of-plane modes. The results related to annealing can commonly be divided into three regimes: low temperature red shift, medium temperature blue shift, and high temperature degradation. A red shift occurs for anneals up to 200°C and is mainly prominent for the Si_3N_4 passivated sample. The blue shift occurs between annealing temperatures 200°C and 400°C. Peak positions changed little between 400°C and 450°C. For high temperature (>400°C) annealing, the behavior of the samples was quite passivation dependent. As previously pointed out, Al_2O_3 and Al_2O_3/Si_3N_4 passivated samples did not illustrate Raman peaks after annealing at 550°C and 500°C, respectively, because of deterioration of the phosphorene. Peaks positions were unpredictable at the highest anneal temperatures for every sample. Most probably this was as a result of the extensive degradation of phosphorene crystal structure. The energy spacing between the A_g^1 and B_{2g} modes declines when the material is stretched; nevertheless, it augments under compression.[38] In this study, the energy spacing between $A2_g$ and B_{2g} modes returns to its low temperature value at higher annealing temperature for Si_3N_4 passivated structure, as demonstrated in Figure 2.3(d). Our finding varies from an earlier report from Late, which specified a linear red shift with rising temperatures up to 400°C on a SiO_2/Si substrate.[38] This could be caused by sample differences. In Late's report, there was no passivation layer on the phosphorene nanosheets, permitting free expansion of phosphorene materials due to annealing.[40] The observed red shift in the spectra is because of stretch in phosphorene in the zigzag direction.[41]

There are two possible reasons for the behavior observed in Figure 2.3(a–d). The first reason is out-diffusion of Zn atoms from the AZO film. The atomic radius of the Zn atoms (1.42 Å) is larger than the radius of P atoms (0.98 Å). Thus, incorporation of Zn in phosphorene can result in a lattice distortion,[40] generating a compressive strain, which compensates already present tensile strain in phosphorene.[38] Since diffusion coefficients increase exponentially with temperature, the amount of Zn incorporation in the phosphorene layer would increase with annealing temperature. The second possible explanation is that the stress in the passivation layers is decreased by thermal annealing, reducing the stress in the phosphorene. To study this, we deposited the passivation layers directly on 4-inch silicon wafers and used the wafer bowing to determine the average film stress. We replicated the anneals done to our phosphorene samples and remeasured the stress after each anneal. The results are shown in Figure 2.4. Because the deposition was done directly on silicon rather than on the sample structure (phosphorene on AZO), the absolute values of the stress in these films may be different in this measurement. However, the annealing trends should show similar effects.

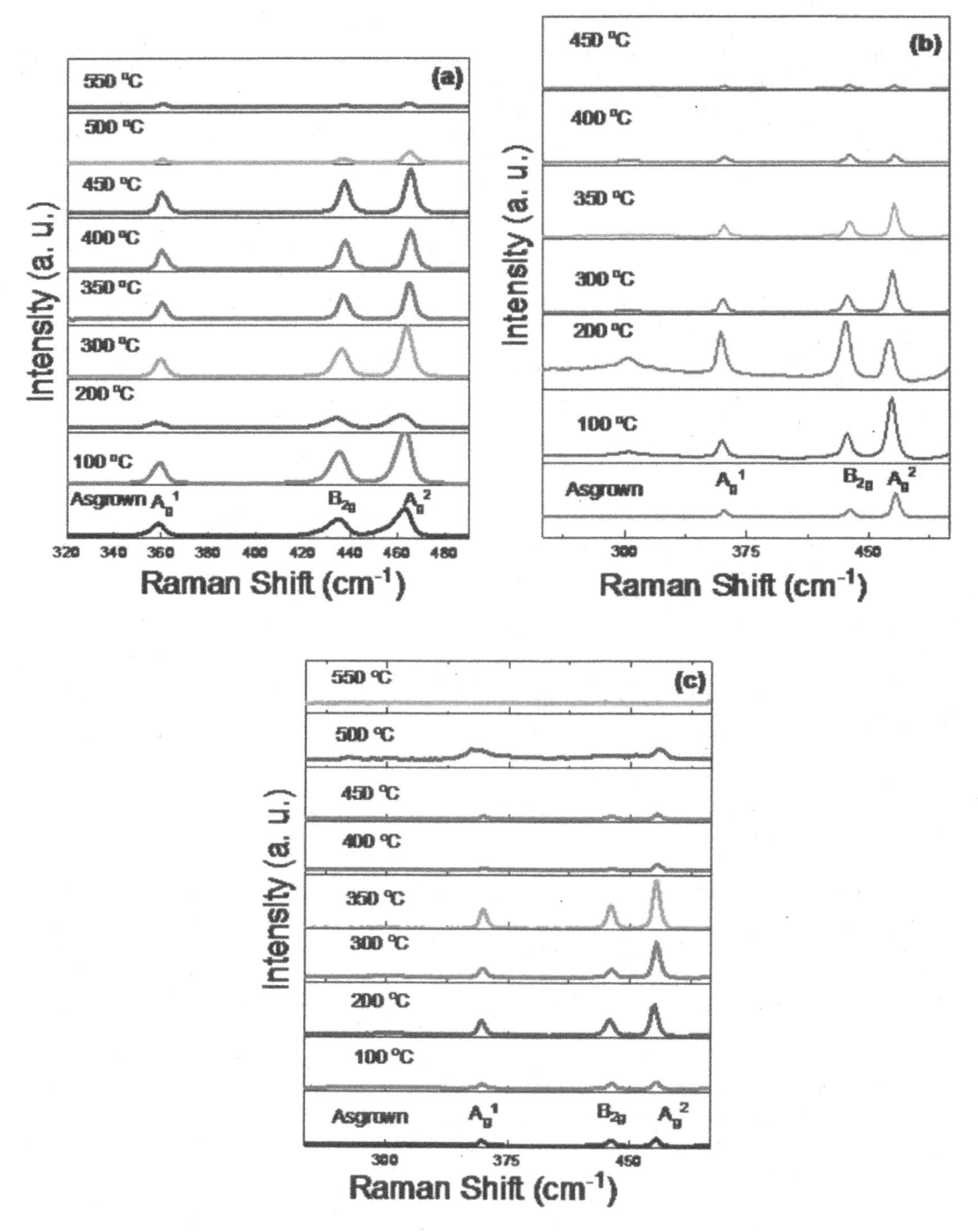

FIGURE 2.2 Raman spectra of the phosphorene/AZO samples passivated by (a) Si_3N_4, (b) Al_2O_3, and (c) Si_3N_4/Al_2O_3 layers annealed at different temperatures. (Adapted from S. K. Pandey et al., RSC Adv. **7**, 46201–46207, 2017.)

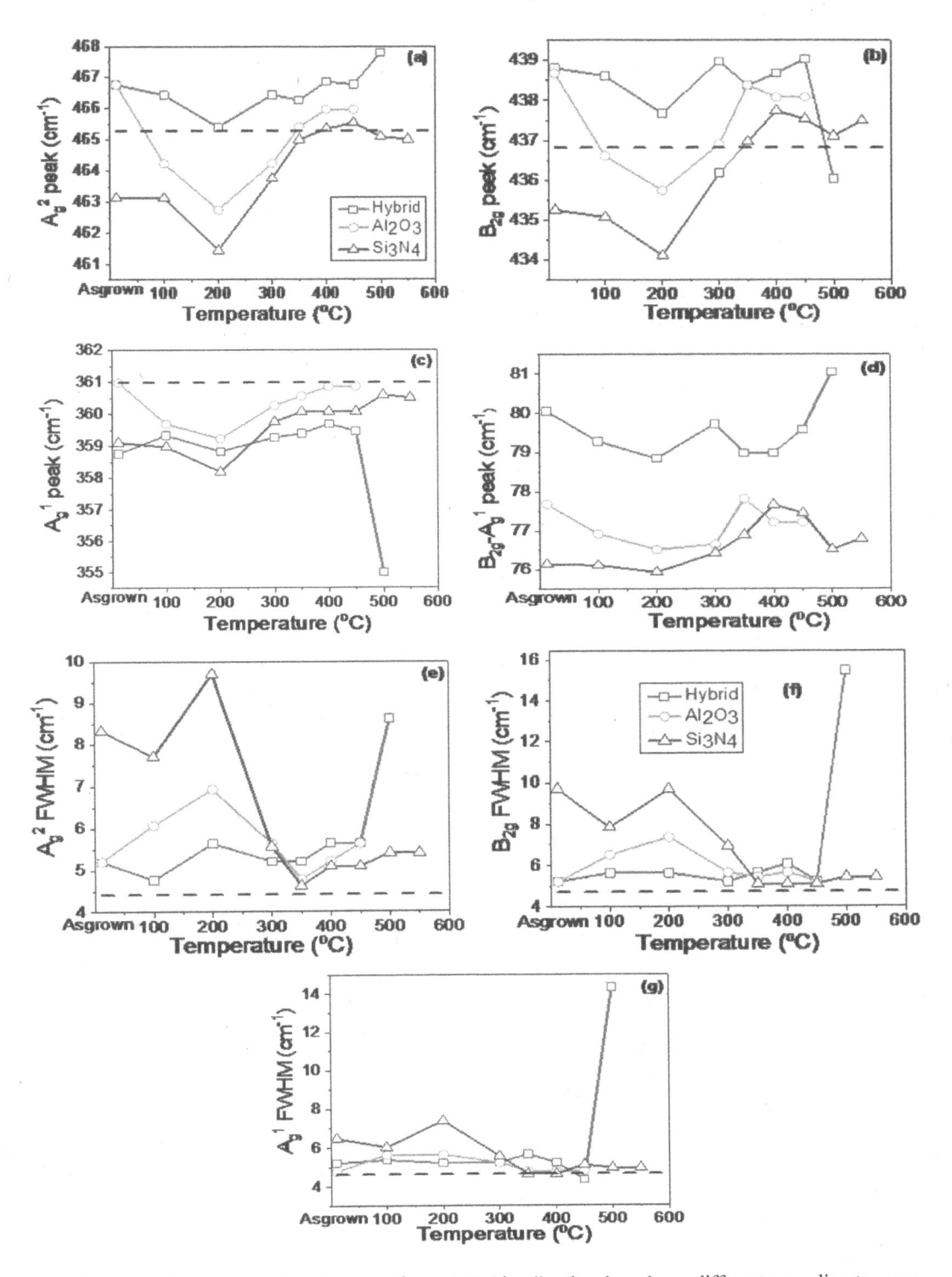

FIGURE 2.3 Raman shifts of (a) A^2_g, (b) B^2_g, and (c) A^1_g vibrational modes at different annealing temperatures. (d) The delta Raman shift between B2g and A1g vibrational modes at different annealing temperatures. The full-width at half-maximum (FWHM) variation of (e) A^2_g, (f) B^2_g, and (g) A^1_g phonon modes with annealing temperature for different samples. (Adapted from S. K. Pandey et al., RSC Adv. **7**, 46201–46207, 2017.)

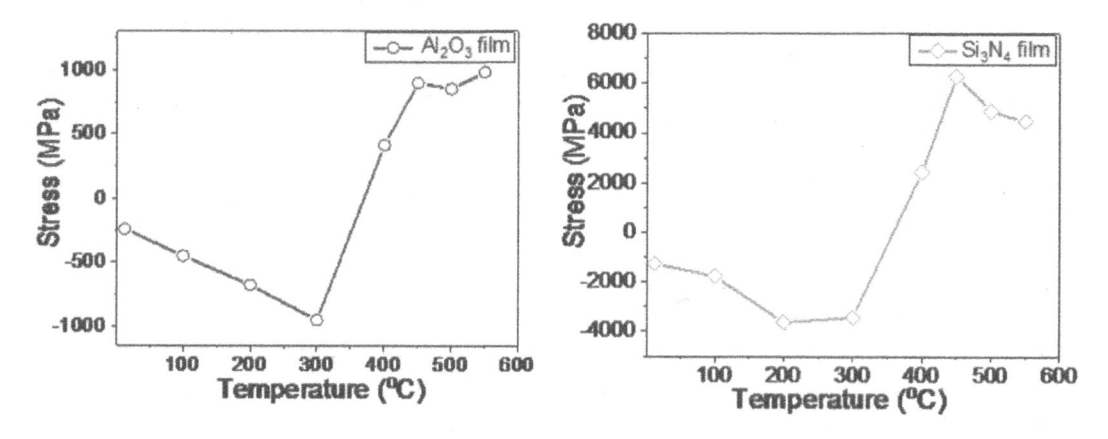

FIGURE 2.4 Stress in the passivation layers as a function of anneal temperature. (Adapted from S. K. Pandey et al., RSC Adv. **7**, 46201–46207, 2017.)

Both films show a pronounced change in stress upon anneals of 300–450°C. This effect has been reported elsewhere[38] and is believed to be due to a densification caused by a loss of hydrogen from the films. This leads to tensile strain in the passivation layer when it is deposited on conventional materials. When deposited on a 2D material, the weak interplane forces will permit significant strain relaxation, but some degree of tensile strain will still exist. Thus, the densification of the passivation layer leads to compressive strain in the phosphorene, producing the blue shift observed in Figure 2.3.

It is tempting to ascribe the medium temperature shift in the Raman position (Figure 2.3), which occurs from 200°C to 400°C, with this H_2 loss in the passivation layers, but for the difference in temperature. However, there are a few reasons to be skeptical of this apparent temperature difference. As described in the experimental section, the samples were annealed in an optically heated rapid thermal processing system. The observed temperature in this system is the read out for a thermocouple in contact with a 4-inch silicon wafer. For the work summarized in Figure 2.4, this is the actual coated wafer, whereas the work described in Figure 2.2 was done on small samples (~1 cm²), which are placed on top of the monitored 4-inch silicon wafer in the rapid thermal process (RTP) system. Poor thermal contact could be responsible for some of this difference. More significant, however, is the fact that we are heating optically. Phosphorene has a very short absorption length in the optical,[38] and the thin layer provides a very small thermal mass. It is likely, therefore, that the temperature of the phosphorene and the adjacent passivation layer significantly exceeds the measured temperature during heating transients due to the poor thermal contact of the system.

The crystalline properties of phosphorene were also investigated for different annealing temperature by calculating the full-width at half-maximum (FWHM) of A^2_g, B_{2g}, and A^1_g phonon peaks, as shown in Figure 2.3(e–g). It is well known that the FWHM of phonon peaks indicates crystallinity quality. It was observed that annealing at 350°C produced phonon peaks with a much lower FWHM. Measurements of bulk BP indicated an FWHM of about 5 cm⁻¹, comparable to the annealed results. Nitride passivated films showed marginally better results than other samples. The decomposition of the Al_2O_3/Si_3N_4 passivated structure at 500°C is clearly visible in this plot. The return of the FWHM to the bulk value is an interesting result. If the strain relaxation was driven by Zn up-diffusion, one would expect an increase in the peak width due to alloy broadening effects related to the disorder in the crystal. The fact that this was not seen suggests that this is not the case for our samples. The small FWHM in the high temperature annealed Si_3N_4-coated samples strongly suggests that the nitride acts as a diffusion barrier for phosphorus, allowing the crystalline layer to not only remain intact but actually improve its crystallinity for anneals that would otherwise

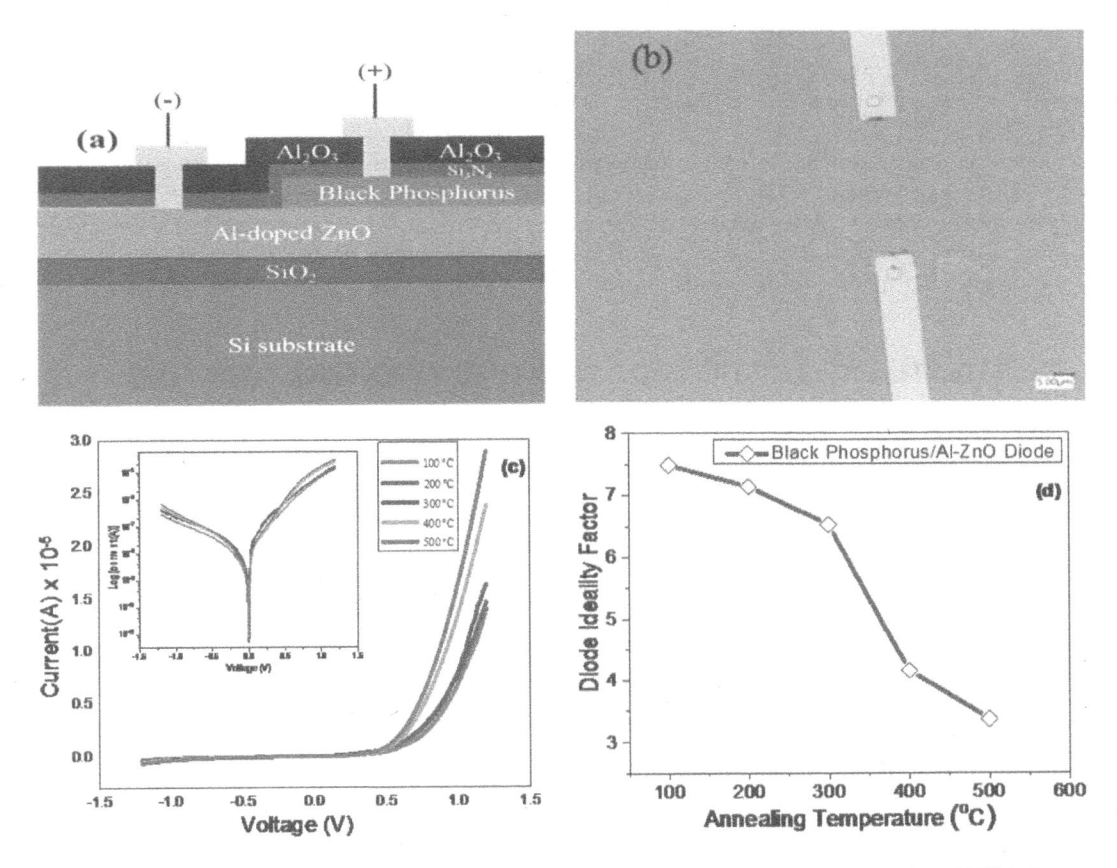

FIGURE 2.5 (a) Schematic of device structure. (b) Microscopic image of fabricated device. (c) I-V characteristics of the phosphorene/AZO heterojunction diode at different annealing temperature. The inset shows the I-V characteristics in semilog scale. (d) Calculated ideality factor at different annealing temperature. (Adapted from S. K. Pandey et al., RSC Adv. **7**, 46201–46207, 2017.)

destroy the material through sublimation. The upper limit for this effect is not known, but likely exceeds the maximum temperature, studied here, of 550°C.

Figure 2.5(a) and (b) shows the schematic and optical image of fabricated device. Figure 2.5(c) illustrates the current-voltage (I-V) characteristics of phosphorene/AZO heterojunction measured at different annealing temperature. The inset of this I-V characteristic shows the semilog plot. Ti/Au electrodes produced ohmic contact on AZO and phosphorene. The diode-like rectifying behavior confirmed the formation of p-n diode using phosphorene and AZO layers. It is well known that AZO has n-type conductivity; this indicates the p-type nature of phosphorene. The forward and reverse currents of this diode increased as annealing temperature increased. The calculated ideality factor of diode measured at different annealing temperature is shown in Figure 2.5(d). The ideality factor decreased with increase of annealing temperature and achieved a minimum value of 3.3 with an annealing temperature of 500°C, indicating improvement in quality of diode. The increase in forward and reverse currents and reduction of ideality factor can be explained by the following phenomenon. The resistance of this device can be divided into three parts: (1) the contact resistances of metal/AZO and metal/phosphorene interfaces, (2) the sheet resistances of AZO and phosphorene layers, and (3) the resistance of the p-n heterojunction near the interface.[38] The increase in annealing temperature reduced the sheet resistance of phosphorene and the contact resistance of metal/phosphorene (and AZO), thus increasing the forward and reverse current. The AZO film has high electron concentration (in the order of 10^{19} cm^{-3}) so its sheet resistance was not reduced significantly by temperature. The diode-rectifying characteristics came from the p-n heterojunction interface region, which was improved by increasing annealing temperature. It was observed in earlier study that phosphorene crystalline quality was improved at higher annealing temperature, resulting in improvement of phosphorene/AZO interface. This improvement in interface property resulted in lower ideality factor of diode, but the rectifying I-V characteristics disappeared and ohmic behavior was observed at 550°C annealing temperature. This might be due to diffusion of metal atoms into the AZO layer through the phosphorene layer for metal/phosphorene contact.

2.6 POTENTIAL APPLICATIONS OF STABLE PHOSPHORENE

Phosphorene material is preferred over other 2D materials due to its feature of high mobility, and optical properties make it one of the most favorable materials for application in the next-generation devices. However, obtaining the stable monolayers and the few layers of the phosphorene will enable their application in devices. Additionally, the phosphorene nanosheet with the semiconducting behavior alleviates the research toward their application in electronic and optoelectronic devices. Moreover, phosphorene attains various fascinating layer-dependent properties, high hole mobility with the anisotropic behavior for the hole and electron along with x- and y-directions, anisotropic optical response, and phonon anisotropy.[7,8] Therefore, phosphorene thickness-dependent bandgap and other properties make phosphorene the most favorable 2D material for the following application areas.

2.6.1 ENERGY HARVESTING

The research is working toward a sustainable future in the area of interest, for which solar power is the most important component for the clean energy harvesting by replacing fossil fuels. Traditional solar cells are based on semiconducting materials, such as silicon, CIGSe, CdTe, etc., which generate electrons under the influence of the incident photons.[7,8] Based on the bandgap and the absorption coefficient of the materials, the amount of incident photon absorption is governed. Recently, the advent of the 2D materials with direct bandgap behavior and intriguing optical and absorption properties make them a high potential candidate for photovoltaic applications.[7,8] However, a fully verified method for the formation of the tunable bandgap-based 2D material fabrication is required for the application in the field of solar cells. These 2D materials have the potential to absorb up to 5%–10% of the incident sunlight with the thickness of 1 nm and can achieve the one order of magnitude higher light absorption compared with the traditional solar absorbers such as Si and CdTe.[7,8] Two-dimensional materials have the advantage of fabricating ultrathin, flexible, highly efficient, and lightweight photovoltaic devices. However, to fabricate these photovoltaic devices it is most important to develop a process to fabricate high-quality, defect-free, highly controlled growth of 2D materials. Among established 2D materials, TMDs and phosphorene have been constantly explored due to their application in the field of photovoltaics. The property of phosphorene, such as tunable bandgap, optical properties, mobility, and electrical conductivity, makes it a potential candidate for energy harvesting applications. Phosphorene compatibility with the various material systems such as dye-sensitized solar cells, perovskite solar cells, etc., makes it an active constituent of the photovoltaic system. Additionally, various combinations of the composites, such as phosphorene/TMDs, Si/SiO$_2$-phosphorene, graphene-phosphorene, etc., are predicted and used in the photovoltaic device structure for extracting the improved performance parameters from the photovoltaic device. Therefore, after resolving the issues associated with the material growth, the phosphorene will be the irreplaceable part of the highly efficient ultrathin photovoltaic systems.

2.6.2 Energy Storage

2.6.2.1 Battery

Lithium-ion batteries (LIBs) are among the new class of batteries that have stable cycling performance, high energy density, and high storage capacity. Currently, LIBs are the source of energy in smartphones, tablets, laptops, etc. The LIB consists of electrodes (anode and cathode), separators, and an electrolyte. In general, graphite is used as an anode, and lithium metal oxides are used as a cathode electrode. For better performance of the battery, the anode material has been improved by replacing with materials having high power density and a long life cycle. Phosphorene has attracted researchers to explore its application in the field of batteries because it has high power density and a long life cycle. In general, phosphorene composite with the carbon nanomaterials has been utilized in LIBs. Such composites provide improvement in the lifetime and charge/discharge capacities of batteries.[7,8]

There are different classes of materials, which are the primary area of research in the field of energy storage applications. Among the new class of battery research, lithium-sulfur is one of the promising areas. The predicted energy density of about 2597 Wh/kg can be extracted from the Li-S battery, which is the biggest motivation behind the research in this class of battery. However, there are various issues associated with the Li-S battery, such as weak ionic and sulfur conductivity, irreversible loss of active species, and during the charging and discharging cycle, large volumetric variation in the S electrode, which results in loss of contact between conductive species and the current collector. These issues can be minimized by utilizing the unique features of phosphorene such as extraordinary physical and chemical properties, high surface-to-volume ratio, high carrier mobility, and appreciable low diffusion energy barrier toward Li, which makes phosphorene an excellent candidate for the anode of Li-S battery.

2.6.2.2 Super Capacitor

Supercapacitors are one of the most promising areas for research in the field of electronics. In general, materials within the high surface area such as graphene, etc., are the major contributors in the field of the electrode for supercapacitors. Recently, phosphorene is proposed as the alternative of the graphene for supercapacitor applications. The development in phosphorene fabrication and stabilized growth process using passivation techniques will contribute to the improvement of the properties such as high power density, long life cycle, fast charge/discharge rate, etc., and the supercapacitor performance parameters. Additionally, the introduction of 2D materials, specifically phosphorene, also has the advantage of the stretchable supercapacitor application for the energy storage devices. These stretchable supercapacitor devices may be utilized in wearable computing applications, flexible mobiles, flexible sensor devices, etc.

2.6.2.3 Field-Effect Transistors

Unique properties of 2D materials have been explored for revolutionizing the electronics domain. The incorporation of phosphorene passivated with Ag+ metal ion in the FET results in improved mobility and ON/OFF ratio, which is due to improved ambient stability of the phosphorene. Comparing TMD-based FETs, phosphorene-based FETs exhibit fivefold higher mobility. Therefore, for high performance, phosphorene-based FETs are more suitable due to its high hole mobility at room temperature and tunable bandgap. Additionally, phosphorene can be the potential candidate that has the ability to extract the best-optimized values of mobility and ON/OFF ratio from the FET.[7,8]

2.6.2.4 Sensing Applications

For sensing applications of phosphorene, the property utilized is its high surface-to-volume ratio, which is due to its puckered lattice structure. The high surface-to-volume ratio leads to the phosphorene application in the field of gas sensing, humidity sensing, photodetection, biosensing, and ion sensing. Gas sensors are essential to ensure public safety, indoor air quality control, industrial chemical sensing, etc. Metal oxide gas sensors have been emerged as the major contributor in the sensor field due to their low cost and ease of synthesis. However, the relatively high operating temperature hinders their application due to high power consumption and potential thermal power safety. Therefore, there is a need for room temperature sensors. Recently, 2D nanomaterials have shown great potential to be good candidates for room temperature sensing layers because of their large surface-to-volume ratio and high surface activities. Among various 2D materials, phosphorene is highly sensitive to the surrounding atmosphere and can be used to detect different gas molecules. Additionally, humidity sensing is another aspect of sensing, which can be performed using stacked phosphorene flakes prepared using the liquid phase epitaxy method.[7,8]

2.7 CONCLUSION

Beyond graphene, phosphorene is the most commended layered material due to its extraordinary properties. This material has potential application in the fields of nanoelectronics, nanophotonics, energy conversion, energy storage, and biomedicals. The major challenges in this materials field are related to its deposition of large-area ultrathin films and stability. In this study, different methods to prepare phosphorene and future strategies to grow large-area crystals and films are discussed. The CVD process can be the potential technique to grow large-area phosphorene crystals and films. Presently, the biggest challenge in the field is obtaining stable phosphorene for the fabrication of phosphorene-based devices. The efficient technique to stabilize phosphorene in air and at high temperatures is discussed in detail. The different properties of passivated phosphorene are examined in detail by Raman spectroscopy, optical microscopy, AFM, and current-voltage measurement of fabricated devices. The potential applications of passivated phosphorene, such as solar cells, batteries, supercapacitors, FETs, and sensing devices, are revealed in detail. The information presented in this chapter will help to make phosphorene a shining star among all 2D materials.

REFERENCES

1. K. Kim, J. Y. Choi, T. Kim, S. H. Cho, and H. J. Chung, Nature **479**, 338 (2011).
2. Q. H. Wang, K. K. Zadeh, A. Kis, J. N. Coleman, and M. S. Strano, Nat. Nanotechnol. **7**, 699 (2012).
3. F. H. L. Koppens, T. Mueller, P. Avouris, A. C. Ferrari, M. S. Vitiello, and M. Polini, Nat. Nanotechnol. **9**, 780 (2014).
4. F. Schwierz, Nat. Nanotechnol. **5**, 487 (2010).
5. G. Fiori, F. Bonaccorso, G. Iannaccone, T. Palacios, D. Neumaier, A. Seabaugh, S. K. Banerjee, and L. Colombo, Nat. Nanotechnol. **9**, 768 (2014).
6. K. S. Novoselov, V. I. Fal'ko, L. Colombo, P. R. Gellert, M. G. Schwab, and K. Kim, Nature **490**, 192 (2012).
7. D. K. Sang, H. Wang, Z. Guo, N. Xie, and H. Zhang, Adv. Funct. Mater. **29**, 1903419 (2019).
8. M. Batmunkh, M. Bat-Erdene, and J. G. Shapter, Adv. Mater. **28**, 8586–8617 (2016).
9. M. K. Jana, A. Singh, D. J. Late, C. R. Rajamathi, K. Biswas, C. Felser, U. V. Waghmare, and C. N. R. Rao, J. Phys. Condens. Matter **27**, 285401 (2015).
10. D. J. Late, S. N. Shirodkar, U. V. Waghmare, V. P. Dravid, and C. N. Rao, ChemPhysChem **15**, 1592 (2014).
11. L. Li, Y. Yu, G. J. Ye, Q. Ge, X. Ou, H. Wu, D. Feng, X. H. Chen, and Y. Zhang, Nat. Nanotechnol. **9**, 372 (2014).
12. Y. Cai, G. Zhang, and Y. W. Zhang, Sci. Rep. **4**, 6677 (2014).
13. H. Liu, A. T. Neal, Z. Zhu, Z. Luo, X. Xu, D. Tomanek, and P. D. Ye, ACS Nano. **8**, 4033 (2014).
14. H. Liu, Y. Du, Y. Deng, and P. D. Ye, Chem. Soc. Rev. **44**, 2732 (2015).
15. J. Dai, and X. Zeng, J. Phys. Chem. Lett. **5**, 1289 (2014).
16. F. Xia, H. Wang, D. Xiao, M. Dubey, and A. Ramasubramaniam, Nat. Photon. **8**, 899 (2014).
17. K. W. Ang, Z. P. Ling, and J. Zhu, IEEE International Conference on Digital Signal Processing, p. 1223 (2015).
18. P. K. Kannan, D. J. Late, H. Morgan, and C. S. Rout, Nanoscale **7**, 13293 (2015).

19. R. Fei, and L. Yang, Nano Lett. **14**, 2884 (2014).
20. Y. Deng, Z. Luo, N. J. Conrad, H. Liu, Y. Gong, S. Najmaei, P. M. Ajayan, J. Lou, X. Xu, and P. D. Ye, ACS Nano. **8**, 8292 (2014).
21. P. Gehring, R. Urcuyo, D. L. Duong, M. Burghard, and K. Kern, Appl. Phys. Lett. **106**, 233110 (2015).
22. D. Grasseschi, D. A. Bahamon, F. C. B. Maia, A. H. Castro Neto, R. O. Freitas, and C. J. S. de Matos, 2D Mater. **4**, 035028 (2017).
23. Y. Wang, B. Yang, B. Wan, X. Xi, Z. Zeng, E. Liu, G. Wu, Z. Liu, and W. Wang, 2D Mater. 3, 3 (2016).
24. W. Luo, D. Y. Zemlyanov, C. A. Milligan, Y. Du, L. Yang, Y. Wu, and P. D. Ye, Nanotechnology 27, 43 (2016).
25. C. Han, Z. Hu, A. Carvalho, N. Guo, J. Zhang, F. Hu, D. Xiang, J. Wu, B. Lei, L. Wang, C. Zhang, A. H. C. Neto, and W. Chen, 2D Mater. 4, 021007 (2017).
26. J. Wu, N. Mao, L. Xie, H. Xu, and J. Zhang, Angew. Chem. 127, 2396 (2015).
27. F. Xia, H. Wang, and Y. Jia, Nat. Commun. 5, 4458 (2014).
28. V. Tran, R. Soklaski, Y. Liang, and L. Yang, Phys. Rev. B. 89, 235319 (2014).
29. T. Low, A. Rodin, A. Carvalho, Y. Jiang, H. Wang, F. Xia, and A. C. Neto, Phys. Rev. B. 90, 075434 (2014).
30. S. C. Dhanabalan, J. S. Ponraj, Z. Guo, S. Li, Q. Bao, and H. Zhang, Adv. Sci. 4, 1600305 (2017).
31. J. W. Jiang, and H. S. Park, J. Phys. D Appl. Phys. **47**, 385304 (2014).
32. E. Flores, J. R. Ares, A. Castellanos-Gomez, M. Barawi, I. J. Ferrer, and C. Sánchez, Appl. Phys. Lett. **106**, 022102 (2015).
33. H. Mu, S. Lin, Z. Wang, S. Xiao, P. Li, Y. Chen, H. Zhang, H. Bao, S. P. Lau, and C. Pan, Adv. Opt. Mater. **3**, 1447 (2015).
34. H. Liu, A. T. Neal, Z. Zhu, Z. Luo, X. Xu, D. Tománek, and P. D. Ye, ACS Nano. **8**, 4033 (2014).
35. A. Castellanos-Gomez, L. Vicarelli, E. Prada, J. O. Island, K. L. Narasimha-Acharya, S. I. Blanter, D. J. Groenendijk, M. Buscema, G. A. Steele, J. V. Alvarez, H. W. Zandbergen, J. J. Palacios, H. S. J. van der Zant, 2D Mater. **1**, 025001 (2014).
36. M. Buscema, D. J. Groenendijk, S. I. Blanter, G. A. Steele, H. S. J. van der Zant, and A. Castellanos-Gomez, Nano Lett. **14**, 3347 (2014).
37. J. B. Smith, D. Hagaman, and H. F. Ji, Nanotechnology **27**, 215602 (2016).
38. S. K. Pandey, N. Izquierdo, R. Liptak, and S. A. Campbell, RSC Adv. **7**, 46201–46207 (2017).
39. R. Fei, and L. Yang, Appl. Phys. Lett. **105**, 083120 (2014).
40. D. J. Late, ACS Appl. Mater. Interfaces. **7**, 5857 (2015).
41. Z. P. Ling, and K.W. Ang, APL Mater. **3**, 126104 (2015).

3 Study of Transition Metal Dichalcogenides in Junctionless Transistors and Effect of Variation in Dielectric Oxide

Prateek Kumar, Maneesha Gupta, Kunwar Singh, and Ashok Kumar Gupta

CONTENTS

3.1 INTRODUCTION

In 1960, Moore suggested that the "number of transistors per unit area will double after every 18 months" [1]. A possible solution was stacking or decreasing the size of transistors; however, stacking is not a physical solution, which left researchers with option of size reduction. Transistors show a major change in properties until they enter nanometer range and short channel effects (SCEs) dominate the performance of the device [2]. SCEs, which substantially affect the performance of the transistors, are drain-induced barrier lowering (DIBL), velocity saturation, hot electron effect, and punch through types [3, 4]. Vital characteristics that SCEs degrade/hamper are subthreshold slope (SS), I_{OFF}, and power dissipation. Failure of the metal oxide semiconductor field-effect transistor (MOSFET) left researchers with no choice but to seek an alternate device that could work positively in a nanometer regime.

Over the last couple of decades, researchers have provided tunnel field-effect transistors (TFETs) [5–7], nanotubes [8–10], and junctionless transistors [11, 12] as possible candidates to replace MOSFET. Each of the possible candidates has its own set of advantages and disadvantages. TFET was discovered by a team of researchers in IBM in the 2004 [13]. The basic structure of the device is similar to MOSFET; however, the charge transport mechanism differs as in MOSFET transport due to thermionic emission, but in the case of TFET it is due to band-to-band tunneling (BTBT). After the introduction various forms of TFETs have been proposed. Bhuwalka et al. in 2004 proposed a vertical TFET [14]. The results for both experimental-band simulation-based tools were calculated. Basic operation of the device was controlled using a vertical gate. Kumar et al. in their article "Doping-Less Tunnel Field Effect Transistor: Design and Investigation" proposed TFET using the charged plasma technique [15]. It was found that the proposed doping-less device is free from random dopant fluctuations (RDFs). In 2011, Adrian M. Ionescu showed that TFETs can be used in the

DOI: 10.1201/9781003126393-3

future to replace MOSFETs as the principal device as it caused 100-fold less power dissipation [16]. Although TFETs have various sets of advantages, they have disadvantages too, such as large Miller capacitances [17], variations in performance due to RDF, and high costs due to complex fabrication [18]. Similar to TFETs, nanotubes have provided the required boost for nano regime semiconductors. A nanotube is a hollow tubelike structure. In 1997, Collins et al. in their article "Nanotube Nanodevice," fabricated single-walled nanotubes and observed the charge transportation using a scanning tunneling microscope [19]. A new era in electronics was boosted with the discovery of carbon nanotube-based single-electron transistors. In the first publication of a single-electron transistor using a nanotube by Tans et al. discussed the effect of multiwall tube structure, Zeeman effect, etc. [20] In the article "Crossed Nanotube Junctions," the effect of metallic/semiconducting contact on a metallic/semiconductor nanotube and creating metallic-metallic, metallic-semiconductor, semiconductor-metallic, semiconductor-semiconductor interfaces is shown [21]. Although the nanotube has been vastly explored, nanotubes still have basic issues like poor fabrication and poor transport characteristics [22]. Junctionless transistors are another possible candidate to replace MOSFET [23]. Junctionless transistors do not have a varying doping profile, which eliminates the issue of complex fabrication. Previously designed junctionless transistors suffer from poor gate control and lower I_{ON}. In this study we have designed a tube-shaped junctionless transistor (as like a nanotube) to overcome the previously mentioned disadvantages.

Discovery of graphene has led to the application of a whole new category of devices called two-dimensional (2D) materials [24]. Unlike a bulk semiconductor, in 2D materials free charge carrier can not transport across a volume but can transport along a plane. Research for a field of novel semiconducting materials was inevitable due to the failure of silicon (Si) on its transformation from bulk to a many-layered configuration. Major 2D materials under the radar of researchers are graphene [25, 26], transition metal dichalcogenides (TMDCs) [27, 28], and black phosphorus (BP) [29].

In 2004, at University of Manchester, HP Boehm et al. discovered the first grapheme sheet [24]. After the discovery, graphene revolutionized the field of physics because of its multipurpose applications in the field of light-emitting diodes (LEDs), solar cells, touch panels or smart windows, etc. Frank Schwierz, in his article "Graphene Transistors" published in 2010, reviewed graphene thoroughly and presented a broad prospective about how graphene is changing from a topic of condensed matter physics to a lifeline of electronics [25]. He also concluded that although graphene is a promising material, to use it to nourish an electronic material, a sizeable bandgap is required. In its exfoliated form, graphene exhibits zero bandgap, which results in poor I_{OFF}. Recent studies have suggested that graphene bandgap can be tuned, but it alters the mobility of the charge carriers [30].

TMDC materials are in the form of MX_2, where M is a transition metal like molybdenum, tungsten, etc., and X is a chalcogenide like telluride, selenide, etc. TMDC materials possess several advantages as they can be tunable bandgap structures, direct bandgap structures, etc. Manzeli et al., in his article "2D Transition Metal Dichalcogenides," thoroughly reviewed TMDC materials. This review suggested that TMDCs are favorable electronic materials as they have atomic-scale thickness and very strong spin coupling. Different applications of TMDC materials like strain engineering and behavior against high-frequency applications were also analyzed [31]. In the article "Electronics and Optoelectronics of Two-Dimensional Transition Metal Dichalcogenides," the authors have thoroughly discussed applications of TMDC materials in the field of electronics and optical electronics [32]. Although TMDC materials are possible candidates, transistors designed using them have very low I_{ON}, which leads to a poor I_{ON}/I_{OFF} ratio due to less mobility of charge carriers [33].

BP is a multilayered configuration of phosphorus. It can be used from a monolayer configuration to a five-layer configuration. BP changes from indirect bandgap material for bulk to direct bandgap material for structure with a few layers. In BP, with an increase in the number of layers, mobility of the charge carrier also increases (350 cm^2/V-s for monolayer to 2755 cm^2/V-s for five layers). Although BP in the five-layer configuration shows excellent characteristics, it is difficult to fabricate; hence, the development of transistors has high costs, which is important in today's industry in which designers try to keep a tight cap on the price range [29].

Hence, in this work, junctionless transistors have been designed using different TMDC materials like molybdenum disulfide (MoS_2), molybdenum diselenide ($MoSe_2$), molybdenum ditelluride, and tungsten disulfide (WS_2). The previously mentioned materials are chosen as they are the most explored and are found to be stable. Also, all-semiconductor devices suffer from junction capacitance problems, so in this study, semiconductor-on-insulator (SOI) technology is used. The TMDC channel is proposed on sapphire (Al_2O_3) and a gate dielectric HfO_2 is used.

3.2 DEVICE STRUCTURE AND MODELS USED

Cross-sectional view and three-dimensional (3D) view of the proposed device is shown in Figure 3.1.

The proposed device is 122 nm in length, where both drain and source spacer are 1 nm each and the channel is 120 nm. Radius of the device is 10 nm. The insulator on which the device is grown is 2 nm. The width of the TMDC materials is 5 nm. The gate dielectric is 3 nm. As the minimum number of conduction band states available are at a minimum for WS_2 (0.73×10^{19}), doping in the case of each material is chosen as 0.5×10^{19} cm^{-3}. Different models used for simulation are conmob, consrh, trap.auger, fldmob, and BGN print. Copper in the <111> plane configuration is used as a metal electrode. All the physical parameters are specified in Table 3.1.

3.3 RESULTS AND DISCUSSION

In this work MoS_2, $MoSe_2$, $MoTe_2$, and WS_2 are used as channel materials. Different device properties like energy (eV), potential (V), and electric field (V/cm) across the channel are studied. Different analog properties studied are drain current-gate voltage characteristics (I_{DS}-V_{GS}), threshold voltage (V_t), SS, etc. Materials are also compared for various linearity properties like first-order transconductance (g_m), second-order transconductance (g_{m2}), third-order transconductance (g_{m3}), transconductance gain factor (TGF), third-order voltage intercept point (VIP3), current intercept point (IIP3), second-order harmonic distortion (HD2), and third-order harmonic distortion (HD3).

Band energy curve for the proposed device is shown in Figure 3.2. The energy curve is shifting downward due to applied drain voltage V_{DS}. In the OFF state (shown in Figure 3.3), the energy curve around the gate terminal shifts upward, which makes charge carriers difficult to transport from source to drain. On applying positive bias on both, that is $V_{GS} = V_{DS} = 1$ V, the curve shifts downward resulting in flow of current. Higher downward shift across the drain terminal indicates a higher concentration of electron required for conduction.

Figure 3.2 supports the fact that MoS_2 has one of the highest bandgaps across TMDC materials, thus, the shift in energy curve is minute, whereas the shift across the drain terminal is much more significant for the remaining materials. Curve for potential across the channel is plotted in Figure 3.4. A rise in potential for $MoSe_2$, $MoTe_2$, and WS_2 is very steep across the drain terminal

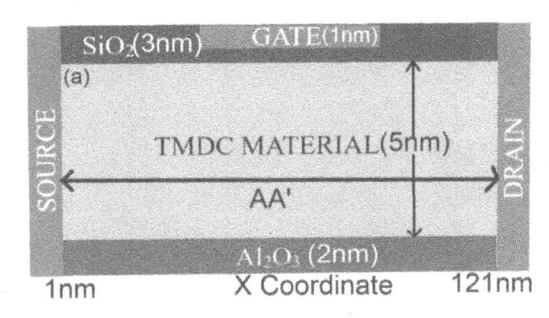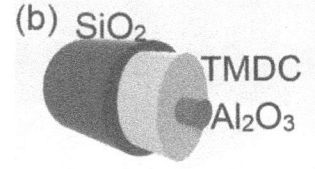

FIGURE 3.1 (a) Cross-sectional view of the device along with physical measurements and (b) a 3D view of the device.

TABLE 3.1

Physical Parameters for Junctionless Transistor

Parameters	TMDC-Based Junctionless Transistor (nm)
TMDC length	122
Source spacer length	1
Drain spacer length	1
Substrate insulator (Al_2O_3)	2
Channel radius	5
Gate dielectric	3
Gate depth	1

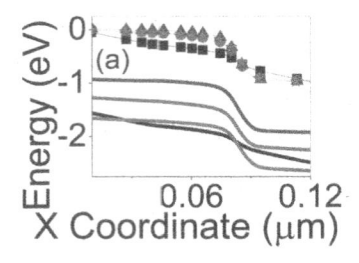

FIGURE 3.2 Energy curve for different TMDC materials is shown. The black, red, blue, and pink lines represent MoS_2, $MoSe_2$, $MoTe_2$, and WS_2, respectively, and will represent the same throughout this chapter. In the curve, symbols represent the conduction band and the solid lines represent the valence band.

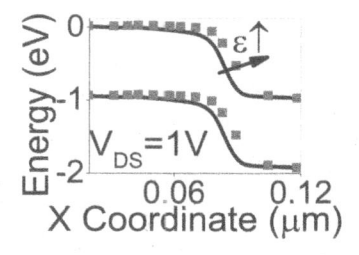

FIGURE 3.3 OFF- and ON-state energy curves are plotted. The black line indicates the OFF state and the ON state is represented by the red line.

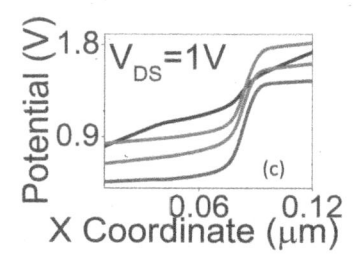

FIGURE 3.4 Potential across the channel along cutline AA′[shown in Figure 3.1(a)] is plotted. Rise in potential curve across the drain terminal is highest for $MoTe_2$ representing higher electron concentration supported by Figure 3.2.

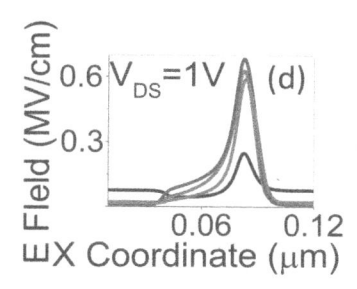

FIGURE 3.5 Curve for electric field across the channel is plotted along the cutline AA′.

and because of the higher electron concentration, which gives rise to a reverse biased potential. As the downward shift is highest in the order $MoTe_2$, $MoSe_2$, and WS_2, the same is supported by the potential curve due to reverse bias formation.

The curve for the electric field across the channel is plotted in Figure 3.5. Electric field is the differential of potential and, hence, follows the curve of potential across the channel (Figure 3.4). As the potential curve is steepest for $MoTe_2$, the peak is highest for $MoTe_2$. Variation in curve is minute for MoS_2, hence, the variation is minimal for MoS_2.

The higher potential curve indicates the presence of higher charge carriers. Higher concentration will result in a higher OFF-state current and the same is supported by I_{DS}-V_{GS} characteristics as plotted in Figure 3.6. OFF state is highest for WS_2, $MoTe_2$, and $MoSe_2$. OFF state is highest for MoS_2 because the presence of charge carriers across most of the channel is highest for MoS_2, as shown in Figures 3.2 and 3.3. Delay in saturation for $MoTe_2$ can be attributed to the lower bandgap available as higher numbers of carriers are available, making it difficult to saturate.

Various characteristics like I_{ON}, I_{OFF}, I_{ON}/I_{OFF}, $V_t =$, and SS for different TMDC materials are summarized in Table 3.2. From Table 3.2, it can be noted that in those applications where faster switching abilities and lowest power dissipation are required, $MoTe_2$ can be used as channel material. MoS_2 should be the least preferred material as it has the highest slope and OFF-state current.

First-order transconductance for TMDC materials is compared in Figure 3.7. The curve of g_m for MoS_2 shows the least variation as slope is the least for the MoS_2-based transistor. Also, g_m directly represents the behavior of the device as an amplifier.

The g_m directly depends on the variation in I_{DS}, hence, g_m follows the curve for I_{DS}-V_{GS}. It is important to note here that delay in saturation of $MoTe_2$ hampers the linearity performance of the material-based device.

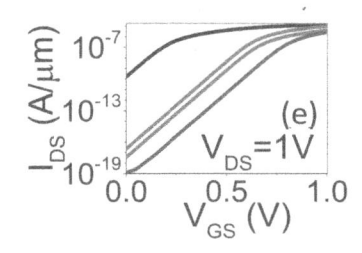

FIGURE 3.6 I_{DS}-V_{GS} characteristics for different TMDC materials as a channel of junctionless transistor are compared.

TABLE 3.2
Values of All the Compared Materials Are Summarized

Materials	I_{ON}	I_{OFF}	I_{ON}/I_{OFF}	V_t	SS
MoS$_2$	1.47×10^{-05}	1.61×10^{-10}	91287.9494	-0.16	60.9
MoSe$_2$	3.95×10^{-06}	3.39×10^{-18}	1.16×10^{12}	0.26	59.47
MoTe$_2$	2.66×10^{-06}	1.21×10^{-19}	2.20×10^{13}	0.34	50.44
WS$_2$	9.86×10^{-06}	2.15×10^{-17}	4.57×10^{-05}	0.22	59.48

TGF is the ratio of first-order transconductance to the drain current and is plotted in Figure 3.8. From Figure 3.7 it can be noted that the initial stage slope for all the materials is less, but it becomes steep after ≈ 0.1 V; hence, there is a sharp drop in the TGF curve initially for all the materials. As slope is constant for all the materials up to 0.6 V, it is constant in TGF. For higher voltages, the value of g_m is lower, which accounts for the steep fall in the curve of TGF.

Before studying the linearity properties, the effect of variation on insulator material on the I_{DS}-V_{GS} characteristics is examined. To study the effect, both inner and outer material is chosen with $k = 25$. The effect of variation in permittivity is only shown on energy across the channel in Figure 3.9(a) and on I_{DS}-V_{GS} characteristics in Figure 3.9(b). Figure 3.9(a) represents that with an increase in dielectric constant, the energy curve shifts upward, which indicates a lack of charge carriers available. It can be noted that with an increase in permittivity I_{OFF} decreases. With a higher k, higher gate voltage will be required for proper channel formation or for particular voltage, there will be less available charge carriers, hence, a lower I_{DS}.

Linearity parameters are analyzed using the following mathematical relations [34]:

$$VIP3 = \sqrt{24 \times g_{m1}/g_{m3}} \tag{3.1}$$

where $g_{m1} = \left(\partial I_d / \partial V_{gs} \right)$, $g_{m2} = \left(\partial^2 I_d / \partial V_{gs}^2 \right)$ and $g_{m3} = \left(\partial^3 I_d / \partial V_{gs}^3 \right)$

$$IIP3 = \frac{2}{3} \frac{g_{m1}}{g_{m3} \times R_S} \tag{3.2}$$

$$HD2 = 0.5V_a \frac{\left(\dfrac{dg_{m1}}{dV_{GT}} \right)}{2g_{m1}} \tag{3.3}$$

$$HD3 = 0.25V_a \frac{\left(\dfrac{d^2 g_{m1}}{dV^2 GT} \right)}{6g_{m1}} \tag{3.4}$$

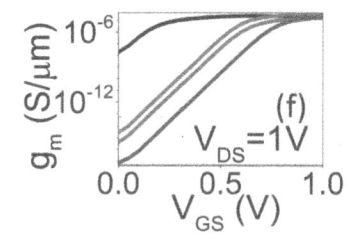

FIGURE 3.7 Variation in g_m against V_{GS} for different TMDC materials is plotted.

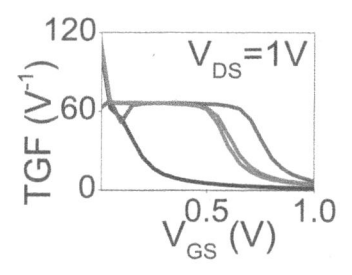

FIGURE 3.8 Variation in TGF curve against gate voltage (V_{GS}) is plotted.

As all the linearity parameters depend on the value of g_{m2} and g_{m3}, before plotting the linearity parameters, curves for g_{m2} and g_{m3} are plotted in Figure 3.10 (a and b, respectively). Both g_{m2} and g_{m3} are nonlinearity parameters, so, for better linearity, their values should be low. At operating point, that is $V_{DS} = V_{GS} = 1$ V, the value of g_{m2} is highest for MoTe$_2$ and as stated while discussing I_{DS}-V_{GS} characteristics, delay in saturation hampers the linearity performance of the device. Third-order transconductance represents the zero cross point (ZCP), which is vital to determine the biasing point of the circuit. ZCP is the value in which g_{m3} goes from positive to negative or negative to positive for the time. Figure 3.10(b) indicates that the value of ZCP is lowest for MoS$_2$ and it is worst for MoTe$_2$. Although the MoS$_2$-based device shows better g_{m2} and g_{m3} characteristics, it is unsuitable for most of the applications because of its poor I_{DS}-V_{GS} characteristics.

For higher linearity, the value of g_{m3} should be low, and from Equation 3.1, it can be related that for higher linearity, the value of VIP3 should be higher. VIP3 gives the point where first- and third-order harmonics of I_{DS} for a particular V_{DS} are equal. The curve for VIP3 is plotted in Figure 3.11. At operating voltage, as value of g_m is nearly the same and variation in g_{m3} is minute, the value of VIP3 is nearly the same. As for most of the range, the value of g_{m3} is lowest for MoS$_2$; hence, the value of VIP3 is highest for MoS$_2$ for most of the range.

IIP3 indicates the point where the third-order input signal becomes equal to third-order harmonic distortion. Like VIP3, IIP3 is also inversely proportional to g_{m3}, so the higher linearity value of IIP3 should also be higher. The curve for IIP3 is plotted in Figure 3.12. It can be noted that at the guiding voltage, the value of IIP3 is the least for MoTe$_2$ and for MoS$_2$, it is high for most of the range.

The curves for HD2 and HD3 are plotted in Figure 3.13(a) and (b), respectively. For better operation of the device, the values of HD2 and HD3 should be low. At operating point, the least values of HD2 and HD3 are achieved for the WS$_2$-based device, whereas the highest value is achieved for MoTe$_2$.

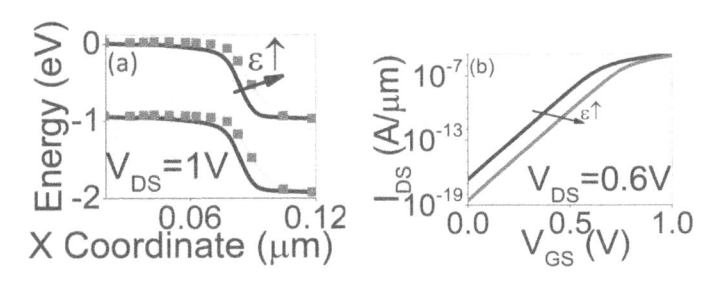

FIGURE 3.9 Effect of variation in dielectric oxide on (a) the energy curve along cutline AA′ [shown in Figure 3.1(a)] and (b) I_{DS}-V_{GS} characteristics are shown.

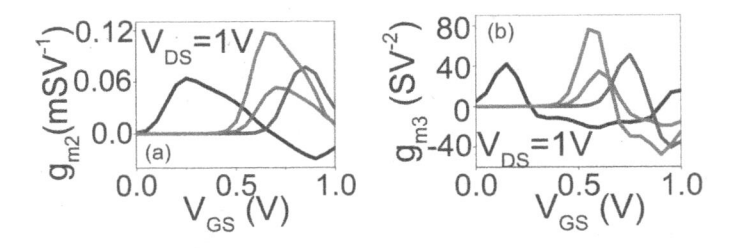

FIGURE 3.10   Variation in (a) g_{m2} and (b) g_{m3} against gate voltage is plotted.

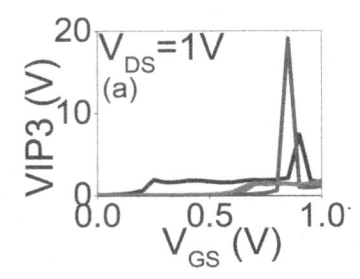

FIGURE 3.11   Variation in VIP3 against gate voltage. Although at operating voltage all of the materials show the same linearity, for most of the range the MoS_2-based device results in the highest linearity.

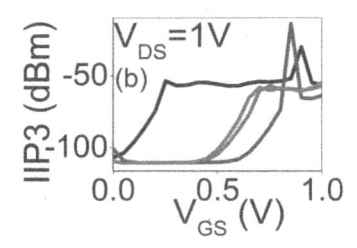

FIGURE 3.12   Variation in third-order IIP is plotted against gate voltage. It shows that at $V_{GS} = 1.0$ V. Linearity achieved is the least for $MoTe_2$.

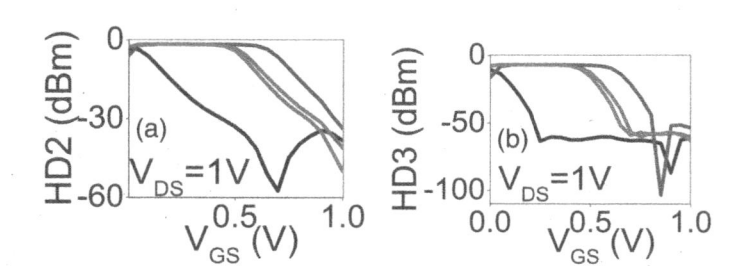

FIGURE 3.13   Variation in (a) HD2 and (b) HD3 against gate voltage is plotted.

3.4 CONCLUSION

In this work, a TMDC materials-based junctionless transistor is designed. It was observed that although the $MoTe_2$-based device provided the least I_{OFF}, due to delayed saturation linearity properties deteriorate. On replacing the transistor materials with higher dielectric materials, performance of the device decreases due to a limited amount of charge carrier availability. The WS_2-based device provides excellent I_{DS}-V_{GS} characteristics and higher linearity; hence, it can be used for the applications in which faster switching and higher linearity are required.

REFERENCES

1. R. R. Schaller, "Moore's Law: Past, Present and Future," IEEE Spectrum, vol. 34, no. 6, pp. 52–59, June 1997, doi: 10.1109/6.591665
2. W. Vandenberghe, "Quantum Transport in Tunnel Field-Effect Transistors for Future Nano-CMOS Applications," Ph.D. thesis, Leuven, Aug 2012.
3. S. M. Sze, "Physics of Semiconductor Devices," 3rd edn. Wiley, 2008.
4. B. G. Streetman, and S. K. Banerjee, "Solid State Electronics Devices," 6th edn. PHI, 2013.
5. O. M. Stuetzer, "Junction Fieldistors," Proceedings of the IRE, vol. 40, no. 11, pp. 1377–1381, Nov. 1952.
6. J. Quinn, G. Kawamoto, and B. McCombe, "Sub-band Spectroscopy by Surface Channel Tunneling," Surface Science, vol. 73, pp. 190–196, 1978.
7. W. Reddick, and G. Amaratunga, "Silicon Surface Tunnel Transistor," Applied Physics Letters, vol. 67, no. 4, pp. 494–496, 1995.
8. S. Iijima, "Helical Microtubules of Graphitic Carbon," Nature, vol. 354, pp. 56–58, 1991.
9. S. J. Tans, A. R. M. Verschueren, and C. Dekker, "Room- Temperature Transistor Based on a Single Carbon Nanotube," Nature, vol. 393, pp. 49–52, May 1998.
10. W. Lu, and C. M. Lieber. "Semiconductor Nanowires," Journal of Physics D, vol. 39, no. 21, pp. R387, 2006.
11. C. W. Lee, A. Afzalian, N. D. Akhavan, R. Yan, I. Ferain, and J. P. Colinge, "Junctionless Multigate Field-Effect Transistor," Applied Physics Letters, vol. 94, no. 5, pp. 053511, 2009.
12. C. W. Lee, A. Borne, I. Ferain, A. Afzalian, R. Yan, N. D. Akhavan, and J. P. Colinge, "High-Temperature Performance of Silicon Junctionless MOSFETs," IEEE Transactions on Electron Devices, vol. 57, no. 3, pp. 620–625, 2010.
13. L. L. Chang, and L. Esaki, "Tunnel Triode—A Tunneling Base Transistor," Applied Physics Letters, vol. 31, no.10, pp. 687–689, 1977.
14. K. K. Bhuwalka, S. Sedlmaier, A. K. Ludsteck, C. Tolksdorf, J. Schulze, and I. Eisele, "Vertical Tunnel Field-Effect Transistor," IEEE Transactions on Electron Devices, vol. 51, no. 2, pp. 279–282, 2004.
15. M. J. Kumar, and S. Janardhanan. "Doping-Less Tunnel Field Effect Transistor: Design and Investigation," IEEE Transactions on Electron Devices, vol. 60, no. 10, pp. 3285–3290, 2013.
16. A. M. Ionescu, and H. Riel, "Tunnel Field-Effect Transistors as Energy-Efficient Electronic Switches," Nature, vol. 479, no. 7373, pp. 329–337, 2011.
17. M. A. Raushan, N. Alam, and M. J. Siddiqui, "Performance Enhancement of Junctionless Tunnel Field-Effect Transistor Using Dual-k Spacers," Journal of Nanoelectronics Optoelectronics, vol. 13, pp. 912–920, 2018.
18. W. Y Choi, and H. K. Lee, "Demonstration of Hetro-Gate-Dielectric Tunneling Field-Effect Transistors," Nano Convergence, vol. 3, pp. 1–15, 2016.
19. P. G. Collins, A. Zettl, H. Bando, A. Thess, and R. E. Smalley, "Nanotube Nanodevice," Science, vol. 278, no. 5335, pp. 100–102, 1997.
20. S. J. Tans, A. R. M. Verschueren, and C. Dekker, "Room-Temperature Transistor Based on a Single Carbon Nanotube," Nature, vol. 393, no. 6680, pp. 49–52, 1998.
21. M. S. Fuhrer, J. Nygård, L. Shih, M. Forero, Y. G. Yoon, H. J. Choi, J. Ihm, S. G. Louie, A. Zettl, and P. L. McEuen, "Crossed Nanotube Junctions," Science, vol. 288, no. 5465, pp. 494–497, 2000.
22. C. Wang, K. Takei, T. Takahashi, and A. Javey, "Carbon Nanotube Electronics–Moving Forward," Chemical Society Reviews, vol. 42, no. 7, pp. 2592–2609, 2013.
23. S. Sahay, and M. J. Kumar. Junctionless Field-Effect Transistors: Design, Modeling, and Simulation. John Wiley & Sons, 2019.

24. H. P. Boehm, A. Clauss, G. O. Fischer, and U. Hofmann, "Das Adsorptionsverhalten sehr dünner Kohlenstoff-folien," Zeitschrift für Anorganische Und Allgemeine Chemie, vol. 316, no. 3–4, pp. 119–127, 1962.

25. F. Schwierz, "Graphene Transistors," Nature Nanotechnology, vol. 5, no. 7, pp. 487, 2010.

26. D. Li, and R. B. Kaner, "Graphene-Based Materials," Nature Nanotechnology, vol. 3, no. 101, 2008.

27. T. Roy, M. Tosun, X. Cao, H. Fang, D. H. Lien, P. Zhao, Y. Z. Chen, Y. L. Chueh, J. Guo, and A. Javey, "Dual-Gated MoS_2/WSe_2 Van Der Waals Tunnel Diodes and Transistors," American Chemical Society NANO, vol. 9, pp. 207–2079, 2015.

28. P. M. Campbell, J. K. Smith, J. Ready, and E. M. Vogel, "Material Constraints and Scaling of 2-D Vertical Heterostructure Interlayer Tunnel Field-Effect Transistors," IEEE Transactions on Electron Devices, vol. 64, no., pp. 2714–2720, 2017.

29. J. Qiao, X. Kong, Z. X. Hu, "Few Layered Black Phosphorus: Emerging 2D Semiconductor with High Anisotropic Carrier Mobility and Linear Dichroism," Nature Communications, vol. 5, pp. 4475, 2014.

30. S. M. Song, J. K. Park, O. J. Sul, and B. J. Cho, "Determination of Work Function of Graphene under a Metal Electrode and Its Role in Contact Resistance," Nano Letters, vol. 12, pp. 3887–3892, 2012.

31. S. Manzeli, D. Ovchinnikov, D. Pasquier, O. V. Yazyev, and A. Kis, "2D Transition Metal Dichalcogenides," Nature Reviews Materials, vol. 2, no. 8, 2017.

32. Q. H. Wang, K. Kalantar-Zadeh, A. Kis, J. N. Coleman, and M. S. Strano, "Electronics and Optoelectronics of Two-Dimensional Transition Metal Dichalcogenides." Nature Nanotechnology, vol. 7, no. 11, pp. 699–712, 2012.

33. R. Yan, S. Fathipour, Y. Han, B. Song, S. Xiao, M. Li, V. Protasenko, D.A. Muller, D. Jena, and H.G. Xing, "Diodes in Van Der Waals Heterojunctions with Broken-Gap Energy Band Alignment," Nano Letters, vol. 15, pp. 5791–5798, 2015.

34. R. Chaujar, R. Kaur, M. Saxena, M. Gupta, and R. S. Gupta, "TCAD Assessment of Gate Electrode Workfunction Engineered Recessed Channel (GEWE-RC) MOSFET and Its Multi-Layered Gate Architecture. Part II: Analog and Large Signal Performance Evaluation," Superlattices Microstructure, vol. 46, no. 4, pp. 645–655, 2009.

4 GNRFET-Based Ternary Repeaters

Prospects and Potential Implementation for Efficient GNR Interconnects

Afreen Khursheed and Kavita Khare

CONTENTS

4.1 INTRODUCTION

During the mid-twentieth century, rapid exponential performance improvement of integrated circuits (ICs) has been observed in accordance with Moore's law [1]. This is possible because of extensive scaling of on-chip technology. Downsizing of technology dimensions results in enhanced transistor performance but at the same time spoils the interconnect performance. Based on International Terrestrial Reference System (ITRS) predictions, interconnect delay dominates gate delay at the deep submicron (DSM) regime [2]. Moreover, parasitic impedance associated with traditional interconnects leads to 50% of total power consumption in ICs. Hence, there is an immediate need to focus on enhancing interconnect performance by adopting smart techniques to reduce signal propagation delay without compromising on-power consumption. One way is to replace the conventional copper (Cu) wires with novel carbon allotropes like carbon nanotubes (CNTs) and graphene nanoribbons (GNRs) as interconnect wire. Based on the selection of the chirality vector and geometry, they can be metallic or semiconducting. For interconnecting wire modeling, only metallic CNT and GNR are considered. Compared with Cu wires, the CNTs, which are concavity cylinders of wrapped graphene sheets, can conduct huge electronic charge flows without signal wear and tear and electron migration [3, 4]. Despite showing promising results, CNTs are the least preferred allotropes compared with GNRs for modeling on-chip interconnects. The main reason behind this is due to CNT's nonplanar nature and complicated fabrication process.

GNR, which is an unzipped version of CNTs due to its ease of fabrication along with commendable electrical properties, better $(m.p)$ mean free path (MFP; 1×10^3 nm), and enhanced current transportation capacity (5–$20 \times 10^8 A/m^2$) emerged as a viable candidate for on-chip interconnects [5]. Based on the layers, GNRs are categorized as solitary layer GNR (SLGNR) and multilayered GNR (MLGNR). SLGNR manifests smaller capacitance due to its small thickness, but its high resistance value can make interconnects lossy. Mitigation in value of resistance can be obtained by increasing the conduction paths as in case of MLGNR. Moreover, with respect to their contact with surrounding devices, MLGNRs are stratified as top-tier-contacted MLGNRs and side-tier-contacted MLGNRs. Despite the fact that top-tier-contacted MLGNRs are easier to fabricate than side-contact MLGNRs, the latter offers significant mitigation in resistance to charge flow. The reason for the reduction is that in the case of top-tier-contacted MLGNRs, only the crest layer is connected with the contacts; whereas in the case of side-tier-contacted MLGNRs, all layers are physically linked with contacts, thereby providing more number of channel paths.

In addition to using a suitable material for modeling interconnect wires, other strategies like repeater insertion and optimized wire sizing can be adopted to further enhance the on-chip interconnect performance. Introducing repeater buffers sandwiched between the long wire lines regenerates the deformed waveform and significantly improves the power-delay product (PDP) [6, 7]. This chapter presents the designing of state-of-the-art repeaters using graphene nanoribbon field-effect transistors (GNRFETs) based on multivalued ternary logic instead of conventional metal oxide semiconductor field-effect transistors (MOSFETs) based on binary logic. The ternary logic has gathered the attention of many researchers because of its ability to operate at reduced power consumption, exhibiting less delay and area to implement the same functions. Furthermore, GNRFET-based repeaters show superiority over Si-based FETs due to replacement of GNRs as channel material on the transistors. Although graphene exhibits zero bandgap, if patterned into nanoscale ribbons, a bandgap opens due to lateral quantum confinement. These nanoribbons overcome short channel effects encountered in Si MOSFET. One of the important touchstones for logic circuit designing is elevated $\frac{I_{ON}}{I_{OFF}}$ ratio and reduced-conductivity off-state for suppressing power dissipation. Studies conducted [8] proved that large-diameter CNTs show evidence of large density of current, but low $\frac{I_{ON}}{I_{OFF}}$ ratio; whereas small-diameter CNTs show evidence of high $\frac{I_{ON}}{I_{OFF}}$ ratio and low current density. Like small-diameter CNTs, the GNR devices with a width of 2- to 3-nm ribbons exhibit a high ON/OFF current ratio [9].

This chapter is organized as follows. Section 4.2 briefly describes the electrical impedance modeling for conventional Cu and novel MLGNR interconnects. The electrical modeling of GNRFET buffers and the designing of GNRFET-based ternary repeaters are explained in Section 4.3. Performance analysis and simulation is carried out in Section 4.4. Finally, this chapter concludes with in a summary in Section 4.5.

4.2 BASICS OF INTERCONNECT

Interconnects are fundamentally the polysilicon wires or metallic strips that provide electrical rivets between transistors or active devices within an application-specific integrated circuit (ASIC) [10]. Interconnects perform a vital task in determining the performance of ASIC as they carry signals within a very large-scale integration (VLSI) chip. Driven by Moore's law, the extensive dimension scaling of technology over the past 10 years has led to various device and interconnect challenges.

The ever-increasing demand for fast portable devices has spurred the focus toward modeling high-performance circuits and interconnect wires carrying signals within the ICs. At the onset of the submicron era, transistors were relatively slow and interconnect wires were wide and exhibited low resistance. In such conditions, wires were treated as equipotential nodes with lumped capacitance. However, as technology steps into the DSM era, transistors switch much faster and at the same time, the interconnect wires become narrower. Thus, at lower technology node, wires no longer behave as mere resistive loads but has parasitic impedances coupled with them [11]. These impedances (RLC) are material reliant and ascertain the value of the Elmore delay. Hence, for accurate estimation of these parasitic wires, a detailed study of interconnect geometry and material property is necessary.

4.2.1 INTERCONNECTS' PARASITIC FORMULATION

This section describes the formulation of Cu and graphene interconnect models.

4.2.1.1 Copper Interconnect Model

The predictive-technology model (PTM) is incorporated to formulate the RLC parasitic of interconnects above the ground plane. The primarily used interconnect wire model for analytical study of Cu wires is illustrated in Figure 4.1. As shown in the figure, the rectangular Cu wires are separated by s, with thickness t, length l, width w, and height h over the ground.

The corresponding conductance $1/G$ of Cu rectangular interconnect wires having area A is given by

$$R = 1/G = \frac{\rho.l}{A} = \frac{\rho.l}{wt} \tag{4.1}$$

Studies show that the electrical behavior of circuits is perturbed by temperature variations, consequently, due to the electron-phonon scattering on the MFP, which computes the interconnect conductance [12]. Hence considering the effect of surface coarseness and granule boundary scattering phenomenon, the conductivity can be formulated using [13] the Sondheim surface roughness resistivity model $\left(\rho_{FS}\right)$ and Mayadas-Shatzkes grain boundary scattering resistivity model $\left(\rho_{MS}\right)$ as

$$\rho_{Cu} = \rho_{FS} + \rho_{MS} \tag{4.2}$$

$$\frac{\rho_{FS}}{\rho_{bulk}} = 1 + 0.75 \left[\frac{\lambda_{Con}\left(1-P\right)}{w} \right]$$

$$\frac{\rho_{MS}}{\rho_{bulk}} = \left[1 - 1.5\Delta + 3\Delta^2 - 3\Delta^3 \ln\left(1+\frac{(1)}{\Delta}\right) \right]$$

$$\Delta = \frac{\lambda_{Con}}{D}\frac{R}{\left(1-P\right)}$$

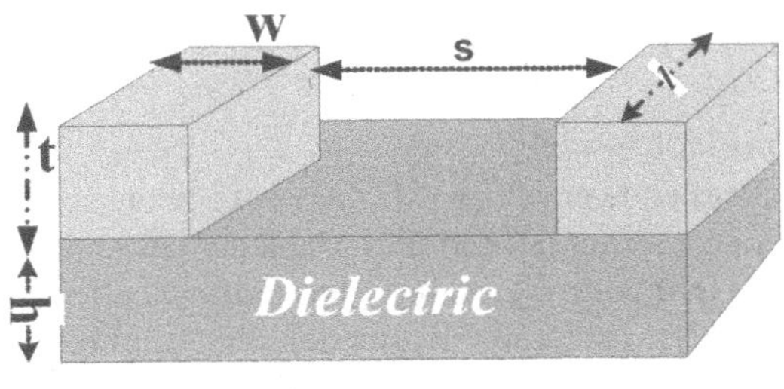

FIGURE 4.1 Model showing Cu as an interconnect.

where bulk material resistivity is ρ_{bulk}, MFP is λ_{Con}, P is Fuchs scattering parameter, the size of mean grain is D, and R is reflection coefficient having values between 0 and 1.

In the second step considering the effect of temperature on resistivity,

$$\rho_{Cu}(T) = \rho_{Cu}(0) + \left(\frac{4\,R\,(\Theta_R)\,T^n}{\Theta_R^n}\right) \int_0^{\frac{\Theta_R}{T}} \frac{Z^p}{(e^Z - 1)(1 - e^{-Z})}\,dZ \qquad (4.3)$$

$$R(\Theta_R) = \frac{h^3}{e^2}\left[\frac{\pi^3 (3\pi^2)^{1/3}}{4n^{2/3}a\,M\,k_B\,\Theta_R}\right]$$

where resistivity of nano Cu interconnects is calculated with Debye temperature Θ_R (320 K), h is Planck's constant, the number of electrons available for current conduction in an atom is n, M is atomic mass, k_B is constant defined by Boltzmann's equation, and e is the electronic charge, whereas p is an integer whose value is the function of the interaction characteristics. Henceforth, resistance of Cu interconnect after taking into consideration all the effects is given by

$$R_{Cu}(T, w_{int}) = (\rho_{Cu} + \rho_{Cu}(T))\frac{l}{w \times t} \qquad (4.4)$$

Inductance $\mathcal{L}$ and mutual inductance $\mathcal{M}$ are expressed by

$$\mathcal{L} = \mu_o \frac{l}{2\pi}\left[\ln\left(\frac{2l}{w+t}\right) + 0.5 + \left(\frac{0.22(w+t)}{l}\right)\right] \qquad (4.5)$$

$$\mathcal{M} = \mu_o \frac{1}{2\pi}\left[\ln\left(\frac{2l}{d}\right) - 1 + \frac{d}{l}\right] \qquad (4.6)$$

Total effective capacitance of Cu interconnect is given by

$$C_{total} = C_g + 2C_c \qquad (4.7)$$

$$C_g = \varepsilon\left[\frac{w}{h} + 2.22\left(\frac{s}{s+0.70h}\right)^{3.19} + 1.17\left(\frac{s}{s+1.51h}\right)^{0.76}\left(\frac{t}{t+4.53h}\right)^{0.12}\right]$$

$$C_c = \varepsilon\left[1.14\frac{t}{s}\left(\frac{h}{h+2.065}\right)^{0.09} + 0.74\left(\frac{w}{w+1.595}\right)^{1.14} + 1.16\left(\frac{w}{w+1.875}\right)^{0.16} + \left(\frac{h}{h+0.985}\right)^{1.18}\right]$$

where permeability $\mu_o = 4\pi10^{-7}$ H/m, d is center-to-center distance between wires, ε is the dielectric permittivity, and s is the inter-conductor spacing. Both inductance and effective capacitance are temperature independent.

4.2.1.2   MLGNR Interconnect Model

Graphene is an allotrope of carbon from which thin layers of ribbons can be obtained. These narrow width conducting ribbons of an ultrathin (<50 nm) graphene layers are known as GNRs. Nakada et al., in 1996, contributed a GNR pedagogical model to study the effect of nanoscale dimensioning of graphene [14]. After this, GNR gathered the attention of several researchers and has eventually emerged as a viable candidate for modeling on-chip interconnects. A unique feature is that a monolithic system is obtained by means of the single-layer GNR for interconnects as well as transistors.

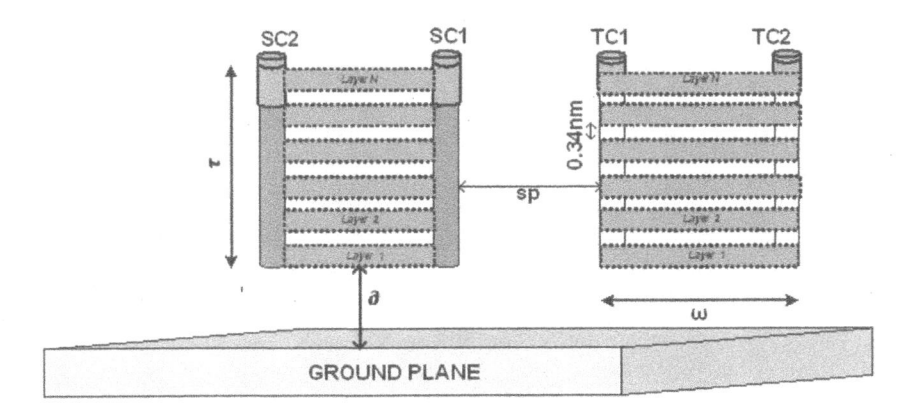

FIGURE 4.2 Model showing MLGNR as an interconnect.

As previously mentioned, the GNR will go one better than the Cu interconnects for narrow widths due to its commendable electrical properties, greater MFP (1×10^3 nm), and superior charge carrying capacity ($5-20 \times 10^8$A/m^2). Figure 4.2 show the schematic of MLGNR structure with top contact and side contact [15]. The figure illustrates that the MLGNR on-chip interconnect has N layers of width ω and height τ, which are held together due to the force exerted by the van der Waals force segregated with δ ($= 0.34$ nm) gap. The space in between the lower interconnect layer and ground is given by ∂, and ε_r is the medium's dielectric constant.

The RLC parameters of the MLGNR interconnect are expressed as in Ref. [16].

The overall count of the MLGNR layers is determined with the help of the following mathematical expression

$$N = 1 + \text{Integer}\left(\frac{\tau}{\delta}\right) \tag{4.8}$$

Every layer of MLGNR is made up of several conducting channels denoted by the variable N_{ch}. The effective number of conducting channels plays a vital role in controlling the MLGNR interconnects' performance as its parasitic impedance (RLC) parameters are a function of N_{ch}. To commensurate the value of N_{ch}, we have to take into consideration the Fermi energy (E_F), temperature (T), width ω, and the impact of spin and sub-lattice degeneracy of *carbon-carbon* atoms.

$$N_{ch} = \sum_{n=0}^{n_c}\left[e^{(E_i - E_F)/kT} + 1\right]^{-1} + \sum_{n=0}^{n_v}\left[e^{(E_i + E_F)/kT} + 1\right]^{-1} \tag{4.9}$$

where k, n_c, and n_v denote the Boltzmann constant, number of conduction bands, and number of valence bands, respectively. The term E_i delineates the smallest/highest energy of ith *conduction/valence* sub-band and can be calculated as

$$E_i = \begin{cases} \dfrac{3\sqrt{3}\, a_0\, ct}{2\left(\omega + \sqrt{3}\, a_0\right)}, & n = 0 \\[2ex] \dfrac{h.v_f}{2.\omega}|n|, & n \neq 0 \end{cases} \tag{4.10}$$

where distance between two carbon-carbon atoms is denoted by a_0 (≈ 0.142 nm), c ($= 0.12$), and t ($= 2.7$eV); h is Planck's constant; and Fermi velocity v_f ($= 8 \times 10^5$) m/s.

The equivalent resistance of MLGNR interconnect has two resistive components, namely lumped resistance $\left(R_{lumpd}\right)$ and per unit length distributed resistance $\left(R_{distributd}\right)$.

$$\left(R_{lumpd}\right) = \frac{1}{\left\{\sum_{i=1}^{N}\left(R_c^i + R_q^i\right)^{-1}\right\}} \tag{4.11}$$

where contact resistance due to imperfect contact between the metal layer and GNR layers is denoted by R_c^i (has values ranging between 1 and 20 kΩ based on the fabrication process), and resistance due to quantum effects of charge carriers in MLGNR layers is denoted by $R_q^i\left(\approx \frac{h}{2e^2 N_{ch}}\, k\Omega\right)$. The integer i has values for the number of layers ranging from 1 to N.

Whereas the distributed resistance $\left(R_{distributd}\right)$, because of phonon and acoustic scattering in MLGNR, is expressed as

$$\left(R_{dis}\right) = \frac{R_s^i}{\{N\}} \tag{4.12}$$

$$R_s^i = \frac{h}{2e^2 \cdot N_{ch} \cdot \lambda_{eff}} = \frac{12.9}{N_{ch} \cdot \lambda_{eff}}\, K\Omega$$

where e denotes electronic charge and λ_{eff} denotes the effective MFP of electrons. Applying Matthiessen's rule to calculate MFP for the nth sub-band, we get

$$\lambda_{eff,n} = \frac{1}{\dfrac{1}{\lambda_d} + \dfrac{1}{\lambda_n}} \tag{4.13}$$

where λ_d is the MFP due to various types of scattering phenomenon of electrons (viz acoustic, optical phonon, impurity, and defect scattering); λ_n is MFP due to edge scattering and is computed as

$$\lambda_n = \frac{\omega}{1-P}\sqrt{\left(\frac{2.\omega.E_f}{n.h.v_f}\right)^2 - 1} \tag{4.14}$$

where P (varies 0 to 1) is the specular constant and determines the level of edge coarseness in GNRs.

The equivalent inductance consists of two components: kinetic inductance L_K and magnetic inductance L_m

$$L = L_K + L_m \tag{4.15}$$

The kinetic inductance L_K is attributable to inertia of the ambulatory charge carrier of each sheet of MLGNR and is expressed as

$$L_k = \frac{L_{k_0}}{2N_{ch}};\ \text{where}\ L_{k_o} = \frac{h}{2e^2\, v_F} \tag{4.16}$$

where v_F is the Fermi velocity $\left(\approx 8\times10^5\,\text{m/s}\right)$ of carriers in graphene. The value of $\left(L_k\right)$ per channel is experimentally calculated as 8 nH/μm.

In addition to this, because of stored energies of carriers in the magnetic field, the MLGNR wires exhibit magnetic inductance denoted by L_m and expressed as

$$L_m = \frac{\mu_o \mu_r d}{w} \qquad (4.17)$$

Each layer of interconnect has quantum capacitance c_q, which is a function of the density of electronic states and given by Equation 4.18

$$c_q = 2c_{q_0} . N_{ch}; \text{ where } c_{q_0} = \frac{2e^2}{hv_F} \qquad (4.18)$$

As a result of coupling of electric field linking the ground plane and bottommost MLGNR layer; it introduces an electrostatic capacitance c_e, which is proportional to the width w of nanoribbons and the distance d above ground and is given by Equation 4.19

$$c_e = \frac{\varepsilon_o \varepsilon_r w}{d} \qquad (4.19)$$

The mutual inductance l_m and coupling capacitance c_m occurring in between adjacent GNR sheets are expressed as

$$l_m = \frac{\mu_o \delta}{w} \text{ and } c_m = \frac{\epsilon_o w}{\delta} \qquad (4.20)$$

4.3 BASICS OF REPEATER BUFFERS

As already mentioned, along with using a suitable material for modeling interconnect wires, several other strategies like repeater insertion and optimized wire sizing can be adopted to further enhance the on-chip interconnect performance. The Elmore delay of wire raises a quadratic ally with augmentation in wire length. Splitting the elongated interconnect into smaller segments (having parasitic impedances such as $R_{SEGMENT}$, $C_{SEGMENT}$, and $L_{SEGMENT}$) and inserting smart buffers in between these segments help reduce the effective wire length, which in turn reduces the RC propagation delay [17].

Apart from making the on-chip network faster, introducing repeaters as a buffer in between the long interconnect lines also regenerates the deformed waveform and significantly improves performance. For instance, with a signal traveling through a very long interconnect wire there are chances of loss of information due to signal distortion. This may result in logic failure. The buffer insertion technique proves beneficial in resolving this issue.

While modeling transistor buffers as repeaters, efforts must be made to design faster logic without affecting power speculations. Technology used for repeater modeling plays a vital role in achieving this objective. Graphene-based transistors emerged as a promising alternative to replace conventional silicon-based transistors for efficient repeater circuit designing. Using multivalued ternary logic for buffer circuits helps to attain faster logic.

4.3.1 GNRFET BUFFERS AS REPEATERS

Graphene is a closely packed sheet of carbon-carbon atoms arranged in a two-dimensional (2D) hexagonal framework. These sheets are etched into one-dimensional (1D) narrow stripes and are termed as GNRs. Subjected to the orientation of carbon atoms on the boundary of the graphene sheet as shown in Figure 4.3, GNRs are grouped into two types: zigzag GNRs (ZGNRs) and armchair GNRs (AGNRs). Normally AGNRs are semiconducting, whereas ZGNRs show metallic properties.

FIGURE 4.3 Graphene lattice structure.

The FETs fabricated using GNRs as channel material are termed as GNRFET. Based on the device structure, GNRFETs are classified as GNR-MOSFETs and GNR-Schottky Barrier FETs (GNR-SBFET). In this work GNR-MOSFETs are used for repeater circuit modeling because, compared with GNR-SBFET, the former has a higher ON/OFF current ratio, more robust process variation, and exhibits thermionic conduction.

To study the simulation behavior of the device, Chen et al. [18] proposed a parameterized, HSPICE-compatible model for GNRFET.

Figure 4.4(a) and (b) illustrates the structure of GNRFET and its equivalent circuit model, respectively.

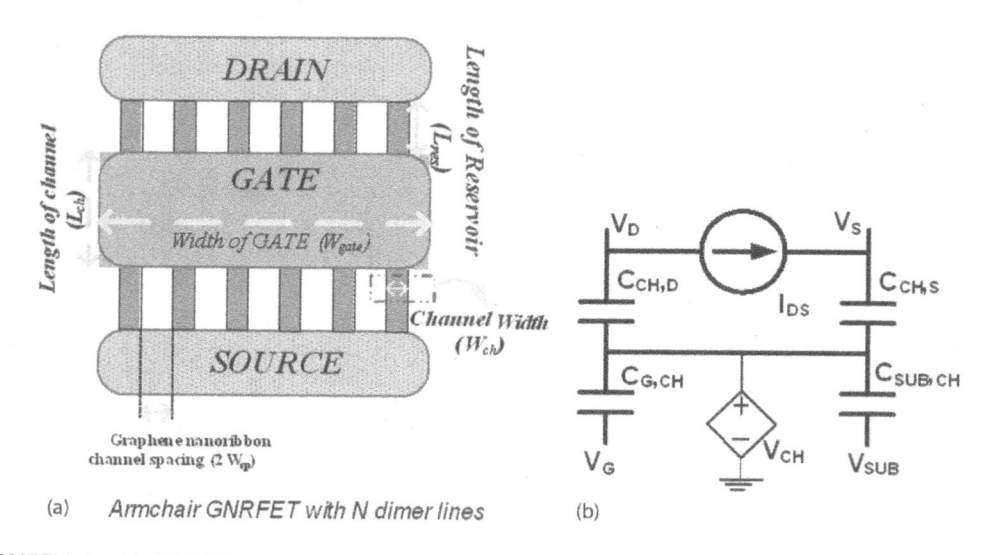

FIGURE 4.4 (a) GNRFET structure and (b) equivalent SPICE circuit model.

As illustrated from the GNRFET structure shown in Figure 4.4(a), the channel region is composed of the semiconducting type of GNRs. The undoped semiconducting ribbons are arranged beneath the gate region, whereas the extremely doped ribbons are placed beneath the drain and source. The device can be turned ON and OFF by applying potential on the gate terminal. The threshold voltage V_t required to switch ON the GNRFET is mathematically given as [19]

$$V_t = \frac{\mathcal{BG}}{3e} = \frac{2|\delta|\Delta E}{3e} \tag{4.21}$$

$$\Delta E = \frac{\pi\, h\, v_f}{W_d} \tag{4.22}$$

$$W_d = \frac{\sqrt{3}a_0(N+1)}{2} \tag{4.23}$$

where $\mathcal{BG}$ represents the bandgap; e denotes charge due to unit electron; and $\delta = 0.27, 0.4$, and 0.066 for $N = 3l$, $3l+1$, and $3l+2$, respectively. N is the number of dimer lines, $h\ (= 6.5821 \times 10^{-16}\,\text{eV})$ is Planck's constant, $v_f\ (= 10^6\,\text{m/s})$, and W_d is thickness of GNR or in other words width of the ribbon. From Equations 4.21 and 4.23 we may infer that the width of ribbon W_d varies directly with the change in dimer lines N and varies inversely with bandgap $\mathcal{BG}$.

The equivalent SPICE circuit model shown in Figure 4.4(b) can be segregated into three parts consisting of channel potential denoted by V_{CH}, current source I_{DS}, and the combination of channel capacitances denoted by variables $C_{G,CH}, C_{SUB,CH}, C_{CH,D}$, and $C_{CH,S}$ [18]. Moreover from Landauer-Buttiker formalism the value of drain-to-source current is given as

$$I_{DS} = \frac{2e}{h} \sum_{\infty} \int_0^\infty \left[f\left(E - E_{FS,C}\right) - f\left(E - E_{FD,C}\right) \right] dE \tag{4.24}$$

where E is the energy level and ∞ is the sub-band index.

4.3.1.1 Modeling of GNRFET-Based Ternary Buffer

In the recent past, ternary logic emerged as a suitable alternative to the conventional binary logic due to some of its salient features like high operating speed and information density, appreciable energy efficiency, and reduced overheads for interconnect and chip area. Ternary logic is basically a type of multivalued threshold logic with three significant logic levels equivalent to 0, $\frac{V_{DD}}{2}$, and V_{DD} voltages and denoted by symbols "0," "1," and "2," respectively. In contrast to the wearisome strategy used in MOS technology to implement multi-V_t FETs by applying different bias voltage to the substrate terminal of MOSFETs, obtaining the variation in threshold voltage for GNRFETs is quite simple.

From the previous section, the value of the threshold voltage of GNRFETs might be controlled by altering the number of dimer lines and the width of GNR layers. Hence, by judicially selecting the different combinations of dimer lines and graphene ribbon widths, multithreshold value logic buffers can be implemented. Apart from this, the performance and the current driving capability of the GNRFET is further improved by introducing more GNR layers.

Figure 4.5 illustrates the schematics of ternary logic-based GNRFET repeater buffer.

The repeater buffer circuit consists of six armchair type GNRFETs (X_1, X_2, and X_3 are N-type FETs and X_4, X_5, and X_6 are P-type FETs) with different widths ($W = 0.86, 1.23$, and 1.59 nm) obtained by selecting the values of dimer lines as $N = 6, 12$, and 18. The corresponding threshold voltages for P-type/N-type FETs obtained are $V_{th} = \pm\,0.43, \pm0.30$, and ±0.23 V, respectively.

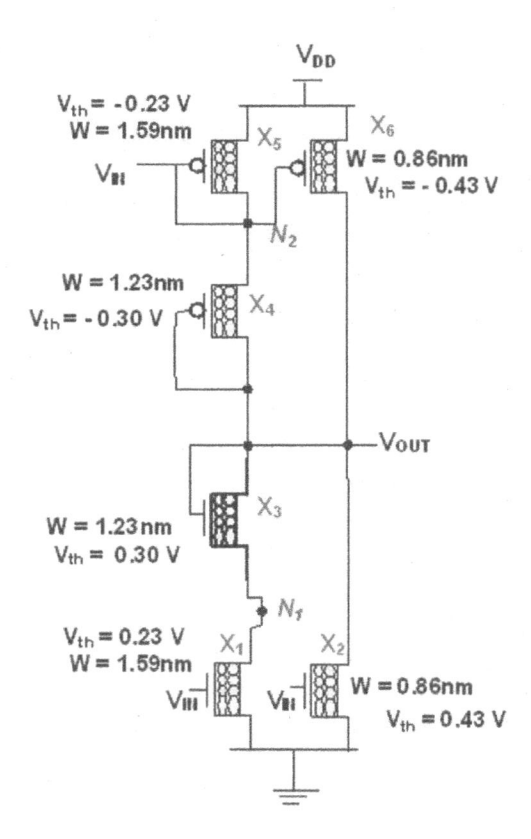

FIGURE 4.5 Ternary GNRFET-based repeater buffer.

Table 4.1 mentions the set of logic levels and voltage ranges considered against different voltage levels in this work.

When applying a potential of $V_{DD} = 0.9$ V the buffer operates in the following manner:

- Initially until the voltage of the buffer is less than 0.23 V, the transistors X_5 and X_6 are in the ON state, whereas X_1 and X_2 are in OFF state. Thus, the value of V_{OUT} is 0.9 V. Hence, the attained output voltage V_{OUT} $(= 0.9$ V$)$ represents logic level **2**.
- Now gradually when the input voltage V_{IN} increases such that it reaches beyond 0.23 V, the transistors X_5 and X_1 are turned ON, whereas X_6 and X_2 are found in the OFF state. In this condition the diode-connected transistors X_3 and X_4 (having their drain terminal and gate terminal shorted) will provide a voltage drop of 0.45 V from node N_2 to the output terminal V_{OUT} and from the output to node N_1 because the threshold voltage V_{th} of X_4 and X_3 are 0.43 V.

TABLE 4.1
Ternary Logic Levels

Voltage Level	Logic Level	Voltage Range
0	0	≤0.3 V
$V_{DD}/2$	1	0.3–0.6 V
V_{DD}	2	≥0.6 V

Hence, the attained output voltage V_{OUT} (= 0.45V), which is half of the applied supply voltage V_{DD}, represents logic level **1**.

- Finally, when the input voltage V_{IN} increases such that it exceeds 0.6 V X_5 and X_6 turn OFF and transistor X_2 turns ON. When X_2 is short-circuited, it brings down the output voltage V_{OUT} to ground potential. Hence, the attained output voltage V_{OUT} (= 0V) represents logic level **0**.

4.4 PERFORMANCE ANALYSIS AND SIMULATION RESULTS

Figure 4.6 shows the driver interconnect load (DIL) model for the HSPICE analysis. The input pulse signal V_{in} is applied with a signal transition time of 10 ns and pulse duration period of 1 μs. The HSPICE tool is used to simulate the DIL. For simulation of the DIL test bench, the SPICE GNRFET model is adapted from Ref. [20].

The coupling capacitance and mutual inductance between interconnect lines, electrostatic capacitance, and magnetic inductance of MLGNR interconnects have been computed using Synopsys Raphael. The interconnects' parasitic (viz resistance per unit length), inductance, and capacitance for Cu interconnect are 61.9 KΩ/m, 2 nH/m, and 120 fF/m, respectively. These values for the MLGNR interconnect are 2.72 KΩ/m, 7 nH/m, and 44 fF/m, respectively.

Here six ribbon layers of graphene as conducting channels are taken and applied and drain-to-source voltage is 0.9 V. Important simulation parameters of the HSPICE model with their values and brief descriptions are listed in Table 4.2.

The inquisitions are accomplished for design metrics such as delay, power, and PDP for different cases.

4.4.1 DELAY ANALYSIS

4.4.1.1 Variation in Delay Time Regarding Interconnect Wire Length without Inserting Buffers as Repeaters

Figure 4.7 shows the variations in propagation delay with variation in length of Cu as well as MLGNR interconnect wires. The interconnect wire length is varied from 200–1200 μm. It is inferred from the graph that with an increase in wire length, the signal delay also rises in a quadratic fashion. Furthermore, time taken for the signal to travel in DIL is greater in the case of the Cu interconnect wire than the MLGNR interconnects because the delay time is a function of the interconnect parasitic.

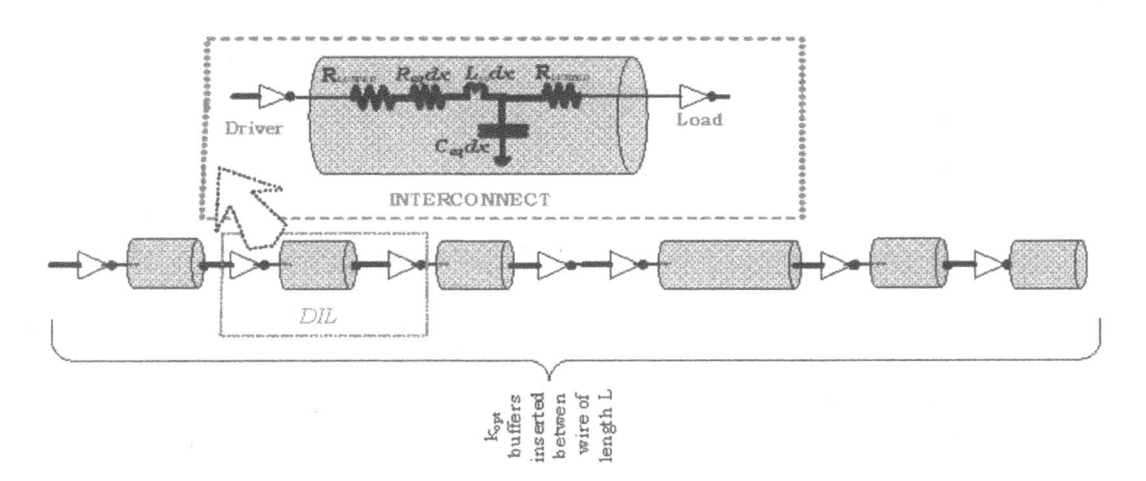

FIGURE 4.6 DIL.

TABLE 4.2
HSPICE Simulation Parameters of GNRFET

Device Parameters	Description	Default Value
L	Physical channel length	32 nm
T_{OX}	Thickness of top-gate dielectric material	0.95 nm
$2W_{sp}$	Spacing between the edges of two adjacent GNRs within the same device	2 nm
N	Number of ribbons in the device	6
P	Edge roughness percentage of the device	0
Dop	Source and drain reservoirs doping	0.001
T_{ox2}	Oxide thickness between channel and substrate/ bottom gate	20 nm
Gate_Tied	Whether gate or substrate hold the same value	0

Thus as values of Cu interconnect, impedance is larger than the MLGNR interconnect, and the time constant in case of the MLGNR is less, which results in a reduced time delay.

4.4.1.2 Variation in Delay Time Regarding Interconnect Length with Insertion of Buffers as Repeaters

Inserting buffers as repeaters in between long interconnect lines helps to minimize the propagation delay in the DIL system, but randomly adding buffers may increase the circuit overhead. Hence, the utmost care must be taken to decide the optimum count of buffers to be inserted as repeaters. Figure 4.8 shows the deviation in total buffer count for subjective interconnect lengths.

Analysis drawn from the graph tells that the number of repeaters requisite for Cu exceeds the number used in the case of MLGNR for a specified wire length. Moreover, the total count of repeaters needed rises monotonically with an increase in the interconnect's wire length.

Figure 4.9 depicts that the increase in number of repeaters shows an inverse relation regarding the time delay. The total amount of repeater buffers inserted is directed by counts equal to 2, 4, 6, 8, and 10. For a specific length of 800 μm, HSPICE simulation shows that delays reduce until the optimum number of buffer counts is reached and later increases.

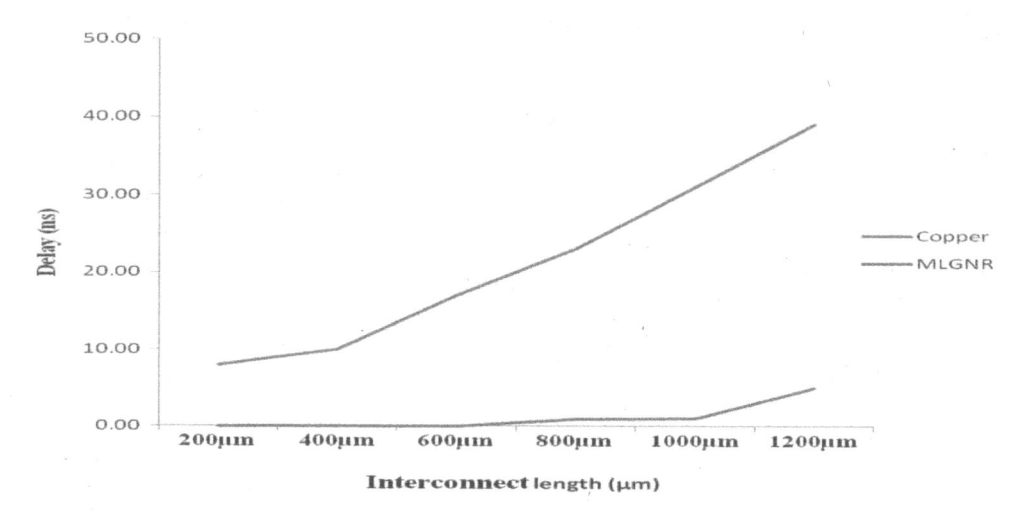

FIGURE 4.7   Delay variations with respect to change in wire length of Cu as well as MLGNR.

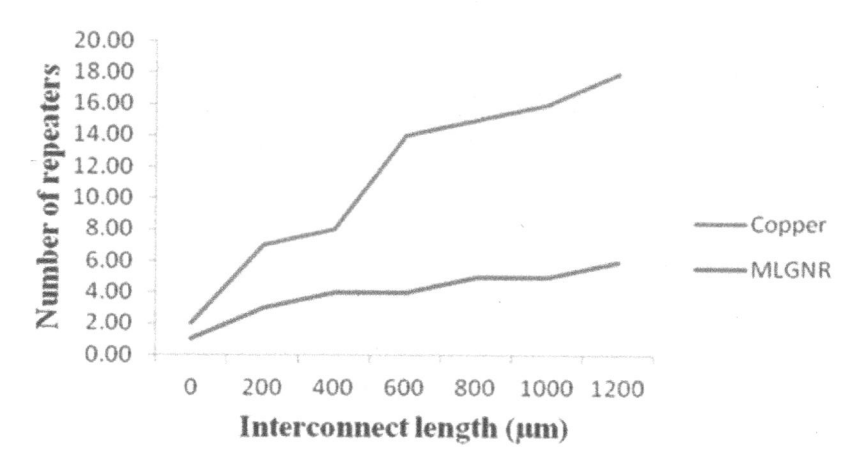

FIGURE 4.8 Variation in number of repeaters as a function of wire length and wire material used for interconnects.

4.4.1.3 Delay Time Variation Regarding Deviation in Interconnect Width

To compute the delay for different interconnect wire widths, it is necessary to know the buffer size and overall count of repeaters inserted in a single network. Bakoglu and Meindl [21] and El-Moursy and Friedman [22] have proposed a model to calculate the value of the optimized number of repeaters $\left(k_{opt} \right)$ and optimized repeater size (h_{opt}).

$$k_{opt} = \left[\frac{1}{1 + 0.16[X]^3} \right]^{0.24} \left(\frac{R_{SEGMENT} C_{SEGMENT}}{2.3 \times R_0 \times C_0} \right)^{\frac{1}{2}} \tag{4.25}$$

$$h_{opt} = \left[\frac{1}{1 + 0.16[X]^3} \right]^{0.3} \left(\frac{R_0 \times C_{SEGMENT\,(w_{int})}}{R_{SEGMENT} \times C_0} \right)^{\frac{1}{2}} \tag{4.26}$$

$$X = \left(\frac{L_{SEGMENT}\left(w_{int} \right)}{R_{SEGMENT} \times R_0 \times C_0} \right)^{\frac{1}{2}} \tag{4.27}$$

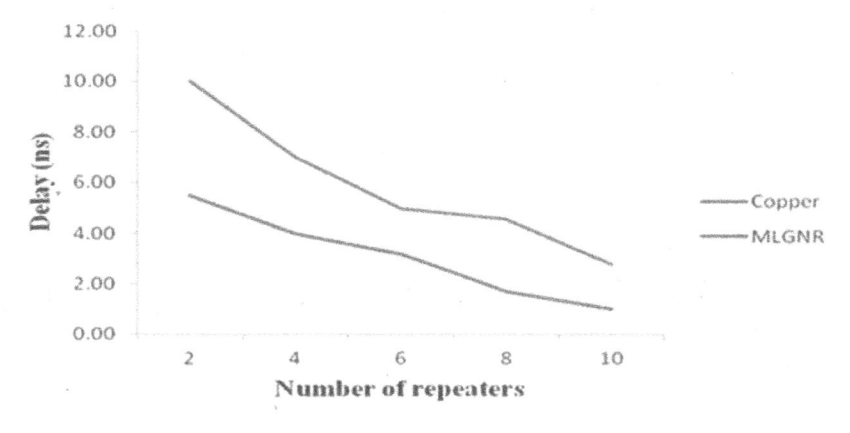

FIGURE 4.9 Delay time variation regarding the number of repeaters.

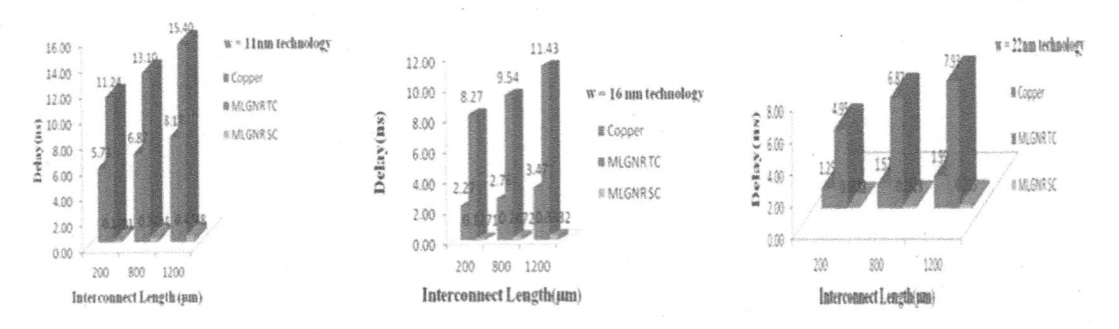

FIGURE 4.10 Propagation delay analysis for different interconnects (Cu, MLGNR TC, and MLGNR SC) at different widths ($w = 11$ nm, 16 nm, and 22nm) for different interconnect wire segment lengths.

where $R_{SEGMENT}$, $C_{SEGMENT}$, and $L_{SEGMENT}$ are impedance parameters of the interconnect segment of width w_{int}, and R_0 and C_0 are the output resistance and capacitance of a minimum-sized repeater. Primitive research shows that raising the ratio of the inductance-to-resistance line reduces the totality of repeaters needed to attain the bare minimum interconnect delay. Therefore, variation in width of wire is the best way to accomplish the minimum interconnect delay with the minimum number of repeaters. Figure 4.10 shows the analysis of wire segment length-dependent signal propagation delay for different types of interconnects (Cu, MLGNR TC, and MLGNR SC) at different widths ($w = 11$ nm, 16 nm, and 22 nm). Due to scaling of the technology node, the interconnect wire width reduces, which in turn reduces the amount of conduction channels and MFP in every MLGNR layer, thereby escalating the effective resistance of MLGNR interconnects.

From the chart, we conclude that with an increase in the width (w) of the interconnect, the wire resistance decreases, reducing the delay of the interconnect wire. Thicker interconnect wires have an extra number of GNR layers, which provides a large conduction path and lesser resistance of that interconnect wire. There is the least amount of delay in the case of side-contact MLGNR as there are more conduction paths compared with the top-contact MLGNR.

4.4.2 Power Analysis

Figure 4.11 illustrates the graphical plot of power dissipation in Cu and MLGNR interconnects for different interconnect lengths. Overall power dissipation is the sum of dynamic power and static

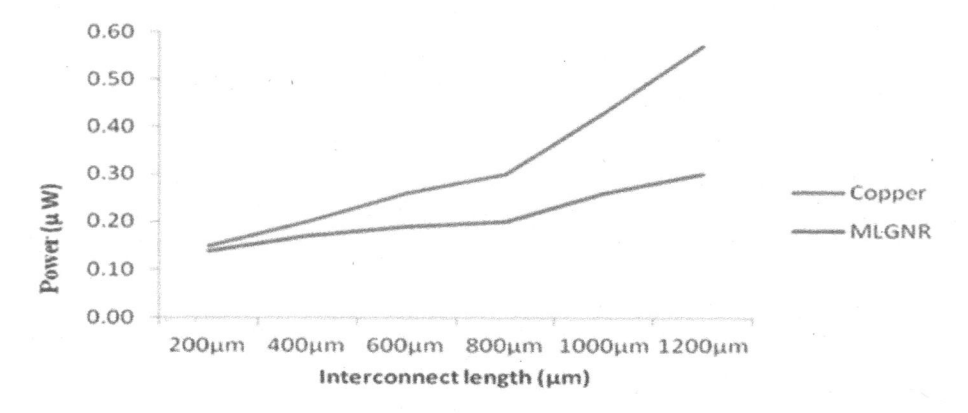

FIGURE 4.11 Power dissipation with respect to change in wire length of Cu and MLGNR.

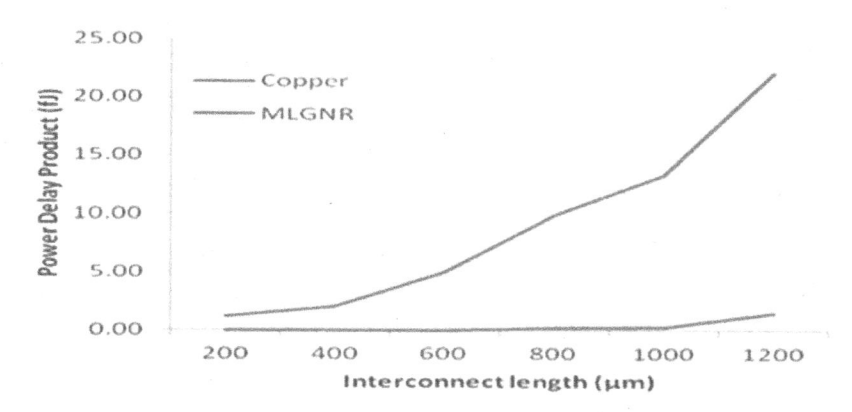

FIGURE 4.12 Variations in PDP regarding change in wire length of Cu and MLGNR.

power dissipation. Dynamic power is mainly due to switching of the device. Here, because of the use of ternary repeaters, which can efficiently operate in the subthreshold region, there is considerable reduction in power dissipation. Moreover, using MLGNR further suppresses the total power consumption. Further analysis shows that the MLGNR interconnect has less power dissipation than a Cu interconnect. Minor power dissipation in the subthreshold region of operation promotes the usage efficiently in ultralow power applications.

4.4.3 Power-Delay Product Analysis

Figure 4.12 shows the PDP variation in DIL as a function of interconnect length. PDP is the figure of merit to evaluate the overall performance efficiency of any DIL network. The lower the value of PDP, the better the network performance. The graph in Figure 4.12 can be interpreted as the MLGNR interconnect using less PDP than the Cu interconnect, indicating that the MLGNR interconnect's performance is superior to Cu wire.

4.5 SUMMARY

This chapter presents a qualitative comparison of novel MLGNR interconnects with conventional Cu interconnects. In conjunction with this, it elucidates the prospects of GNRFET-based ternary buffers as repeaters and investigates their potential implementation on a high-performance on-chip GNR interconnecting wire network.

For efficient analyses and comparison, HSPICE simulation is carried out on the DIL test bench structure. Result analysis obtained from the graphs points out that MLGNR is the superior interconnect in terms of delay, power dissipation, and PDP compared with the Cu interconnect. Using different technology widths of interconnect wire, propagation delay is analyzed at different interconnect lengths. In the previously mentioned argument, an imperative insight is also obtained by applying the approach of repeater insertion in between long interconnect lines. Moreover, modeling repeaters using ternary logic further enhance the performance because of their ability to operate at reduced power consumption, exhibiting less delay and area to implement the same functions. We observed that by inserting the optimum count of repeaters, signal propagation delay is further reduced, although such an exercise is an iterative process. In the end, the authors called for supplementary reading of the various approaches and other strategies proposed by authors in Refs. [4, 7, 8, 11, and 19] for greater understanding of designing efficient buffers for high-performance on-chip interconnects.

REFERENCES

1. Moore GE. Cramming more components onto integrated circuits. IEEE Solid-State Circuits Society Newsletter. 2006;38(8):33–35.
2. Wilson L. "International Technology Roadmap for Semiconductors (ITRS)." 2013. http://www.itrs.net/
3. Khursheed A, Khare K, Malik MM, Haque FZ. Performance tuning of very large scale integration interconnects integrated with deep sub micron repeaters. J Nanoelectron Optoelectron. 2018;13(12):1797–1806.
4. Khursheed A, Khare K. Optimized buffer insertion for efficient interconnects designs. Int J Numer Model Electron Netw Devices Fields. 2020;33(4):e2748.
5. Xu C, Li H, Banerjee K. Modeling, analysis, and design of graphene nano-ribbon interconnects. IEEE Trans Electron Devices. 2009;56(8):1567–1578.
6. Khursheed A, Khare K, Haque FZ. Designing high-performance thermally stable repeaters for nano-interconnects. J Comput Electron. 2019;18:53–64. https://doi.org/10.1007/s10825-018-1271-0
7. Khursheed A, Khare K, Haque FZ. Designing of ultra-low-power high-speed repeaters for performance optimization of VLSI interconnects at 32 nm. Int J Numer Model. 2019;32:e2516. https://doi.org/10.1002/jnm.2516
8. Khursheed A, Khare K. Designing dual-chirality and multi-V_t repeaters for performance optimization of 32 nm interconnects. Circuit World. 2020;46(2):71–83. https://doi.org/10.1108/CW-06-2019-0060
9. Biswas C, Young H. Graphene versus carbon nanotubes in electronic devices, Adv Funct Mater. 2011;21(20):3806–3826, https://doi.org/10.1002/adfm.201101241
10. Rabaey J. Digital Integrated Circuits: A Design Perspective, 2nd ed. Berkeley: Prentice-Hall, 2004.
11. Banerjee K, Mehrotra A. A power-optimal repeater insertion methodology for global interconnects in nanometer designs. IEEE Trans Electron Dev. 2002;49(11):2001–2007.
12. Magen N, Kolodny A, Weiser U, Shamir N. "Interconnect-power dissipation in a microprocessor," presented at the International Workshop System Level Interconnect Prediction, Paris, France, 2004;7–13.
13. nanoHub. "Tool Development." 2021. http://www.nanohub.org/tools
14. Nakada K, Fujita M, Dresselhaus G, Dresselhaus MS. Edge state in graphene ribbons: nanometer size effect and edge shape dependence. Phys Rev B. 1996;54(24):17954–17961.
15. Nishad AK, Sharma R. Analytical time-domain models for performance optimization of multilayer GNR interconnects. IEEE J Sel Top Quantum Electron. 2014;20(1):17–24.
16. Das D, Rahaman H. Analysis of crosstalk in single- and multiwall carbon nanotube interconnects and its impact on gate oxide reliability. IEEE Trans Nanotechnol. 2011;10(6):1362–1370.
17. Ismail YI, Friedman EG. Effects of inductance on the propagation delay and repeater insertion in VLSI circuits. IEEE Trans Very Large Scale Integr Syst. 2000;8(2):195–206.
18. Chen YY, Sangai A, Rogachev A, Gholipour M, Iannaccone G, Fiori G, Chen D. A SPICE-compatible model of MOS-type graphene nano-ribbon field-effect transistors enabling gate- and circuit-level delay and power analysis under process variation. IEEE Trans Nanotechnol. 2015;14(6):1068–1082.
19. Kumar VR, Majumder MK, Kukkam N, Kaushik BK. Time and frequency domain analysis of MLGNR interconnects. IEEE Trans Nanotechnol. 2015;14(3):484–492.
20. ES/CAD: CAD for Emerging Systems. 2021. http://dchen.ece.illinois.edu/tools.html
21. Khursheed, A, & Khare, K. Nano Interconnects: Device Physics, Modeling and Simulation, 1st ed. CRC Press, 2021. https://doi.org/10.1201/9781003104193
22. El-Moursy, MA, Friedman EG. Optimum wire sizing of RLCinterconnect with repeaters. Integr VLSI J. 2004;38(2):205–225.

5 An Effective Study on Particulate Matter (PM) Removal Using Graphene Filter

Katyayani Bhardwaj, Aryan, and R.K. Yadav

CONTENTS

5.1 INTRODUCTION

Technology evolves, leading to major breakthroughs and remarkable new inventions or discoveries. One such notable discovery or invention that has the potential to change and revolutionize the future is graphene, which was discovered by Professor Kostya Novoselov and Professor Andre Geim of the University of Manchester. They discovered a single and isolated atomic layer of carbon with sp^2 bonded carbon atoms. The name "graphene" is given to a flat and tightly packed monolayer of carbon atoms, transformed into a two-dimensional (2D) honeycomb lattice. For the graphitic materials of all other dimensionalities, whether it is fullerenes, which are zero-dimensional (0D) wrapped up in the form of graphene, or one-dimensional (1D) nanotubes that look like graphene and are rolled into a cylinder, or three-dimensional (3D) graphite materials in which graphene sheets are arranged in stacks (Figure 5.1), graphene is the basic building block. Its properties include high thermal and electrical conductivity, along with incredibly high strength and flexibility. Up to the level of 10 layers, any structure is called a 2D crystal, and beyond this limit the crystal is referred to as 3D. Graphene has been studied by various scientists for more than 60 years. Before then, it was not available in the free state, so it was not considered as practically usable; rather, it was referred to as academic material [1]. It was later discovered that graphene has a zero band gap and does exist in a free state. The potential of this material as a revolutionized discovery was understood. Scotch tape played yet another crucial role in its discovery. Using scotch tape, graphene was peeled off from the highly oriented pyrolytic graphite (HOPG) form of graphite [2, 3]. Currently, graphene can be produced by a number of ways: mechanical exfoliation, liquid phase and thermal exfoliation, chemical vapor deposition (CVD), and synthesis on silicon carbide.

DOI: 10.1201/9781003126393-5

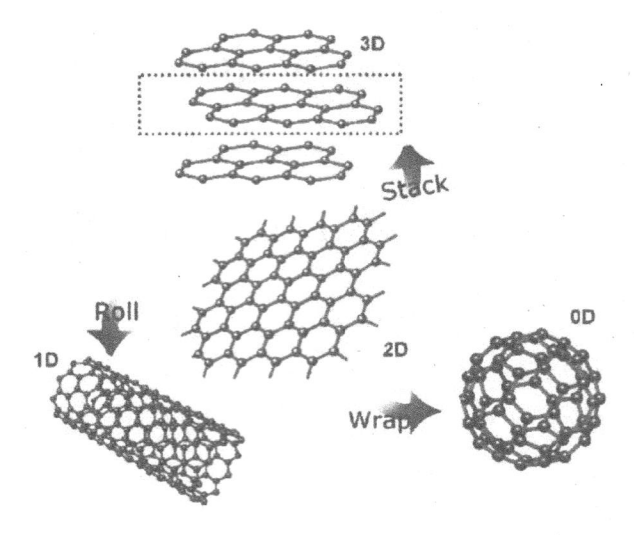

FIGURE 5.1 Different dimensionalities of graphene: 0D (wrapped up form), 1D (rolled form), and 3D (stack form).

TABLE 5.1
Intrinsic Properties of Graphene

Saturation Velocity (cm/s)	**$4-5 \times 10^2$**
Carrier mobility (cm²/Vs)	>100,000
Current density (A/cm²)	~10^9
Thermal conductivity (W/m-K)	4800
Optical opacity (%)	2.3% per layer
Young's modulus (Pa)	0.5-1 TPa

The graphene components were integrated with silicon-based electronics, and a substantial performance improvement is observed resulting in the production of an entirely new product. The intrinsic properties of graphene are displayed in Table 5.1.

5.2 PROPERTIES OF GRAPHENE

The 2D crystal of graphene possesses many properties that make it unique from other materials and make it potentially capable to revolutionize the existing application areas, including being impermeable to gases, chemically inert, and various other areas (refer to Figure 5.2 for more information). The detailed properties are as follows.

5.2.1 MECHANICAL PROPERTIES

As per the studies done by Geim [4], pristine graphene is a very strong material with very high elasticity, intrinsic strength, tensile strength, and brittleness. Additionally, the studies conducted by Lee et al. [5] yielded the results that the Young's modulus of graphene is 1.0 TPa, the third-order elastic stiffness is −2.0 TPa, and ultimate tensile strength is 130,000,000,000 Pa (or 130 GPa). All of these properties, when combined, make this material superior to many other commonly used materials.

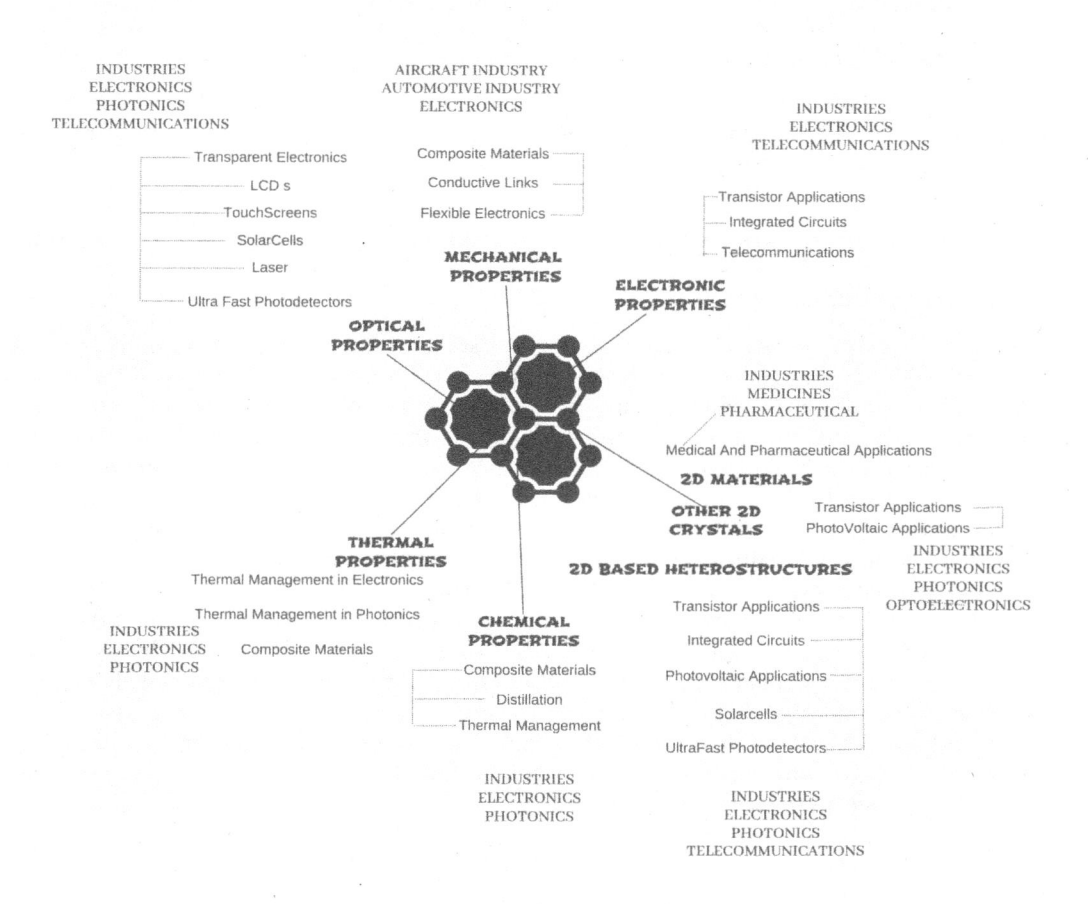

FIGURE 5.2 Unique properties possessed by graphene and their respective application areas.

5.2.2 Chemical Properties

Graphene is a zero-overlap semimetal with each carbon atom connected to three other carbon atoms. They have four electrons in the N shell (the outer shell), and when they are bonded with three other carbon atoms with the same number of outer shell electrons, one electron is left freely available, which is the cause of conduction. Referred to as π-electrons (pi-electrons), these are located in between two graphene ultrathin layers, thus resulting in overlapping of π-orbitals, making the carbon-carbon bond in graphene much stronger [2, 6].

5.2.3 Electrical Properties

In graphene, the freely available pi-electrons are the reason for the electrical conduction, which make it behave as a semiconductor similar to that of silicon, germanium, or gallium arsenide (GaAs). Because the conduction in graphene is different from that of other semiconductor materials, it exhibits many unique properties; one such unique property is electron mobility. In its pristine form, the electron mobility of graphene is more than 200,000 cm^2/Vs. The sheet resistance of graphene is about 30 Ω [1, 6]. The graphene atoms behave similar to photons because they are considered massless.

5.2.4 OPTICAL PROPERTIES

Although it has a thickness of just 1 atom in pristine form, graphene has the ability to absorb 2.3% of white light, which is very appreciable considering the size of graphene. It proves that graphene has a good absorbing capacity [7].

5.3 COMPUTATIONAL GRAPHICAL MODELING OF 2D BILAYERED GRAPHENE

5.3.1 BACKGROUND

Due to the scarcity of energy and environmental issues, a variety of cost-effective and pollution-free technologies have attracted a great deal of attention. Thermoelectricity is one of the new ways by which the problem of waste of pure heat energy can be solved. The thermoelectric effect is the direct conversion of temperature differences to electric voltage and vice versa via a thermocouple. Thermoelectric devices create a voltage when there is a different temperature on each side. Conversely, when a voltage is applied to it, heat is transferred from one side to the other, creating a temperature difference. At the atomic scale, an applied temperature gradient causes the charge carrier in the material to diffuse from the hot side to the cold side. In this phenomenon, both electric and thermal currents contribute to the total current. The thermoelectric effect described is introduced in a natural way by several characteristic coefficients of the material, namely the thermal conductivity κ, the electrical conductivity σ, and the Seebeck coefficient S. These coefficients relate thermal and electrical currents with thermal and electrical gradients, generally referred to as *transport coefficients*.

Accordingly, materials for thermoelectric applications are evaluated in terms of their figure of merit (FOM) value, which, in turn, is determined by their thermal conductivities (appearing in the FOM denominator) and the power factor $P \equiv \sigma S^2$, which appears in the FOM numerator.

$$Z = \sigma S^2 / \mathcal{K} \tag{5.1}$$

Thus, large FOM values require both small thermal conductivity values and large Seebeck coefficient values. Therefore, in searching for promising thermoelectric materials, one must focus on materials exhibiting strong couplings between the electrical and thermal currents.

Computational modeling permits the modeling and simulation of complex nanometer-scale structures. The predictive and analytical power of computation is critical to success in nanotechnology. Nature required several hundred million years to evolve a functional "wet" nanotechnology; the insight provided by computation should allow us to reduce the development time of a working "dry" nanotechnology to a few decades, which will have a major impact on the "wet" side as well.

The wide variety of carbon allotropes and their related physical properties are largely due to the flexibility of carbon's valence electrons and resulting dimensionality of its bonding structures. Among carbon-only systems, 2D hexagonal sheet *graphene* forms the basis of other important carbon structures such as graphite and carbon nanotubes.

Although the pure graphene is a good conductor of heat, its thermal conductivity has been measured to be around 4000–6000 W/(mK), which makes it unsuitable for the thermoelectric material. Still, defected graphene provides a method to limit thermal conductivity by scattering the photons, while maintaining sufficient charge transport to produce more efficient thermoelectric devices. Graphene nanostructures are, therefore, interesting potential thermoelectric materials. Hence, computational modeling can create a great breakthrough in this field of research.

5.3.2 Objective

The objective is to model the bilayer of graphene graphically, its Brillouin zone, and the band structure using Python. This can prove to be a great candidate as a thermoelectric material.

5.3.3 Methodology

In this section we are going to use the Jupyter notebook and, most importantly, pybinding (a Python library) will be used to plot a 2D graph of a bilayer graphene. The computer programming code that has been used for plotting the graph along with the obtained graph as an output are mentioned below in Figures 5.3 and 5.4.

5.4 POLLUTION, ITS TYPES, AND ITS CONTRIBUTORS

Air pollution is a very harmful problem in today's society, which has been recognized for many years. In recent years, information about air quality (measured through the Air Quality Index) and knowledge of why and how air pollution levels increase and how air becomes polluted has increased our knowledge about the atmosphere. The approximate mass of the atmosphere, which is about 99%, lies in the first 50 km of the Earth's surface, which includes the troposphere and the stratosphere, and is where major air pollution occurs. The various types of air pollution are given in Figure 5.5.

We can even detect the contribution of a single molecule in pollution in the environment by using the concept of the Fourier transform. The studies conducted by Kim et al. demonstrate the same. The $F(x)$ part of the Fourier transform is used to select the molecule, and the contribution is further calculated [8].

The harmful materials introduced into the environment are known as pollutants, and they are the major cause of any type of pollution. Figure 5.6 shows the most common air pollutants.

Particulate matter, which is also referred to as PM or particle pollution, is a combination of liquid droplets and solid particles present in the atmosphere. There are two types of PM, namely filterable particulate matter (FPM) and condensable particulate matter (CPM).

The sources of the generation of particles belonging to the category of PM 10–PM 2.5 are given in Figure 5.7. The spread of PM 2.5 and PM 10 occurs more often because of construction work,

```
In [1]: import pybinding as pb
        import numpy as np
        import matplotlib.pyplot as plt
        from math import sqrt
        from matplotlib.artist import allow_rasterization
        from pybinding.repository import graphene
        import matplotlib.collections
        matplotlib.collections.allow_rasterization=allow_rasterization
        pb.pltutils.use_style()
        pi=3.14
        def monolayer_graphene():
            a=0.24595
            r=0.142
            h=-2.8
            lat=pb.Lattice(a1=[a,0],a2=[a/2,a/2*sqrt(3)])
            lat.add_sublattices(('A',[0,-r/2]),('B',[0,r/2]))
            lat.add_hoppings(([0,0],'A','B',h),([1,-1],'A','B',h),([0,-1],'A','B',h),([1,0],'A','B',h))
            return lat

        lattice=monolayer_graphene()
        lattice.plot()
        plt.show()

        lattice=graphene.bilayer()
        plt.savefig("bilayer.png")
        lattice.plot()
        model=pb.Model(graphene.bilayer(),pb.translational_symmetry())
        solver=pb.solver.lapack(model)
        r=graphene.a_cc
        Gamma=[0,0]
        K1=[-4*pi/(3*sqrt(3)*r),0]
        M=[0,2*pi/(3*r)]
        K2=[2*pi/(3*sqrt(3)*r),2*pi/(3*r)]
        #bands.plot(point_labels=['K',r'$\Gamma$','M','K'])
        plt.savefig("Band diagram.png")
        plt.show()
        model=pb.Model(graphene.bilayer(),pb.translational_symmetry())
        solver=pb.solver.lapack(model)
        r=graphene.a_cc
        Gamma=[0,0]
        K1=[-4*pi/(3*sqrt(3)*r),0]
        M=[0,2*pi/(3*r)]
        K2=[2*pi/(3*sqrt(3)*r),2*pi/(3*r)]
        bands = solver.calc_bands(K1, Gamma, M, K2)
        bands.plot(point_labels=['K', r'$\Gamma$', 'M', 'K'])

        plt.savefig("Band diagram.png")
        plt.show()
```

FIGURE 5.3 Computer programming code used to plot a 2D graph of a bilayer graphene.

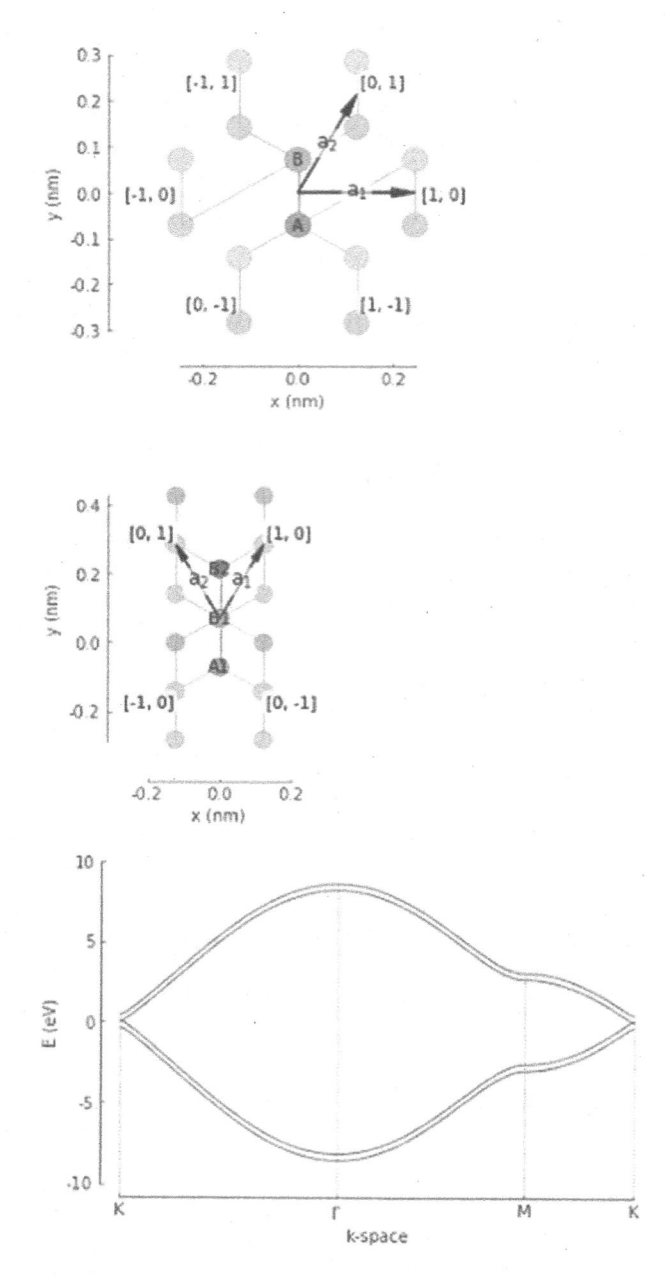

FIGURE 5.4   Graphs obtained by running the computer programming code mentioned in Figure 5.3.

dust, and burning of various forms of waste. These particles cannot be seen by the naked eyes, and an electron microscope is generally used to detect them. Particle pollution consisting of PM 10 and PM 2.5 is comparatively dangerous. The atmospheric PM with a diameter of 2.5 μm or less, which is about 3% of the diameter of a human hair, can be detected using the electron microscope. PM 2.5 is small when compared with PM 10 particles. The PM 10 particles have a diameter of 10 μm and are referred to as fine particles. PM 10 is also called respirable PM according to an environmental expert. Due to their small size, both PM 10 and PM 2.5 particles act as gases. These particles penetrate into the lungs and can cause cough or even asthma attacks. Heart attack, high blood pressure, irritation

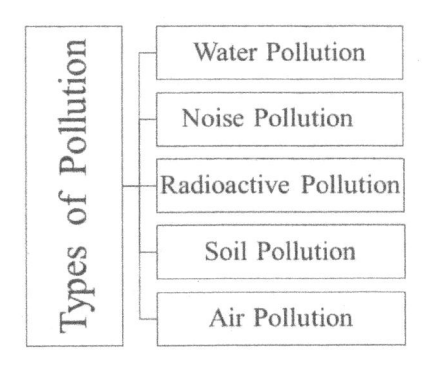

FIGURE 5.5 Different types of pollution leading to an ecological imbalance.

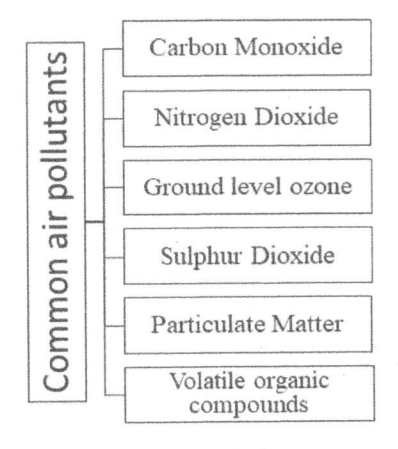

FIGURE 5.6 Most common air pollutants present in the atmosphere.

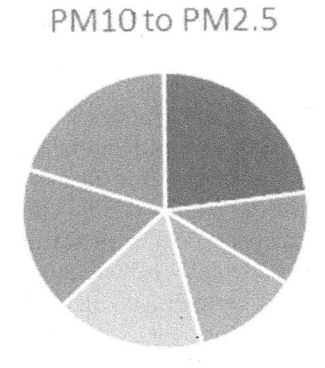

FIGURE 5.7 Major sources of the generation of particles of category (PM 10 to PM 2.5).

in the nose, eyes, and throat, stroke, etc., are some of the serious issues they cause, and can even lead to premature death. Visibility is also affected when high levels of PM 2.5 cause high fog and air mist. Younger and older age groups are adversely affected by these particles [9].

After the discovery of graphene, scientists began to apply it to address and resolve different kinds of problems. To combat air pollution, they prepared graphene oxide, which was used to filter and purify air that was polluted with PMs of various diameters (PM 10, PM 2.5, or PM 1.0). We observed that filters with the addition of graphene oxide were able to purify polluted air containing PM at even higher concentrations, which was not done by any of the commercially available filters. The efficiency of these filters was higher compared with that of other filters available commercially. They also proposed a mechanism by which the amount of PM available in the air can be diminished effectively.

5.5   THE MECHANISM OF FILTERING

Basically, graphene is a 2D material with a honeycomb-like structure that is utilized for solar cells, optical absorbers, and thermal collectors. As per the studies conducted by Jung et al. [10], humans are adversely affected by air pollution leading to various other environmental problems. Under the category of PM, it is PM 2.5 that is the most dangerous for humans because its diameter is less than 2.5 μm. As mentioned earlier, there is a need to remove both types of PM, i.e., FPM and CPM, because both contribute to pollution during emission of coal boilers, respectively, in quite large amounts. Thus began the requirement of any filter to remove both FPM and CPM from the source of pollution. For the fabrication process, chemical stability needs to be analyzed even after the emission of gas at high temperature and toxic acids into the atmosphere. Commercial markets require any filter that is designed in such a manner to be easily installed directly at the pollution source. The solution is the fabrication of a type of filter that can remove both CPM and FPM from the source of pollution at the same time. Unlike other PM removal filters available commercially, the improved chemical resistance and heat resistibility of this type of filter made it capable and suitable enough for the purpose of filter installation, which will be done directly at pollution sources. As per the studies, a condenser and oil heater can be used to generate both types of PM. The CPM is similar to that of FPM except for its small size average diameter; both of them exist in either the liquid or solid phase and contain ions soluble in water, which makes the PM particles polar. The electrically conductive rGO, because of its porous structure, could absorb the dust particles. The electrostatic force of attraction is used by a negative ionizer when a higher voltage (positive) is applied to that of the rGO filter. While summarizing our literature study regarding rGO filter design systems, FPM was removed from the very first membrane followed by the elimination of CPMs using a condenser. The second membrane of the rGO filter has electrical conductivity and moderate pore size unlike the other PM removal filters, thus making it an electrostatic precipitator, which works effectively while removing the CPM with a diameter of 2.5 μm or less. Thus, this study proved to have removed both CPM and FPM from the source of pollution simultaneously by ensuring the durability and usability of the rGO filter, and, hence, verifying its long-term operational ability. As per our survey, this filter is reusable and its properties of durability and reusability have also been assured. While extending this study, we decided to add a sensor to receive the notifications regarding the update of filter performance. One major challenge is to define sensors because the technical literature contains contradictory approaches. A sensor is a major part of the equipment in all measurement systems and devices and is widely used in fields like medicine, science, automation, and so on. The term sensor can be replaced by various other terms like transducer, detector, meter, and gauge. At present, no uniform standard definition has been agreed on universally. According to the American National Standards Institute (ANSI), transducers should be treated as sensors, and it states that "a sensor is a device that provides usable output in response to a specified quantity which is measured." On the other hand, the National Standard of China states that sensors consist of sensing components and sends the information to a device and electronic circuits. The output of a sensor can be an electrical, optical, or mechanical signal. Hence, we can conclude that the output of a mechanical sensor will be a mechanical signal. Both transducer and sensor are used in the context of a measurement system and can be often interchangeable. Transducer is more often used in the United States, whereas the term sensor is used in China and Europe. In this study, we will be using a differential pressure sensor, because it will provide a measurement that can be compared between two points. One such example is the two conditions, before and after, of a valve in a pipe. The pressure will be the same on both the sides when the valve is fully open, and it is possible that either the valve is not opened fully or there is a blockage when there is a difference in the pressure and the differential pressure sensors come into play. There are generally two sensors to which the pipes are attached. The pipes are then connected to a system and measurements are taken. Integration of industrial differential sensors into a standardized fitting allows them to be built into existing pipework. Unlike gauge sensors, here the measurements made are not at all dependent on the atmospheric pressure. The differential pressure sensors can even be used in industrial environments to determine the flow of gases or liquids using the difference in pressure, which can include offshore and sub-sea gas and oil processing, effluent treatment plants, and remote heating systems that utilize the heated water or steam, as well as many more application areas. All in all, we can use one sensor across all filter elements. Hence, FPM and CPM are both captured using rGO filters. According to the European Standard, there are four groups of filters, which have further filter classes (refer to Table 5.2). Each class has a certain value of arrestance, in which arrestance refers to the filter efficiency. The value of arrestance shows the percentage of pollutants that can be filtered out. The first group of filters is G, which represents the filters for coarse dust (where particles are 10 μm). The second group of filters is F, which represents the filters for fine size (where particles are 0.01 μm). The third and fourth groups of filters are H (HEPA) and U (ultra-low penetration air [ULPA]), which represent the filters for microparticles (where particles are 10 μm). Table 5.2 contains detailed information of international classification of air filters of atmospheric dust.

5.6 CONCLUSION

In this study, the level of PM and its contribution to air pollution was analyzed, and we have linked the latest air filters to those pollutants and have shown how a particular filter can effectively act on those pollutants. Until now, the majority of the current research was focused on dealing with FPM. In this study, we have successfully dealt with CPM too.

TABLE 5.2
International Classification of Air Filters of Atmospheric Dust

Group of Filters	Filter Class	Arrestance A_m (%), E_m (%)
	*Average Weight: Arrestance A_m (%)	
Filters for coarse dust (efficiency for particles 10 μm) G	G1	60
	G2	70
		80
	G3	90
	G4	
	**Average Atmospheric Dust Spot Efficiency E_m (%)	
Filters for fine dust (efficiency for particles 1 μm) F		40
	F5	50
		60
	F6	70
		80
	F7	90
	F8	
	F9	

(Continued)

TABLE 5.2 *(Continued)*
International Classification of Air Filters of Atmospheric Dust

Group of Filters	Filter Class	Arrestance A_m (%), E_m (%)
Filters of microparticles (efficiency for particles 0.01 μm)	Initial Penetration (%) / Efficient	
	MMPS	85
	H10	90
	H11	95.0
		97.0
		99.0
H (HEPA)	H12	99.5
		99.7
		99.9
	H13	99.95
		99.97
		99.99
	H14	99.995
		99.997
		99.999
	U15	99.9995
		99.9997
		99.9999
U (ULPA)	U16	99.99995
		99.99997
	U17	99.999995

* Arrestance represents the filter efficiency.

A_m represents the arrestance measure through the weight of molecules.

** E_m represents the arrestance measure through the number of particles of molecules.

REFERENCES

1. Geim, A. K. & Novoselov, K. S. (2007). "The Rise of Graphene." Nature Mater. 6: 183–191.
2. Novoselov, K. S. et al. (2004). "Electric Field Effect in Atomically Thin Carbon Films." Science. 306: 666.
3. Jayasena, B., & Subbiah S. (2011). "A Novel Mechanical Cleavage Method for Synthesizing Few-Layer Graphenes." Nanoscale Res Lett., 6: 95.
4. Geim, A. (2009). "Graphene: Status and Prospects." Science. 324 (5934): 1530–1534.
5. Lee, C., Wei, X., Kysar, J. W., & Hone, J. (2008). "Measurement of the Elastic Properties and Intrinsic Strength of Monolayer Graphene." Science. 321: 385. https://www.science.org/doi/10.1126/science.1157996
6. Novoselov, K. S., Falqko, V. I., Colombo, L., Gellert, P. R., Schwab, M. G., & Kim, K. (2012). "A Roadmap for Graphene." Nature. 490: 192–200. https://www.nature.com/articles/nature11458
7. Kinaret, J., Falko, V., Ferrari, A., Helman, A., Kivioja, J., Neumaier, D., Novoselov, K. S., Palermo, V., & Roche, S. (2012). "Publishable Flagship Proposal Report on the GRAPHENE-CA – Coordination Action for Graphene-Driven Revolutions in ICT and Beyond." https://cordis.europa.eu/docs/projects/cnect/8/284558/080/deliverables/001-DeliverableD63.pdf
8. Kim, E., Lee, S., Kim, J., Kim, C., Byun, Y., Kim, H., & Lee, T. (2012). "Pattern Recognition for Selective Odour Detection with Gas Sensor Arrays." Sensors. 12 (12): 16262–16273. doi:10.3390/s121216262
9. Jawaid, M., Ahmad, A., Ismail, N., & Rafatullah, M. (2021). Environmental Remediation through Carbon Based Nano Composites. Singapore: Springer. doi:10.1007/978-981-15-6699-8
10. Jung, W., Lee, J. S., Han, S., Ko, S. H., Kim, T., & Kim, Y. H. (2018). "An Efficient Reduced Graphene-Oxide Filter for PM2.5 Removal." J Mater Chem A. 6 (35): 16975–16982. doi:10.1039/c8ta04587

6 Recent Trends in Fabrication of Graphene-Based Devices for Detection of Heavy Metal Ions in Water

Avik Sett, Monojit Mondal, Santanab Majumder, and Tarun Kanti Bhattacharyya

CONTENTS

DOI: 10.1201/9781003126393-6

6.1 INTRODUCTION

Due to extensive industrialization and detrimental human activities, toxic heavy metal ion pollutants such as cadmium, lead, mercury, and arsenic are becoming easily attached in the environmental food chain. These heavy metal ions from various sources such as combustion of coal, batteries, lead and tanning industries, fertilizers, and pesticides pose a serious threat to the environment and cause serious health hazards. The introduction of these toxic ions in the food chain may cause threats such as lung cancer, osteoporosis, hypertension, Alzheimer's disease, etc. The safety limit of cadmium, lead, mercury, and arsenic in drinking water as per the Environmental Protection Agency (EPA) is given as 5, 15, 2, and 10 parts per billion (ppb), respectively. Hence, there is significant demand to develop heavy metal ion detectors to determine the numerous toxic analytes, like metal ions, in water. Conventional techniques like atomic emission/absorption spectroscopy (AAS/AES), anodic stripping voltammetry (ASV), inductively coupled plasma mass spectrometry (ICP-MS), and chromatographic separation require proper training and are very time-consuming as they undergo complicated measuring steps. Several sensors based on different working principles have been developed including electrochemical, fluorescent, and calorimetric principles; field-effect transistors (FETs); surface-enhanced Raman scattering (SERS); and chemiresistor-based sensors. These fabricated sensors are designed keeping in mind the aspects of high sensitivity, good selectivity, and stability under adverse atmospheric perturbations. The developed sensors are aimed at fabrication simplicity and portability to carry out real-time monitoring of water. Graphene and graphene-based functional nanomaterials in this context are attracting researches due to the unique properties related to their electrical, optical, and mechanical features. Graphene, with a thickness in the regime of one atomic layer, exhibits high carrier mobility, exceptionally low electronic noise, and chemical stability; moreover, it can be used for miniaturization of devices. Various technologies for synthesizing graphene have been discovered, for instance, sublimating SiC at high temperature, chemical vapor deposition (CVD), mechanical exfoliation, etc. However, to ensure the wide availability of graphene-based sensor devices, the cost of these sensors must be as low as possible and production of graphene on a larger scale should be feasible. Reducing graphene oxide is an economical process in graphene production and requires very user-friendly and low-cost equipment and resources. The modified Hummers method-based synthesis of graphene is well known. Various graphene oxide-reducing techniques ensure the attachment of different functional groups to reduce graphene oxide. Chemically derived graphene induces several defects on its matrix compared with pristine graphene. These defect sites act as electroactive and optical centers for sensing applications. Tuning mechanical, optical, and electrical properties of graphene for various applications can be done quite efficiently by (1) metal doping, (2) decorating it with nanoparticles, (3) selectively reducing the functional groups, (4) making nanocomposites with metal oxide nanoparticles or polymers, and so forth.

This chapter discusses the working principles of conventional techniques for heavy metal ion detection and their limitations. Various synthesis techniques of graphene have been analyzed, with special focus on chemical exfoliation of graphite and fabrication of functionalized graphene nanocomposites. Optical methods, which include simple design and sensitive recognition of toxic metals, are elucidated. The evolution of zero-dimensional quantum dots based on graphene and several modified graphene-based fluorescent quenchers paves the way for efficient determination of toxic metal ions in water. The electrical detection technique of toxic metal ions by functionalized graphene-based resistive sensors is discussed. Graphene-based FETs that are developed as highly sensitive detectors of these heavy metals by functionalizing the graphene sheets with sensitive probes are also demonstrated. Finally, challenges faced by graphene-based sensors and the outlook to future research and developments are discussed. The advantage of this work lies in the integration of graphene-based semiconductor devices with the existing complementary metal oxide semiconductor (CMOS) technology. Sensors that work within 2 V are easily integrated with advanced existing CMOS circuits and readily available for real-time detection of toxic elements in the environment. As per the knowledge of the author, there are hardly any reviews available on the various techniques of heavy metal detection with graphene as a sensing material, which summarizes the potential of this novel and fascinating material.

6.2 CONVENTIONAL METHODS FOR HEAVY METAL ION DETECTION

The determination of toxic metal ions like Hg^{2+}, Pb^{2+}, Cu^{2+}, As^{5+}, and Cd^{2+} in water is highly challenging. It is even more challenging when there are several heavy metal ions present together in different proportions. To mitigate this issue, several sensitive and highly accurate techniques for measurements of toxic heavy metal ions are developed and have achieved great attention worldwide. Such traditional or conventional techniques such as hydride generation atomic spectroscopy (HG-AFS), ASV, chromatographic separation, ICP-MS, and AAS/AES are used worldwide. These methods are highly sensitive and accurate but suffer few real-time limitations. These conventional techniques need highly expensive instruments, complex measuring steps, and a trained skilled workforce, and they are very time-consuming. Some of the conventional methods for detecting heavy metal ions in water are discussed in the following sections.

6.2.1 HYDRIDE GENERATION ATOMIC FLUORESCENCE SPECTROSCOPY (HG-AFS)

HG-AFS is a technique in which toxic metals are vaporized by transforming them into volatile hydrides. It separates the metals by forming hydrides from a range of matrices. Traditionally this method of HG-AFS, a technique of microwave digestion by high-pressure and HG-AFS, was utilized for the detection of mercury and arsenic concentration in fucoidans. This concept was also extended to detecting metals in water. Liu et al. [1] optimized the detecting conditions, which included the working parameters of the instrument, such as lamp current, potential of the photomultiplier tube, and atomic temperature. The proper conditions for generation of hydrides, such as speed of the carrying gas, concentration of acid in the loading fluid, and the KBH_4 concentration, were optimized. The experiments showed promising results, where the relation between the concentrations of arsenic and mercury and the fluorescent intensity was found to be linear. For conducting the analysis precisely, the samples were continuously detected and standard deviation of mercury and arsenic were obtained as 1.69% and 1.09%, respectively. Despite high accuracy and efficiency, the bulkiness of the instruments along with their complexity, restrict their use in real-time on-site monitoring of water quality by detecting these ions.

6.2.2 ANODIC STRIPPING VOLTAMMETRY (ASV)

ASV is a technique used to determine ionic species specifically and quantitatively. During the step of deposition, the analyte of interest is used to electroplate the working electrode, and during stripping the ions are oxidized from the electrode. The stripping step involves the measurement of current, and the current peak is allotted to the oxidation of ionic species at a specific potential. A linear, square wave, pulse, or staircase type of stripping is generally allowed. The ASV technique uses three types of electrodes, namely working, counter (also called auxiliary electrodes), and reference. An electrolyte is added to the solution of interest, and usually the working electrode used is a mercury or bismuth film electrode. The configuration of these electrodes can be planar strip or a disk type. The metal ions in the solution form an amalgam with the mercury film, which when oxidized results in a sharp peak in current, facilitating the selectivity between different toxic heavy metals. The glassy carbon electrodes may be made of mercury film, or a mercury drop electrode can be used as the working electrode. Inert metal electrodes such as platinum or gold may be used when the oxidizing potential of the ions is higher than that of mercury; in that case the mercury electrodes are not suitable.

The technique usually comprises four valuable steps when the working electrode is a mercury drop/film electrode, and continuous stirring of the solution is achieved during the first two steps. After the second step, stirring is completely stopped. The first step is the cleaning step, in which the oxidizing potential is kept higher than that of the analytes for a particular duration, for removing the analytes completely from the electrode. The second step (deposition) is crucial where the potential is deliberately kept low to allow reduction of the ions and deposition fully on the electrode. The first two steps combined can be termed as preconcentration steps. The third step also involves the potential to be kept low to allow the heavy metal ions to distribute more thoroughly in mercury. If the electrode used is an inert one, then the third step is not required. Raising the working electrode to a higher anodic potential in the last step involves oxidizing or stripping the analyte from the electrode. Electrons are released as observed by the rise in the current peak at this particular potential. ASV has the capability to detect heavy metals at concentrations as low as $\mu g/L$ [2–4].

In a typical work, Kiekens et al. [5], used glassy carbon as the working electrode due to high electrical conductivity, chemical inertness, and an easily renewable surface. Determination of very small amounts of mercury by the ASV technique, through deposition of this analyte in a rotating glassy carbon electrode, was carried out. To enhance the detection capabilities, a secondary ion (Cd^{2+}) was added to the solution of interest. The experiments related to stripping are carried out on electrolyte solutions such as 0.1 M $HClO_4$ and 0.1 M $NaSCN$-0.1 M $HClO_4$. The advantage of co-deposition of cadmium ions is to allow mercury to be determined at very low concentrations of about 10^{-8} M. This mechanism is called "support effect," which is explained by enhancing reversibility to electrodeposition of mercury at these metal nuclei, which are found to be deposited from the easily reduced auxiliary ion to the glassy carbon surface.

6.2.3 INDUCTIVELY COUPLED PLASMA ATOMIC EMISSION SPECTROMETRY (ICP-AES)

Inductively coupled plasma atomic emission spectrometry (ICP-AES) is an analytical method used to determine the presence of chemical hazards. ICP is used to excite ions and atoms present in the solution of interest, which when reaches its ground state emits electromagnetic radiations at specific frequencies (or wavelengths). These specific frequencies of radiation are associated with certain ions or atoms. The emission intensity at a particular frequency (or wavelength) directly corresponds to the number of chemical analytes present in the solution. The plasma is created by a heated source of ionized argon gas. Inductive coupling maintains this plasma at megahertz frequencies. This technique facilitates determining toxic metal ions in soil and water; however, the complexity in measurements and requirement of skilled and trained personnel limits its use in real-time monitoring in water quality. Apart from the detection of toxic elements in water, ICP-AES is also used to determine the presence of metals in wine, metals such as arsenic and lead in food, and trace protein bound elements [6–8].

6.3 GRAPHENE AS A TRANSDUCING MATERIAL DUE TO ITS OUTSTANDING PROPERTIES AND ITS CHARACTERIZATION USING RAMAN STUDIES

Graphene is a sp^2 hybridized [9] honeycomb-like structure with an array of carbon atoms [10] featuring a bond length of 1.42 Å and 120° angles. It is a monolayer structure established as the thinnest form of carbon nanomaterials and probably the thinnest material discovered [11]. The symmetry of this two-dimensional (2D)lattice gives rise to some fascinating electronic and mechanical

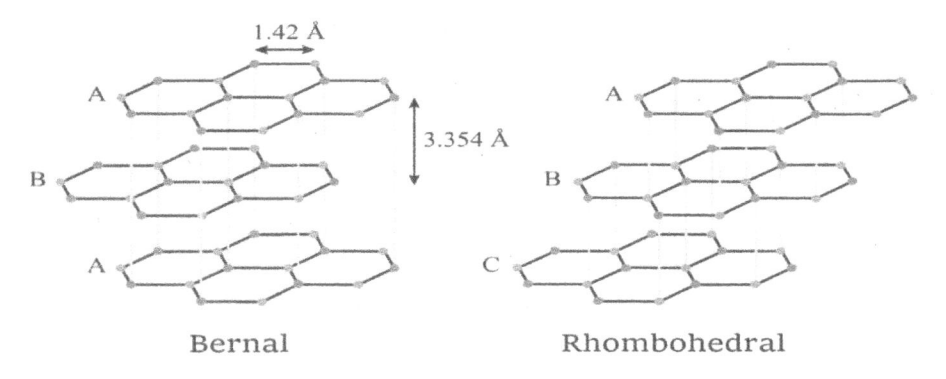

FIGURE 6.1 Stable allotropes of graphite, Bernal (ABA) and rhombohedral (ABC).

properties [12]. Graphene is known to be one of the hardest materials found [13] with an ability to be bent without affecting its internal structure and properties [14]. Graphite is the parent material and comprises stacking of several layers of single-layer graphene on top of each another having a separation of 3.354 Å between layers [15, 16]. The graphene layers could be stacked in two different ways, as shown in Figure 6.1. Eighty percent of graphite observed in nature is the Bernal graphite, which is an allotrope of graphite comprising ABA stacking configuration. Fourteen percent of the graphite found is rhombohedral with an ABC stacking configuration, whereas the last 6% is disordered in nature with no such stacking configuration. Monolayer graphene is very difficult to obtain and has very low yield. Hence, the few-layered graphene is considered in most of the cases for device fabrication for heavy metal sensors, as its properties are in between that of monolayer graphene and graphite in bulk form, with high mechanical and electrical stability. However, both single-layer and few-layered graphene may be used for sensing applications after proper functionalization.

6.3.1 ELECTRONIC PROPERTIES

The formation of a perfect lattice, which is honeycomb in nature, from the distribution of carbon atoms results in electronic properties of graphene. Hybridized orbitals that are sp^2 in nature is the backbone of graphene. The p-orbitals that are normal to the lattice attribute to the conductivity in the graphene matrix. These orbitals are conjugated to form conduction bands (CBs) (π^*) and valence bands (VBs) (π), which are significant to sensing of various analytes. The transduction of the interaction between the graphene-based materials and the analytes to a readable electrical signal is facilitated by these orbitals. The level of Fermi energy of pristine graphene is located exactly in the position where the CB meets with the VB. Together, in the Brillouin zone, there are six such points where CB meets the VB, as shown in Figure 6.2(a). These points are termed as neutrality points or Dirac points and in the reciprocal space, it is labeled as K and K' [12]. The ability of electrons in graphene to travel submicron lengths without scattering (travel without electrical resistance) is termed ballistic transport. This type of transport in graphene is observed at room temperature and is not affected by the adsorbates present in the graphene matrix or the substrate's topography [17]. Even though at room temperature, impurity scattering limits the mobility of charge carriers, and the mobility of carriers in graphene stays very high even in the presence of chemical and electrical dopants at high concentrations [18]. Electrical robustness of graphene-based sensors is attributed to these types of transport phenomena. Tuning of charge carriers between holes and electrons for doping graphene is possible because of the effect of a strong ambipolar electric field. Facile modulation of graphene FETs allows detection of n-type and p-type analytes with very high sensitivity, as will be discussed later. At the level of Fermi energy, graphene's conductivity does not vanish due to quantization [19]. Eventually, at this level where the density of states approaches zero, there

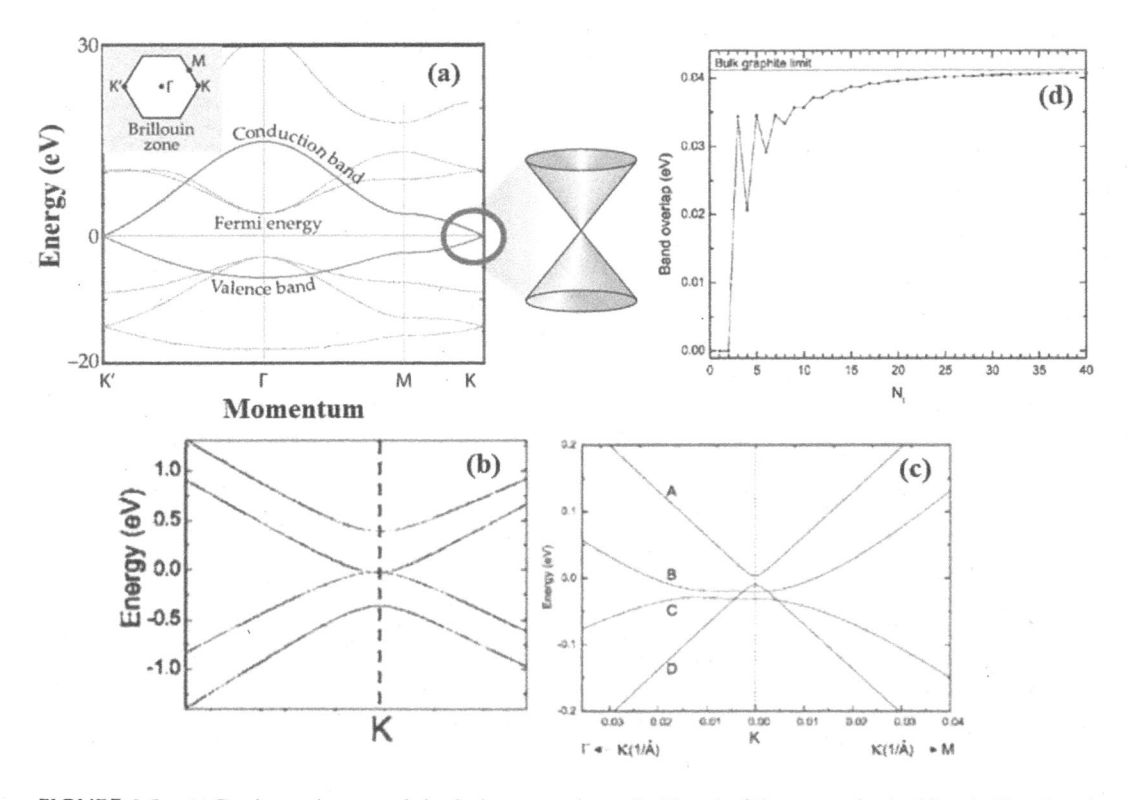

FIGURE 6.2 (a) Reciprocal space of single-layer graphene. In blue the Dirac cone is elucidated. (Reprinted with permission from [16]. Copyright 2007, American Institute of Physics.) (b) Graphene bilayer and (c) graphene trilayer in the ABA (Bernal) stacking sequence. (d) Evolution of the band overlap from graphene to graphite. (Adapted with permission from [12] Copyright 2007, The American Physical Society.)

still exists finite conductivity. The last charge carrier provides e^2/h as the minimum conductivity [12]. Hence, graphene is found to have no transition of the metal insulator while measurement is in σ_{min}. The experimentally obtained value for σ_{min} is observed as $4e^2/h$ [18]. The electronic property of graphene changes with layer thickness (number of layers). At one horizon, monolayer graphene is semiconducting with no band gap (zero overlap semimetal) and on the other horizon, graphite is found to be a semimetal with a band overlap of 41 meV [16]. With respect to the properties of graphene, the threshold between bulk and single-layer graphene is not clear. With an increase in the layer thickness, the band overlap starts increasing until it reaches that of the bulk graphite, as depicted in Figure 6.2(b and c). The amount of overlap seems to be a decent approximation of the threshold between the behavior of 2D and three-dimensional (3D) graphene. Figure 6.2(d) shows that, for graphene with more than 11 layers, the overlap is seen to vary only 10% compared with that of bulk graphite. Hence, 10 layers are suggested to be the limit to few-layer graphene [16]. Therefore, single-layer graphene as well as few-layer graphene (up to 10 layers) can be used for water contaminant and heavy metal detection.

6.3.2 MECHANICAL PROPERTIES

Investigation of the fracture strength of a single-layer graphene along with its Young's modulus is performed by simulations based on molecular dynamics [20–22]. Measurements related to force displacements by atomic force microscopy (AFM) were conducted to measure the Young's modulus of multilayer graphene. The AFM measurements were conducted on strips of graphene after being suspended over trenches [23]. Measurements using AFM by nanoindentation were conducted to measure the intrinsic breaking strength and properties of graphene related to elasticity [24]. Monolayer graphene that was defect free was found to have a 130-GPa fracture strength and 1.0-TPa Young's modulus.

6.3.3 RAMAN STUDIES

Characterization of graphene sheets is mainly done by Raman spectroscopy. Graphene correlates to Raman spectra consisting of a 2D peak around 2700 cm^{-1}, and the 1580 cm^{-1} peak attributes to the G peak. These spectral peaks are caused due to the boundary phonons of second-order zone and optical vibrations in the plane, respectively. The location of D peak for first-order phonons is at 1350 cm^{-1} in graphene, which contains a defect; however, this peak is not visible in defect-free graphene. The position and shape of the 2D peak can facilitate determination of the number of layers in graphene (n-layer up to five), and the quality of the graphene can be analyzed [25]. The shift in the 2D peak position to a higher wave number indicates that the number of layers has increased. The increment layer numbers can also be evaluated by broadening of the 2D peak. Mechanical strain in the graphene matrix can be analyzed by the splitting and shifting of Raman modes. The Raman spectrum shows no change for various substrates incorporating graphene, i.e., Raman spectra are substrate independent. This is mainly true for micromechanically cleaved graphene sheets as the interactivity between the transferred graphene sheets and substrate is minimum. Hence, proper transfer of monolayer graphene onto the desired substrate is quite tedious work and requires certain skills. The range of frequency for 2D and G bands can also be tuned by introducing dopants through changes due to electron-phonon coupling [26]. The ratios in intensity of D and G peaks [27] allow the calculation of the disorder in graphene matrix due to edges, impurities, defects, and domain boundaries.

6.4 SYNTHESIS OF GRAPHENE FACILITATING HEAVY METAL DETECTION

The synthesis of graphene can be subdivided in two different techniques, top-down and the bottom-up. Approaches related to the top-down technique deal with graphite exfoliation into single-layer graphene (or few-layer graphene). Approaches related to the bottom-up technique include pyrolysis, epitaxial growth, plasma synthesis, and CVD.

6.4.1 TOP-DOWN TECHNIQUES

The top-down technique deals with attacking the raw bulk graphite and separating the layers to obtain sheets of graphene. Chemical and mechanical exfoliation of bulk graphite powder leads to either a single-layer or a few-layer graphene sheet characterized by Raman spectroscopy.

6.4.1.1 Mechanical Exfoliation

Mechanical exfoliation of bulk graphite to obtain a monolayer or few-layer graphene over desired substrates is a well-known method. The superficial part of the graphite experiences longitudinal or transverse stress during mechanical exfoliation. The interlayer distance between graphene layers is 3.354 Å and the bond energy is observed to be 2 eV/nm^2 [28]. The external force required for mechanical cleaving of graphite layers to obtain a single atomic layer graphene sheet is approximately 300 nN/mm^2 [29]. When the partially filled p-orbitals overlap perpendicularly on a plane sheet, the sheets stack due to van der Waals forces. The higher lattice spacing in vertical direction and poor bonding forces between the layers causes the reverse step of stacking and exfoliation to occur when external forces are applied. This mechanical exfoliation is carried out through peeling different graphitic substances such as highly oriented pyrolytic graphite (HOPG), natural graphite, and monocrystal graphite. Mechanical exfoliation is generally carried out using different agents such as ultrasonication [30], electric field [31], and scotch tape [32].

6.4.1.2   Chemical Reduction of Graphite Oxide

One of the top-down approaches for producing enormous volumes of graphene is through reduction of graphite oxide. Synthesizing graphite oxide requires oxidation of graphite, which is usually done by utilizing certain oxidants such as potassium permanganate, concentrated nitric acid, and sulfuric acid. In 1860, graphene oxide was first produced by Brodie [33], Staudenmaier [34], and Hummers [35]. After this discovery, the modified Hummers method and the improved Hummers method were created. The major differences between these methods are the types of oxidants, the number of steps, and degree of toxicity. With progress, the toxic nature of synthesis could be eliminated with a decrease in reduction steps. Once the oxidation is complete, various reducing agents are employed to remove the oxidizing groups from graphene oxide. The reducing agents may be sodium borohydride, hydroxylamine, glucose, pyrrole, ascorbic acid, phenyl hydrazine, and many more. However, utilizing these reducing agents does not suffice for full reduction as there are some leftover oxidizing groups. Graphite oxide or graphene oxide is a material with high potential due to the hydrophilic nature of its layers. However, to achieve single-layer graphene, first graphene oxide needs to be suspended in water. Then, it should undergo extreme sonication, which results in single-layer sheets by overcoming the weak interlayer forces. The deposition of these suspended graphene oxide sheets onto substrates is then carried out through spin coating or filtration. Once transferred to the substrate, graphene oxide undergoes reduction treatments to produce defect-induced single- or few-layer graphene films. Formation of reduced graphene oxide (rGO) suspensions in organic solvents is common practice, because graphene oxide is hydrophilic in nature, but rGO is hydrophobic, which leads to graphene sheets becoming agglomerated. Graphene oxide reduction through a chemical process is a very famous technique among the research community due to the (1) large quantity of graphene sheets; (2) desired functionalization of graphene sheets for various applications including sensing, transparent conducting electrodes for solar cells, electrochemical high-energy electrode material for energy storage, and many more; and (3) tuning defects or the presence of functional groups by various approaches of reduction [36].

6.4.2   BOTTOM-UP TECHNIQUES

Bottom-up techniques deal with the formation of graphene from carbonaceous species in the vapor state. The formation of graphene from atoms of carbon results in defect-free pure graphene sheets. However, the yield of graphene, which is low in such cases, should be improved. Pyrolysis, epitaxial growth, CVD, and plasma synthesis are the bottom-up strategies discussed in this chapter.

6.4.2.1   Pyrolysis

The pyrolysis technique comprises a formation of graphene sheets chemically by the solvo-thermal method. For example, one may use sodium and ethanol in a 1:1 molar ratio in a container vessel during the entire duration of thermal treatment. Another approach is pyrolization of sodium ethoxide through the process of sonication. This results in better performance during the detachment of graphene sheets.

6.4.2.2   Epitaxial Growth

One of the methods for growing graphene on substrate surfaces is epitaxial growth. Heating and cooling down a crystal of silicon carbide can be used to prepare graphene at an optimum pressure. Graphene can grow on both silicon and carbide faces. Growth over the silicon face of the crystal leads to monolayer or bilayer graphene formation. Few-layer graphene is formed when growth takes place on the carbide face of the crystal [37]. The parameters used during the growth process, i.e., heating rate (ramp), pressure, and temperature, facilitate the quality and nature of graphene produced. When high temperature and pressure is applied, the growth of carbon nanotubes is favored compared with graphene. Hence, by varying the conditions, different morphologies and materials can be obtained such as carbon nanotubes, graphene sheets, graphene nanoribbons, etc. The lattice of the Ni(III) surface is relatively similar to that of graphene with a dissimilarity of only 1.3%. Hence, by using nickel diffusion [38], a thin layer of nickel might be evaporated onto the SiC crystal surface. After heat is applied, graphene will be found on the surface due to diffusion of carbon through the nickel layer. However, the entire process depends on the temperature and heating rate. This additional layer of nickel allows better separation of graphene that is grown on the SiC surface. However, due to defect formation along with grain boundaries, the quality of graphene is not the best and the formed graphene lacks homogeneity.

6.4.2.3 Chemical Vapor Deposition (CVD)

CVD is the best technique to synthesize high-quality graphene sheets. The technique involves reaction in the gaseous phase and deposition onto a desired substrate [39]. The combination of gaseous species in a reaction chamber occurs at a desired optimum pressure and temperature. When the reactant gases meet the substrate in a heated reaction chamber, a material film is seen to grow over the substrate. There are certain by-products in the chamber that are removed during the process. The temperature of the substrate plays a significant role in facilitating proper reaction mechanisms. The speed of deposition of material over the substrate is very low and in the range of a few microns per hour. Two types of CVD are widely used, namely ultrahigh vacuum CVD (UHVCVD) and low-pressure CVD (LPCVD).

6.4.2.4 Plasma Synthesis

Apart from UHVCVD and LPCVD, graphene synthesis using plasma is another bottom-up approach. It consists of both plasma doping to form graphene sheets [40] and plasma-enhanced CVD (PECVD). The PECVD techniques are categorized into three types, inductively coupled plasma-enhanced CVD (ICP-PECVD), direct current plasma enhanced CVD (DC-PECVD), and microwave plasma enhanced CVD (MW-PECVD). ICP-PECVD includes wave heating and plasma is generated without the presence of electrodes, thus, impurities could be eliminated [41]. Temperature, gas flow rate, and pressure are the parameters The time period of film growth and the power of plasma can be adjusted in ICP-PECVD. This approach is environmentally friendly and highly scalable, but it is less effective compared with other processes. In case of DC-PECVD, the occurrence of plasma and its sustainability depends on secondary electrons, which are created by impinging accelerated ions among the electrodes [42]. When the cathode is impinged by both atoms and plasma ions, the atoms tend to release. The atoms that are sputtered may diffuse through the plasma causing impurity atoms to spread over the produced graphene film. Due to the wave heating approach of MW-PECVD, wet etching could be avoided leading to no pollution of polymers or metal impurities on the produced graphene films [43]. Plasma doping is a technique to dope graphene with other substances to create variations in electrical properties. Introduction of heteroatoms into graphene by treating it with plasma to modify the electrical characteristics is one such approach. Nitrogen doping on graphene sheets is an appropriate example of such a process [44]. These plasma processes involve low energy, faster production, less contamination, and retaining the main properties of graphene [40].

6.5 FUNCTIONALIZATION OF GRAPHENE TO ATTAIN SELECTIVITY TO DIFFERENT ANALYTES

Functionalization of graphene is conducted while preserving the electron transport properties to a certain extent. The surface functionalization of graphene is very significant while detecting specific target analytes in water. Pristine graphene can be doped easily with various chemical species, and different functional moieties may be attached to graphene sheets to achieve selectivity in heavy metal ion detection. Water consists of various chemical species of unknown composition, where various organic compounds, living organisms, different cations, and anions coexist simultaneously forming a complex matrix. Hence, attachment of a selective probe onto the surface of the transducing material, i.e., graphene, is of utmost importance. This attachment or functionalization of the surface can be done in many different ways, and the various strategies for functionalization are provided extensively in the reviews [45–50]. We limit ourselves to certain relevant approaches used during fabricating solid-state sensing devices. Formation of chemical bonds by covalent functionalization can be obtained between reactive species (carbenes, arynes, free radicals, etc.) and sp^2 backbone of graphene sheet or graphene oxide [51–54]. The attachment of functional groups to graphene oxide is facilitated by the dangling oxygenated bonds. After the desired functionalization step, graphene oxide reduction is initiated to obtain rGO [55, 56]. The stable nature of covalent functionalization is an advantage; however, it may lead to disruption of the conjugated π system, causing the electron transport properties to degrade. This approach in fabrication of FET-based sensors and chemiresistive sensors are not very popular due to degradation in the transport phenomenon. Rather, a less invasive approach that will help to preserve electron transport is functionalization by the noncovalent approach achieved through π-π interactions.

Graphene is the most promising transducing material to undergo noncovalent functionalization due to its extensive conjugated π system [57]. The device needs to be incubated into the probe containing a solution for surface functionalization. This strategy is quite famous due to its effectiveness and simplicity. Chemiresistive and FET-based sensors actively detect heavy metal ions by conductivity change when exposed to water containing toxic ions. Interaction of the transducing material (graphene) and analytes is the simplest approach; however, to eliminate cross-sensitivity to undesired analytes, selectivity is achieved by functionalization of the channel containing graphene. This approach enables various strategies of sensing, namely protonation, complexation of the probe introduced, catalytic reactions, etc. [58]. Exposing the selective probe to graphene is one of the easiest approaches to functionalizing graphene. This is established through strong π-π interactions of the selective probes with the conjugated π system of graphene. The selective probes sometimes do not establish enough interaction with graphene; hence, functionalization is not feasible. In those cases, the common strategy is to use a linker and noncovalently bond (functionalize) it to graphene. These linkers now can be covalently bonded to the probes used for selectivity purposes. Hence, this linker acts as a mediator between the graphene sheet and the selective probes. Sometimes, instead of using a linker, graphene is decorated with gold nanoparticles, which is then covalently bonded to a selective probe. These functionalization strategies allow us to selectively determine the concentration of heavy metal ions present in the solution.

6.6 OPTICAL DETECTION OF HEAVY METAL IONS BY GRAPHENE-BASED FLUORESCENT PROBES

The detection of heavy metal ions optically by graphene is segmented into two parts: (1) graphene or graphene oxide and (2) graphene quantum dots.

6.6.1 GRAPHENE OXIDE OR GRAPHENE

6.6.1.1 Loading Nanoparticles on Graphene or Graphene Oxide

Graphene oxide produced by the Hummers method contains various groups containing oxygen such as phenol, hydroxyl, carboxylic, epoxide, and carbonyl, whereas the graphite produced by mechanical exfoliation does not contain these chemical groups. Hence, treatment of both types of graphene to load nanoparticles would be different. The ability of graphene or its derivatives to incorporate other nanoparticles in its matrix is significant for improving the performance of heavy metal ion sensors. Doping of pristine graphene by certain heteroatoms such as nitrogen or phosphorus may lead to an increase in active sites for sensing and improve the hydrophilicity of pristine graphene. Adding various functional groups and selective probes to graphene oxide is efficient due to unsaturation in the structure, thus, providing selectivity for various toxic metal ions.

6.6.1.2 Raman Enhancement by Different Graphene Nanocomposites

Electromagnetic mechanism (EM) or chemical mechanism (CM) is responsible for the increase in SERS. These two mechanisms are responsible for SERS signal amplification when noble metal nanocomposites with graphene are used. Chemical enhancement may take place in graphene sheets decorated with Raman probe molecules because of the transfer of charge between them. However, the noble metal nanoparticles result in local surface plasmon resonance causing a local electric field, leading to the electromagnetic effect. Toxic heavy metal ion detection is applicable using this overall enhancement in SERS. The assembly between gold nanoparticles and graphene was facilitated by a molecular glue and a local SERS reporter, cucurbit[7]uril (CB[7]), which was employed by Shi et al [59]. The immobilization of lead ions (Pb^{2+}) allows desorption of CB[7] from the surface of gold nanoparticles, which leads to "turning off" the SERS signal. This is due to stronger interaction of CB[7] with Pb^{2+} ions compared with gold nanoparticles. In situ fabrication of gold nanoparticles/rGO heterojunction with enhanced SERS activity was conducted through a seed-assisted growth mechanism by Ding et al. [60]. The nanocomposite was modified by thymine-rich DNA, which responded specifically to Hg^{2+}, as the T-Hg^{2+}-T complex was formed. Strong SERS signals were generated by the work proposed by Zhang et al. [61], where they synthesized gold triangular nanoarrays/n-layer graphene/gold nanoparticles forming a structure much like a sandwich in nature. The structure was found to be uniform with large sub-nanometer gaps, providing higher electromagnetic hot spots, enhancing the Raman signal to a greater extent. Here also T-Hg^{2+}-T complexes were formed in the SERS substrate, which facilitated SERS signal generation.

6.6.1.3 Fluorescence Detection by Graphene Nanocomposites

Loading fluorophore onto graphene or graphene oxide by direct assembly of nanoparticles that are fluorescent in nature is a technique for heavy metal detection in solution. Liu et al. applied functionalization to the rGO/Fe_3O_4 nanocomposite by utilizing Hg^{2+}-specific thymine-rich DNA [62]. Simultaneous detection and removal of Hg^{2+} ions by double stranded DNA and sensitive SYBR Green I was facilitated by high surface area of graphene sheets. Graphene oxide and double gold nanocluster composite (GO/D-GNCs) were prepared by Wu et al. with the help of electrostatic interactions for detection of Hg^{2+}, where D-GNCs facilitated enhancement in sensitivity and graphene oxide enhanced the reaction rate owing to its large specific surface area [63]. Another work was performed by Yu et al., in which they decorated graphene oxide with BSA functionalized gold nanoclusters (AuNCs/BSA) for separation and detection of Hg^{2+} [64]. The detection mechanism was based on the enhanced metal binding nature of AuNCs/BSA and very large specific surface area of graphene oxide. The highly improved efficiency of separation of this nanocomposite is based on the interaction between Hg^{2+} and Au^+. Cu^{2+} detection using graphene oxide modified with Ag-In-Zn-S quantum dots was performed by Liu et al. [65]. Here, the immobilization of Cu^{2+} initiated aggregation of the quantum dots leading to fluorescence quenching. The amount of quenching was attributed to the concentration of Cu^{2+} ions in solution.

6.6.1.4 Graphene Nanocomposites as Fluorescent Quenchers

It is known that dye labeled ssDNA can be easily functionalized to graphene oxide through pep stacking and hydrophobic interactions. This leads to fluorescence quenching of dyes by the fluorescence resonance energy transfer (FRET) mechanism; however, dsDNA does not interact and bind easily to the graphene oxide surface. Hence, graphene oxide can be used as a DNA-based platform for sensing heavy metals with more signal-to-noise ratio. The Hg^{2+} ion detection by functionalization with thymine-rich ssDNA based on FRET/SIMNSEF (shell-insulated metal nanoparticle surface-enhanced fluorescence) spectroscopy on graphene oxide and shell-insulated metal nanoparticles (SIMNs) was proposed by Kong et al. [66]. Easy adsorption of the labeled ssDNA attached to SIMNs takes place over graphene oxide when the Hg^{2+} ions are not present, which leads to quenching of fluorescence by the FRET mechanism. The interaction between ssDNA and graphene oxide is weakened by the formation of T-Hg^{2+}-T coordination when Hg^{2+} is present. This leads to desorption of ssDNA functionalized SIMNs, and recovery of fluorescence is observed. Here, amplification of the fluorescence signal takes place through the surface-enhanced fluorescence (SEF) effect. The change in fluorescence allowed many researchers to construct similar sensing platforms for Ag$^+$ and Pb^{2+} and many more by using the mechanism of conformation evolutions of DNA when combined with metal ions. These led to the formation of the Pb^{2+}-G quadruplex and T-Hg^{2+}-T and C-Ag$^+$-C base pairs [67–70]. Many other research groups have worked on heavy metal ion detection depending on quenching of fluorescence of the fluorescent nanoparticles like carbon dots, nanophosphors, etc. Also, fluorescence recovery is the measure of heavy metal ions: the greater the concentration is, the greater the recovery of fluorescence [71, 72]. Bridging graphene oxide with graphene quantum dots by G-rich DNA as a linker is another strategy used to form the Pb^{2+}-G quadruplex, which may lead to a variation in the length of the DNA chain. The shortening in the chain length between graphene quantum dots and graphene oxide may lead to quenching in fluorescence of graphene quantum dots. Efficient energy transfer between the two nanostructures is responsible for quenching, hence, the greater the concentration is, the greater the energy transfer and quenching [73].

6.6.2 GRAPHENE QUANTUM DOTS

6.6.2.1 Graphene Quantum Dots Directly as Fluorescence Probes

As graphene oxide shows conditional fluorescence depending on the size and the surface functional groups, the quantum yield of fluorescence is very low. Because of the absence of bandgap in graphene oxide, the full-width half-maximum (FWHM) is wider and the emission peaks are asymmetric in nature. Hence, many efforts are made to synthesize graphene quantum dots by bottom-up or top-down approaches. The intrinsic characteristics of the graphene quantum dot may be tuned by doping certain heteroatoms such as boron, phosphorus, and nitrogen into the graphene matrix. Optical characteristics may be manipulated through changing surface features and allowing fluorescent probes to be functionalized onto the surface for the purpose of heavy metal ion sensing. The synthesis of sulfur and nitrogen co-doped graphene quantum dots by Shen et al. [74] is done via the hydrothermal approach through a one-step bottom-up molecular fusion reaction. This doped graphene quantum dot showed simultaneous detection of Ag$^+$, Cu^{2+}, and Fe^{3+} ions with low limits of detection. Simultaneous detection of Fe^{3+} and Hg^{2+} by utilizing dual excitation of graphene quantum dots was conducted by Xu et al. by developing a masking agent free method [75]. Blue fluorescence was emitted by the prepared graphene quantum dots when excited by two different wavelengths. Fe^{3+} responded to change in one of the fluorescent channels and Hg^{2+} allowed quenching of the other fluorescence. Hence, both metal ions can be detected in a single sensing matrix. Direct interaction strategies between metal ions and graphene quantum dots is dependent on the mechanism of "turn off." This suffers from lack of selectivity and leads to false-positive results. The single signal mechanism of turning off was overcome by Hua et al. [76], where the self-assembled graphene quantum dots were electrostatically immobilized on the surface of silica and acted as active Hg^{2+} sites. The CdTe quantum dots, which emitted red fluorescence, were entrapped in the silica nanosphere and acted as a shield to environmental effects. This ratio-metric probe was used for quantification of Hg^{2+} ions. The fluorescence of the graphene quantum dot was quenched when Hg^{2+} ions were immobilized, whereas the fluorescence of red CdTe was maintained. Hence, with the addition of Hg^{2+}, the solution's fluorescence color transformed from blue to red and it can be visualized by our eyes. This sensitive detection technique is depicted in Figure 6.3. However, a dual probe of fluorescence was proposed by Sun et al. [77], where graphene quantum dots functionalized with glutathione were used to act as internal standards. CdTe quantum dots were utilized as a fluorescence probe of response on detection of Cu^{2+}. When the Cu^{2+} ions were immobilized, the red fluorescence due to CdTe was quenched, whereas the blue fluorescence of the graphene quantum dot was maintained. Hence, with the addition of Cu^{2+} in solution, the fluorescence turned from red to blue.

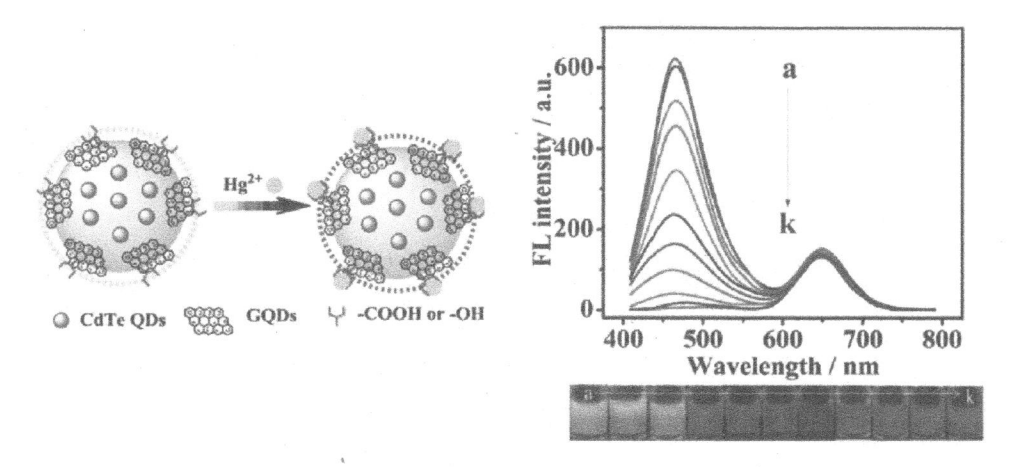

FIGURE 6.3 Ratio-metric fluorescence probe for the detection of Hg^{2+} and fluorescence intensity spectra when exposed to various concentrations of Hg^{2+}. (Reproduced with permission from Ref. [76].)

6.6.2.2 Graphene Quantum Dots Modified with Large Biomolecules

Heavy metal ion detection is facilitated by functionalization of graphene quantum dots with large biomolecules, such as T-rich DNA and C-rich DNA to impart selectivity. Park et al. used Pb^{2+} aptamers based on DNA to functionalize graphene quantum dots to detect lead ions specifically [78]. The electron transfer mechanism from graphene quantum dots to Pb^{2+} leads to fluorescence quenching of graphene quantum dots, which provides quantitative measure of the concentration of Pb^{2+} ions. The sensor demonstrated a limit of detection (LOD) of 0.64 nM. In a similar work Zhao et al. [79] used thymine-rich DNA to functionalize graphene quantum dots. A T-Hg^{2+}-T hairpin-like structure with DNA is formed, which allows electron transfer from graphene quantum dots to Hg^{2+}, leading to a decrease in fluorescence from the quantum dots. Cu^{2+} ion sensing was done by Ha et al. [80], where a Cu^{2+}-specific aptamer was used. Here the up-conversion photoluminescence (UCPL) quenching took place due to electron transfer between graphene quantum dots and Cu^{2+}. The quenching provided the quantitative information about the presence of Cu^{2+} ions in solution. Functionalization of graphene quantum dots with derivatives of copper-zinc superoxide dismutase (E_2Zn_2SOD) which is Cu free, was performed by Zhu et al. [81]. These functionalized dots were used as receptors and Nile Blue dye was used as probes for reference. Two photon excitation caused dual emission bands and the immobilization of Cu^{2+} ions caused the blue fluorescence of graphene quantum dots to get quenched. However, the red fluorescence of Nile blue dye was found to remain the same. Therefore, a change in color to yellow from blue was observed when Cu^{2+} ions were added, leading to the formation of a Cu^{2+} ratio-metric fluorescent sensor.

6.6.2.3 Graphene Quantum Dots Assembled with Nanoparticles

Graphene has an outstanding property to host various metal nanoclusters or nanoparticles in its matrix to form a highly stable assembly. Many research teams have conducted synthesis of graphene quantum dots decorated by noble metal nanoparticles and studied its effect on the fluorescence of graphene quantum dots. Ran et al. studied that silver nanoparticles can form hybrids with graphene quantum dots, as Ag^+ can get easily attached to graphene quantum dots electrostatically [82]. The fluorescence of graphene quantum dots becomes quenched when functionalized with silver nanoparticles. However, with the addition of bio-thiols, fluorescence of graphene quantum dots gets completely quenched due to the reducing nature of bio-thiols which acts as bridge between the silver nanoparticles. Hence, by utilizing the hybrids, detection of Ag^+ and bio-thiols are possible. When gold nanoparticles were decorated over graphene quantum dots by Niu et al [83], it resulted in quenching of fluorescence. Functionalizing gold nanoparticles with DNA leads to separation of graphene quantum dot and gold nanoparticles, when Pb^{2+} ions are immobilized. This separation is initiated by the Pb^{2+} ions; hence, the fluorescence recovery of graphene quantum dots is possible. The fluorescence recovery can be a quantitative analysis of the concentration of Pb^{2+} ions in solution.

6.7 ELECTRICAL DETECTION OF HEAVY METAL IONS BY GRAPHENE-BASED DEVICES

6.7.1 Working Mechanism of Graphene FETs and Chemiresistive Devices

FETs consist of source and drain contacts, a gate dielectric, and a channel comprising transducing material and gate contacts. When gate dielectric and gate electrodes are removed from this configuration, it is simplified to form a chemiresistor device. Figure 6.4(a–c) schematically depicts the three different types of sensor devices, namely back-gated FET, solution-gated FET, and chemiresistors. The source and drain electrodes are frequently covered with a dielectric, so that there is negligeable current leakage through the solution. Electrostatic gating effect and charge transfer are the two mechanisms of sensing that are used to explain the change in conductance of the sensing layer in FETs and chemiresistors. However, these two effects are contradictory in nature. For adsorbates that are negative in nature, the positively charged holes are attracted by the electrostatic gating effect, making the transducing material p-type in nature. In case the working mechanism is a charge transfer, then the adsorbates would lead to n-type doping by donating electrons to the channel. The reverse phenomenon is observed when positively charged adsorbates are immobilized in the channel region. Both effects can occur simultaneously; however, one will dominate. The two effects are elucidated in Figure 6.5(a). The device performance is affected by an electrical double layer (EDL) when electrostatic gating dominates in nature.

Spontaneous formation of EDL will occur at the interface of liquid and channel. Dropping surface potential and reaching a minimum at the boundary (the distance termed as Debye length) takes place. Screening of the surface at this point takes place by the ions present in the solution and the charge particles in the solution have no influence, which affects the ability to detect toxic ions in the solution. The event of sensing needs to take place within the EDL of the channel. If it does not occur with the double layer, then these ions are screened and will not be able to affect the channel transport properties, restricting the sensing mechanism. The details of the scenario are elucidated in Figure 6.5(b). Hence, performance of the sensing devices depends on the thickness of the ELD. If

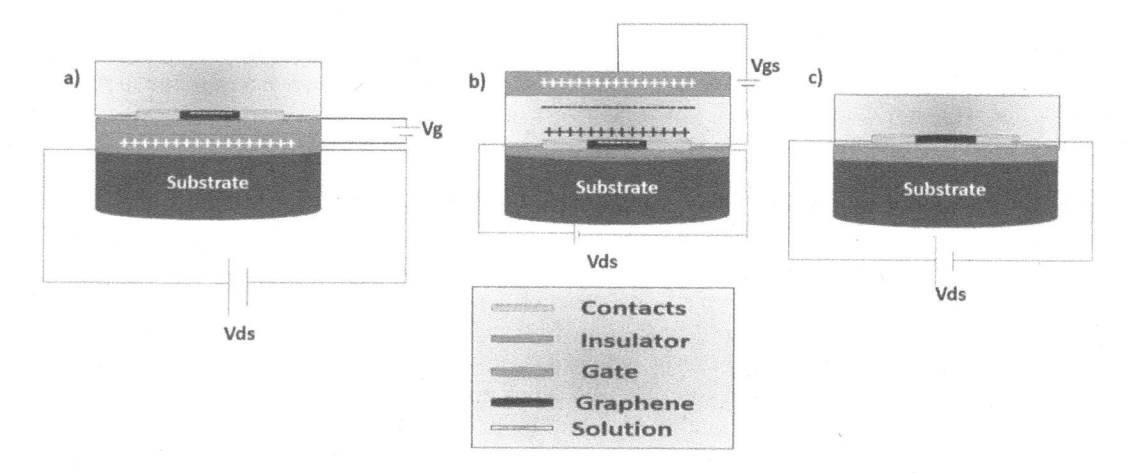

FIGURE 6.4 (a) Back-gated FET, (b) solution-gated FET, and (c) chemiresistor. Working FETs with positive bias on the gate when the channel is doped intrinsically is depicted in the figures. If negative bias is applied, the opposite behavior is observed.

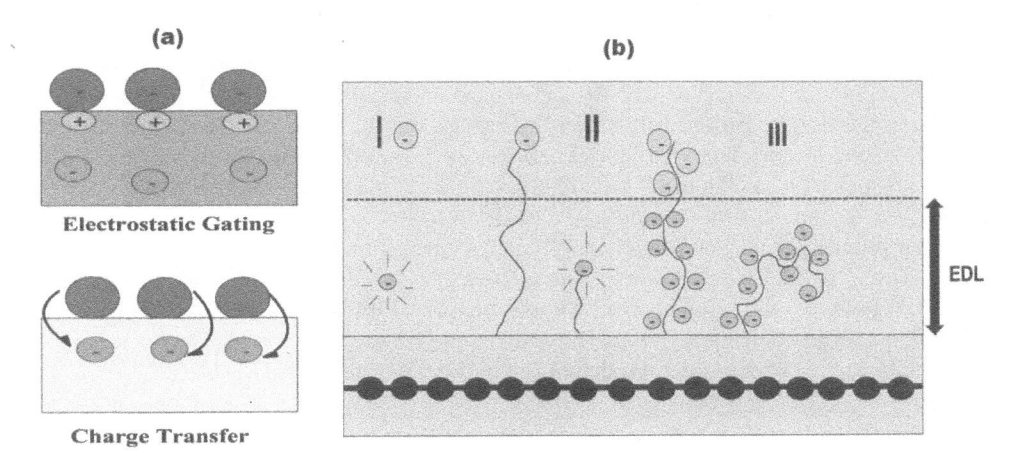

FIGURE 6.5 Mechanism of sensing in chemiresistors and FETs: (a) Electrostatic gating effect and mechanism of charge transfer on channel doping when exposed to negative adsorbates. (b) Sensitivity of the device based on EDL thickness (Debye length): (I) determining the distance from the channel, (II) determining the length of the probe, and (III) modulation of the channel conductance is determined by the conformational changes.

the Debye length is short, then the sensing event must take place much closer to the channel, which in turn decreases the device sensitivity. The solution's ionic strength is the determining factor of the Debye length, where the Debye length gets shortened with increase in ionic strength. Therefore, in other words, the composition and pH of the electrolyte plays a major role in the sensing mechanism. However, with the charge transfer mechanism these considerations are less significant and the main contributing factor is the change in electrical distribution of the sensing probes reaching the channel.

The doping of the graphene-based channel in the FET device can be found by recording the I_{sD}-V_G curves. The doping may be n-type or p-type, I_{sD} is the source-to-drain current, and V_G is the applied voltage at gate terminal. The minima of the plot should appear at zero bias of gate for pristine graphene. When graphene is p-doped, the excess holes need to be neutralized by a greater number of electrons; hence, the minima appears at positive biases. On the other hand, graphene with n-type doping will have its minima at negative gate voltages. Chemiresistors are not modulated like FETs, but they work on the same sensing mechanisms.

6.7.2 BACK-GATED FETS FOR HEAVY METAL ION SENSING

The gate is built with heavily doped silicon, and a dielectric (mostly silicon dioxide) separating it from the graphene-based transducing channel. The operation of FETs can be both in static and dynamic mode. In static mode the source-to-drain current is measured with change in gate voltage keeping the source-to-drain voltage fixed. However, in the dynamic mode, the source-to-drain current is measured with respect to time at a constant source-to-drain voltage. The static mode will allow us to plot the gate voltage of the Dirac point with respect to the analyte concentration. Static plots facilitate characterization of devices (i.e., whether they are n-type doped or p-type). The dynamic mode will measure the change in current or resistance with change in concentration of the analyte.

Functionalization of graphene layers by 1-octadecane-thiol by Zhang et al. was conducted [84], where self-assembly into nanostrips takes place along the in-plane axis of graphene. The sensitivity of the device to Hg^{2+} was rendered because of interaction with thiol groups. The shifting of the Dirac point to more positive values took place when the device was immobilized with 10 ppm of Hg^{2+}.

This indicated p-type doping of graphene layers. Chen et al. functionalized the rGO channel with benzyl-triethyl ammonium chloride (TEBAC) to attain selectivity of nitrates [85]. A reduction in channel conductance was observed with an increase in analyte concentration. The graphene-based device was observed to be p-doped intrinsically through the I_{sD}-V_G curves. Here the interaction is basically due to charge transfer, where the charge is transferred from the complex of TEBAC-NO_3- to the rGO sensing layer. Wang et al. functionalized the channel with DNAzyme with a substrate strand [86]. The graphene film was attached to the selective probe via noncovalent interactions using a linker pyrene group. Intrinsically p-type doping was revealed from characterization of the devices. However, when Pb^{2+} ions were immobilized, the Dirac point shifted to negative values. This showed n-type doping of the channel when exposed to Pb^{2+} ions. Thus, the positively charged ion enforced the electrostatic gating effect leading to a shift in the Dirac point. The device expressed good specificity to Pb^{2+} ions in the presence of Ca^{2+}, Mg^{2+}, K^+, and Na^+.

6.7.3 Solution-Gated FETs for Heavy Metal Ion Sensing

Solution-gated FETs possess higher transconductance when compared with back-gated FETs and has more sensitivity when compared with back-gated FETs. This may be attributed to the shorter Debye length of the EDL when compared with the gate dielectric's thickness. This type of configuration uses a reference electrode, i.e., Ag/AgCl, to be precise. This electrode is used to apply gate bias through the solution.

An intrinsically p-type rGO-based sensor device was fabricated by Park et al. [87], where poly-furan nanotubes were embedded in the rGO film to selectively monitor changes due to Hg^{2+} ion concentrations. When the concentration of Hg^{2+} ions was increased, the conductivity of the device was seen to increase, due to interaction of the lone pair of electrons of the oxygen atom present in furan with the immobilized metal ions. The sensor expressed enhanced selectivity in the presence of other metal ions like Ce^{2+}, Na^+, Pb^{2+}, Ni^{2+}, Zn^{2+}, Li^{2+}, and Cu^{2+}. Another similar work was conducted by Park et al. where rGO films were decorated by polypyrrole nanotubes [88]. This work was intended to detect H_2O_2 in water. The sensor device was intrinsically found to be p-doped, and with an increase in concentration of the analyte, current was seen to increase. The sensor was selective to H_2O_2 in the presence of ascorbic acid, uric acid, and glucose. Single-layer graphene was functionalized noncovalently by linker 1,5-diaminonapthalene (DAN) and was then attached to glutaraldehyde with the help of a Schiff-base reaction. The composite was then capped with a Hg^{2+} aptamer [89]. The two thymine bases binds with metal ions by displacing the protons, which results in a duplex structure. With increase in concentration, the conductance was found to be increased. The sensor device was highly selective in presence of various ions such as Cd^{2+}, Na^+, Zn^{2+}, Li^+ and many more. Instead of an aptamer, Tu et al. introduced ssDNA to the composite [90]. Here a duplex structure also was formed when Hg^{2+} ions were exposed. The sensor worked with negative gate potential and on immobilization of mercury ions, the conductance was observed to decrease. The presence of various other cations did not interfere with the sensing performance.

6.7.4 Chemiresistors for Heavy Metal Ion Sensing

Chemiresistor-based sensor devices are also very common where they work similar to that of solution-fated FETs (the gate potential is absent). The device can be thought of as a FET-based device with source and drain with zero gate bias. Functionalization of graphene films are done so that there is direct interaction of the functionalized sensing channel with the target analytes. Similar to other approaches, selectivity also is obtained through functionalizing the transducing material with a selective probe.

In a typical work, Tan et al. [91] fabricated a chemiresistive sensor device by functionalizing rGO with a linker 1-Pyrenebutanoic acid succinimidyl ester (PASE) and then immobilized and attached a Hg^{2+} specific aptamer with the help of a nucleophilic substitutional reaction. The sensor's

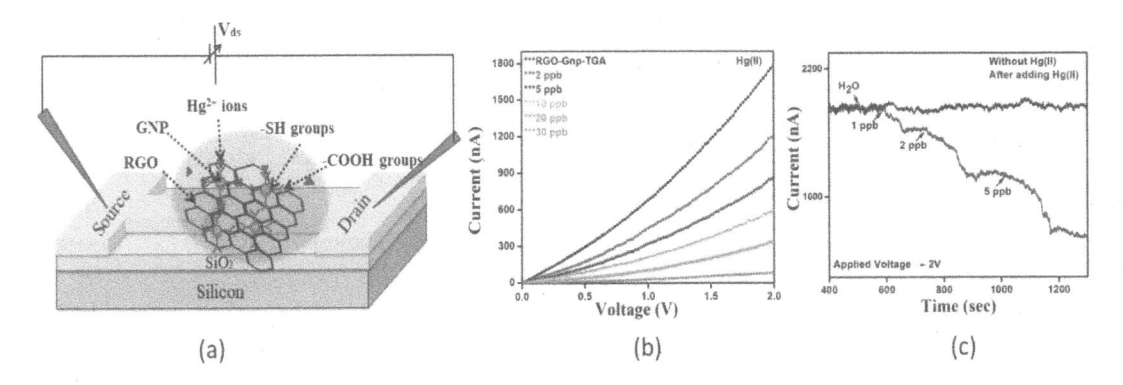

FIGURE 6.6 Chemiresistor device based on the RGT sensing layer: (a) schematic of the device, (b) i-v characteristics when exposed to various concentration of Hg^{2+} ions, and (c) i-v-t characteristics (dynamic response) of the sensor at 2 V.

conductivity is enhanced with an increase in the analyte's concentration. The authors proposed the sensing mechanism. As the positively charged Hg^{2+} ions are exposed, the mechanism repels the holes from the rGO channel, leading to the change in conductance. The device showed high selectivity to Hg^{2+} ions in the presence of Li^+, Na^+, Zn^{2+}, Cu^{2+}, Pb^{2+}, K^+, and Mg^{2+} ions. In another work, Sett et al. fabricated a Hg^{2+} chemiresistor sensor based on gold nanoparticles functionalized with thiol groups and attached to rGO sheets [92]. The gold nanoparticles were functionalized with thiol groups which were specific to Hg^{2+} ions. The gold nanoparticles acted as a mediator (linker) between the rGO sheets and the thiol molecules as depicted in Figure 6.6(a). Figure 6.6(b) shows the plot for current-voltage characteristics at different concentrations of mercury ranging from 2 to 30 ppb. Figure 6.6(c) illustrates the i-v-t (dynamic) plot of the RGT-based mercury sensor, where the resistance is seen to increase with an increase in mercury ion concentration. The sensing is conducted at 2 V and is highly selective to Hg^{2+} ions in the presence of Pb^{2+}, Cd^{2+}, and As^{3+}.

6.8 FUTURE SCOPE AND AREAS OF DEVELOPMENT

After reviewing the literature, it can be revealed that FET-based sensor geometry is renowned in the area of water-quality sensors. Water quality includes sensing of heavy metal ions as well as pH sensing and sensing of various anions such as nitrates and fluorides, and many other organic compounds. Seventy-five percent of the water-quality monitoring devices are FET-based devices. The in-depth analysis of the electronic properties and device characterization makes FET-based sensors advantageous. The chemiresistor-based configuration is very popular due to its simplicity and direct interaction with the analytes; hence, a shift to a chemiresistive configuration is expected. Covalent attachment of probes is found to be very complicated and it affects the intrinsic electrical aspects of graphene, allowing it to deteriorate the overall performance of the device. Noncovalent functionalization of probes on graphene substrate is dominant; however, π-π functionalization strategies might also suffer stability issues. This is because resetting the sensor is problematic in this case; however, for long usage, resetting the sensor is required.

The most dominating feature of sensing is functionalizing the channels to detect the target analytes in water. The analytes and detecting units may be heavy metals, different cations and anions, pathogens, or organic compounds. The most common among all the units is heavy metal ion sensing due to its toxic and fatal effects on human life. Until now only 14 FET-based heavy metal ion sensors are reported, among which most of them are mercury and lead ion detectors. Cadmium, arsenic, copper, and cobalt ions also need more focus due to their detrimental effects on human health. Moreover, apart from sensing heavy metal ions, other contaminants present along with these heavy metals that need to be analyzed as well. Selectivity to different analytes is achieved through attachment of different selective probes onto the graphene structures. Inspiration may be drawn from various optical methods in which different proteins and enzymes are utilized; however, use of such proteins may lead to stability issues of such sensors. Hence, utilization of protein- or enzyme-free probes may be a new area of research that will in turn improve the lifetime (stability) of the sensors.

6.9 CONCLUSION

In this work, for the first time, different detection strategies for heavy metals are elaborated using graphene as the transducing material. First, the conventional methods of heavy metal ion sensing are discussed along with their working principles and drawbacks. The astonishing properties of graphene are highlighted as it is used as the transducing layer in most of the sensitive heavy metal ion detectors. Various synthesis techniques are discussed with more focus on chemical exfoliation of graphene. Functionalization of graphene sheets to introduce selectivity among various analytes are elucidated. Functionalization with various proteins, ssDNA, DNAzyme, and chemical groups are discussed. Optical methods for detection are discussed in two parts, one in which graphene works as a matrix for various nanoparticles and acts as a component in nanocomposites, and the other where graphene quantum dots are directly utilized as fluorescent probes (the quantum dots are functionalized by various biomolecules to impart selectivity in sensing target analytes). Electrical detection including three different types of configuration is discussed. Back-gate, solution-gate, and chemiresistor architecture is discussed in detail. Chemiresistors seem to be the newest prospect due to their simplicity in fabrication and direct interaction of the analytes with the functionalized channel. FET-based sensors seem to be the future of heavy metal ion sensing; however a great deal of work is still left to determine these ions in complex environments, which may consist of various anions, cations, and organic compounds. The readers may go through a few recent publications related to this work in Refs. [93–95].

ACKNOWLEDGMENT

The authors would like to thank the DST NNetra project for its extensive support.

REFERENCES

1. Liu, Y. X., Y. P. Wu, L. Wang, and Z. Y. Huang. "Determination of total arsenic and mercury in the fucoidans by hydride generation atomic fluorescence spectroscopy." Guang Pu Xue Yu Guang Pu Fen Xi 28, no. 11 (2008): 2691–2694.
2. Dai, Xuan, Olga Nekrassova, Michael E. Hyde, and Richard G. Compton. "Anodic stripping voltammetry of arsenic (III) using gold nanoparticle-modified electrodes." Analytical Chemistry 76, no. 19 (2004): 5924–5929.
3. Wehmeyer, Kenneth R., and R. Mark Wightman. "Cyclic voltammetry and anodic stripping voltammetry with mercury ultramicroelectrodes." Analytical Chemistry 57, no. 9 (1985): 1989–1993.
4. Hwang, Gil Ho, Won Kyu Han, Joon Shik Park, and Sung Goon Kang. "Determination of trace metals by anodic stripping voltammetry using a bismuth-modified carbon nanotube electrode." Talanta 76, no. 2 (2008): 301–308.
5. Kiekens, P., E. Temmerman, and F. Verbeek. "Voltammetric study of the redox behaviour of the Hg (II)/Hg (I)/Hg system at a rotating metal-ring/glassy-carbon disc electrode." Talanta 31, no. 9 (1984): 693–701.
6. Karami, Hassan, Mir Fazlollah Mousavi, Yadollah Yamini, and Mojtaba Shamsipur. "On-line preconcentration and simultaneous determination of heavy metal ions by inductively coupled plasma-atomic emission spectrometry." Analytica Chimica Acta 509, no. 1 (2004): 89–94.
7. Rui, Y. K., et al. "Application of ICP-MS/ICP-AES to detection of 22 trace elements in fruits of elm." Guang pu xue yu Guang pu fen xi= Guang pu 27, no. 10 (2007): 2111–2113.

8. Liang, Pei, Taqing Shi, Hanbing Lu, Zucheng Jiang, and Bin Hu. "Speciation of Cr (III) and Cr (VI) by nanometer titanium dioxide micro-column and inductively coupled plasma atomic emission spectrometry." Spectrochimica Acta Part B: Atomic Spectroscopy 58, no. 9 (2003): 1709–1714.

9. Chatterjee, Shyamasree Gupta, Somenath Chatterjee, Ajoy K. Ray, and Amit K. Chakraborty. "Graphene–metal oxide nanohybrids for toxic gas sensor: a review." Sensors and Actuators B: Chemical 221 (2015): 1170–1181.

10. Allen, Matthew J., Vincent C. Tung, and Richard B. Kaner. "Honeycomb carbon: a review of graphene." Chemical Reviews 110, no. 1 (2010): 132–145.

11. Li, Dan, and Richard B. Kaner. "Graphene-based materials." Nature Nanotechnology 3 (2008): 101.

12. Geim, Andrey, and Allan McDonald. "Graphene: exploring carbon flatland." Physics Today 60, no. 8 (2007): 35.

13. Lee, Changgu, Xiaoding Wei, Jeffrey W. Kysar, and James Hone. "Measurement of the elastic properties and intrinsic strength of monolayer graphene." Science 321, no. 5887 (2008): 385–388.

14. Kim, Keun Soo, Yue Zhao, Houk Jang, Sang Yoon Lee, Jong Min Kim, Kwang S. Kim, Jong-Hyun Ahn, Philip Kim, Jae-Young Choi, and Byung Hee Hong. "Large-scale pattern growth of graphene films for stretchable transparent electrodes." Nature 457, no. 7230 (2009): 706–710.

15. Lipson, Henry Solomon, and A. R. Stokes. "The structure of graphite." Proceedings of the Royal Society of London. Series A. Mathematical and Physical Sciences 181, no. 984 (1942): 101–105.

16. Partoens, Bart, and F. M. Peeters. "From graphene to graphite: electronic structure around the K point." Physical Review B 74, no. 7 (2006): 075404.

17. Geim, Andre Konstantin. "Graphene: status and prospects." Science 324, no. 5934 (2009): 1530–1534.

18. Geim, Andre K., and Konstantin S. Novoselov. "The rise of graphene." In Nanoscience and Technology: A Collection of Reviews from Nature Journals (2010): 11–19. https://doi.org/10.1142/9789814287005_0002

19. Novoselov, K. S., A. K. Geim, S. V. Morozov, and D. Jiang, M. I. Katsnelson, I. V. Grigorieva, S. V. Dubonos, and A. A. Firsov. "Two-dimensional gas of massless Dirac fermions in graphene." Nature 438, no. 7065 (2005): 197.

20. Van Lier, Gregory, Christian Van Alsenoy, Vic Van Doren, and Paul Geerlings. "Ab initio study of the elastic properties of single-walled carbon nanotubes and graphene." Chemical Physics Letters 326, no. 1–2 (2000): 181–185.

21. Reddy, C. D., S. Rajendran, and K. M. Liew. "Equilibrium configuration and continuum elastic properties of finite sized graphene." Nanotechnology 17, no. 3 (2006): 864.

22. Kudin, Konstantin N., Gustavo E. Scuseria, and Boris I. Yakobson. "C 2 F, BN, and C nanoshell elasticity from ab initio computations." Physical Review B 64, no. 23 (2001): 235406.

23. Frank, I. W., David M. Tanenbaum, Arend M. van der Zande, and Paul L. McEuen. "Mechanical properties of suspended graphene sheets." Journal of Vacuum Science & Technology B 25, no. 6 (2007): 2558–2561.

24. Lee, Changgu, Xiaoding Wei, Jeffrey W. Kysar, and James Hone. "Measurement of the elastic properties and intrinsic strength of monolayer graphene." Science 321, no. 5887 (2008): 385–388.

25. Ferrari, A. C., J. C. Meyer, V. Scardaci, C. Casiraghi, M. Lazzeri, F. Mauri, S. Piscanec, D. Jiang, K. S. Novoselov, S. Roth, and A. K. Geim. "Raman spectrum of graphene and graphene layers." Physical Review Letters, 97, no. 18 (2006): 187401.

26. Wang, Y. Y., Z. H. Ni, T. Yu, Z. X. Shen, H. M. Wang, Y. H. Wu, W. Chen, and A. T. Shen Wee. "Raman studies of monolayer graphene: the substrate effect." Journal of Physical Chemistry C 112, no. 29 (2008): 10637–10640.

27. Robinson, Joshua A., Maxwell Wetherington, Joseph L. Tedesco, Paul M. Campbell, Xiaojun Weng, Joseph Stitt, Mark A. Fanton et al. "Correlating Raman spectral signatures with carrier mobility in epitaxial graphene: a guide to achieving high mobility on the wafer scale." Nano Letters 9, no. 8 (2009): 2873–2876.

28. Bhuyan, Md Sajibul Alam, Md Nizam Uddin, Md Maksudul Islam, Ferdaushi Alam Bipasha, and Sayed Shafayat Hossain. "Synthesis of graphene." International Nano Letters 6, no. 2 (2016): 65–83.

29. Zhang, Yuanbo, Joshua P. Small, William V. Pontius, and Philip Kim. "Fabrication and electric-field-dependent transport measurements of mesoscopic graphite devices." Applied Physics Letters 86, no. 7 (2005): 073104.

30. Ci, Lijie, Li Song, Deep Jariwala, Ana Laura Elias, Wei Gao, Mauricio Terrones, and Pulickel M. Ajayan. "Graphene shape control by multistage cutting and transfer." Advanced Materials 21, no. 44 (2009): 4487–4491.

31. Liang, Xiaogan, Allan S. P. Chang, Yuegang Zhang, Bruce D. Harteneck, Hyuck Choo, Deirdre L. Olynick, and Stefano Cabrini. "Electrostatic force assisted exfoliation of prepatterned few-layer graphenes into device sites." Nano Letters 9, no. 1 (2009): 467–472.

32. Chen, J.-H., Masa Ishigami, Chaun Jang, Daniel R. Hines, Michael S. Fuhrer, and Ellen D. Williams. "Printed graphene circuits." Advanced Materials 19, no. 21 (2007): 3623–3627.

33. Brodie, B. C. "XXIII—Researches on the atomic weight of graphite." Quarterly Journal of the Chemical Society of London 12, no. 1 (1860): 261–268.

34. Staudenmaier, L. "Process for the preparation of graphitic acid." Reports of the German Chemical Society 32, no. 2 (1899): 1394–1399.

35. Hummers Jr, William S., and Richard E. Offeman. "Preparation of graphitic oxide." Journal of the American Chemical Society 80, no. 6 (1958): 1339–1339.

36. Sett A., S. Majumder, and T. K. Bhattacharyya, "Flexible room temperature ammonia gas sensor based on low-temperature tuning of functional groups in graphene." IEEE Transactions on Electron Devices 68, no. 7 (2021): 3181–3188.

37. Chaste, Julien, Amina Saadani, Alexandre Jaffre, Ali Madouri, José Alvarez, Debora Pierucci, Zeineb Ben Aziza, and Abdelkarim Ouerghi. "Nanostructures in suspended mono-and bilayer epitaxial graphene." Carbon 125 (2017): 162–167.

38. Fogarassy, Zsolt, Mark H. Rümmeli, Sandeep Gorantla, Alicja Bachmatiuk, Gergely Dobrik, Katalin Kamarás, László Péter Biró, Károly Havancsák, and János L. Lábár. "Dominantly epitaxial growth of graphene on Ni (1 1 1) substrate." Applied Surface Science 314 (2014): 490–499.

39. Tetlow, H., J. Posthuma De Boer, I. J. Ford, D. D. Vvedensky, Johann Coraux, and L. Kantorovich. "Growth of epitaxial graphene: theory and experiment." Physics Reports 542, no. 3 (2014): 195–295.

40. Wang, Qi, Xiangke Wang, Zhifang Chai, and Wenping Hu. "Low-temperature plasma synthesis of carbon nanotubes and graphene based materials and their fuel cell applications." Chemical Society Reviews 42, no. 23 (2013): 8821–8834.

41. Wu, Angjian, Xiaodong Li, Jian Yang, Changming Du, Wangjun Shen, and Jianhua Yan. "Upcycling waste lard oil into vertical graphene sheets by inductively coupled plasma assisted chemical vapor deposition." Nanomaterials 7, no. 10 (2017): 318.

42. Fray, Derek, and Ali Kamali. "Method of producing Graphene." U.S. Patent 10,458,026, issued October 29, 2019.

43. Bo, Zheng, Mu Yuan, Shun Mao, Xia Chen, Jianhua Yan, and Kefa Cen. "Decoration of vertical graphene with tin dioxide nanoparticles for highly sensitive room temperature formaldehyde sensing." Sensors and Actuators B: Chemical 256 (2018): 1011–1020.

44. Li, Nan, Zhiyong Wang, Keke Zhao, Zujin Shi, Zhennan Gu, and Shukun Xu. "Large scale synthesis of N-doped multi-layered graphene sheets by simple arc-discharge method." Carbon 48, no. 1 (2010): 255–259.

45. Kuila, Tapas, Saswata Bose, Ananta Kumar Mishra, Partha Khanra, Nam Hoon Kim, and Joong Hee Lee. "Chemical functionalization of graphene and its applications." Progress in Materials Science 57, no. 7 (2012): 1061–1105.

46. Georgakilas, Vasilios, Michal Otyepka, Athanasios B. Bourlinos, Vimlesh Chandra, Namdong Kim, K. Christian Kemp, Pavel Hobza, Radek Zboril, and Kwang S. Kim. "Functionalization of graphene: covalent and non-covalent approaches, derivatives and applications." Chemical Reviews 112, no. 11 (2012): 6156–6214.

47. Sett, Avik, et al. "Graphene and Its Nanocomposites Based Humidity Sensors: Recent Trends and Challenges." (2021).

48. Qing, Zhen Zhou, and Zhongfang Chen. "Graphene-related nanomaterials: tuning properties by functionalization." Nanoscale 5, no. 11 (2013): 4541–4583.

49. Georgakilas, Vasilios, Jitendra N. Tiwari, K. Christian Kemp, Jason A. Perman, Athanasios B. Bourlinos, Kwang S. Kim, and Radek Zboril. "Noncovalent functionalization of graphene and graphene oxide for energy materials, biosensing, catalytic, and biomedical applications." Chemical Reviews 116, no. 9 (2016): 5464–5519.

50. Sett, Avik, Kunal Biswas, Santanab Majumder, Arkaprava Datta and Tarun Kanti Bhattacharyya. (2021). Graphene and Its Nanocomposites Based Humidity Sensors: Recent Trends and Challenges [Online First]. Available from: https://www.intechopen.com/online-first/graphene-and-its-nanocomposites-based-humidity-sensors-recent-trends-and-challenges

51. Park, Jaehyeung, and Mingdi Yan. "Covalent functionalization of graphene with reactive intermediates." Accounts of Chemical Research 46, no. 1 (2013): 181–189.

52. Chua, Chun Kiang, and Martin Pumera. "Covalent chemistry on graphene." Chemical Society Reviews 42, no. 8 (2013): 3222–3233.

53. Hsiao, Min-Chien, Shu-Hang Liao, Ming-Yu Yen, Po-I. Liu, Nen-Wen Pu, Chung-An Wang, and Chen-Chi M. Ma. "Preparation of covalently functionalized graphene using residual oxygen-containing functional groups." ACS Applied Materials & Interfaces 2, no. 11 (2010): 3092–3099.

54. Wang, Aijian, Wang Yu, Zhipeng Huang, Feng Zhou, Jingbao Song, Yinglin Song, Lingliang Long et al. "Covalent functionalization of reduced graphene oxide with porphyrin by means of diazonium chemistry for nonlinear optical performance." Scientific Reports 6, no. 1 (2016): 1–12.

55. He, Fuan, Jintu Fan, Dong Ma, Liming Zhang, Chiwah Leung, and Helen Laiwa Chan. "The attachment of Fe_3O_4 nanoparticles to graphene oxide by covalent bonding." Carbon 48, no. 11 (2010): 3139–3144.

56. Song, Weina, Chunying He, Wang Zhang, Yachen Gao, Yixiao Yang, Yiqun Wu, Zhimin Chen, Xiaochen Li, and Yongli Dong. "Synthesis and nonlinear optical properties of reduced graphene oxide hybrid material covalently functionalized with zinc phthalocyanine." Carbon 77 (2014): 1020–1030.

57. Teo, Ellie Yi Lih, Gomaa AM Ali, H. Algarni, Wilairat Cheewasedtham, Thitima Rujiralai, and Kwok Feng Chong. "One-step production of pyrene-1-boronic acid functionalized graphene for dopamine detection." Materials Chemistry and Physics 231 (2019): 286–291.

58. Mohtasebi, Amirmasoud, and Peter Kruse. "Chemical sensors based on surface charge transfer." Physical Sciences Reviews 3, no. 2 (2018).

59. Shi, Xinhao, Wei Gu, Cuiling Zhang, Longyun Zhao, Li, Weidong Peng, and Yuezhong Xian. "Construction of a graphene/Au-nanoparticles/cucurbit 7. uril-based sensor for Pb^{2+} sensing." Chemistry 22, no. 16 (2016): 5643–5648.

60. Ding, Xiaofeng, Lingtao Kong, Jin Wang, Fang, Dandan Li, and Jinhuai Liu. "Highly sensitive SERS detection of Hg2+ ions in aqueous media using gold nanoparticles/graphene heterojunctions." ACS Applied Materials & Interfaces 5, no. 15 (2013): 7072–7078.

61. Zhang, Xingang, Zhigao Dai, Shuyao Si, Xiaolei Zhang, Wei Wu, Hongbing Deng, Fubing Wang, Xiangheng Xiao, and Changzhong Jiang. "Ultrasensitive SERS substrate integrated with uniform subnanometer scale "hot spots" created by a graphene spacer for the detection of mercury ions." Small 13, no. 9 (2017): 1603347.

62. Liu, Yanchen, Xiangqing Wang, and Hui Wu. "Reusable DNA-functionalized-graphene for ultrasensitive mercury (II) detection and removal." Biosensors and Bioelectronics 87 (2017): 129–135.

63. Xiaofei, Wu, Li Ruiyi, Li Zaijun, Liu Junkang, Wang Guangli, and Gu Zhiguo. "Synthesis of double gold nanoclusters/graphene oxide and its application as a new fluorescence probe for Hg^{2+} detection with greatly enhanced sensitivity and rapidity." RSC Advances 4, no. 48 (2014): 24978–24985.

64. Yu, Xiaoqing, Wei Liu, Xiaolu Deng, Shuying Yan, and Zhiqiang Su. "Gold nanocluster embedded bovine serum albumin nanofibers-graphene hybrid membranes for the efficient detection and separation of mercury ion." Chemical Engineering Journal 335 (2018): 176–184.

65. Liu, Yongfeng, Ming Deng, Xiaosheng Tang, Tao Zhu, Zhigang Zang, Xiaofeng Zeng, and Shuai Han. "Luminescent AIZS-GO nanocomposites as fluorescent probe for detecting copper (II) ion." Sensors and Actuators B: Chemical 233 (2016): 25–30.

66. Kong, Lingtao, Jin Wang, Guangchao Zheng, and Jinhuai Liu. "A highly sensitive protocol (FRET/SIMNSEF) for the determination of mercury ions: a unity of fluorescence quenching of graphene and enhancement of nanogold." Chemical Communications 47, no. 37 (2011): 10389–10391.

67. Zhang, Jian Rong, Wei Tao Huang, Wan Yi Xie, Ting Wen, Hong Qun Luo, and Nian Bing Li. "Highly sensitive, selective, and rapid fluorescence Hg^{2+} sensor based on DNA duplexes of poly (dT) and graphene oxide." Analyst 137, no. 14 (2012): 3300–3305.

68. Zhai, Kun, Yonghong Liu, Dongshan Xiang, Guangguang Guo, Tianying Wan, and Hongqing Hu. "Dual color fluorescence quantitative detection of mercury in soil with graphene oxide and dye-labeled nucleic acids." Analytical Methods 7, no. 9 (2015): 3827–3832.

69. Zhang, Huan, Sisi Jia, Min Lv, Jiye Shi, Xiaolei Zuo, Shao Su, Lianhui Wang, Wei Huang, Chunhai Fan, and Qing Huang. "Size-dependent programming of the dynamic range of graphene oxide–DNA interaction-based ion sensors." Analytical Chemistry 86, no. 8 (2014): 4047–4051.

70. Zhou, Ying, Xiao-Jing Xing, Dai-Wen Pang, and Hong-Wu Tang. "An exonuclease III-aided "turn-on" fluorescence assay for mercury ions based on graphene oxide and metal-mediated "molecular beacon." RSC Advances 5, no. 17 (2015): 12994–12999.

71. Chen, Xia, Huina Zhou, Niu Zhai, Pingping Liu, Qiansi Chen, Lifeng Jin, and Qingxia Zheng. "Graphene oxide-based homogeneous fluorescence sensor for multiplex determination of various targets by a multifunctional aptamer." Analytical Letters 48, no. 12 (2015): 1892–1906.

72. Lv, Hua, Shuang Li, Yumin Liu, Gongke Wang, Xiang Li, Yan Lu, and Jianji Wang. "A reversible fluorescent inhibit logic gate for determination of silver and iodide based on the use of graphene oxide and a silver–selective probe DNA." Microchimica Acta 182, no. 15–16 (2015): 2513–2520.

73. Sun, Xiangying, Yan Peng, Youlan Lin, Lifen Cai, Fang Li, and Bin Liu. "G-quadruplex formation enhancing energy transfer in self-assembled multilayers and fluorescence recognize for Pb^{2+} ions." Sensors and Actuators B: Chemical 255 (2018): 2121–2125.

74. Shen, Chao, Shuyan Ge, Youyou Pang, Fengna Xi, Jiyang Liu, Xiaoping Dong, and Peng Chen. "Facile and scalable preparation of highly luminescent N, S co-doped graphene quantum dots and their application for parallel detection of multiple metal ions." Journal of Materials Chemistry B 5, no. 32 (2017): 6593–6600.

75. Xu, Fengzhou, Hui Shi, Xiaoxiao He, Kemin Wang, Dinggeng He, Xiaosheng Ye, Jinlu Tang, Jingfang Shangguan, and Lan Luo. "Masking agent-free and channel-switch-mode simultaneous sensing of Fe^{3+} and Hg^{2+} using dual-excitation graphene quantum dots." Analyst 140, no. 12 (2015): 3925–3928.

76. Hua, Mengjuan, Chengquan Wang, Jing Qian, Kan Wang, Zhenting Yang, Qian Liu, Hanping Mao, and Kun Wang. "Preparation of graphene quantum dots based core-satellite hybrid spheres and their use as the ratiometric fluorescence probe for visual determination of mercury (II) ions." Analytica Chimica Acta 888 (2015): 173–181.

77. Sun, Xiangying, Pengchao Liu, Lulu Wu, and Bin Liu. "Graphene-quantum-dots-based ratiometric fluorescent probe for visual detection of copper ion." Analyst 140, no. 19 (2015): 6742–6747.

78. Park, Minsu, Hyun Dong Ha, Yong Tae Kim, Jae Hwan Jung, Shin-Hyun Kim, Do Hyun Kim, and Tae Seok Seo. "Combination of a sample pretreatment microfluidic device with a photoluminescent graphene oxide quantum dot sensor for trace lead detection." Analytical Chemistry 87, no. 21 (2015): 10969–10975.

79. Zhao, Xin, Jinsuo Gao, Xin He, Longchao Cong, Huimin Zhao, Xiaoyu Li, and Feng Tan. "DNA-modified graphene quantum dots as a sensing platform for detection of Hg 2+ in living cells." RSC Advances 5, no. 49 (2015): 39587–39591.

80. Ha, Hyun Dong, Min-Ho Jang, Fei Liu, Yong-Hoon Cho, and Tae Seok Seo. "Upconversion photoluminescent metal ion sensors via two photon absorption in graphene oxide quantum dots." Carbon 81 (2015): 367–375.

81. Zhu, Anwei, Changqin Ding, and Yang Tian. "A two-photon ratiometric fluorescence probe for cupric ions in live cells and tissues." Scientific Reports 3 (2013): 2933.

82. Ran, Xiang, Hanjun Sun, Fang Pu, Jinsong Ren, and Xiaogang Qu. "Ag nanoparticle-decorated graphene quantum dots for label-free, rapid and sensitive detection of Ag+ and biothiols." Chemical Communications 49, no. 11 (2013): 1079–1081.

83. Niu, Xiaofang, Yuanbo Zhong, Rui Chen, Fei Wang, Yanjun Liu, and Dan Luo. "A "turn-on" fluorescence sensor for Pb2+ detection based on graphene quantum dots and gold nanoparticles." Sensors and Actuators B: Chemical 255 (2018): 1577–1581.

84. Zhang, Tao, Zengguang Cheng, Yibing Wang, Zhongjun Li, Chenxuan Wang, Yibao Li, and Ying Fang. "Self-assembled 1-octadecanethiol monolayers on graphene for mercury detection." Nano Letters 10, no. 11 (2010): 4738–4741.

85. Chen, Xiaoyan, et al. "Real-time and selective detection of nitrates in water using graphene-based field-effect transistor sensors." Environmental Science: Nano 5, no. 8 (2018): 1990–1999.

86. Wang, Chenyu, Xinyi Cui, Ying Li, Hongbo Li, Lei Huang, Jun Bi, Jun Luo et al. "A label-free and portable graphene FET aptasensor for children blood lead detection." Scientific Reports 6 (2016): 21711.

87. Park, Jin Wook, Seon Joo Park, Oh Seok Kwon, Choonghyen Lee, and Jyongsik Jang. "High-performance Hg 2+ FET-type sensors based on reduced graphene oxide–polyfuran nanohybrids." Analyst 139, no. 16 (2014): 3852–3855.

88. Park, Jin Wook, Seon Joo Park, Oh Seok Kwon, Choonghyeon Lee, and Jyongsik Jang. "Polypyrrole nanotube embedded reduced graphene oxide transducer for field-effect transistor-type H_2O_2 biosensor." Analytical Chemistry 86, no. 3 (2014): 1822–1828.

89. An, Ji Hyun, Seon Joo Park, Oh Seok Kwon, Joonwon Bae, and Jyongsik Jang. "High-performance flexible graphene aptasensor for mercury detection in mussels." ACS Nano 7, no. 12 (2013): 10563–10571.

90. Tu, Jiawei, Ying Gan, Tao Liang, Qiongwen Hu, Qian Wang, Tianling Ren, Qiyong Sun, Hao Wan, and Ping Wang. "Graphene FET array biosensor based on ssDNA aptamer for ultrasensitive Hg^{2+} detection in environmental pollutants." Frontiers in Chemistry 6 (2018): 333.

91. Tan, Feng, Longchao Cong, Nuvia Maria Saucedo, Jinsuo Gao, Xiaona Li, and Ashok Mulchandani. "An electrochemically reduced graphene oxide chemiresistive sensor for sensitive detection of Hg^{2+} ion in water samples." Journal of Hazardous Materials 320 (2016): 226–233.

92. Sett, Avik, and Tarun Kanti Bhattacharyya. "Functionalized gold nanoparticles decorated reduced graphene oxide sheets for efficient detection of mercury." IEEE Sensors Journal 20, no. 11 (2020): 5712–5719.
93. Maity, A., Sui, X., Pu, H., Bottum, K.J., Jin, B., Chang, J., Zhou, G., Lu, G., and Chen, J. "Sensitive field-effect transistor sensors with atomically thin black phosphorus nanosheets." Nanoscale 12, no. 3 (2020): 1500–1512.
94. Tan, Z., W. Wu, C. Feng, H. Wu, and Z. Zhang. "Simultaneous determination of heavy metals by an electrochemical method based on a nanocomposite consisting of fluorinated graphene and gold nanocage." Microchimica Acta 187, no. 7 (2020): 1–9.
95. Kumar, A., A. K. Yadav, A. S. Kushwaha, and S. K. Srivastava. "A comparative study among WS_2, MoS_2 and graphene based surface plasmon resonance (SPR) sensor." Sensors and Actuators Reports, 2, no. 1, (2020): 00015.

7 Vertical Tunnel FET Having Dual MOSCAP Geometry

Vandana Devi Wangkheirakpam, Brinda Bhowmick, and Puspa Devi Pukhrambam

CONTENTS

7.1 INTRODUCTION

One of the hallmarks of the semiconductor industry is the continued reduction of device dimensions through scaling to increase the transistor's density on a chip (i.e., metal oxide semiconductor field-effect transistors [MOSFETs]) [1]. This was predicted by Gordon Moore, which is popularly known as Moore's law, and his philosophy has been the guiding principle for the exponential growth of the electronic industry. In using Moore's law, there is a trade-off between the objectivity and the relevance. The reduction of the device dimensions of MOSFETs led to some issues relating to performance degradation, or short channel effects (SCEs) [2–4]. These effects come into the picture because in MOSFETs with a short channel, charges are shared between the source and drain causing phenomena like velocity saturation, hot electron effects, drain-induced barrier lowering (DIBL) effect, and so on. These issues have become a hindrance to the growth of MOSFETs as a low-power device. Thus, researchers are on the lookout for novel devices that can operate at low voltage and consume less power [5, 6].

In the recent past, researchers explored many devices whose working principles were different from the MOSFET's principles and that could overcome the previously mentioned problems [7]. These devices include carbon nanotube FETs, graphene FETs, nanowire FETs, tunnel FETs (TFETs), negative capacitance FETs, and so on [8]. Among these recent emerging devices, TFETs have acquired more attention because they can attain a steeper subthreshold swing (SS) contradicting the 60-mV/dec limit of MOSFETs at room temperature. This type of transistor enables an aggressive scaling in nanometer dimensions, and their elementary procedure of fabrication is close to that of MOSFETs [9]. The TFET is a reverse biased insulator-gated p-i-n diode with a three-terminal structure, namely source, gate, and drain. Unlike MOSFETs, the current conduction mechanism of TFETs does not depend on thermionic emission [10]. Instead, it conducts through the band-to-band tunneling (BTBT) technique ensuring a lower SS and low OFF-state current [11]. However, some unwanted features of TFETs like lower current at the ON state (I_{ON}) and ambipolar nature render it to be a less acceptable device [12]. Here ambipolarity refers to the conduction of current by the application of both positive and negative gate voltage. Exploring different techniques to overcome these major drawbacks are indispensable to making TFET a superior technique [13]. Until now, different architectural designs of the TFET have been proposed to enhance its performance. Some of the widely used structures include vertical TFET (V-TFET), gate-all-around structure, L-shaped gate TFET, nanowire TFET, heterojunction TFET, silicon-on-insulator (SOI) TFET, circular gate TFET, etc. [14–17]. V-TFET rectifies the ON-state current problem associated with conventional TFET to some extent. Such TFET architecture is designed to have an additional BTBT orthogonal to the gate-stack along with point tunneling occurring parallel to the gate length [18–20].

This chapter presents the V-TFET with dual MOS capacitor (MOSCAP) geometry and analyses of various aspects related to it [21, 22]. Section 7.2 explains the geometry and concepts of the conventional TFET. The design approach and simulation setup of dual MOSCAP (D-MOS) V-TFET based on Synopsys technology computer-aided design (TCAD) are discussed in Section 7.3. The comparative analyses of the I_D-V_{GS} characteristics between the devices when there is no spacing between the sources (L_{Gap}) and the proposed device structure are reviewed in Section 7.5. Linearity performance and intermodulation (IM) distortions analyses are performed in Section 7.5. This chapter concludes with Section 7.6.

7.2 CONVENTIONAL TUNNEL FET: GEOMETRY AND CONCEPT

7.2.1 DEVICE GEOMETRY

The conventional TFET has an insulator-gated reverse biased p-i-n geometry as presented in Figure 7.1. It is a device with source, gate, and drain terminals. The source is highly doped p-type and the drain is highly doped n-type; whereas the channel remains intrinsic or doped lightly with p/n-type material. The channel region is sandwiched between the source and drain areas creating a p-i-n geometry. This arrangement provides a narrow depletion layer and forms a favorable condition

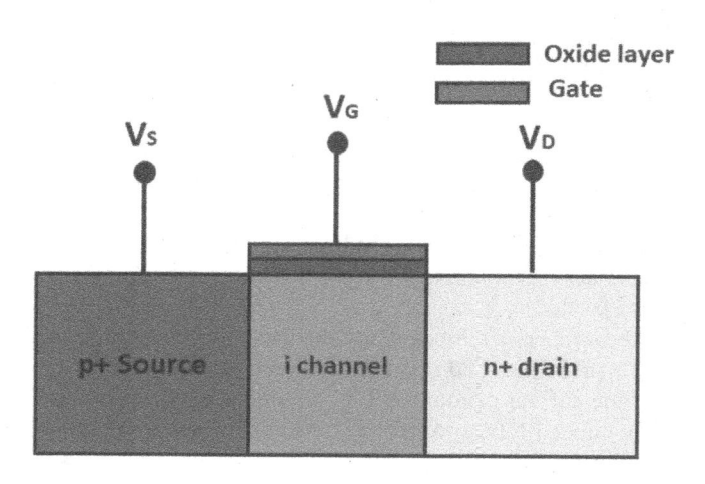

FIGURE 7.1 Two-dimensional (2D) schematic of conventional TFET.

for the carriers to tunnel from the source valence band (E_V) to the channel conduction band (E_C). The degree of the channel being intrinsic varies from device to device.

7.2.2 OPERATING PRINCIPLE

TFETs are one of the recent emerging devices that work under the principle of Zener tunneling. The electron transport mechanism in TFET is BTBT in contrast to that of thermionic emission in MOSFET. Depending on the polarity of the applied voltages, a TFET can be classified into p channel and n channel modes. In n-TFET, the source terminal is connected to the ground and the drain terminal is positively biased (V_{DS}). The energy bands of the channel region are reduced and pulled down when the gate terminal is applied with a positive voltage (V_{GS}). Further increase in this gate voltage causes more and more band bending resulting in the reduction of the tunneling width, i.e., the area between the E_V of the source and E_C of the channel where the bending is taking place. As V_{GS} increases beyond a particular limit known as the threshold voltage (V_{th}), carrier tunneling starts facilitating from the filled state E_V to the empty state E_C of the source and channel, respectively. This is depicted in Figure 7.2 where the nature of E_C and E_V of the conventional TFET is plotted.

The tunneling probability of the energy barrier governs the current conduction of TFET. This probability is modeled through Wentzel-Kramers-Brillouin (WKB) approximation assuming a triangular barrier and is given by

$$T(E) = \exp\left(\frac{-4\lambda\sqrt{2m^*}E_g^{3/2}}{3q\hbar\left(E_g + \Delta\varnothing\right)}\right) \tag{7.1}$$

where m^* represents the effective mass; E_g denotes the forbidden energy gap; $\Delta\varnothing$ represents the energy of the area at the tunnel junction where bands are overlapped; λ is the tunneling width; and q and $\hbar$ are the value of an electronic charge and reduced Planck's constant, respectively. The variable λ is defined as

$$\lambda = \sqrt{\varepsilon_s/\varepsilon_{ox}}\,t_{ox}t_s \tag{7.2}$$

where ε_s and ε_{ox} correspond to the dielectric constants of substrate and oxide layer, respectively, and t_{ox} and t_s, respectively, represent the oxide and substrate thickness.

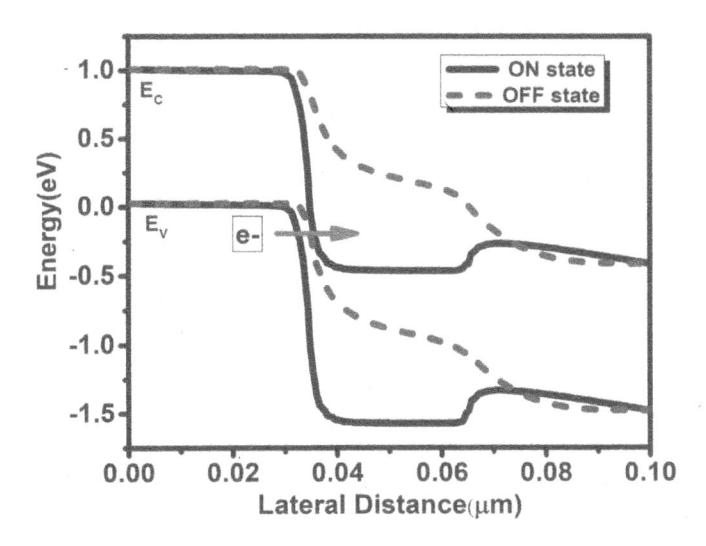

FIGURE 7.2 Energy band diagram of conventional TFET.

7.3 DUAL MOS CAPACITOR (MOSCAP) VERTICAL TFET

7.3.1 DEVICE STRUCTURE AND WORKING PRINCIPLE

This section describes the structure of D-MOS capacitor (MOSCAP) V-TFET, which has a δ-doped SiGe layer at the source-channel junction.

The two-dimensional (2D) schematic of the D-MOS V-TFET structure is presented in Figure 7.3. It shows that the two MOS capacitors sandwich the channel and drain regions, and form a shape that resembles an inverted T. The MOSCAP present in this device is comprised of the gate-stack, a narrow intrinsic layer, and the source region. This narrow intrinsic layer resides in the region between the gate-stack and the source and is known as the epi-layer. This structure provides the whole region of epi-layer lying under the gate-stack to perform the BTBT operation normal to gate-oxide thickness, thus giving rise to an enhanced ON-state current and a steeper SS. A δ-doped layer of SiGe with 40% germanium concentration and 2-nm thickness is deposited at the source/channel junction to further improve the BTBT rate and boost the electrical performance of this D-MOS V-TFET because SiGe has an energy bandgap lower than silicon. An intrinsic layer, with a length L_{Gap} = 20 nm, separates the source region into two. The elevated channel has a thickness of 5 nm. The effective oxide thickness (EOT) of the uniform parallel and normal SiO_2 layer is 1 nm. L_{epi} and T_{epi} represent the length and thickness of the epitaxial layer with values of 30 nm and 4 nm, respectively. Gate metal has a work-function ϕ_m = 3.8 eV. The source, channel and drain are, respectively, p-type doped (1×10^{20} cm^{-3}), intrinsic (1×10^{15} cm^{-3}), and n-type doped (1×10^{18} cm^{-3}). The I_D-V_{GS} characteristics graph of the D-MOS V-TFET when drain bias V_{DS} = 0.7 V is shown in Figure 7.4(a). The corresponding energy band diagram under the application of positive V_{GS}, i.e., ON state, is observed in Figure 7.4(b). An I_{ON}/I_{OFF} ratio of 8.46×10^8 and an average SS (SS_{avg}) of 18.67 mV dec^{-1} is obtained, which indicates the effectiveness of the proposed D-MOS V-TFET when used in low-power applications. The formula for SS_{avg} is given by

$$SS_{avg} = \frac{\left(V_{th} - V_{off}\right)}{\left[log\left(I_{th}\right) - log\left(I_{off}\right)\right]} mV \ dec^{-1} \tag{7.3}$$

where V_{th} is the magnitude of V_{GS} when drain current I_{th} = 10^{-7} A, and I_{off} provides the amount of I_D at V_{off} i.e., V_{GS} = 0 V.

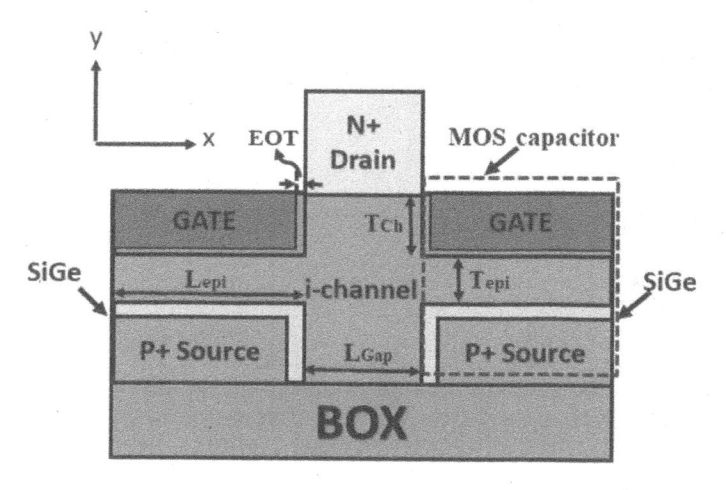

FIGURE 7.3 Two-dimensional (2D) schematic of dual MOSCAP (D-MOS) V-TFET.

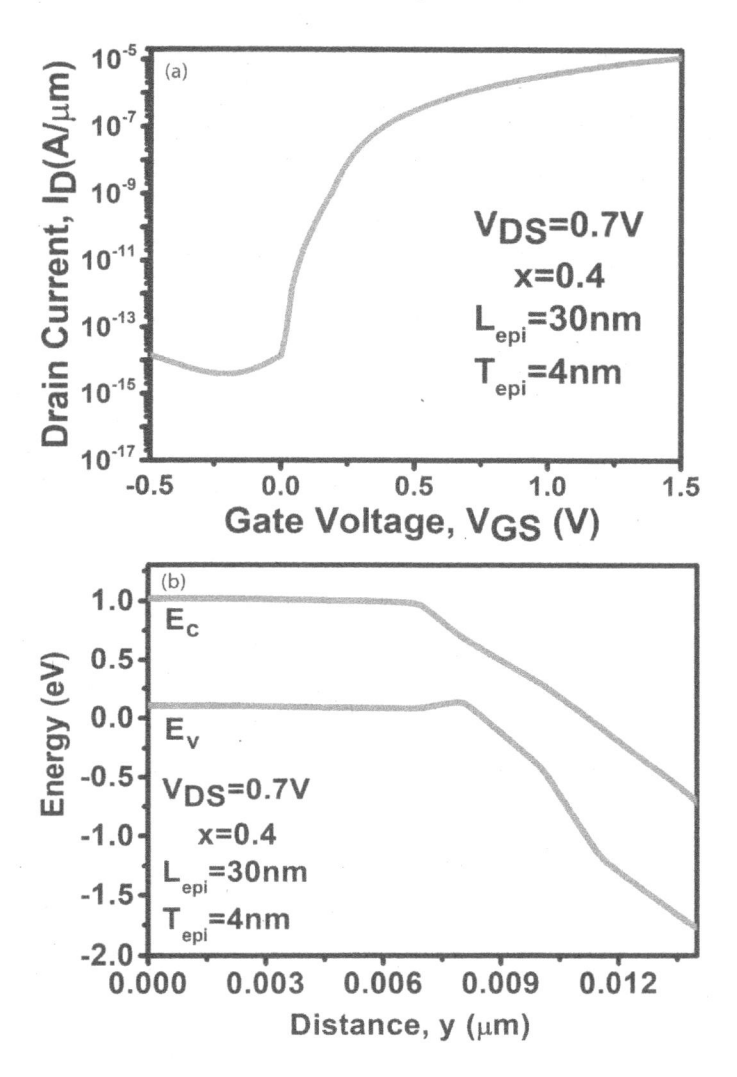

FIGURE 7.4 (a) I_D-V_{GS} characteristics and (b) ON-state energy band of D-MOS V-TFET with L_{Gap}.

This chapter presents an enhanced variant of TFET with a D-MOS geometry and an elevated channel/drain region. This novel device has an improved vertical BTBT with better I_{ON}. The reason behind these enhanced characteristics is because of the presence of the dual orthogonal gate, which controls the epi-layers and channel region independently. Unlike the conventional TFET, the elevated channel/drain regions of D-MOS V-TFET provide better electrostatics at the 2D source edge. Along with this, a reduced parasitic leakage is obtained by mitigating the unwanted parasitic tunneling paths.

7.3.2 SIMULATION SETUP

In this section, the design approach and simulation setup of D-MOS V-TFET based on Synopsys TCAD are discussed including various simulation models. The doping-dependent model and nonlocal BTBT model are activated in device physics to establish the BTBT process at the tunneling junction [23]. The BTBT generation rate given in Equation (7.4) is evaluated over the area where the tunneling takes place. To perform this, the barrier width (L_{BW}) is measured at the point where the electric field is maximum using the energy band diagram [23].

$$G_{BTBT}\left(E_{avg}\right) = AE_{avg}^2 \exp\left(-\frac{B}{E_{avg}}\right) \tag{7.4}$$

where E_{avg} is the average electric field and is calculated using the following equation:

$$E_{avg} = \frac{\left(\int E_{Total}\, dy\right)}{L_{BW}} \tag{7.5}$$

where $E_{Total} = \left(\left|E_X\right|^2 + \left|E_Y\right|^2\right)^{1/2}$. The parameters for the nonlocal path band-to-band model are available in the parameter set `Band2BandTunneling`. To make the parameter set of the model consistent with the existing BTBT models, the prefactor and the exponential factor are chosen as the input parameters. Up to three different tunneling paths can be specified in the parameter file, and they are activated by setting `Model=NonlocalPath1 | NonlocalPath2 | NonlocalPath3` in the `Band2Band` option of the command file. `NonlocalPath1` selects the first tunneling path, `NonlocalPath2` selects the first and second tunneling paths, and `NonlocalPath3` selects all three tunneling paths. The effects of a degenerated source and drain on the device performance are accounted by the bandgap narrowing model. The Shockley-Read-Hall (SRH) recombination model looks after the quantization of the carriers' density gradient. Recombination through deep defect levels in the gap is usually labeled the SRH recombination. In the Sentaurus device, the following form is implemented [23]

$$R_{net}^{SRH} = \frac{np - n_{i,eff}^2}{\tau_p\left(n + n_1\right) + \tau_n\left(p + p_1\right)} \tag{7.6}$$

where $n_1 = n_{i,eff}\exp\left(\frac{E_{Trap}}{KT}\right)$ and $p_1 = n_{i,eff}\exp\left(\frac{-E_{Trap}}{KT}\right)$

where E_{Trap} is the difference between the defect level and intrinsic level. The variable is accessible in the parameter file. The silicon default value is $E_{Trap} = 0$. The doping-dependent SRH model is activated by specifying the additional argument `DopingDependence` for the SRH keyword in the `Recombination` statement. In addition to these, the mobility model and Fermi-Dirac statistics model are also considered in the simulation process. In Figure 7.5, the plots for the model's calibration utilized in the simulation against the experimentally available data [20] are presented. Because the experimentally fabricated data of D-MOS V-TFET is not available, the calibration is carried out using the experimental data of a V-TFET in which the channel is sandwiched between two lightly doped Si.

7.4   IMPACT OF THE PRESENCE OF L_{Gap}

This section analyses the advantages of the presence of an intrinsic layer (L_{Gap}) separating the p+ source into two regions (Figure 7.1). The 2D schematic structures of the proposed device with and without L_{Gap} is given in Figure 7.6. The importance of geometrical differences on the electrical performance and tunneling components of D-MOS V-TFET is investigated and their transfer characteristics studies. I_D-V_{GS} characteristics of these two devices are compared in this section and are plotted in Figure 7.7. From this figure, it is clearly understood that the OFF-state current (I_{OFF}) as well as the SS_{avg} increases when L_{Gap} is not present. The reason behind this effect is the rise in the effect of

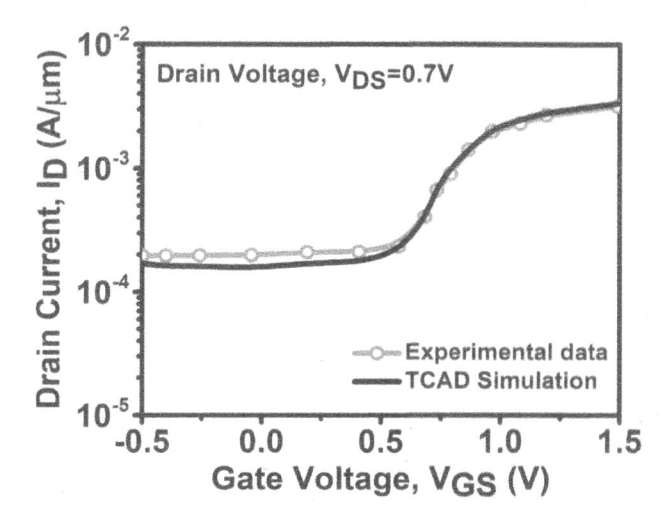

FIGURE 7.5 Model calibration using the fabricated data available in Ref. [20].

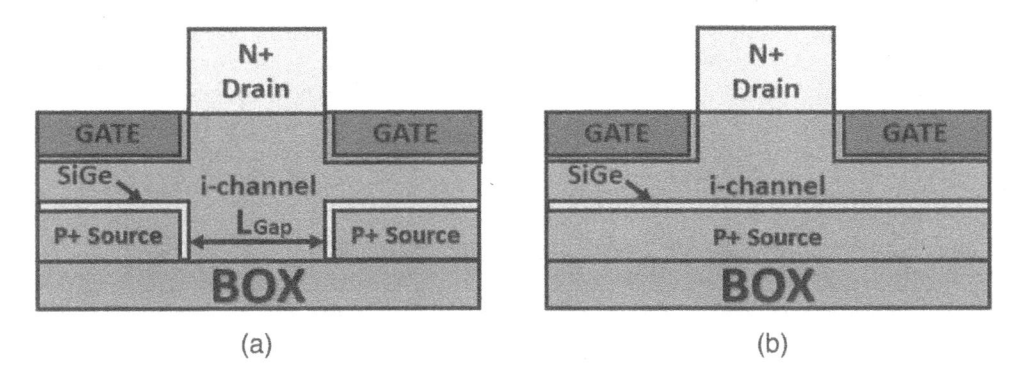

FIGURE 7.6 Two-dimensional (2D) schematic structures of (a) D-MOS V-TFET with L_{Gap} and (b) D-MOS V-TFET without L_{Gap}.

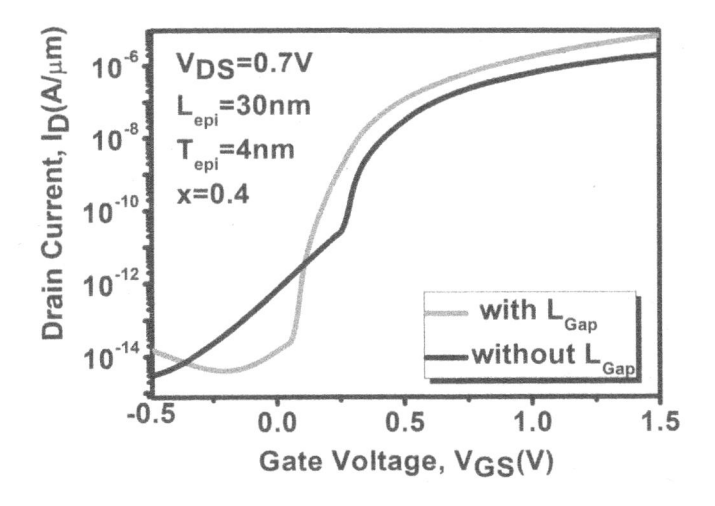

FIGURE 7.7 I_D-V_{GS} characteristics compared with D-MOS V-TFET with and without L_{Gap}.

TABLE 7.1

Comparison of I_{ON}, I_{OFF}, I_{ON}/I_{OFF}, and SS_{avg} at $V_{DS} = 0.7V$

Parameter	D-MOS V-TFET Without L_{Gap}	D-MOS V-TFET with L_{Gap}
I_{ON} (A/μm)	1.98×10^{-6}	6.91×10^{-6}
I_{OFF} (A/μm)	6.64×10^{-13}	1.42×10^{-14}
I_{ON}/I_{OFF}	2.98×10^{6}	4.86×10^{8}
SS_{avg} (mV/dec)	27.93	18.67

the parasitic leakage components on the conduction of the device operation. The ON-current (I_{ON}) is also observed to be similar to the case with L_{Gap} because of the additional tunneling that occurs parallel to the gate-stack when L_{Gap} is present. The comparison of I_{ON}, I_{OFF}, I_{ON}/I_{OFF}, and SSa_{vg} are shown in Table 7.1. It can be concluded from this observation that the proposed D-MOS V-TFET outperforms the one without L_{Gap} and it can be a superior device for low-power applications.

7.5 LINEARITY PERFORMANCE AND INTERMODULATION DISTORTION

Linearity performance and IM distortion analyses are also explored in this section considering the feasibility of the device for practical use. Today, radiofrequency integrated circuits (RFICs) require a device with low IM distortions so that the linear operation continues even if the receiving signal is poor [24]. The desired weak signal can be interfered with by the strong unwanted signal and cross-modulation might take place. The nonlinear behavior of any RFICs depends on the nonlinearity of a device [25]. The presence of IM distortions at various frequencies disturbs the linear behavior of the device. Hence, reliability prediction and minimization of IM distortions play significant roles in communication systems like mobile communication and wireless systems [26]. The sensitivity of the receiver needs to be maximized, whereas minimization of inference from the transmitter is essential for an efficient communication system with minimal signal distortion. In communication systems, IM distortions are considered to be one of the major sources of noise [27]. Optimization of linearity can be done through the fabrication process; however, this effort consumes more time and it is expensive. Hence, linearity analysis is done in an efficient manner by considering some figures of merit (FOMs) [28–32]. This section presents the linearity performance and IM distortion analyses for D-MOS V-TFET with and without L_{Gap} based on FOMs like g_{m2}, g_{m3}, second-order voltage intercept point (VIP_2), third-order voltage intercept point (VIP_3), third-order input intercept point (IIP_3), third-order IM distortion (IMD_3), and 1-dB compression point.

7.5.1 SECOND-ORDER (g_{m2}) AND THIRD-ORDER HARMONICS (g_{m3})

To maintain a better linearity of a device over an intended input voltage limit, transconductance should remain constant. However, transconductance is found to vary with input signal in both MOSFET and TFET showing nonlinear characteristics. Harmonic distortions in MOS circuits are mainly caused by the nonlinear nature of the higher-order transconductances. The perfect description of nonlinearity requires infinite power terms in series expansion. Practically, the characterization of a circuit can be sufficiently made with proper precision by considering only the first three terms. Equations (7.7) and (7.8) provide the formulae for calculating g_{m2} and g_{m3}, respectively.

$$g_{m2} = \frac{\partial^2 I_D}{\partial V_{GS}^2} \tag{7.7}$$

$$g_{m3} = \frac{\partial^3 I_D}{\partial V_{GS}{}^3} \tag{7.8}$$

It is known that the application of balanced topologies cancels out the even harmonics; thus, distortions due to g_{m2} can be controlled to some extent, but g_{m3} is uncontrollable and acts as a dominant parameter in the nonlinear behavior of a device. The lower limit of the distortion is determined by the amplitude of g_{m3}; therefore, reducing g_{m3} and improving the transconductance (g_m) becomes a viable key to enhancing device linearity.

Figure 7.8(a–c) plots the transconductance g_m, g_{m2}, and g_{m3} variations in response to the applied gate voltage, respectively, of the proposed D-MOS V-TFET with and without L_{Gap}. For optimum

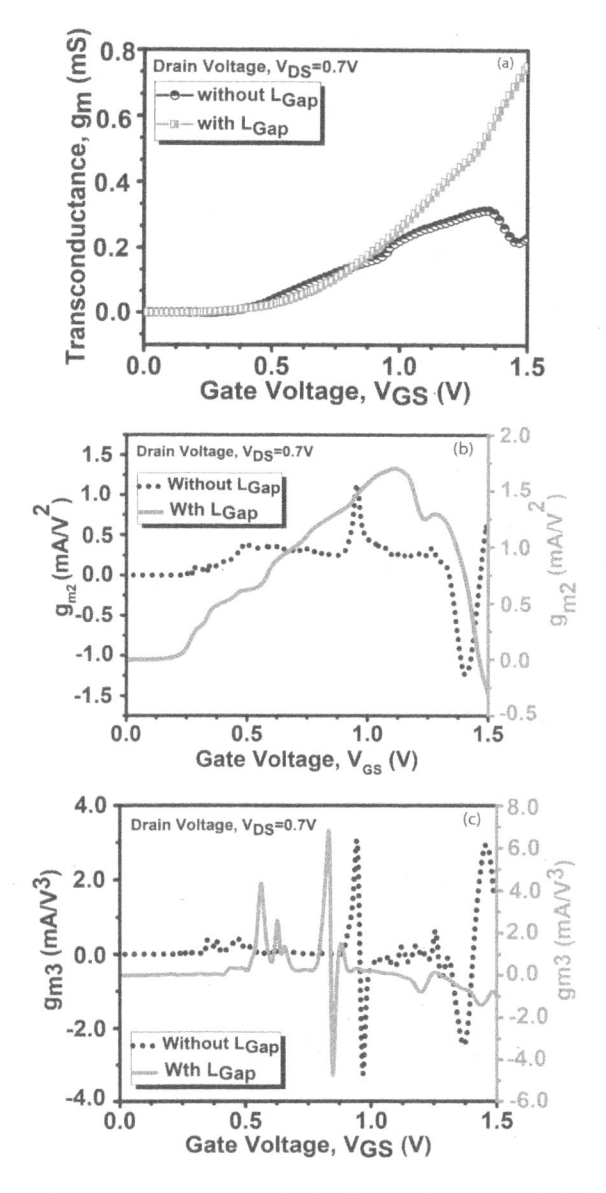

FIGURE 7.8 (a) Transconductance (g_m), (b) second-order harmonic (g_{m2}), and (c) third-order harmonic (g_{m3}) of D-MOS V-TFET with and without L_{Gap}.

device operation, the biasing point is obtained by the zero-crossover point (ZCP) of g_{m3}. Thus, the nonlinearity of a device can be reduced to the smallest possible degree by keeping the DC bias near this ZCP. In other words, the ZCP of g_{m3} with lower V_{GS} is desirable for the linearity of a semiconductor device. Figure 7.8(c) shows that the D-MOS V-TFET with L_{Gap} has ZCP of g_{m3} toward lower V_{GS} (0.79 V), whereas the one without L_{Gap} has the ZCP at $V_{GS} = 0.94$ V. The reason for this is the incorporation of an intrinsic layer separating the source region into two in the proposed D-MOS V-TFET. Thus, the proposed device requires smaller gate voltage to suppress the harmonic distortions.

7.5.2 SECOND-ORDER VOLTAGE INTERCEPT POINT (VIP_2)

The second-order voltage intercept point (VIP_2) is the extrapolated input voltage where the first harmonic voltage is equal to the second-order harmonic voltage. Second-order nonlinearity results in second-order IM distortion, whereas third-order nonlinearity is responsible for third-order IM distortion. VIP_2 is evaluated using Equation (7.9).

$$VIP_2 = 4 \times \frac{g_m}{g_{m2}} \tag{7.9}$$

where g_m and g_{m2} are the transconductance and second-order derivative of I_D-V_{GS} characteristics. As mentioned earlier, the application of balanced topologies cancels out the even harmonics. Consequently, RF distortion is mainly defined by the third-order derivative of I_D with regard to V_{GS}, i.e., g_{m3}. Figure 7.9 illustrates the comparative results of VIP_2 for the proposed D-MOS V-TFET in the absence and presence of L_{Gap}. It is known that the higher-order harmonics (g_{m2}, g_{m3}) should be minimized. Therefore, the amplitude of VIP_2 is expected to be more to obtain a distortion-less output. Figure 7.9 shows that VIP_2 is greater for D-MOS V-TFET with L_{Gap}. This indicates that the presence of an intrinsic between the two source regions provides a great advancement in conduction current. Consequently, the device becomes highly efficient and enhances the transconductance, thereby exhibiting higher linearity and lower distortion behavior compared with the one without L_{Gap}.

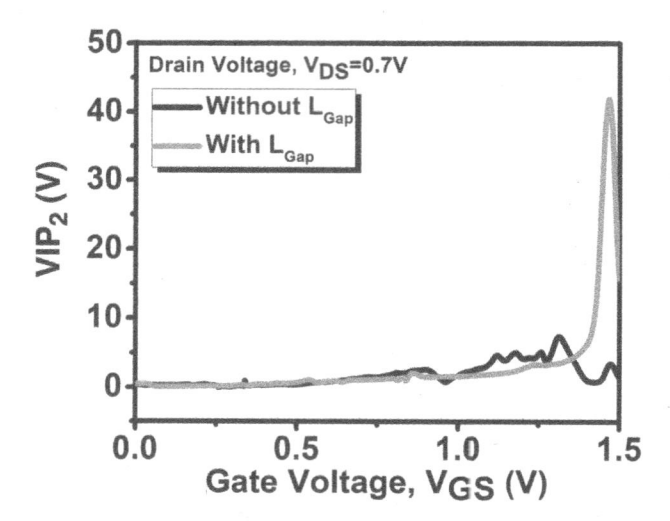

FIGURE 7.9 VIP_2 for D-MOS V-TFET with and without L_{Gap}.

7.5.3 THIRD-ORDER VOLTAGE INTERCEPT POINT (VIP_3)

The third-order voltage intercept point (VIP_3) represents the extrapolated input voltage where the first-order harmonic voltage is equal to the third-order harmonic voltage. It is expressed in volt (V). VIP_3 is mathematically evaluated using Equation (7.10).

$$VIP_3 = \sqrt{24 \times \frac{g_m}{g_{m3}}} \tag{7.10}$$

where g_m and g_{m3} are, respectively, the transconductance and third-order derivative of I_D with regard to V_{GS}. It is evident from Equation (7.10) that VIP_3 is dependent on the third-order of transconductance (g_{m3}), which is a dominant parameter out of the other nonlinear sources. As discussed earlier, the value of g_{m3} should be minimized. Hence, the amplitude of VIP_3 is expected to be as high as possible because a larger value of VIP_3 indicates a more linear device. Figure 7.10 shows the comparative analysis of VIP_3 plots for the proposed D-MOS V-TFET with and without L_{Gap}. Figure 7.10 shows that the device with L_{Gap} between the two source regions has a comparatively higher value of VIP_3 than the counterpart without L_{Gap}. This result specifies that the proposed D-MOS V-TFET with L_{Gap} has the ability to drive drain current with great enhancement leading to the improvement of its efficiency and, hence, exhibiting higher linearity.

7.5.4 THIRD-ORDER INPUT INTERCEPT POINT (IIP_3)

The third-order input intercept point (IIP_3) represents the extrapolated input power in which the first-order harmonic is equal to the third-order harmonic. IIP_3 is a unique parameter because it can by itself serve as a factor for linearity comparison in various devices and circuits. The detailed analysis of this device parameter is necessary in a communication system to study the linearity behavior. IIP_3 is mathematically measured using the formula given in Equation (7.11).

$$IIP_3 = \frac{2}{3} \times \frac{g_m}{g_{m3} \times R_s} \tag{7.11}$$

where g_m and g_{m3} are, respectively, the transconductance and third-order derivative of I_D with regard to V_{GS}. For most RF applications, R_s is assumed to be 50 Ω. IIP_3 is directly proportional to g_m and

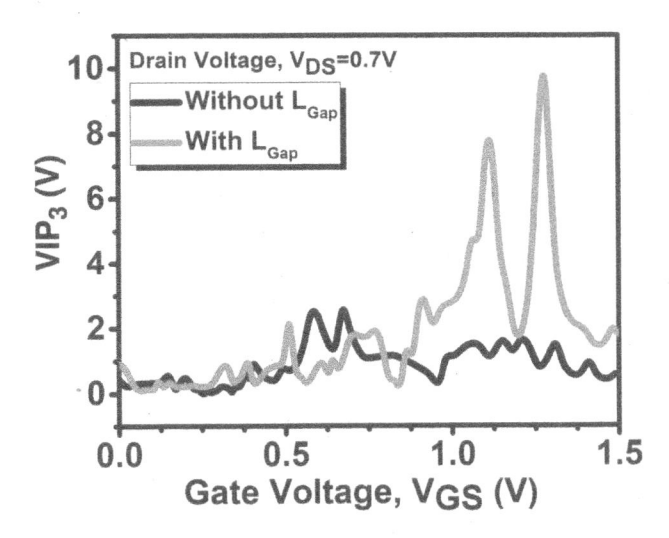

FIGURE 7.10 VIP_3 for D-MOS V-TFET with and without L_{Gap}.

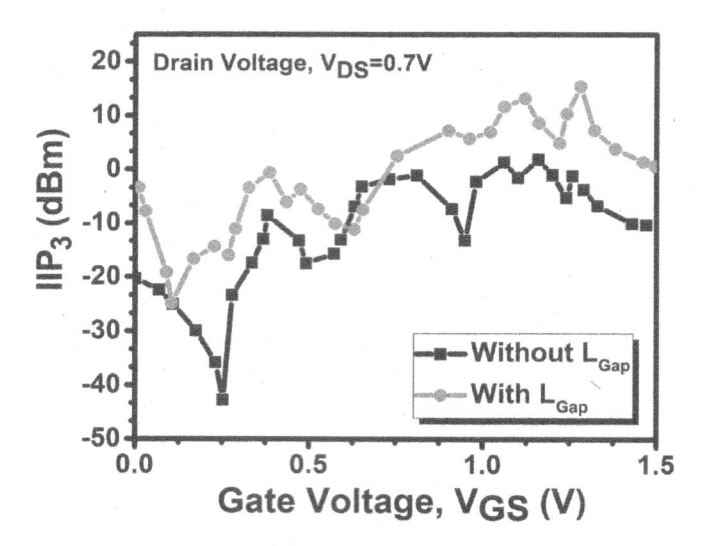

FIGURE 7.11 IIP_3 for D-MOS V-TFET with and without L_{Gap}.

inversely dependent on g_{m3}. As discussed in the previous section, g_m should be maximized and g_{m3} should be minimized. Therefore, to achieve better linearity, the parameter IIP_3 is expected to be as high as possible. The unit of IIP_3 is dBm, which is the power ratio in decibels (dB) of the measured power with respect to 1 mW. It is expressed as:

$$IIP_3\,(\text{dBm}) = 10\log_{10}\left(IIP_3\right) + 30 \tag{7.12}$$

A comparison has been made with the IIP_3 of the proposed D-MOS V-TFET with and without L_{Gap} (Figure 7.11). Because the D-MOS V-TFET with L_{Gap} has better characteristics in terms of g_m and g_{m3}, it is obvious that this particular device will have a higher IIP_3 compared with one without L_{Gap}. This is clearly seen in Figure 7.11. Thus, the proposed D-MOS V-TFET can be an efficient device with better linearity, which can replace MOSFET in the near future.

7.5.5   Third-Order Intermodulation Distortion (IMD_3)

The third-order IM distortion (IMD_3) is the extrapolated IM current in which the first-order harmonic current is equal to the third-order harmonic current. In a wireless communication system, this factor IMD_3, which is derived from the transfer characteristics of a device, has the capability to destruct the information signals in the neighboring channel. Hence, it is essential to perform a detailed study of this parameter for better understanding of the linearity behavior of the device. IMD_3 is measured using the relation provided in Equation (7.13).

$$IMD_3 = \left[\frac{9}{2}\times\left(VIP_3\right)^2\times g_{m3}\right]^2\times R_S \tag{7.13}$$

where VIP_3 is the third-order voltage intercept point and g_{m3} represents the third-order derivative of I_D with regard to V_{GS}. For RF applications, R_S is assumed as 50 Ω. Because IMD_3 is directly proportional to $gm3$, it is desired to have a lower value. It is also expressed in dBm, which is given by:

$$IMD_3\,(\text{dBm}) = 10\log_{10}\left(IMD_3\right) + 30 \tag{7.14}$$

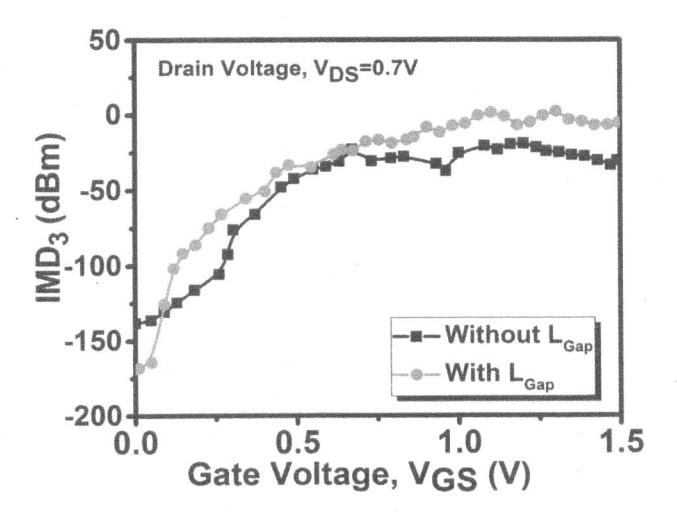

FIGURE 7.12   Third-order intermodulation distortions (IMD_3) for D-MOS V-TFET with and without L_{Gap}.

Figure 7.12 shows the plot of IMD_3 versus gate voltage for the proposed D-MOS V-TFET in the absence and presence of L_{Gap}. It is evident from Figures 7.11 and 7.12 that IIP_3 of the proposed D-MOS V-TFET with L_{Gap} is significantly higher than the IMD_3 of the same device architecture. The difference is also greater than the device without L_{Gap}. This is due to the incorporation of an intrinsic layer between the two source regions (L_{Gap}) of the proposed device ensuring a higher ON current, better SS, and reduced V_{th}. Therefore, the proposed D-MOS V-TFET with L_{Gap} is a superior device in terms of linearity and distortion compared with the one without L_{Gap}.

7.5.6   1-dB Compression Point

The 1-dB compression point is also one of the most important parameters for measuring the upper limit of linearity performance. The 1-dB compression point can be defined as the value of input power where the gain shifts from linearity by 1 dB. The linear behavior of an amplifier prevails up to this point, and beyond this the performance of the amplifier gain starts degrading. Under such a nonlinear situation, the IM products and signal distortions come into the picture and amplifier gain saturates producing a condition known as compression. Hence, a device with a higher than 1-dB compression point is desirable for a better linearity performance and lower IM distortions. It is mathematically calculated using Equation (7.15).

$$1-\text{dB compression point} = 0.22 \times \sqrt{\frac{g_m}{g_{m3}}} \tag{7.15}$$

Figure 7.13 plots the 1-dB compression point with regard to V_{GS} for D-MOS V-TFET with and without L_{Gap}. It can be inferred from Figure 7.13 that the proposed D-MOS V-TFET with L_{Gap} that has a higher 1-dB compression point has better linear characteristics and lower IM distortion compared with its counterpart without L_{Gap}.

7.6   CONCLUSION

This chapter presented a comprehensive study of a D-MOS V-TFET geometry using extensive Sentaurus TCAD simulation data. A precis of the principle of operation of the conventional TFET was also discussed to provide better understanding of its pros and cons. The D-MOS V-TFET

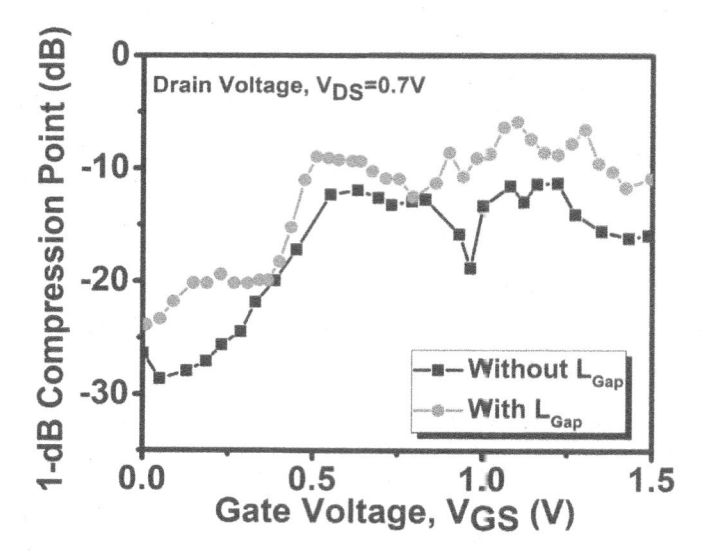

FIGURE 7.13   A 1-dB compression point for D-MOS V-TFET with and without L_{Gap}.

with a vertically elevated channel/drain region provides an enhanced vertical tunneling due to the existence of an intrinsic layer (L_{Gap}) separating the source region into two. The inclusion of a SiGe δ-layer also enhances the device performance because SiGe has lower bandgap energy than silicon. A comparison was made with the impact of the presence of L_{Gap} by considering another device without the presence of L_{Gap}. The purposed device outperforms the one without L_{Gap} in terms of I_{ON}/I_{OFF} and SS_{avg}. The simulation results of the proposed device manifest a very high I_{ON}/I_{OFF} ratio of 8.46 × 10^8 and a steeper average SS (SS_{avg}) of 18.67 mV dec^{-1} making it a superior candidate for low-power applications. Furthermore, linearity performance and IM distortion analyses were also explored in this chapter considering the feasibility of the device for its practical use. These analyses were carried out for the proposed device with and without L_{Gap} using different FOMS such as $gm2$, $gm3$, VIP_2, VIP_3, IIP_3, IMD_3, and 1-dB compression point. Through these analyses it is concluded that the proposed D-MOS V-TFET with L_{Gap} is comparatively linear where minimum IM distortions and a lower value of higher-order harmonics are guaranteed. Due to the incorporation of L_{Gap}, D-MOS V-TFET possesses significantly superior values of VIP_2, VIP_3, IIP_3, 1-dB compression point, and a lower value of IMD_3, which is an extremely important criteria for high linearity and low distortion operation. Therefore, the status of the proposed D-MOS V-TFET is highlighted as an efficient device for RF circuit applications. The results based on Sentaurus TCAD tool can be validated using appropriate physics-based models.

REFERENCES

1. Frank, D. J., R. H. Dennard, E. Nowak, P. M. Solomon, Y. Taur and H. S. P. Wong. 2001. Device scaling limits of Si MOSFETs and their application dependencies. *Proc. IEEE* 89:259–288.
2. Loan, S. A., S. Qureshi, and S. S. K. Iyer. 2010. A novel partial-ground-plane-based MOSFET on selective buried oxide: 2-D simulation study. *IEEE Trans. Electron Devices.* 57:671–680.
3. Ehteshamuddin, M., S. A. Loan, and M. Rafat. 2018. Planar junctionless silicon-on-insulator transistor with buried metal layer. *IEEE Electron Device Lett.* 39:799–802.
4. Ionescu, A. M., and H. Riel. 2011. Tunnel field-effect transistors as energy-efficient electronic switches. *Nature* 479:329–337.
5. Lu, H., and A. Seabaugh. 2014. Tunnel field-effect transistors: state-of-the-art. *IEEE J. Electron Devices Soc.* 2:44–49.
6. Choi, W. Y., B. G. Park, J. D. Lee, and T. J. K. Liu. 2007. Tunneling field-effect transistors (TFETs) with subthreshold swing (SS) less than 60 mV/dec. *IEEE Electron Device Lett.* 28:743–745.

7. Avci, U. E., D. H. Morris, and I. A. Young. 2015.Tunnel field-effect transistors: prospects and challenges. *IEEE J. Electron Devices Soc.* 3:88–95.

8. Ehteshamuddin, M., S. A. Loan, and M. Rafat. 2018. A vertical-gaussian doped SOI-TFET with enhanced DC and analog/RF performance. *Semicond. Sci. Technol.* 33:075016.

9. Verhulst, A. S., W. G. Vandenberghe, K. Maex, and G. Groeseneken. 2017. Tunnel field-effect transistor without gate-drain overlap. *Appl. Phys. Lett.* 91:053102.

10. Devi W. V., and B. Bhowmick. 2019. Optimisation of pocket doped junctionless TFET and its application in digital inverter. *Micro Nano Lett.* 14:69–73.

11. Rooyackers, R. *et al.* 2014. Ge-source vertical tunnel FETs using a novel replacement-source integration scheme. *IEEE Trans. Electron Devices.* 61:4032–4039.

12. Panda, S., S. Dash, S. K. Behera, and G. P. Mishra. 2016. Delta-doped tunnel FET (D-TFET) to improve current ratio (I_{ON}/I_{OFF}) and ON current performance. *J. Comput. Electron.* 15:857–864.

13. Ehteshamuddin, M., S. A. Loan, A. G. Alharbi, A. M. Alamoud, and M. Rafat. 2019. Investigating a dual MOSCAP variant of Line-TFET with improved vertical tunneling incorporating FIQC effect. *IEEE Trans. Electron Devices* 66:4638–4645.

14. Kim, S. W., W. Y. Choi, M. C. Sun, H. W. Kim, and B. G. Park. 2012. Design guideline of Si-based L-shaped tunneling field-effect transistors. *Jpn. J. Appl. Phys.* 51:06FE09-1–06FE09-4.

15. Kim, S. W., W. Y. Choi, M. C. Sun, and B. G. Park. 2013. Investigation on the corner effect of L-shaped tunneling field-effect transistors and their fabrication method. *J. Nanosci. Nanotechnol.* 13:6376–6381.

16. Yang, Z. 2016. Tunnel field-effect transistor with an L-shaped gate. *IEEE Electron Device Lett.* 37:839–842.

17. Wang, W. *et al.* 2014. Design of U-shape channel tunnel FETs with SiGe source regions. *IEEE Trans. Electron Devices* 61:193–197.

18. Bagga, N., A. Kumar, and S. Dasgupta. 2017. Demonstration of a novel two source region tunnel FET. *IEEE Trans. Electron Devices* 64:5256–5262.

19. Vanlalawmpuia, K., and B. Bhowmick. 2019. Investigation of a Ge-source vertical TFET with delta-doped layer. *IEEE Trans. Electron Devices* 66:4439–4445.

20. Kim, J. H., S. Kim, and B. G. Park. 2019. Double-gate TFET with vertical channel sandwiched by lightly doped Si. *IEEE Trans. Electron Devices* 66:1656–1661.

21. Wangkheirakpam, V. D., B. Bhowmick, and P. D. Pukhrambam. 2020. Investigation of a dual MOSCAP TFET with improved vertical tunneling and its near-infrared sensing application. *Semicond. Sci. Technol.* 35:065013: 1–8.

22. Wangkheirakpam, V. D., B. Bhowmick, and P. D. Pukhrambam. 2021. Investigation of Temperature Variation and Interface Trap Charges in Dual MOSCAP TFET. *Silicon* 13:2971–2978.

23. TCAD Sentaurus User Guide. 2013. Mountain View, CA: Synopsys Inc.

24. Dassi, M., J. Madan, R. Pandey, and R. Sharma. 2020. A novel Source Material-Engineered DG-TFET for RFIC Applications *Semicond. Sci. Technol.* 35:105013.

25. Vanlalawmpuia, K., and B. Bhowmick. 2020. Linearity performance analysis due to lateral straggle variation in heterostacked TFET. *Silicon* 12:955–961.

26. Saha, R., B. Bhowmick, and S. Baishya. 2020. Impact of lateral straggle on linearity performance in gate-modulated (GM) TFET. *Appl. Phys. A.* 126:1–5.

27. Narwal, S., and S. S. Chauhan. 2019. Investigation of RF and linearity performance of electrode work-function engineered HDB vertical TFET. *Micro Nano Lett.* 14:17–21.

28. Datta, E., A. Chattopadhyay, and A. J. Mallik. 2020. Relative study of analog performance, linearity, and harmonic distortion between junctionless and conventional SOI FinFETs at elevated temperatures. *J. Electron. Mater.* 49:3309–3316.

29. Paras, N., and S. S. Chauhan. 2019. Insights into the DC, RF/Analog and linearity performance of vertical tunneling based TFET for low-power applications. *J. Microelectron. Eng.* 216:111043.

30. Vanlalawmpuia, K., and B. Bhowmick. 2021. Optimization of a hetero-structure vertical tunnel FET for enhanced electrical performance and effects of temperature variation on RF/linearity parameters. *Silicon* 13:155–166.

31. Ashish, K. S., M. R. Tripathy, K. Baral, P. K. Singh, and S. Jit. 2020. Investigation of DC, RF and linearity performances of a back-gated (BG) heterojunction (HJ) TFET-on-selbox-substrate (STFET): Introduction to a BG-HJ-STEFT based CMOS inverter. *Microelectron. J.* 102:104775.

32. Emona, D., A. Chattopadhyay, A. Mallik, and Y. Omura. 2020. Temperature dependence of analog performance, linearity, and harmonic distortion for a Ge-Source tunnel FET. *IEEE Trans. Electron Devices* 67:810–815.

8 Leakage Current and Capacitance Reduction in CMOS Technology

Ajay Somkuwar and Laxmi Singh

CONTENTS

8.1 INTRODUCTION

Recently technological growth and microprocessors have played an important role in the advancement of technology. Continuous research has led to the development of a faster chip. A number of techniques have been used to enhance the performance of the microprocessor chip. Complementary metal oxide semiconductor (CMOS) transistors are used to reduce the low-power application and to increase the speed. According to Moore's law, the number of transistors increase exponentially and presently the performance is increased every year. A multiplexer is used in most digital and analog designs. A multiplexer is used to control particular data and create the selection path for these data. In the electronics system a multiplexer is used for specific selection of data or a channel. A multiplexer is also called a data selector due to its communication advantage. The arithmetic logic unit (ALU) also plays an important role in the design of microprocessor chips and adder circuitry. It performs a different type of function like addition, subtraction, division, or multiplication and logically performs a function like AND, OR, NOT, XOR, XNOR, etc., by using CMOS technology. A number of other parameters are also considered like leakage current, power consumption, bandwidth and ON resistance, etc. To design a memory element, the basic structure consists of logic gates or a flip-flop and a structure like data conversion is used by a digital component in the CMOS design. CMOS is implemented by using the NMOS (N-channel metal-oxide semiconductor) and PMOS (P-channel metal-oxide semiconductor) structure combinations, which give better result comparisons to the BJT (Bipolar Junction Transistor) technology. Leakage is basically the problem of the CMOS, and it is important to use all of its benefits the best way possible. Subthreshold conduction is also part of the leakage. This problem arises when the transistor leaks current when it is in the OFF condition. The subthreshold conduction develops the carriers due to the potential barrier. Different types of leakage are available like gate leakage, junction leakage, dielectric leakage, and gate leakage produced by the quantum effects through tunneling. Junction leakage is produced by the junction diffusion between the source and drain with substrate. By using the 180-nm technology, a number of parameters are reduced like leakage current, power consumption, subthreshold conduction, and capacitances that affect the different types junctions. To implement these things, a layout must be developed to create better performance and results. To design the layout of the any digital circuit, first draw the transistor-level diagram and then implement the design by different rules. Layout design provides a number of advantages for the analysis and performance.

8.2 HISTORY

Earlier technology consists of transistor-transistor logic and emitter-coupled logic. A comparison of these technologies shows that CMOS technology has much less power and CMOS consumes less power when there is no switching. With an increase in device speed and chip density, consumption of power increases. Roy et al. showed that as threshold voltage, channel length, and gate oxide thickness is reduced, which is a significant contributor to power dissipation of CMOS circuits, there is high leakage current in the deep sub-micrometer regimes [1].

Abdollahi et al. presented sequential circuits by modifying the scan chains in leakage current reduction. Reducing the leakage current of combinational very large-scale integration (VLSI) circuit input vector control is an effective technique when these circuits are in the sleep mode [2]. Abdollahi et al. reviewed a gate oxide leakage current analysis made by Lee et al., which discusses the growing issue of gate oxide leakage current at the circuit level [3]. Their analysis "Gate Oxide Leakage Current Analysis and Reduction for VLSI Circuits" is seen in Ref. [4]. Blaauw et al., in "Static Leakage Reduction Through Simultaneous V_t/T_{ox} and State Assignment," reported the pressing concern for mobile applications that rely on standby modes to extend battery life by minimizing standby leakage current [5]. Morgenshtein et al. in their review discussed propagation delay and area while maintaining a less complex design. The gate diffusion input (GDI), which allows reduction in power consumption, with respect to the layout area, number of devices, delay and power dissipation, is different for each method (dual-Vth, multi-Vth, optimal standby input vector selection, transistor stacking, and body bias. Multiple thresholds) compared. The technology compatibility issues are top-down design and precomputing synthesis. GDI compared with other methods with advantages and drawbacks is discussed in Ref. [6]. The GDI technique that furnished a novel low-power and area efficient carry look ahead adder. This technique uses full adder as the essential component in the digital design [7]. In the article "Capacitance and Power Modeling at Logic-Level," Martins et al. stated that precision and optimization of the CMOS circuit with fast power used to meet stringent power specifications during the design phase is required to guide power optimization techniques [8].

Joao Baptista Martins et al. discussed the accurate global and local circuit leakage current analysis in Ref. [9]. However, capacitance and delay modeling is crude. For better delay and power performance the implementation of five different modified GDI full adders and their performance issues results in the proposed modified GDI full adders. It is compared with the existing GDI technique carried out by Uma and Dhavachelvan [10]. Using the Tanner simulator Taiwan Semiconductor Manufacturing Company Ltd. (TSMC) Berkeley Short-Channel IGFET Model (BSIM), 0.250 μm technologies were used to analyze the GDI techniques [11]. The purpose of the work is to explore the close logic-level power estimates needed to obtain circuit-level simulators.

8.3 LEAKAGE REDUCTION TECHNIQUE

8.3.1 POWER DISSIPATION IN CMOS CIRCUITS

To analyze the power dissipation in CMOS, circuits are divided into different parts like leakage power, short-circuit power, and dynamic power. In the power dissipation techniques, dynamic power is the main leakage and other power does not affect it. The dynamic power and the leakage power play very important roles in CMOS technology if the deep submicron scale is used. To charge and discharge the load capacitances, dynamic power is used when the transistor is switched ON or OFF. Assume that in the CMOS inverter with load capacitance (C_L) one cycle involves a rising and a falling transition at the gate output. Charge $Q = CV_{DD}$ is required on a low-to-high transition at the gate output, and the charge is dumped to ground (GND) during the high-to-low transition at the gate output. Charging and discharging presents the T_{fsw} times, where T presents the interval and fsw presents the frequency of the input signal.

So the dynamic power is

$$P_{dynamic} = \alpha C_L V^2{}_{DD} f \tag{8.1}$$

where
α = Switching activity
f = Clock frequency

It is not possible for the NMOS and PMOS transistors to be ON at the same time, so an error is seen due to the short circuit current.

Now the short circuit power is given by:

$$P_{Short\ circuit} = I_{mean}.V_{DD} \tag{8.2}$$

Static power is also called leakage power because the of the OFF transistor in the process.
Assume

N = Number of transistors in a circuit
I_{offi} = OFF-state current of the ith transistor

Then total leakage power of the circuit is represented by the following equation:

$$P_{leakage} = V_{DD} \sum_{i=1}^{N} I_{offi} \tag{8.3}$$

8.3.2 PROCESS VARIATION AND LEAKAGE MINIMIZATION

Post-manufacturing, transistor parameters like effective length, width of MOS, and thickness of gate oxide have spread across the chip and different regions within the chip. This leads to variations in the transistors' ON and OFF currents, which in turn lead to variations in performance and power of the manufactured chips.

In the process variation technique, dynamic power is less affected due to the linear parameter. In this technique leakage is very sensitive and is contingent on the process variation. To reproduce the effect of process variation, gate delay and leakage current are used as random variables. Subthreshold leakage is also dependent on the process variation. Statistical leakage current variations increase the presence of process variations. Subthreshold leakage affects the threshold voltage (V_{th}). The subthreshold current with threshold voltage is given by the following formula

$$I_{sub=\mu_0 C_{ox}} \frac{W}{L_{eff}} V_T^2 e^{1.8} \exp\left(\frac{V_{gs} - V_{th}}{nV_T}\right) \cdot \left(1 - \exp\left(\frac{-V_{ds}}{V_T}\right)\right) \tag{8.4}$$

The leakage power, for a static CMOS circuit presents

$$P_{leak} = V_{dd} \sum_i I_{leaki} \tag{8.5}$$

In this chapter, we projected a leakage reduction technique with GDI for the reduction power leakage dynamic power. The GDI presents the design of fast, low-power circuits, using a reduced number of transistors. Here we have used a circuit diagram to discuss different types of multiplexers, such as 2×1, 4×1, 16×1, and ALU. A simple transistor-level cell is used for the GDI.

GDI inputs include:

1. G-common gate input of NMOS and PMOS
2. P-input to the source/drain of PMOS
3. N-input to the source/drain of NMOS

Substrates of NMOS represent aN, and substrates of PMOS represent aP. Here we analyzed the probability density function (PDF) for calculating the leakage of an individual gate. T analyzes the function I on L, that is, $I = h(L)$. To find L as a role of I, find the inverse function $g(I)$ that expresses

$$L = h^{-1}(I) = g(I) \tag{8.6}$$

L is differentiated by the current, which is used to calculate the PDF of the leakage. The connection between leakage current and channel length complexity satisfies these two conditions. The $h(L)$ does not allow for the derivation of $g(I).h(L)$ fit for the approximate. The required inverse functions are calculated while maintaining good accuracy.

The expression and the PDF of $g(I)$ in the closed form is

$$PDF(I) = fy(I) = \frac{fxg(I)}{h'(L)} \tag{8.7}$$

where

$$h'(L) = \text{First derivative of the function } h(L)$$

Let us draw the gate length as a Gaussian distribution with a fixed mean μ and standard deviation σ; then the PDF is defined as

$$PDF = \left(\frac{1}{h'(L)}\right)\left(\frac{1}{\sqrt[q]{2\pi}}\right) \times \exp\left(\frac{-(g(I) - \mu)2}{2\sigma 2}\right) \tag{8.8}$$

Subthreshold current is represented by

$$I = I_0 \exp\left(\frac{(V_{gs} - V_{th})}{nV_T}\right)\left(1 - \exp\left(\frac{-V_{ds}}{V_T}\right)\right) \tag{8.9}$$

where $V_T = KT/q$.

$$I_0 = \mu_0 C_{ox}\left(W_{eff}/L_{eff}\right)V_T^2 \tag{8.10}$$

since V_{ds} is much greater than the thermal voltage V_T.

8.4 MOSFET CAPACITANCE MODEL

The metal oxide semiconductor field-effect transistor (MOSFET) capacitance model presents a number of capacitances that are automatically generated by the process. The associated parasitic capacitance is the gate-to-diffusion overlap capacitance, C_{gol}, and the diffusion area and perimeter capacitances are C_{jb} and C_{jbsw}. The perimeter capacitance is C_{jbswg} with the gate side. The capacitance is voltage dependent for diffusion capacitance, with gate capacitance extract switching transition for delay estimation and for effective averaged capacitance.

Figure 8.1 shows parasitic device capacitances with a two-input gate. The output node and the ground are connected with lumped capacitance. The load capacitance value, C_{load}, is

$$C_{load} = C_{gd,A} + C_{gd,B} + C_{gd,load} + C_{db,A} + C_{db,B} + C_{sb,load} + C_{wire} \tag{8.11}$$

There is output load capacitance present for single-input switching. The load capacitance C_{load} is present if only one input is active and all other inputs are low at the output node. A number of parasitic capacitances are present in circuit design:

1. Gate-to-source capacitance (C_{GS})
2. Gate-to-drain capacitance (C_{GD})
3. Gate-to-bulk capacitance (C_{GB})
4. Source-to-bulk capacitance (C_{SB})
5. Drain-to-bulk capacitance (C_{DB})

The size of the transistor (W/L), a value of the parasitic capacitances, is presented as

$$C_{GD} = C_{GS} = \frac{1}{2}C_{ox}WL + C_0 \tag{8.12}$$

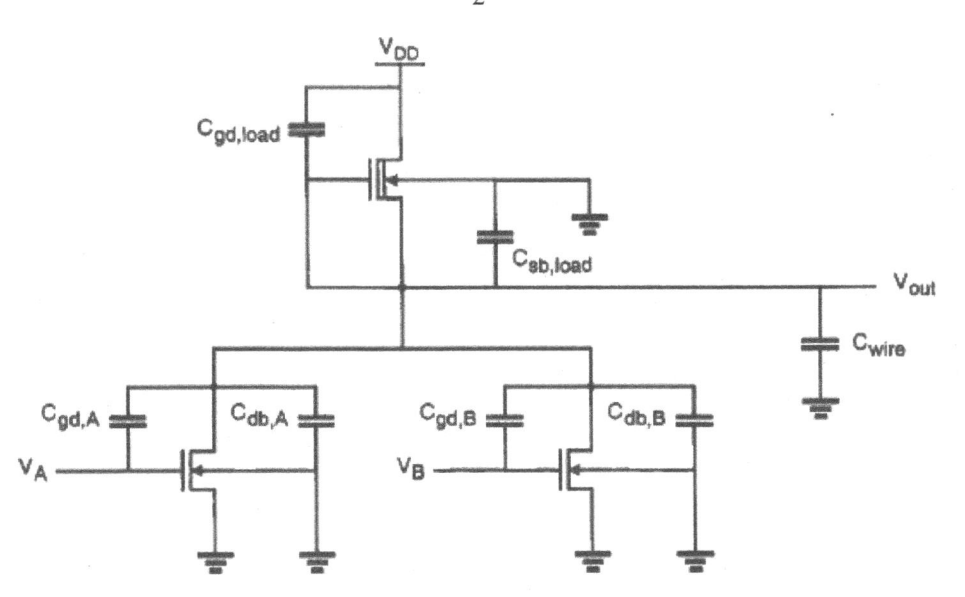

FIGURE 8.1 Parasitic device capacitances in circuits.

In a layout, the capacitance is calculated as

$$C_{DB} = C_{DB,bot} + C_{DB,SW} \tag{8.13}$$

$$C_{SB} = C_{SB,bot} + C_{SB,SW} \tag{8.14}$$

$$C_{DB,bot} = K(V_l)C_{j0}A_D \tag{8.15}$$

$$C_{DB,SW} = K_{1/3}(V_l)C_{jsw}l_D \tag{8.16}$$

$$C_{SB,bot} = K(V_l)C_{j0}A_S \tag{8.17}$$

$$C_{SB,SW} = K_{1/3}(V_l)C_{jsw}l_S \tag{8.18}$$

where L= channel length, W = channel width, Co = overlap capacitance, C_{j0} = zero bias capacitance per unit area, C_{jsw} = zero-bias sidewall capacitance per unit perimeter, A_D = area of drain, l_D = perimeter of drain, A_S = area of source, and L_S = perimeter of source.

These concepts are all implemented in 2×1 MUX, 4×1 MUX, 8×1 MUX, 16×1 MUX; ALU with GDI technique; and CMOS technique and compared with each other.

8.5 PERFORMANCE AND CIRCUIT SIMULATION RESULT

8.5.1 DESIGN OF MULTIPLEXER AND ALGORITHM LOGIC UNIT

The simulation of different multiplexers and ALU is shown in following figures. Implementing the 2-to-1 multiplexer by GDI technique in which one NMOS and one PMOS are used with two inputs is applied in two terminals and a selection line as a thread terminal (Figure 8.2). When the selection line S is low input I0 will select, otherwise I1 will select; in other words the selection line controlling the whole circuit produces the output. Figure 8.3 represents the transistor-level diagram of the 4-to-1 multiplexer with GDI technique with four inputs (I0, I1, I2, and I3) and two selection lines (S0 and S1).

An 8-to-1 multiplexer, which corresponds to every minterm, and three variable switches that select one input depending on its state are illustrated in Figure 8.4. The circuit provides the ability to select 1 bit of data from up to eight sources, and Figure 8.5 shows the implementation of the 16-to-1 multiplexer.

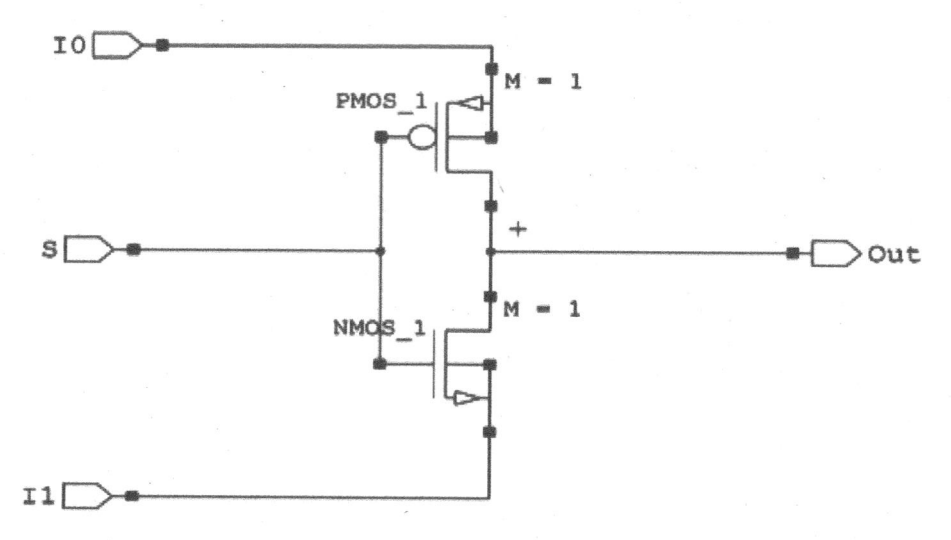

FIGURE 8.2 Transistor-level representation of 2-to-1 MUX with GDI technique.

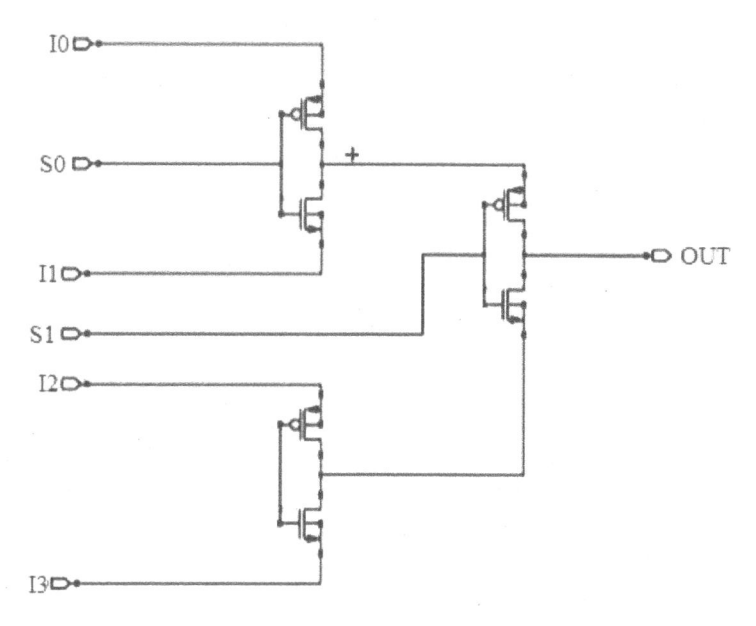

FIGURE 8.3 Transistor-level representation of 4-to-1 MUX with GDI technique.

Figure 8.6 shows the transistor-level representation of a 1-bit ALU with GDI technique. In this circuit a total of 54 transistors (27 for NMOS and 27 PMOS) has been used. The design is used for 1-bit operation in real-time applications with better hardware reduction, which in turn minimizes the number of gates over binary logic. The design can be extended to n-bit operation for real-time applications.

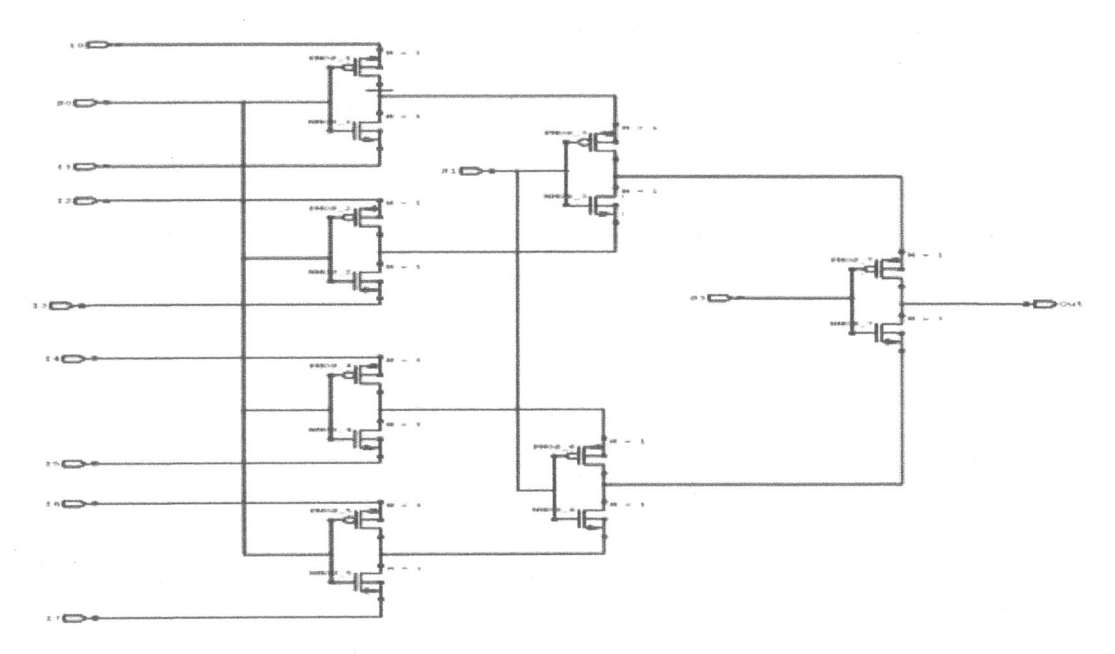

FIGURE 8.4 Transistor-level representation of 8-to-1 MUX with GDI technique.

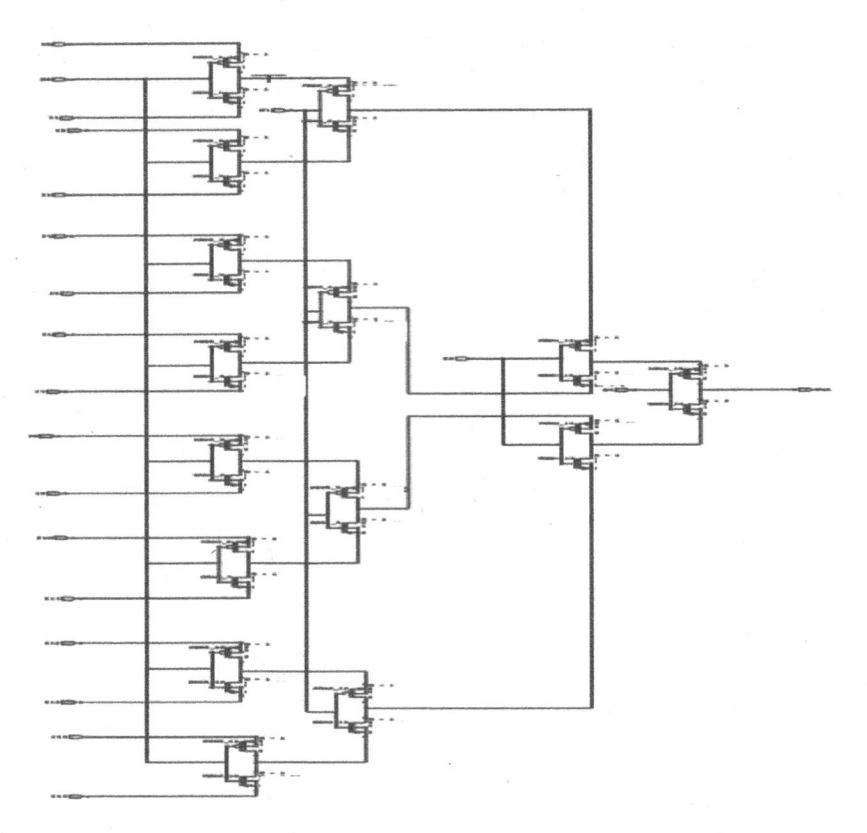

FIGURE 8.5 Transistor-level representation of 16-to-1 MUX with GDI technique.

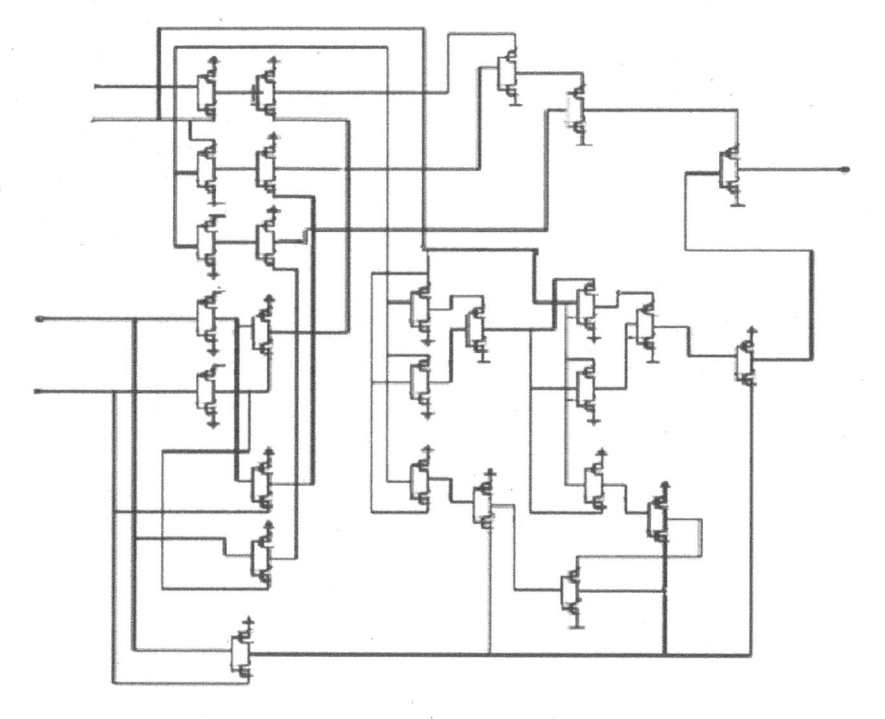

FIGURE 8.6 Transistor-level representation of 1-bit ALU with GDI technique.

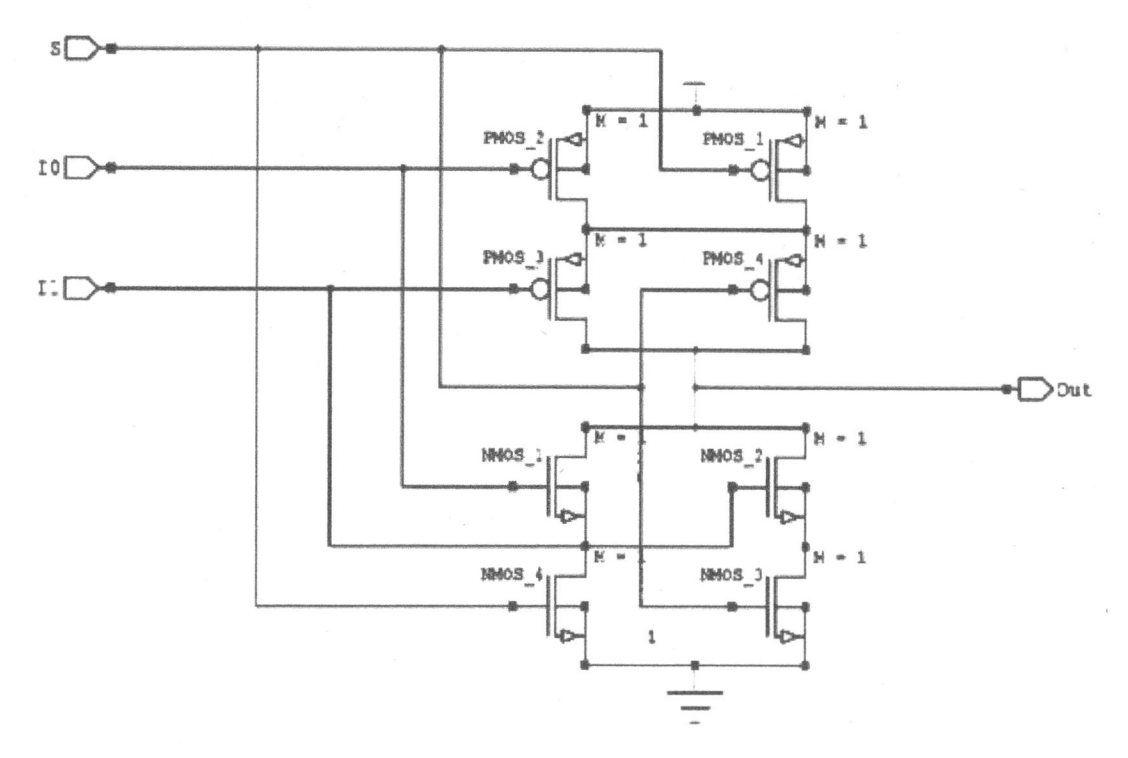

FIGURE 8.7   Transistor-level implementation of the 2-to-1 MUX with CMOS technique.

Figures 8.7–8.10 represent transistor-level implementation of the 2-to-1, 4-to-1, 8-to-1, and 16-to-1 multiplexers with CMOS technique. The ALU implemented using CMOS is illustrated in Figure 8.11 clearly showing that there is a greater number of transistors required compared with the GDI techniques.

Table 8.1 lists the number of transistors required in the GDI and CMOS technology for MUX and ALU designs; whereas Tables 8.2 and 8.3 show the comparison of both techniques in terms of leakage current and dynamic power consumption, respectively.

8.5.2   CMOS Power Consumption Over GDI Technique

Comparison of the GDI technique and the CMOS technique shows that CMOS devices have very low static power consumption, which is the result of leakage current. This power consumption occurs when all inputs are held at some valid logic level and the circuit is not in the charging state,

TABLE 8.1

Comparison Between GDI and CMOS

Device	No. of Transistors in CMOS	No. of Transistors in GDI
2×1 MUX	8	2
4×1 MUX	24	6
8×1 MUX	64	14
16×1 MUX	160	30
1-Bit ALU	118	54

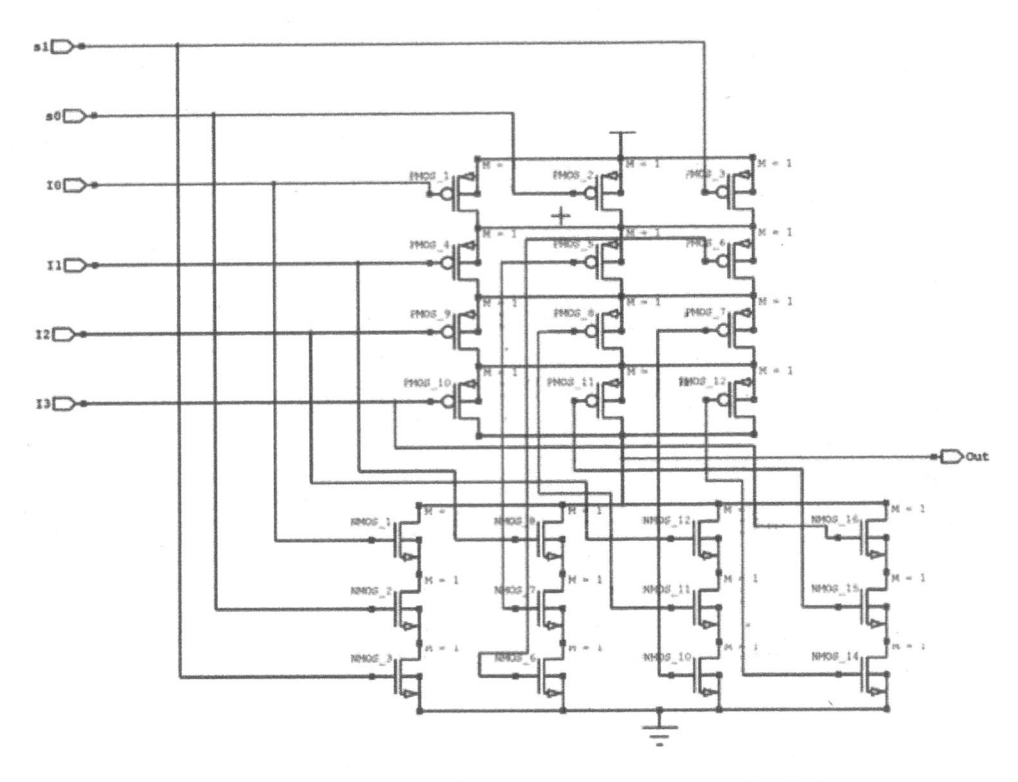

FIGURE 8.8 Transistor-level implementation of the 4-to-1 MUX with CMOS technique.

but switching at a high-frequency dynamic power consumption can contribute significantly to overall power consumption. Charging and discharging a capacitive output load further increases this dynamic power consumption. This chapter addresses possible solutions to minimize power consumption in a CMOS system. The simplified model of a CMOS circuit consisting of several gates can be viewed as one large capacitor that is charging and discharging between the power supply rails. Therefore, the power dissipation capacitance (C_{pd}) is often specified as a measure of this equivalent capacitance and is used to approximate the dynamic power consumption. It includes both internal parasitic capacitance and through currents that are present while a device is switching and both n-channel and p-channel transistors are momentarily conducting. Historically, CMOS design operated at supply voltages much larger than their threshold voltages tunneling current through gate-oxide SiO_2, which is a very good insulator, but at very small thickness levels electrons can tunnel across the very thin insulation. The probability drops

TABLE 8.2
Simulation Result in Terms of Leakage Current

Device	Leakage Current in CMOS (mA)	Leakage Current in GDI (mA)
2×1 MUX	0.137	0.007
4×1 MUX	1.43	0.212
8×1 MUX	2.96	0.122
16×1 MUX	4.43	1.13
1-Bit ALU	8.98	3.77

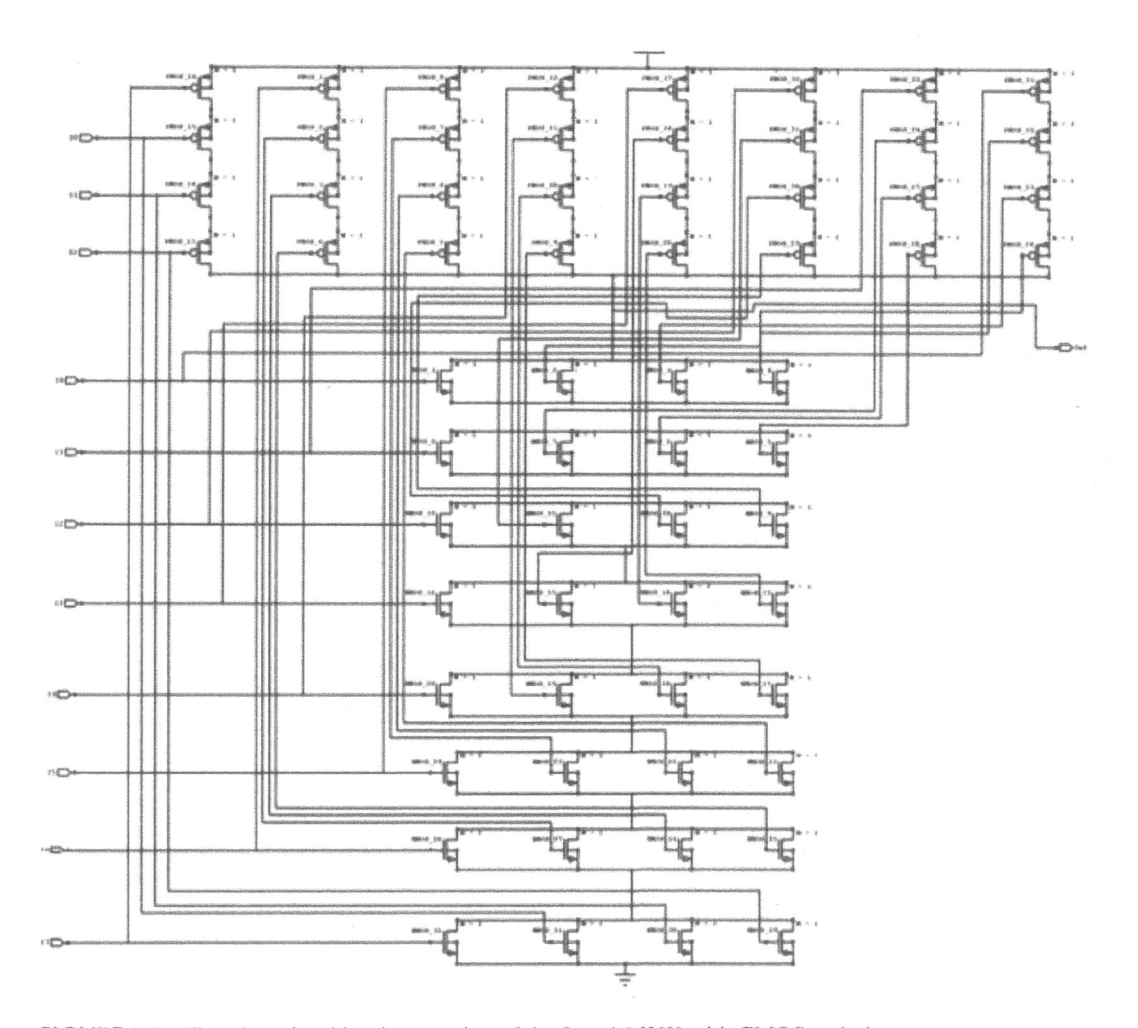

FIGURE 8.9 Transistor-level implementation of the 8-to-1 MUX with CMOS technique.

TABLE 8.3
Simulation Result in Terms of Dynamic Power Consumption

Device	Dynamic Power Consumption in CMOS (μw)	Dynamic Power Consumption in GDI (μw)
2×1 MUX	9.88	2.56
4×1 MUX	16.09	3.23
8×1 MUX	24.16	5.134
16×1 MUX	74.55	15.01
1-Bit ALU	112.5	40.3

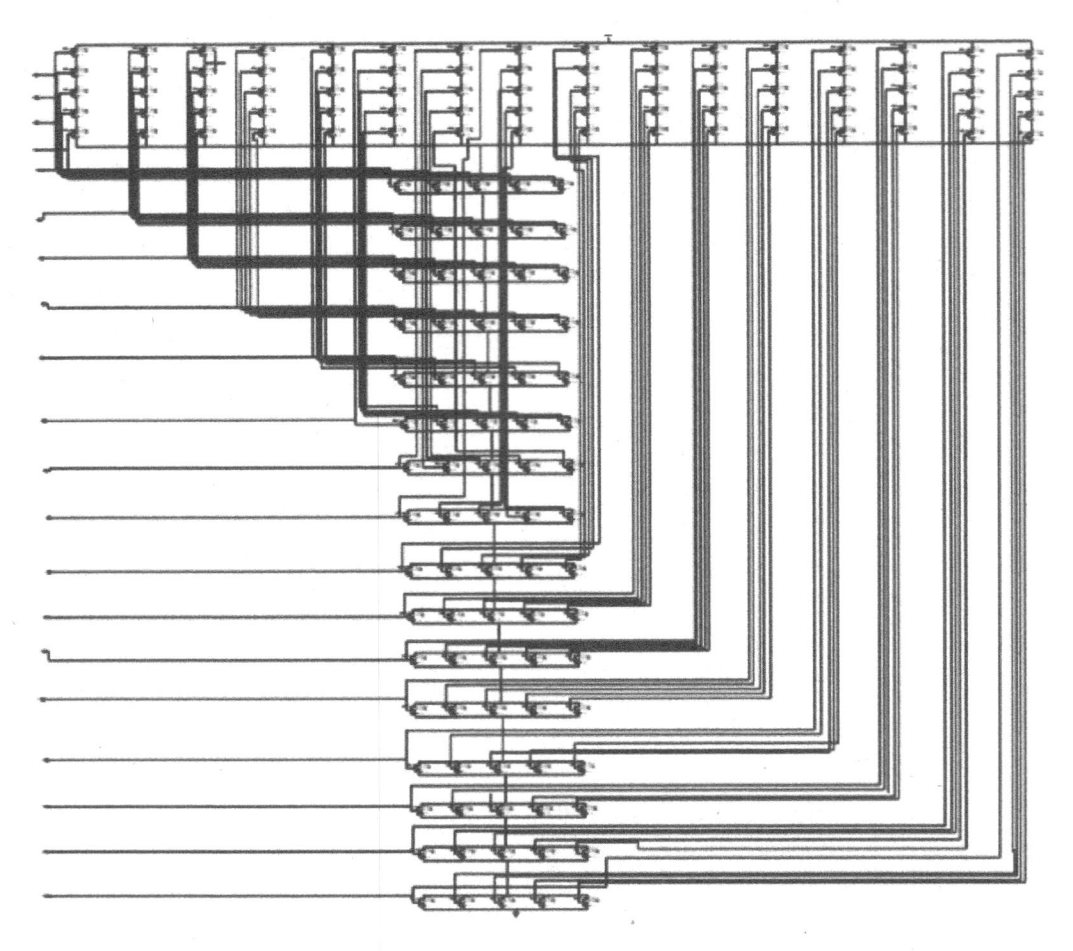

FIGURE 8.10   Transistor-level implementation of the 16-to-1 MUX with CMOS technique.

exponentially with oxide thickness. Tunneling current becomes very important for transistors below 130-nm technology with gate oxides of 20 Å or thinner. Leakages of current occur through reverse-biased diodes. Small reverse leakage currents are formed due to formation of reverse bias in the diffusion region between the wells (e.g., p-type diffusion vs. n-well) and substrate (e.g., n-well, p-substrate). In the modern process, diode leakage is very small compared with subthreshold and tunneling current, so these may be neglected during power calculations. To speed up the current, design manufacturers have switched to constructions that have lower voltage thresholds, but because of this a modern NMOS transistor with V_{th} of 200 mV has a significant subthreshold leakage current. Designs that include vast numbers of circuits that are not actively switching still consume power because of leakage.

8.6   CAPACITANCE REDUCTIONS

The on-chip capacitances found in MOS circuits are in general complicated functions of the layout geometries and the manufacturing process. Most of these capacitances are not lumped, but distributed, and their exact calculations would require complex three-dimensional nonlinear charge-voltage models.

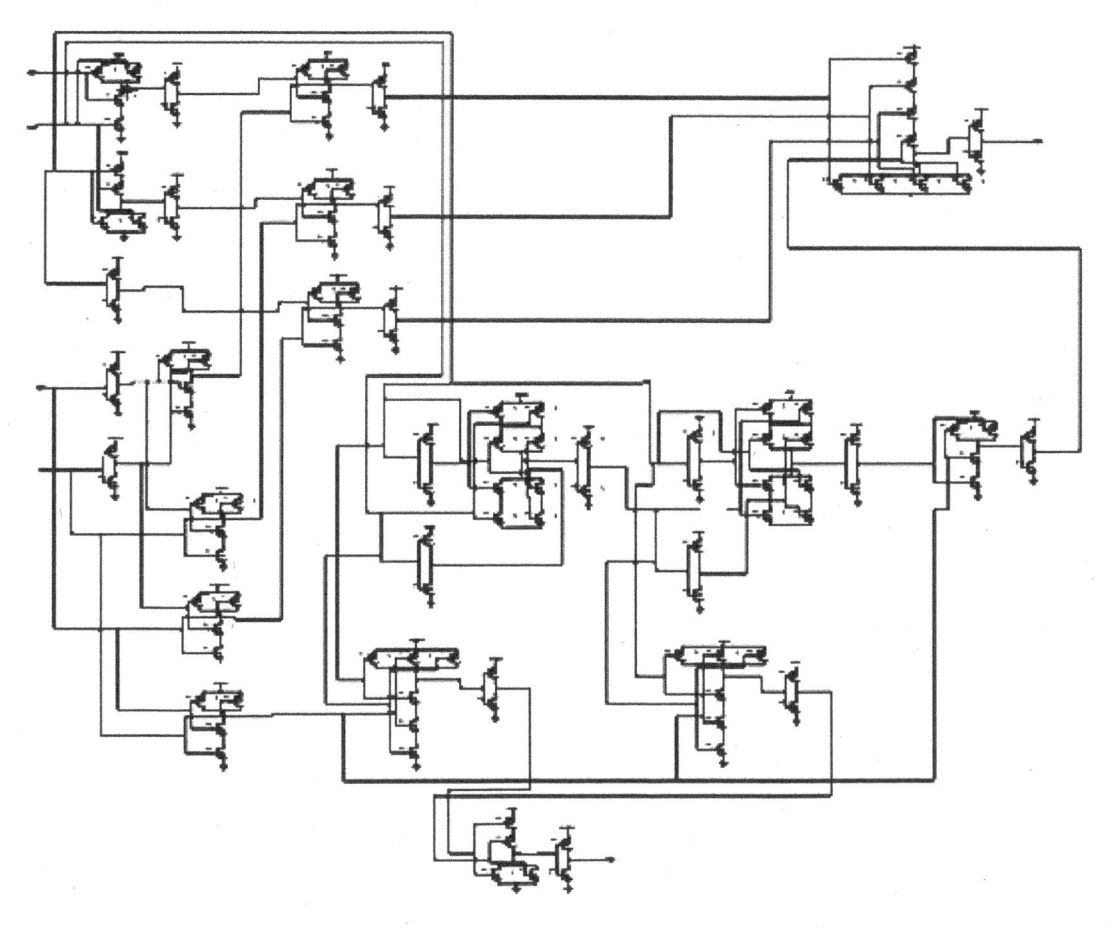

FIGURE 8.11 Transistor-level implementation of 1-bit ALU with CMOS technique.

8.6.1 PARASITIC CAPACITANCE

The parasitic capacitance of a transistor has two components. The junction capacitance C_{ja}, expressed in F/area in μm^2, and the periphery capacitance C_{jp} of the periphery length expressed in F/μm. Each transistor has a parasitic capacitance from the layout

$$C_p = C_{ja}.W.L_{diff} + C_{jp}\left(2W + L_{diff}\right) \tag{8.19}$$

where W is channel width and L is channel length.

The bulk/drain capacitance C_{BD}, bulk/source capacitance C_{BS}, gate-to-source overlap capacitance C_{GSOV} gate-to-drain overlap capacitance C_{GDOV} bulk/gate overlap capacitance C_{GBOV} and MOSFET gate capacitances (C_{GB}, C_{GS}, and C_{GD}) are calculated as per the BSIM [12].

8.6.2 CALCULATION OF PARASITIC CAPACITANCE

The specific values for the simulated results are given in Table 8.4.

TABLE 8.4

Calculations of Parasitic Capacitances of the MOS Transistor in ALU, with GDI Technique

	Equation	Values for $V_{GS} = 0$ V, $V_{DS} = 4$ V, $V_{SB} = 0$ V
C_{BD}	$C_{BD} = C_{BDJ} + C_{BDSW}$	$C_{BD} = 1.855 \times 10^{-13} + 2.04 \times 10^{-16} = 1.857 \times 10^{-13}$ F
	$C_{BDJ} + A_D\, C_J\, (1 + V_{DB}/\varnothing_B)^{-\text{mJ}}\ (\varnothing_B = P_B)$	$C_{BDJ} = (3.032 \times 10^{-15})(1 + (3/1))^{-0.56} = 1.86 \times 10^{-15}$ F
	$C_{BDSW} = P_D\, C_{JSW}\, (1 + V_{DB}/\varnothing_B)^{-\text{mJSW}}$	$C_{BDSW} = (3.2 \times 10^{-16})(1 + (3/1))^{-0.5} = 2.13 \times 10^{-16}$ F
	(P_D **may or may not include channel edge**)	
C_{BS}	$C_{BS} = C_{BSJ} + C_{BSSW}$	$C_{BS} = 3.032 \times 10^{-15} + 4.1 \times 10^{-16} = 3.44 \times 10^{-15}$ F
	$C_{BSJ} + A_S\, C_J\, (1 + V_{SB}/\varnothing_B)^{-\text{mJ}}$	$A_S\, C_J = (6.2 \times 10^{-15})(5.7 \times 10^{-4}) = 3.53 \times 10^{-18}$ F
	$C_{BSSW} = P_S\, C_{JSW}\, (1 + V_{SB}/\varnothing_B)^{-\text{mJSW}}$	$P_S\, C_{JSW} = (8.1 \times 10^{-6})(4 \times 10^{-11}) = 3.24 \times 10^{-16}$ F
C_{GSOV}	$C_{GSOV} = W_{EFF}\, C_{GSO};\ W_{EFF} = W - 2\,W_D$	$C_{GSOV} = (6 \times 10^{-6})(4 \times 10^{-10}) = 2.4 \times 10^{-16}$ F
C_{GDOV}	$C_{GDOV} = W_{EFF}\, C_{GSO}$	$C_{GDOV} = (6 \times 10^{-6})(4 \times 10^{-10}) = 2.4 \times 10^{-15}$ F
C_{GBOV}	$C_{GBOV} = L_{EFF}\, C_{GBO};\ L_{EFF} = L - 2\,L_D$	$C_{GDOV} = (0.5 \times 10^{-6})(3 \times 10^{-10}) = 1.5 \times 10^{-16}$ F
C_{GS}	$C_{GS}/C_O = 0$ (OFF), 0.5 (lin.), 0.66 (sat.)	$C_O = (6 \times 10^{-6})(0.5 \times 10^{-6})(0.00345) = 1.03 \times 10^{-14}$ F
	C_O **(oxide capacitance)** $= W_{EF}\, L_{EFF}\varepsilon_{ox}/T_{ox}$	$C_{GS} = 0.0$ F
C_{GD}	$C_{GD}/C_O = 0$ (OFF), 0.5 (lin.), 0 (sat.)	$C_{GD} = 0.0$ F
C_{GB}	$C_{GB} = 0$ (ON), $= C_O$ in series with C_{GS} (OFF)	$C_{GB} = 3.18 \times 10^{-16}$ F, $C_S =$ **depletion capacitance**

8.7 CONCLUSIONS

In this chapter the analysis of the reduction of leakage current and leakage capacitance was presented. To implement all these things we applied an electronic design automation (EDA) environment that allows implementing and integrating in a single framework. We applied a GDI technique that was implemented by using a CMOS technique in multiplexers and ALU. Then comparisons between the GDI and CMOS techniques were made. To implement these things, we used Cadence Virtuoso 4.1 software. When a particular technology is selected a set of configuration- and technology-related files are employed to customize the environment. The resulting layout verified geometric rules dependent on the technology. When comparing CMOS and GDI techniques, the number of transistors to implement 2×1, 4×1, 8×1, 16×1 was 8, 24, 64 and 160 in CMOS technology. Whereas the number of transistors was reduced in the GDI technique, resulting in only 2, 6, 14, and 30 transistors required for the same circuits. In addition to the number of transistors using the GDI technique, for ALU implementation the amount is reduced to 54 compared with 118 transistors using CMOS technology. Similarly, the leakage current produced 81% of the benefits, and the power consumption provided a 75% benefit. The number of gates in the GDI technique generated a 73% advantage, whereas the capacitance calculation produced a 75% reduction in leakage current.

REFERENCES

1. Kaushik Roy, Saibal Mukhopadhyay, and Hamid Mahmoodi-Meimand, "Leakage Current Mechanisms and Leakage Reduction Techniques in Deep-Submicrometer CMOS Circuits," Proceedings of the IEEE, Vol. 91, No. 2, February 2003, pp. 305–327.
2. Afshin Abdollahi, Farzan Fallah, and Massoud Pedram, "Leakage Current Reduction in Sequential Circuits by Modifying the Scan Chains," Proceedings of the Fourth International Symposium on Quality Electronic Design (ISQED'03), Vol. 05, 2003, pp. 213–218.

3. Afshin Abdollahi, Farzan Fallah, and Massoud Pedram, "Leakage Current Reduction in CMOS, VLSI Circuits by Input Vector Control," IEEE Transactions on Very Large Scale Integration (VLSI) Systems, Vol. 12, No. 2, February 2004, pp. 140–154.

4. Dongwoo Lee, David Blaauw, and Dennis Sylvester, "Gate Oxide Leakage Current Analysis and Reduction for VLSI Circuits," IEEE Transactions on Very Large Scale Integration (VLSI) Systems, Vol. 12, No. 2, February 2004, pp. 155–166.

5. Dongwoo Lee, David Blaauw, and Dennis Sylvester, "Static Leakage Reduction Through Simultaneous Vt/Tox and State Assignment," IEEE Transactions on Computer-Aided Design of Integrated Circuits and Systems, Vol. 24, No. 7, July 2005, pp. 1014–1029.

6. Arkadiy Morgenshtein, Alexander Fish, and Israel A. Wagner, "An Efficient Implementation of D-Flip-Flop Using the GDI Technique," IEEE International Symposium on Circuits and Systems, Vol. 10, No. 05, 2004, pp. 566 -581.

7. Pakkiraiah Chakali, Adilaxmi Siliveru, and Jagadeesh Y, "A Novel Low Power Gray to Binary Code Converter Uses Gate Diffusion Input (GDI)," International Journal of Advanced Research in Computer Engineering and Technology (IJARCET), Vol. 1, No. 6, August 2012, pp. 225–230.

8. M. Yap San Min, O. Thomas, A. Valentian, and F. de Crecy, "Accurate Global and Local Circuit Leakage Current Analysis Based on Design of Experiment Method," IEEE International Conference on IC Design and Technology, 2009.

9. João Baptista Martins, Ricardo Reis, and José Monteiro, "Capacitance and Power Modeling at Logic-Level," UFSM and CAPES/Brazil. https://www.researchgate.net/publication/228974501_ Capacitance_ and Power_ Modeling_at_ Logic-Level

10. R. Uma and P. Dhavachelvan, "Modified Gate Diffusion Input Technique: A New Technique for Enhancing Performance in Full Adder Circuits," 2nd International Conference On Communication, Computing & Security [ICCCS], 2012.

11. Balakrishna Batta, Manohar, Choragudi, Mahesh Varma D, "Energy Efficient Full-Adder Using GDI Technique," International Journal of Research in Computer and Communication Technology, IJRCCT, Vol. 1, No. 6, November 2012.

12. B.J. Sheu, D.L. Scharfetter, P.-K. Ko, and M.-C. Jeng, "BSIM: Berkeley Short-Channel IGFET Model for MOS Transistor," IEEE Journal for Solid-State Circuits, Vol. 22, No. 4, August 1987, pp. 558–566.

9 Design of Gate-All-Around TFET with Gate-On-Source for Enhanced Analog Performance

Navaneet Kumar Singh, Rajib Kar, and Durbadal Mandal

CONTENTS

9.1 INTRODUCTION

Energy-efficient electronic appliances are essential for the latest development of the Internet of Things (IoT) technology [1]. The present mobile industry, including growing IoT technologies like wearable and smart sensor networks, require low power and high package density of integrated circuits. The most expedient method to lower the active power is to decrease the V_{th} and the supply voltage to retain the same performance [2]. The advancement of the complementary metal oxide semiconductor (CMOS) is the key to the electronics manufacturing industry. However, metal oxide semiconductor field-effect transistor (MOSFET) scaling enables a single chip to contain numerous compact devices [3]. After the invention of integrated circuits in the 1950s, the continuous scaling of MOSFETs is done with the changing technology to attain the reduced technology nodes. Moore's law governs the CMOS scaling, but due to leakage current, heat dissipation, and channel length modulation, the scaling has deviated from Moore's prediction [2–4]. Therefore, new technology is needed that can resolve these problems, including cost, reliability, speed, etc. Now is the time to switch from a single-gate MOSFET to a multigate MOSFET for superior electrostatic control by the gate [5–6]. In this regard, researchers have developed the concept of double-gate MOSFET, triple-gate MOSFET, nanowire MOSFET, and many more. The nanowire, when compared with planer devices based on bulk material, shows a large surface to volume ratio and smaller channel length [6–8].

Furthermore, the gate-all-around (GAA) or cylindrical gate-based nanowire FET (NWFET) has enhanced electrostatic control on charge carriers. Thus, the GAA NWFET offers further CMOS scaling with controlled short-channel effects (SCEs) [6]. The FinFET is a structure in which scaling is possible up to a 10-nm or even to a 7-nm technology node. Due to the similarity of GAA NWFET with FinFET, the GAA is a promising structure in sub-7-nm technology to overcome the scalability limits of the FinFET [2–4]. The least gate length required to reduce the SCEs in NWFET is governed by Equation 9.1 [2].

$$\lambda_{gate-all-around} = \sqrt{\frac{2\varepsilon_{si}t_{si}^2 \ln\left(1 + \dfrac{2t_{ox}}{t_{si}}\right) + \varepsilon_{ox}t_{si}^2}{16\varepsilon_{ox}}} \qquad (9.1)$$

where t_{ox}, ε_{ox}, t_{si}, and ε_{si} are the thickness and permittivity of gate dielectric and silicon, respectively. The cylindrical structure confirms the inverse dependence of gate capacitances on the channel thickness in the logarithmic scale. Thus, the channel length in this cylindrical structure is scaled down with nanowire diameter without any decrease in the dielectric gate thickness.

The tunnel field-effect transistor (TFET) has become an admirable structure to attain a lesser power consumption due to the steeper subthreshold slope (SS) [7–10]. The projected OFF-state current is considerably high in the traditional MOSFET and is limited to the nanoampere range. Because of the fantastic specialty of TFET, such as negligible power consumption and increased operating speed [11–13], it has been drawing the attention of researchers for the past few years. By using an appropriate biasing and monitoring band bending in the vicinity of both junctions, the desired tunneling of carriers in TFET can be attained [14]. Despite the smaller SS, TFET suffers from a small drain current. Several methods have been proposed to avoid this problem, e.g., using high-K gate oxide material, dual material gate, gate stacking, etc. [15–25]. The other way to boost the ON current is by utilizing SiGe in the source, which is a low bandgap material. Due to an extensive energy bandgap of silicon, a drain current is observed in Si-sourced TFET that is lower than that of the MOSFET. The current in the ON state of TFET is enhanced by improving the tunneling probability within the source region. In this regard, a material with a low bandgap (SiGe) is utilized within the region of the source; however, a material with a high bandgap within the drain region is used to mitigate the leakage in the device. The evolution of SiGe-sourced nanowire TFET (NWTFET) may be regarded as an extension of SiGe MOSFET [26–32]. However, the majority of the devices have some drawbacks. The low band-to-band tunneling efficiency in silicon results in low I_{ON}. TFET, similar to MOSFET, rather than having the P-I-N structure in TFET, permits the low dissipation of static power and reduced operating voltage. These features are responsible for achieving a superior subthreshold swing (<60 mV/dec) [15–17].

SiGe-sourced NWTFET already exists, but the novelty of this work subsists in the extended gate-on-source (GOS) of NWTFET-GOS and its impact on the analog performance. The analog performance and electrical performance of NWTFET, NWTFET-GOS, NWTFET-gate-on-drain (GOD) and NWTFET-gate-on-source-and-drain (GOSD) are also compared for the first time in this work.

The goal of this chapter is to propose and study the NWTFET-GOS with SiGe as a source material. The analog performance of the proposed structure is analyzed, and compared with NWTFET, NWTFET-GOD, and NWTFET-GOSD. The results achieved after employing the extended GOS are reported. Various design parameters, e.g., extended GOS length, SiGe with various mole fractions, etc., are also examined, and their effects are compared among the considered devices as mentioned earlier.

9.2 DEVICE STRUCTURES AND SIMULATION

The two-dimensional view of NWTFET, NWTFET-GOS, NWTFET-GOD, and NWTFET-GOSD are presented in Figure 9.1(a–d), respectively. The length of the gate used in each device is 20 nm. The gate electrode is extended toward the source, drain, and both drain and source sides in NWTFET-GOS, NWTFET-GOD, and NWTFET-GOSD, respectively. Each device's extended gate is 2, 4, 6, 8, and 10 nm, respectively. The high-k dielectric (hafnium) is used with a 1-nm thickness

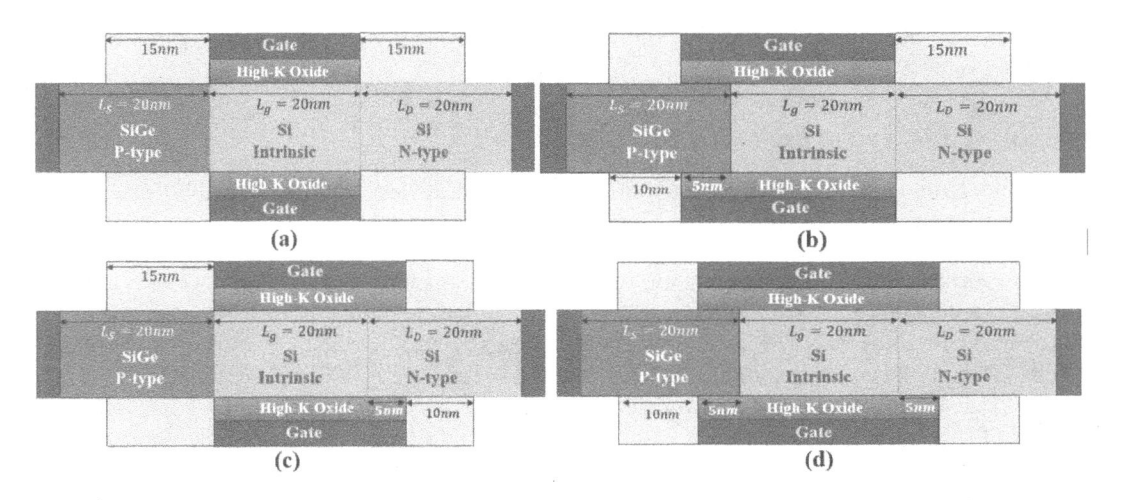

FIGURE 9.1 Two-dimensional view of proposed structures: (a) NWTFET, (b) NWTFET-GOS, (c) NWTFET-GOD, and (d) NWTFET-GOSD.

for all the devices. The material chosen within source bulk is SiGe with a p-type material with a concentration of 1×10^{20}/cm³. The type of channel is intrinsic, and the n-type drain region has a doping concentration of 5×10^{18}/cm³. Molybdenum ($\phi = 4.75$ eV) is used as gate material for the design of all the devices. Other specifications associated with the device structure are given in Table 9.1, in which the high-k dielectric is used as a spacer, which suppresses the leakage in the device.

The Silvaco TCAD tool is used to simulate and analyze the structures. The model used for nonlocal band-to-band tunneling is bbt.kane. Other models used for the simulation are conmob, auger, bgn, srh, fldmob, and print. A lower bandgap material $Si_{1-x}Ge_x$ is used in the bulk of the source to improve the ON current. Different mole fractions (x) of 0.45, 0.5, 0.55, and 0.6 are used in SiGe to analyze its impact on the proposed structure's DC and analog performance. The Wentzel-Kramer-Brillouin approximation in Equation 9.2 states that material in the bulk of the source region with a lower bandgap increases the probability of tunneling, resulting in an improved ON-state current [28].

$$T_{WKB} \approx \exp\left(-\frac{4\lambda\sqrt{2m^*}\sqrt{E_g}^3}{3qh\left(E_g + \Delta\varphi\right)}\right) \tag{9.2}$$

TABLE 9.1

Device Parameters Used for Simulation

Device Parameters	NWTFET	NWTFET-GOS	NWTFET-GOD	NWTFET-GOSD
Gate length, L_g (nm)	20	20	20	20
Extended gate lengths (nm)	-	2, 4, 6, 8, and 10	2, 4, 6, 8, and 10	2, 4, 6, 8, and 10
Oxide thickness, t_{Hfo2} (nm)	1	1	1	1
Source doping (cm⁻³)	1e + 20	1e + 20	1e + 20	1e + 20
Channel doping (cm⁻³)	1e + 15	1e + 15	1e + 15	1e + 15
Drain doping (cm⁻³)	5e + 18	5e + 18	5e + 18	5e + 18
Work function of gate (eV)	4.75	4.75	4.75	4.75
Spacer length toward source/drain side (nm)	15/15	5, 7, 9, 11, and 13/15	15/5, 7, 9, 11, and 13	5, 7, 9, 11, 13/5, 7, 9, 11, and 13

The tunneling rate can be improved by diminishing the device body thickness, reducing the oxide thickness, or using high-k dielectric.

9.3 RESULTS AND DISCUSSION

The comparison of I_d-V_{gs} curves for each of the mentioned devices with the proposed structure at $V_{gs} = 1.8$ V and $V_{ds} = 1$ V is presented in Figure 9.2. The ON current is the highest in the NWTFET-GOS with a reduced leakage in the device. However, the least leakage current is observed in the conventional NWTFET. The comparison of electrical parameters, such as ON current, OFF current, threshold voltage, current ratio, and SS of all the devices is shown in Table 9.2. Among all the considered structures, NWTFET-GOS yields the highest I_{ON}/I_{OFF} and the lowest threshold voltage. NWTFET-GOS has a comparatively faster-switching speed than NWTFET-GOD and NWTFET-GOSD because of the improved subthreshold swing. The NWTFET-GOS, when compared with NWTFET, also shows an improvement in I_{ON}, I_{ON}/I_{OFF}, and threshold voltage. The poorest SS is observed for the device NWTFET-GOD.

The comparative investigation of electrical parameters of the devices is depicted in Figure 9.3(a–e). Figure 9.4(a–d) represents the variations of the electric field, center potential, recombination rate, and band diagram of each device at $V_{gs} = 1.8$ V and $V_{ds} = 1$ V. The devices' center potential become almost constant in the channel along the source region, and NWTFET-GOS has the lowest center potential. The electric field magnitudes in the drain and channel region become almost equal to zero due to the constant center potential in those regions. The rate of recombination in NWTFET-GOS is less than that of NWTFET and NWTFET-GOD. The lowest rate of recombination is achieved for NWTFET-GOSD. The comparison of the energy band diagram shown (EBD) in Figure 9.4(d) shows that the worst band bending occurs in NWTFET-GOSD, and an uneven band bending is observed in NWTFET-GOD. A superior tunneling rate is perceived in NWTFET and NWTFET-GOS. The capacitances C_{gs}, C_{gd}, and overall gate capacitance C_{gg} are presented in Figure 9.5(a–c), respectively.

The capacitance, C_{gs}, is found to be greater than C_{gd}, and the overall gate capacitance is the summation of these two capacitances. Transconductance is the rate of change in drive current with respect to V_{gs} at constant drain voltage. The transconductance and its derivatives are depicted in Figure 9.6(a–c), respectively. Transconductance shows the magnitude of the current developed after the application of V_{gs}. Transconductance represents the gain, and its derivatives signify the amount

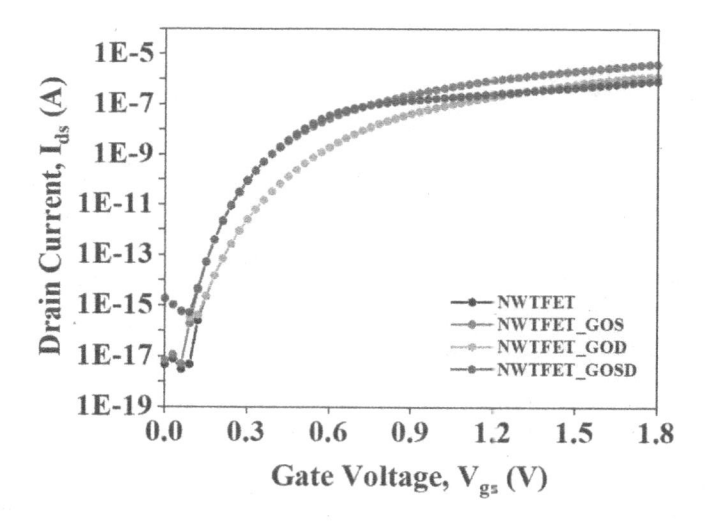

FIGURE 9.2 Comparison of I_d-V_{gs} of various proposed structures at $V_{gs} = 1.8$ V and $V_{ds} = 1$ V.

TABLE 9.2

Comparison of Electrical Performances of NWTFET, NWTFET-GOS, NWTFET-GOD, and NWTFET GOSD

Parameters	NWTFET	NWTFET-GOS	NWTFET-GOD	NWTFET-GOSD
I_{ON} (A)	7.54×10^{-8}	3.96×10^{-7}	7.54×10^{-8}	1.71×10^{-7}
I_{OFF} (A)	4.74×10^{-18}	6.96×10^{-18}	1.83×10^{-15}	1.83×10^{-15}
I_{ON}/I_{OFF}	1.59×10^{10}	5.68×10^{10}	4.12×10^{7}	9.30×10^{7}
V_{th} (V)	1.055	0.76	1.05	0.78
SS (mV/dec)	17.51	19.08	37.81	28.45

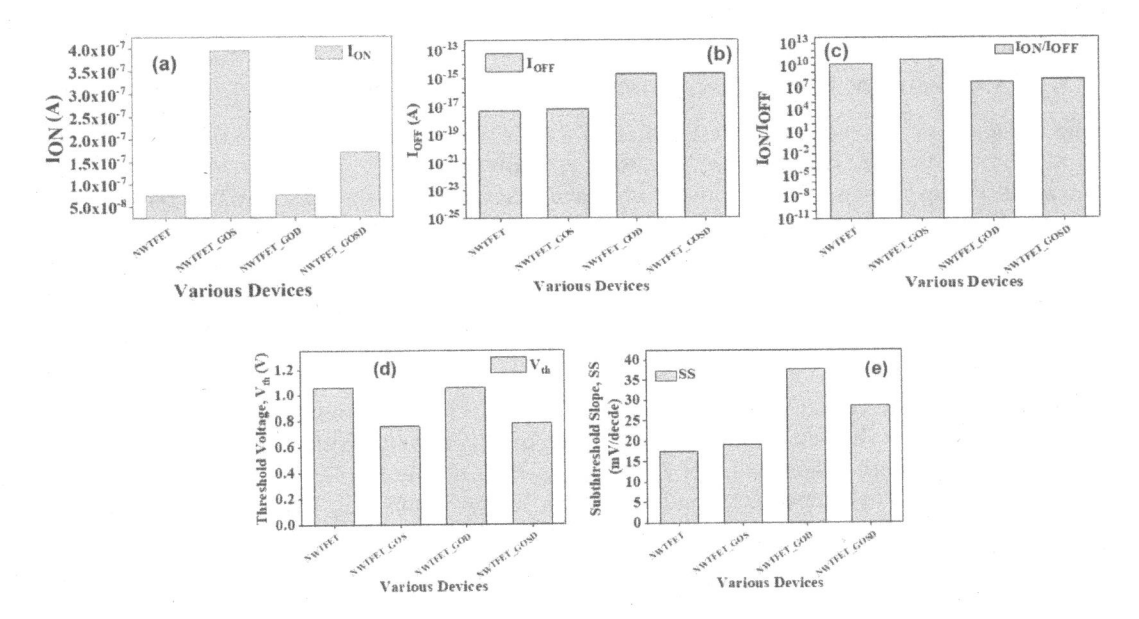

FIGURE 9.3 Comparison of electrical parameters: (a) I_{ON}, (b) I_{OFF}, (c) I_{ON}/I_{OFF}, (d) V_{th}, and (e) subthreshold slope at $V_{gs} = 1.8$ V and $V_{ds} = 1$ V.

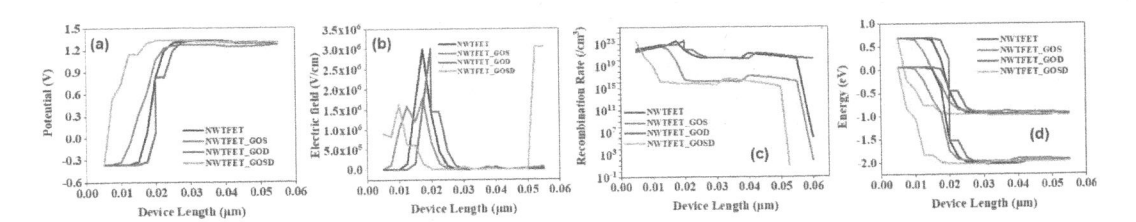

FIGURE 9.4 Plot of (a) center potential, (b) electric field, (c) recombination rate, and (d) energy band diagram along the device length (z-axis) for the NWTFET, NWTFET-GOS, NWTFET-GOD, and NWTFET-GOSD at $V_{gs} = 1.8$ V and $V_{ds} = 1$ V.

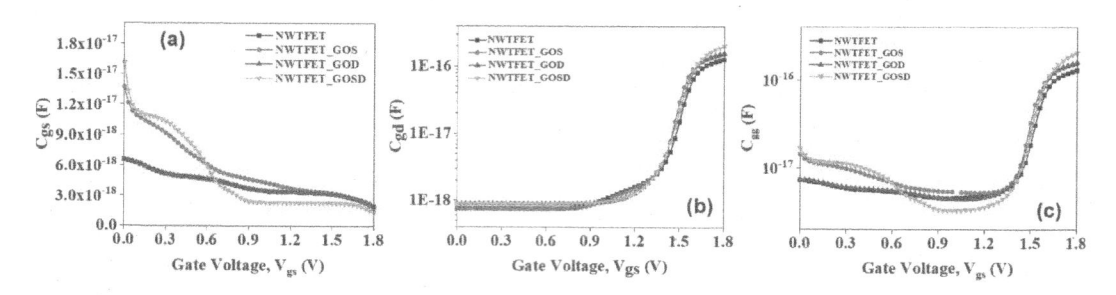

FIGURE 9.5 Plot of (a) C_{gs}, (b) C_{gd}, and (c) overall gate capacitance (C_{gg}) with the gate voltage for NWTFET, NWTFET-GOS, NWTFET-GOD, and NWTFET-GOSD at $V_{gs} = 1.8$ V and $V_{ds} = 1$ V.

of nonlinearity that exists in the device. The higher-order derivatives of transconductance must be minimized. Transconductance is compared for all the devices, and the highest transconductance found is that for NWTFET-GOS.

9.3.1 THE IMPACT OF DESIGN CONSTRAINTS ON ELECTRICAL PERFORMANCES

The influence of the extended gate length and a considerable magnitude of mole fractions "x" in $Si_{1-x}Ge_x$ is observed on electrical factors, e.g., drain current (I_d), threshold voltage, and sub-threshold swing. The graphs of transfer characteristics for the proposed NWTFET-GOS with extended gate lengths of 2, 4, 6, 8, and 10 nm at $V_{gs} = 1.8$ V and $V_{ds} = 1$ V are plotted in Figure 9.7. The plot shows that ON current increases with the increase in extended length gate electrodes in NWTFET-GOS. Simultaneously, the device's leakage also increases; however, I_{ON}/I_{OFF} improves with the extended gate length. The subthreshold swing, which is responsible for faster switching, increases with increasing extended gate length. Still, the smaller subthreshold swing is observed for the extended lengths of the gate electrode of 4 nm, and its magnitude is calculated as 15.25 mV/dec. The electrical performances of NWTFET-GOS for extended gate lengths of 2, 4, 6, 8, and 10 nm are shown in Table 9.3. The value of V_{th} improves with an increasing extended gate on source length. The graphs of I_{ON}, I_{OFF}, I_{ON}/I_{OFF}, the threshold voltage (V_{th}), and SS of NWTFET-GOS for extended gate lengths of 2, 4, 6, 8, and 10 nm at $V_{gs} = 1.8$ V and $V_{ds} = 1$ V are represented in Figure 9.8(a–e), respectively.

Figure 9.9(a–e) shows the center potential, electric field, recombination rate, EBD, and carrier concentration, respectively, along the device length (z-axis) of NWTFET-GOS for extended gate lengths of 2, 4, 6, 8, and 10 nm at $V_{gs} = 1.8$ V and $V_{ds} = 1$ V. A considerable variation of center potential is observed toward the source side, whereas it changes slightly in other regions. On the source side, the electric field changes due to the center potential variations because the electric field is negative to the potential gradient. Additionally, in this region, the recombination rate is found to

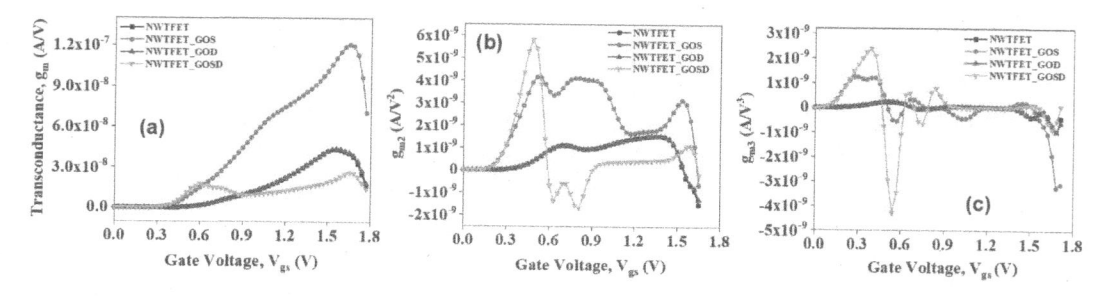

FIGURE 9.6 Plot of (a) transconductance, g_m; (b) g_{m2}; and (c) g_{m3} with the V_{gs} for NWTFET, NWTFET-GOS, NWTFET-GOD, and NWTFET-GOSD at $V_{gs} = 1.8$ V and $V_{ds} = 1$ V.

TABLE 9.3

Electrical Performance of NWTFET-GOS for Extended Lengths of the Gate Electrode of 2, 4, 6, 8, and 10 nm

Parameters	2 nm	4 nm	6 nm	8 nm	10 nm
I_{ON} (A)	2.11×10^{-7}	3.27×10^{-7}	4.53×10^{-7}	5.41×10^{-7}	6.13×10^{-7}
I_{OFF} (A)	5.86×10^{-18}	6.59×10^{-18}	7.39×10^{-18}	8.52×10^{-18}	1.00×10^{-17}
I_{ON}/I_{OFF}	3.60×10^{10}	4.97×10^{10}	6.13×10^{10}	6.35×10^{10}	6.08×10^{10}
V_{th} (V)	0.87	0.79	0.73	0.70	0.67
SS (mV/dec)	17.90	15.25	21.25	21.84	21.59

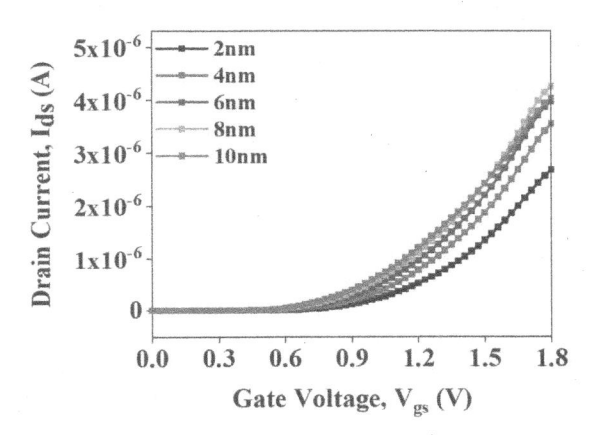

FIGURE 9.7 I_d-V_{gs} graph for the proposed NWTFET-GOS for extended gate lengths of 2, 4, 6, 8, and 10 nm at $V_{gs} = 1.8$ V and $V_{ds} = 1$ V.

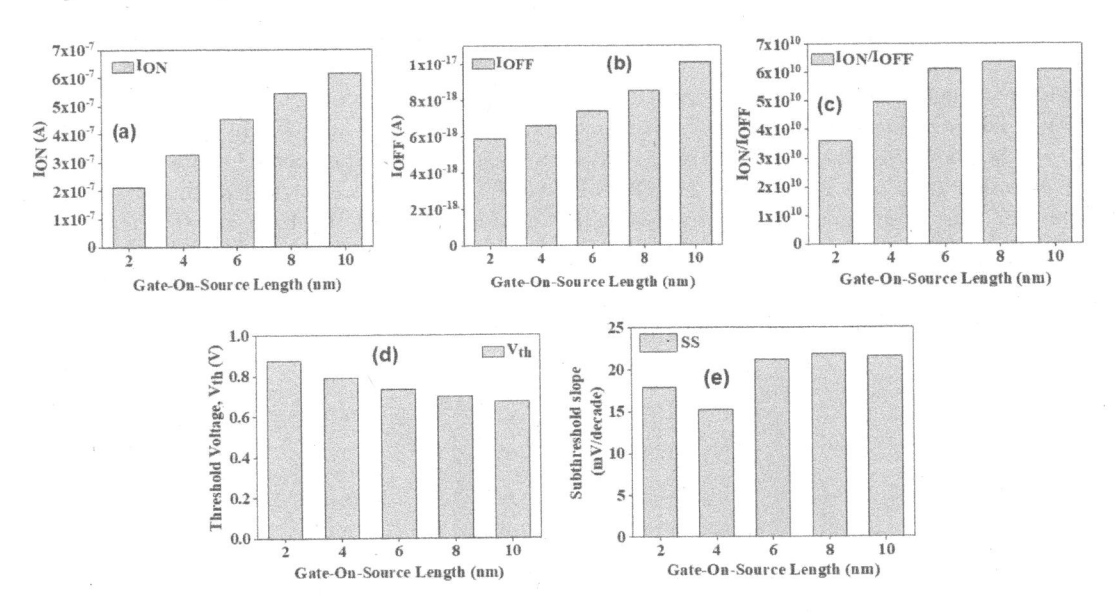

FIGURE 9.8 Comparison of device electrical factors: (a) I_{ON}, (b) I_{OFF}, (c) I_{ON}/I_{OFF}, (d) threshold voltage, and (e) subthreshold slope of NWTFET-GOS for extended gate lengths of 2, 4, 6, 8, and 10 nm at $V_{gs} = 1.8$ V and $V_{ds} = 1$V.

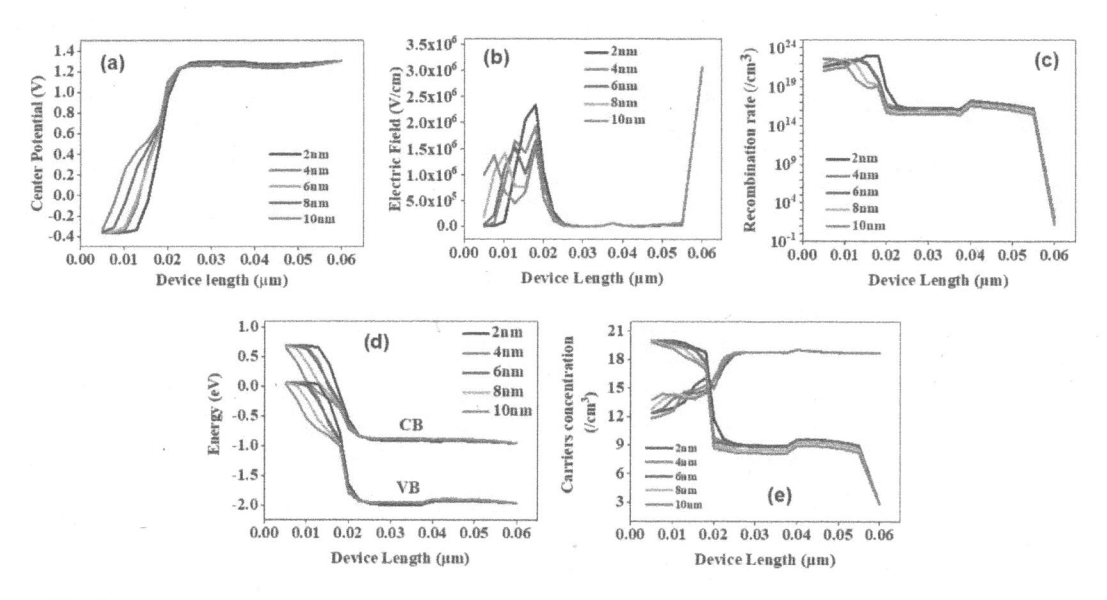

FIGURE 9.9 Graph of (a) center potential, (b) electric field, (c) recombination rate, (d) energy band diagram, and (e) carrier concentration along the device length (z-axis) of NWTFET-GOS for extended lengths of gate of 2, 4, 6, 8, and 10 nm at V_{gs} = 1.8 V and V_{ds} = 1 V.

be higher. The EBD signifies the maximum changes toward the source region with device length. The graph of carrier concentration with respect to the device length is shaped like scissors. With an increase in the extended gate's length on the source, a decrease in energy (eV) of valance and conduction band is observed at the source side. In Figure 9.10(a–c), the capacitances C_{gs}, C_{gd}, and overall gate capacitance with V_{gs} of NWTFET-GOS for extended lengths of the gate electrode of 2, 4, 6, 8, and 10 nm at V_{gs} = 1.8 V and V_{ds} = 1 V are shown, respectively. Higher C_{gs} is observed for the larger extended gate on source lengths. The C_{gs} reduces with the rise in V_{gs}. The C_{gd} and overall gate capacitance shapes are almost similar and are found to be higher for the higher extended gate on source lengths. Figure 9.11(a–c) shows the plots of transconductance (g_m), g_{m2}, and g_{m3} with V_{gs} of NWTFET-GOS for the gate on source extended lengths of 2, 4, 6, 8, and 10 nm at V_{gs} = 1.8 V and V_{ds} = 1 V. The transconductance is high for the higher extended gate on source lengths. The impact of the magnitude of x in $Si_{1-x}Ge_x$ for NWTFET-GOS on different electrical parameters is also analyzed. The magnitudes of mole fraction values used are x = 0.45, 0.5, 0.55, and 0.6. The material with a low bandgap within the bulk of the source region enhances the tunneling probability. As the mole fraction is increased, the current (I_d) is enhanced. With the rise in x value, an increase in

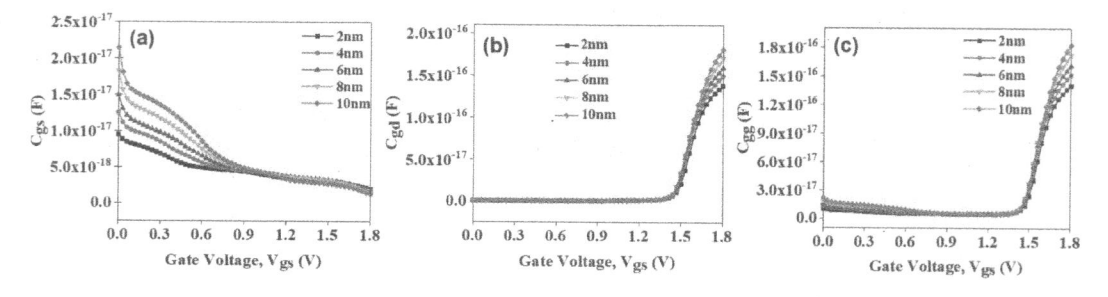

FIGURE 9.10 Graph of (a) C_{gs}, (b) C_{gd}, and (c) overall gate capacitance (C_{gg}) with V_{gs} of NWTFET-GOS for extended lengths of the gate electrode of 2, 4, 6, 8, and 10 nm at V_{gs} = 1.8 V and V_{ds} = 1 V.

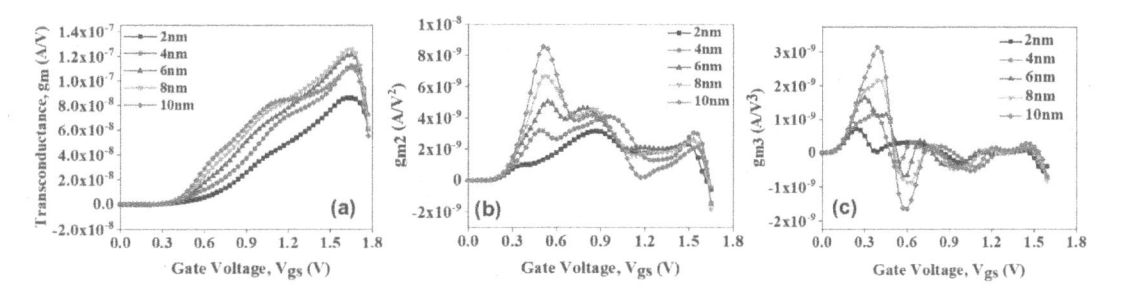

FIGURE 9.11 Graph of (a) transconductance, g_m; (b) g_{m2}; and (c) g_{m3} with V_{gs} of NWTFET-GOS for extended lengths of the gate electrode of 2, 4, 6, 8, and 10 nm at $V_{gs} = 1.8$ V and $V_{ds} = 1$ V.

leakage and the downfall of the current ratio are observed. The threshold voltage is defined as the minimum V_{gs} required for the channel formation, and this value is improved with an increase in mole fraction. The minimum subthreshold swing of 12.18 mV/dec is observed for $x = 0.5$. The effect of mole fraction on the electrical constraints is included in Table 9.4.

9.3.2 ANALYSIS OF ANALOG PERFORMANCE

Among various structures, NWTFET-GOS leads to better electrical performances. The analog performance indicators are also compared for the considered devices. The various analog parameters considered here are unity gain cutoff frequency (f_T), intrinsic device delay (IDD), transconductance frequency product (TFP), and gain-bandwidth product (GBWP). The individual influence of the extended length of the gate electrode and mole fraction (x) on the analog behavior of NWTFET-GOS is also studied. Figure 9.12(a) represents the plot of f_T with V_{gs} up to 1.8 V. A single peak of f_T is detected in all the devices except for NWTFET-GOSD, where a double peak is observed. Among all the devices, f_T is found to be highest in NWTFET-GOS. The highest value of (f_T) in NWTFET-GOS is calculated as 2.34 GHz. The graph of IDD is shown in Figure 9.12(b), which decreases with the gate voltage.

Comparison of NWTFET-GOS with the other considered devices shows that the lowest delay is achieved for the higher gate-to-source voltage. IDD for NWTFET-GOS is found to be 16.03 ps at $V_{gs} = 1.5$. TFP is demonstrated in Figure 9.12(c), and its highest value for NWTFET-GOS is 2.44×10^8 Hz/V. GBWP is plotted in Figure 9.12(d). This figure shows that NWTFET-GOS yields the highest value of GBWP. The highest value of GBWP in NWTFET-GOS is 8.91 GHz. The analog performance indicators are compared at $V_{gs} = 1.8$ V and $V_{ds} = 1$ V and are given in Table 9.5.

TABLE 9.4
Impact of Mole Fraction of SiGe in NWTFET-GOS on Electrical Performances

Parameters	X = 0.45	X = 0.5	X = 0.55	X = 0.6
I_{ON} (A)	2.14×10^{-7}	2.91×10^{-7}	3.96×10^{-7}	5.41×10^{-7}
I_{OFF} (A)	9.47×10^{-19}	2.50×10^{-18}	6.96×10^{-18}	1.91×10^{-17}
I_{ON}/I_{OFF}	2.26×10^{11}	1.16×10^{11}	5.68×10^{10}	2.83×10^{10}
V_{th} (V)	0.86	0.81	0.76	0.71
SS (mV/dec)	14.56	12.1826	19.08	16.33

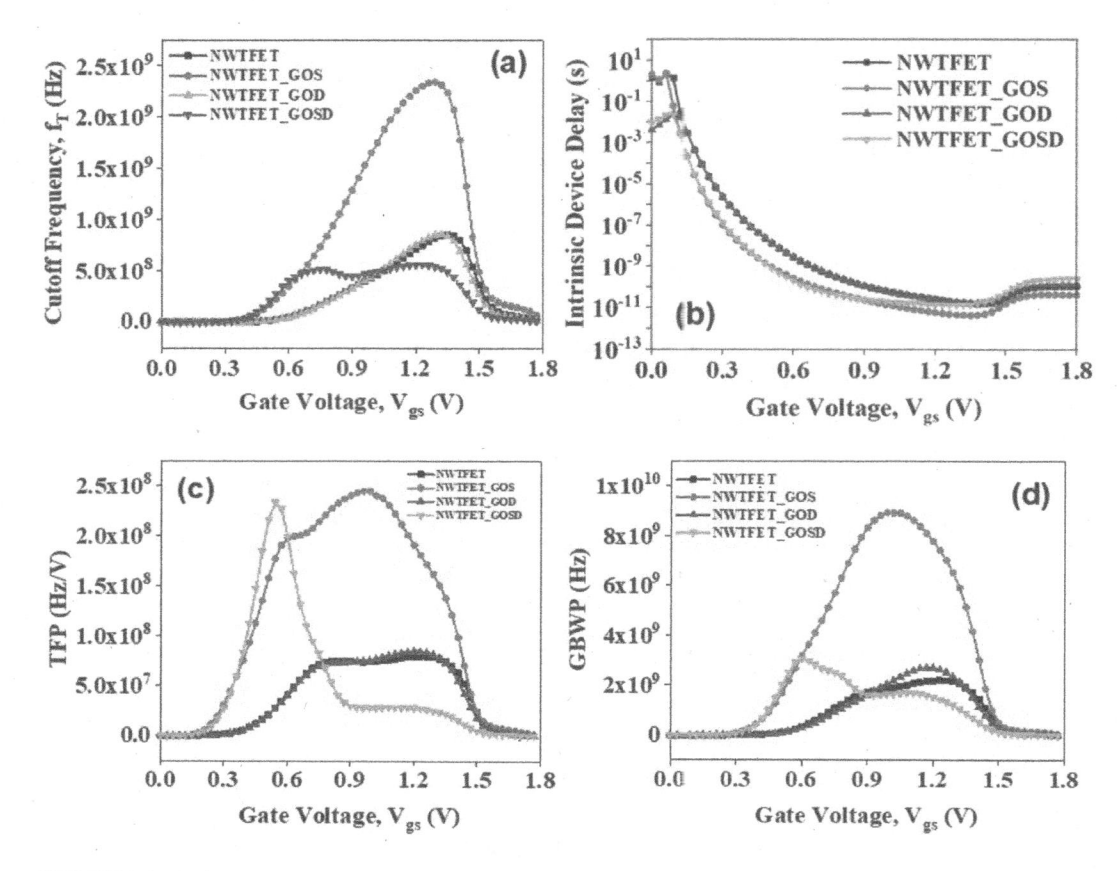

FIGURE 9.12 Comparison of the analog performance indicators: (a) Unity gain cutoff frequency, (b) intrinsic device delay, (c) transconductance frequency product, and (d) gain-bandwidth product with the gate-to-source voltage for the NWTFET, NWTFET-GOS, NWTFET-GOD, and NWTFET-GOSD at $V_{gs} = 1.8$ V and $V_{ds} = 1$ V.

9.3.2.1 The Impact of the Extended Length of the Gate Electrode on the Analog Performance of NWTFET-GOS

The analog performance of the proposed NWTFET-GOS structure is analyzed, and the effect of the extended gate length is observed for a length of 2, 4, 6, 8, and 10 nm for the proposed NWTFET-GOS. The various analog parameters, like unity gain cutoff frequency (f_T), IDD, TFP, and GBWP for various extended gate lengths are presented in Figure 9.13(a–d), respectively, at $V_{gs} = 1.8$ V and

TABLE 9.5
Analysis of the Analog Performance Indicators of NWTFET, NWTFET-GOS, NWTFET-GOD, and NWTFET-GOSD

Different Devices	f_T (max) (Hz)	IDD (ps) at $V_{gs} = 1.5$V	TFP (max) (Hz/V)	GBWP (max) (Hz)
NWTFET	8.57×10^8	29.81	7.93×10^7	2.21×10^9
NWTFET-GOS	2.34×10^9	16.03	2.44×10^8	8.91×10^9
NWTFET-GOD	8.56×10^8	40.31	8.41×10^7	2.71×10^9
NWTFET-GOSD	5.59×10^8	52.92	2.33×10^8	3.06×10^9

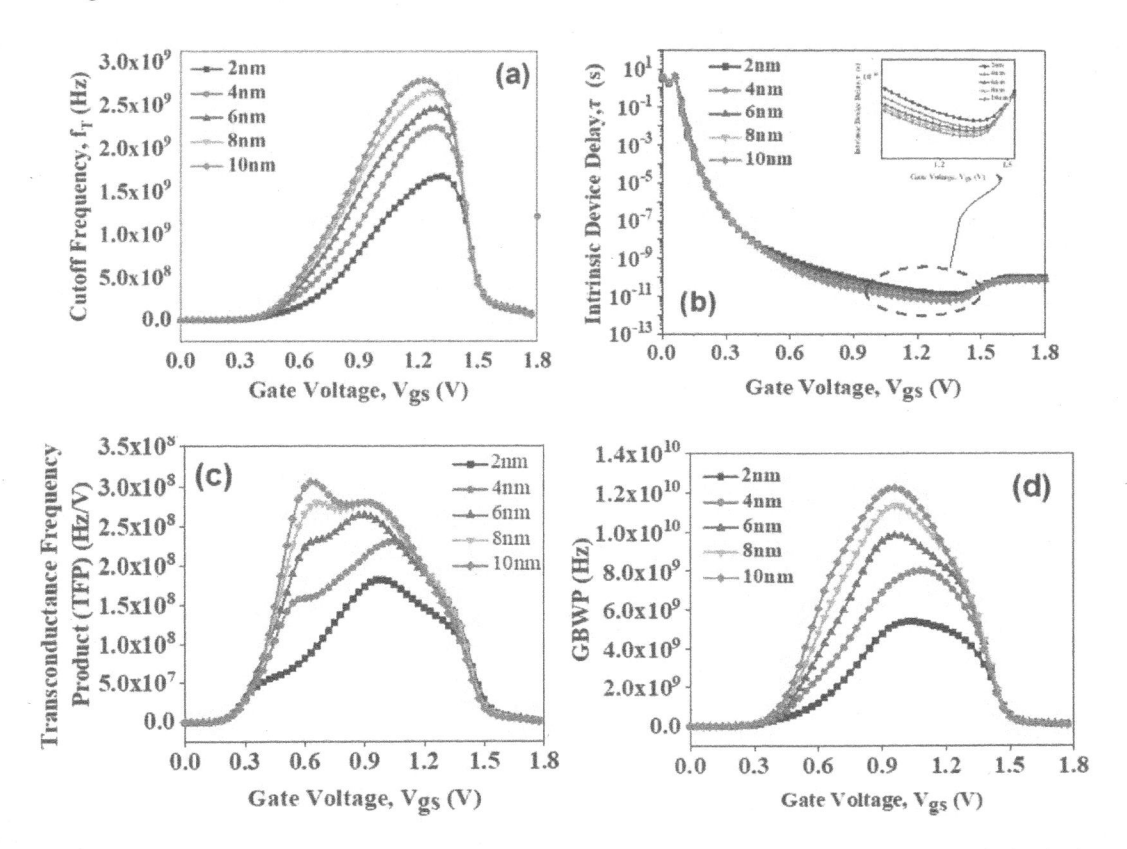

FIGURE 9.13 Comparison of analog performance indicators: (a) unity gain cutoff frequency, (b) intrinsic device delay (c) transconductance frequency product, and (d) gain-bandwidth product with V_{gs} of NWTFET-GOS for extended lengths of the gate electrode of 2, 4, 6, 8, and 10 nm at $V_{gs} = 1.8$ V and $V_{ds} = 1$ V.

$V_{ds} = 1$ V. The unity gain cutoff frequency increases with the increase in extended lengths of the gate electrode. The highest value of f_T for an extended length of the gate electrode of 10 nm is 2.77 GHz. The IDD for NWTFET-GOS falls as the extended gate length increases. The magnitude of IDD for the extended length gate electrode of 10 nm is the lowest with a magnitude of 27.82 ps at $V_{gs} = 1.5$ V. The analog performance indicators of NWTFET-GOS for extended gate lengths of 2, 4, 6, 8, and 10 nm are listed in Table 9.6. The TFP in NWTFET-GOS is found to be higher for the larger values of extended gate lengths. The maximum value of TFP is 3.05×10^8 Hz/V for the proposed structure for an extended length of the gate electrode of 10 nm.

From the plots, it is evident that the GBWP initially increases to attain a peak, and then decreases with the gate-to-source voltage. GBWP is improved as the extended gate length increases, and the maximum value of GBWP for an extended length of the gate electrode of 10 nm is calculated as 12.2 GHz. From the results shown from the analog behavior of NWTFET-GOS, it may be concluded that with the extended gate toward the source region, the analog performance is improved.

9.3.2.2 The Effect of Mole Fractions of SiGe on the Analog Performance of NWTFET-GOS

The effect of mole fractions (x) on the analog performance indicators is also studied. Silicon is the most considered bandgap material. The tunneling rate is lower; however, choosing the material of a lower bandgap in NWTFET-GOS improves the tunneling in the device, enhancing the drive current. The unity gain cutoff frequency is represented mathematically by $f_T = g_m/2\pi C_{gg}$. $C_{gg} = C_{gs} + C_{gd}$

TABLE 9.6

Analog Performance Indicators of NWTFET-GOS for Extended Gate Lengths of 2, 4, 6, 8, and 10 nm

Extended Gate Lengths (nm)	f_T (max) (Hz)	IDD (ps) at $V_{gs} = 1.5$ V	TFP (max) (Hz/V)	GBWP (max) (Hz)
2	1.67×10^9	32.88	1.82×10^8	5.41×10^9
4	2.23×10^9	29.12	2.30×10^8	8.01×10^9
6	2.45×10^9	28.78	2.64×10^8	9.83×10^9
8	2.64×10^9	28.19	2.81×10^8	1.13×10^{10}
10	2.77×10^9	27.82	3.05×10^8	1.22×10^{10}

gives the overall gate capacitance. The analog performance indicators, like unity gain cutoff frequency, IDD, TFP, and GBWP with the gate-to-source voltage of NWTFET-GOS for different mole fraction values of 0.45, 0.5, 0.55, and 0.6 in SiGe material in the bulk of source region at $V_{gs} = 1.8$ V and $V_{ds} = 1$ V are portrayed in Figure 9.14. The parameter f_T first increases due to the rise in transconductance (g_m), attains the maximum value, and decreases due to the increase in C_{gg}. The cutoff frequency increases with an increase in the magnitude of x in $Si_{1-x}Ge_x$. The most significant value of f_T observed for mole fraction of $x = 0.6$ is 3.06 GHz. The IDD is defined as IDD $= C_{gg}V_{ds}/I_d$.

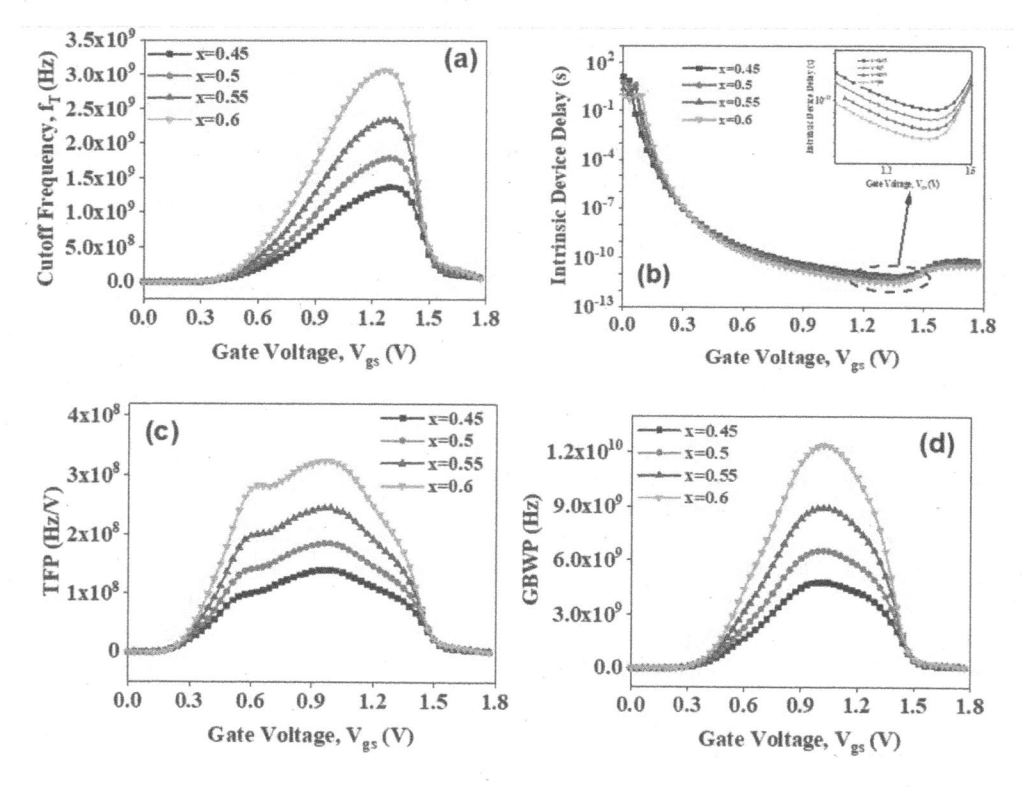

FIGURE 9.14 Comparison of the analog performance indicators: (a) unity gain cutoff frequency, (b) intrinsic device delay, (c) transconductance frequency product, and (d) gain-bandwidth product with the gate-to-source voltage of NWTFET-GOS for multiple mole fraction values of 0.45, 0.5, 0.55, and 0.6 in SiGe material in the source at $V_{gs} = 1.8$ V and $V_{ds} = 1$ V.

TABLE 9.7

Analysis of Analog Behavior of the NWTFET-GOS for the Mole Fractions of $x = 0.45$, 0.5, 0.55, and 0.6

Mole Fraction of SiGe in Source Side	f_T (max) (Hz)	IDD (ps) at $V_{gs} = 1.5$ V	TFP (max) (Hz/V)	GBWP (max) (Hz)
$x = 0.45$	1.36×10^9	20.51	1.39×10^8	4.77×10^9
$x = 0.5$	1.79×10^9	17.77	1.85×10^8	6.52×10^9
$x = 0.55$	2.34×10^9	16.03	2.44×10^8	8.91×10^9
$x = 0.6$	3.06×10^9	15.16	3.23×10^8	1.23×10^{10}

IDD decreases with V_{gs} and can be further decreased if the value of mole fraction value x in $Si_{1-x}Ge_x$ is increased. The magnitude of IDD for $x = 0.6$ is the lowest at 15.16 ps. The expression of TFP is TFP $= (g_m/I_d) \times f_T$. TFP initially increases then decreases. The peak value of TFP is increased as the mole fraction value x in $Si_{1-x}Ge_x$ is increased, and the peak value for $x = 0.6$ is calculated as 3.23×10^8 Hz/V. The analog performance indicators of NWTFET-GOS for considerable magnitude of mole fraction values $x = 0.45$, 0.5, 0.55, and 0.6 are listed in Table 9.7. The GBWP is represented mathematically by GBWP $= g_m/2\pi C_{gd}$. GBWP increases with the increase in mole fractions of SiGe. The peak value of GBWP for a mole fraction of $x = 0.6$ is found to be 12.3 GHz. From the observation and analysis of analog performance indicators of NWTFET-GOS, it may be inferred that with an increase in mole fractions of SiGe, a substantial improvement in the analog performance may be achieved.

9.4 CONCLUSION

This chapter proposes the NWTFET-GOS structure. The structure is simulated, and the device parameters are compared with those of NWTFET, NWTFET-GOD, and NWTFET-GOSD. The highest value of I_{ON}, higher I_{ON}/I_{OFF}, and the lowest threshold voltage achieved for the proposed NWTFET-GOS confirm its acceptability for the analog applications. The proposed device also possesses the lower SS (SS < 60 mV/dec) of 19.08 mV/dec, helping in the faster transition from the OFF state to the ON state and vice versa. The impact of mole fraction and an extended gate length of SiGe are investigated for analog performance in NWTFET-GOS devices. All the analog performance parameters considered in this chapter are improved with the increase in mole fractions and the increase in the extended gate on source lengths. Hence, the NWTFET-GOS becomes a competitive candidate among all the structures considered in this chapter.

REFERENCES

1. Lu, Wei, Xie, Ping, Lieber, Charles M. 2008. Nanowire Transistor Performance Limits and Applications. IEEE Transactions on Electron Devices. 55: 2859–2876.
2. Zhu, Hao. 2017. Semiconductor Nanowire MOSFETs and Applications, ed. Maaz Khan. New Insights. Intech Open. DOI: 10.5772/67446
3. Kwong, D.-L., Li, X., Sun, Y. et al. 2011. Vertical Silicon Nanowire Platform for Low Power Electronics and Clean Energy Applications. Journal of Nanotechnology. 2012: 1–21.
4. Lind, Erik, Memisevic, Elveden, Dey, Anil W, Wernersson, Lars-Erik. 2015. III-V Heterostructure Nanowire Tunnel FETs. IEEE Journal of the Electron Devices Society. 3: 96–102.
5. Sharma, A. K., Zaidi, S. H., Lucero, S., Brueck, S. R. J., Islam, N. E. 2004. Mobility and Transverse Electric Field Effects in Channel Conduction of Wrap-Around-Gate Nanowire MOSFETs. IEE Proceedings – Circuits, Devices and Systems. 151: 422–430.
6. Singh, Sarabdeep, Raman, Ashish. 2018. Gate-All-Around Charge Plasma-Based Dual Material Gate-Stack Nanowire FET for Enhanced Analog Performance. IEEE Transactions on Electron Devices. 65: 3026–3032.

7. Ravindran, Ajith, George, Abraham, Praveen C. S., Kuruvilla, Nisha. 2017. Gate All Around Nanowire TFET with High ON/OFF Current Ratio. Materials Today: Proceedings. 4: 10637–10642.

8. Saurabh, Sneh, Kumar, M. Jagadesh. 2011. Investigation of the Novel Attributes of a Dual Material Gate Nanoscale Tunnel Field Effect Transistor. IEEE Transactions on Electron Devices. 58: 404–410.

9. Chen, Z. X., Yu, H. Y., Singh, N. et al. 2009. Demonstration of Tunnelling FETs Based on Highly Scalable Vertical Silicon Nanowires. IEEE Electron Device Letters. 30: 754–756.

10. Singh, P. K., Baral, K., Kumar, S. et al. 2020. Source Pocket Engineered Underlap Stacked-Oxide Cylindrical Gate Tunnel FETs with Improved Performance: Design and Analysis. Applied Physics A. 126: 166. https://doi.org/10.1007/s00339-020-3336-8

11. Verhulst, A. S., Vandenberghe, W. G., Maex, K. et al. 2008. Complementary Silicon-Based Heterostructure Tunnel-FETs with High Tunnel Rates. IEEE Electron Device Letters. 29: 1398–1401.

12. Jhaveri, R., Nagavarapu, V., Woo, J. C. S. 2011 Effect of Pocket Doping and Annealing Schemes on the Source-Pocket Tunnel Field-Effect Transistor. IEEE Transactions on Electron Devices 58(1): 80–86.

13. Kumar, Naveen, Mushtaq, Umar, Amin, S., Anand, Sunny. 2018. Design and Performance Analysis of Dual-Gate All Around Core-Shell Nanotube TFET. Superlattices and Microstructures. 125: 356–364.

14. Anand, S., Amin, S. I., Sarin, R. K. 2016. Analog Performance Investigation of Dual Electrode-Based Doping-Less Tunnel FET. Journal of Computational Electronics. 15: 94–103.

15. Biswal, Sudhansu Mohan, Baral, Biswajit, De, Debashis, Sarkar, Angsuman. 2016. Study of the Effect of Gate-Length Downscaling on the Analog/RF Performance and Linearity Investigation of InAs-Based Nanowire Tunnel FET. Superlattices and Microstructures. 91: 319–330.

16. Gautam, Rajni, Saxena, Manoj, Gupta, R. S., Gupta, Mridula. 2012. Effect of Localized Charges on Nanoscale Cylindrical Surrounding Gate MOSFET: Analog Performance and Linearity Analysis. Microelectronics Reliability. 52: 989–994.

17. Abhinav, Rai Sanjeev. 2017. Reliability Analysis of Junction-Less Double Gate (JLDG) MOSFET for Analog/RF Circuits for High Linearity Applications. Microelectronics Journal. 64: 60–68.

18. Djeffal, F., Ferhati, H., Bentrcia, T. 2016. Improved Analog and RF Performances of Gate-All-Around Junctionless MOSFET with Drain and Source Extensions. Superlattices and Microstructures. 90: 132–140.

19. Narang, Rakhi, Saxena, Manoj, Gupta, R. S., Gupta, Mridula. 2011. Linearity and Analog Performance Analysis of Double Gate Tunnel FET: Effect of Temperature and Gate Stack. International Journal of VLSI Design & Communication Systems (VLSICS). 197: 466–475.

20. Mohapatra, S. K., Pradhan, K. P., Sahu, P. K. 2014. Linearity and Analog Performance Analysis in GSDG-MOSFET with Gate and Channel Engineering. Annual IEEE India Conference. DOI: 10.1109/INDICON.2014.7030435

21. Rawat, Akash Singh, Gupta, Santosh Kumar. 2017. Potential Modelling and Performance Analysis of Junction-Less Quadruple Gate MOSFETs for Analog and RF Applications. Microelectronics Journal. 66: 89–102.

22. Patel, Sapna, Kumar, Dushyant, Chaurasiya, Nitesh Kumar, Tripathi, Shweta. 2019. Analytical Modelling of Surface Potential and Drain Current of Hetero-Dielectric DG TFET and Its Analog and Radio-Frequency Performance Evaluation. Physics Of Semiconductor Devices. 53: 1797–1803.

23. Lim, Tao Chuan, Bernard, Emilie, Rozeau, Olivier et al. 2009. Analog/RF Performance of Multichannel SOI MOSFET. IEEE Transactions on Electron Devices. 56: 1473–1482.

24. Sharma, Rupendra Kumar, Bucher, Matthias. 2012. Device Design Engineering for Optimum Analog/RF Performance of Nanoscale DG MOSFETs. IEEE Transactions on Nanotechnology. 11: 992–998.

25. Tiwari, Pramod Kumar, Kumar, Mukesh, Sakru Naik, Ramavathu, Saramekala, Gopi Krishna. 2016. Analog and Radio-Frequency Performance Analysis of Silicon-Nanotube MOSFETs. Journal of Semiconductors. 37(6): 064003.

26. Anand, S., Sarin, R. K. 2016. Analog and RF Performance of Doping-Less Tunnel FETs with Si0.55Ge0.45 Source. Journal of Computational Electronics. 15: 850–856.

27. Singh, A., Amin, S. I. Anand, S. 2020. Label-Free Detection of Biomolecules Using SiGe Sourced Dual Electrode Doping-Less Dielectrically Modulated Tunnel FET. Silicon. 12: 2301–2308.

28. Kumar, Apoorva, Amin, Naveen, Intekhab, S, Anand, Sunny. 2020. Design and Performance Optimization of Novel Core-Shell Dopingless GAA-Nanotube TFET with $Si_{0.5}Ge_{0.5}$-Based Source. IEEE Transactions on Electron Devices. 67: 789–795.

29. Kumar, Naveen, Raman, Ashish. 2019. Design and Investigation of Charge-Plasma-Based Work Function Engineered Dual-Metal-Heterogeneous Gate Si-Si0.55Ge0.45 GAA-Cylindrical NWTFET for Ambipolar Analysis. IEEE Transactions on Electron Devices. 66: 1468–1474.

30. Patel, Jyoti, Sharma, Dheeraj, Yadav, Shivendra, Lemtur, Alemienla, Suman, Priyanka. 2019. Performance Improvement of Nanowire TFET by Hetero-Dielectric and Hetero-Material: At Device and Circuit Level. Microelectronics Journal. 85: 72–82.
31. Rahimian, Morteza, Fathipour, Morteza. 2017. Improvement of Electrical Performance in Junctionless Nanowire TFET Using Hetero-Gate-Dielectric. Materials Science in Semiconductor Processing. 63: 142–152.
32. Singh, Sankalp Kumar. Kakkerla, Ramesh Kumar. Joseph, H. Bijo. et al. 2019. Optimization of InAs/GaSb Core-Shell Nanowire Structure for Improved TFET Performance. Materials Science in Semiconductor Processing. 101: 247–252.

10 Solving Schrodinger's Equation for Low-Dimensional Nanostructures for Understanding Quantum Confinement Effects

Amit Kumar

CONTENTS

10.1 BACKGROUND AND NEED

The advent of technologies and techniques made it possible to derive the materials into low-dimensional structures so that their classical bulk properties can be tailored to desired levels. The material structures can now be fabricated in quasi-two-dimensions as a *quantum well*, in a quasi-one-dimension as *quantum wires,* and even in a quasi-zero-dimension as *quantum dots.* Today, a great deal of research is concentrated on the materials in low dimensions as these offer tailor-driven property management for different fields like electrocatalytic applications [1], quantum information technology [2], device modeling for modern very large-scale integration (VLSI) applications [3], etc. The special focus is directed to the incorporation of low-dimensional structures into the next level of device fabrication techniques. This needs a proper understanding of quantum confinement effects and prior theoretical modeling [4]. The traditional ways to shrink size of the devices in an integrated circuit by reducing channel lengths of the metal oxide semiconductor field-effect transistors (MOSFETs; scaling theory) are also finding difficulties as we reached the classical physics limitations; thus, it becomes important to study quantum

DOI: 10.1201/9781003126393-10

confinement effects in small-scaled crystals. Now, device modeling at such a low dimensions requires highly dedicated computing machines to accomplish the task. These small isolated crystallites, called nanocrystallites, exhibit substantially different properties compared with bulk materials even with the same local structure and with the same composition. When the semiconductor particles are reduced in scale to nanometer dimensions, their linear and nonlinear optical, and electro-optical properties start deviating from those of bulk materials with the same composition but with small dimensions. Such deviations are expected because quasi-particles, like electrons, excitons, phonons, etc., become confined in small crystalline boundaries. The study of quantum confinement effects in nano-sized semiconductor systems has attracted considerable attention, especially in the domains of electronic and optical properties. This is attributed to the controllable flexibility for the structural designs of electronic and optical devices. When the size of such crystalline materials becomes less than the mean free path of the quasi-particle, the quantization of the quasi-particle wave vector becomes important. The quantization of electronic states also has profound effects on the behavior of quasi-particles. From a scientific viewpoint, a quantum dot can be considered as a laboratory to test the approximations to the quantum theories. Some quantum confinement effects observed at room temperatures are the emergence of sharp exciton peaks, blue shifts in the absorption spectra of quantum dots, etc. The thermodynamics of quantum dots is also important; as the size of quantum dots decreases, there is a large depression in the melting point of quantum dots. This phenomenon is attributed to the large surface to volume ratio that quantum dots have. In two dimensions, the ground-state binding energy of the Wannier exciton is four times more than that in the three-dimensional material. Therefore, the exciton effects are more easily observed in the quasi-two-dimensional material systems, even at room temperature. The exciton resonance becomes modified by the application of external fields or by the optical excitations; therefore, quantum well structures show room-temperature optical nonlinearity. These encouraging results have provided a new direction for the investigations of optical, electrical, and vibrational properties of systems with lower dimensionalities, so that the material may be tailored to obtain the desired optical, electro-optical, and nonlinear applications. The recent experimental investigations of these mesoscopic systems are motivated by the device application capabilities of such structures in communications, computing, and optics.

The initial quantum dots were made hundreds of years ago by people who created colored glasses by melting a certain amount of semiconducting materials, such as ZnSe or ZnS, with the usual glass materials. Such a diluted system of semiconductor nano-crystallites, in the glass matrix, absorbs light at the characteristic wavelengths resulting in the coloring of glass. The size of these earlier nanocrystallites was not well controlled, which lead to greater size distribution and also distributed compositions. During the 1970s, more controlled methods were attempted to create quantum dots inside another matrix. Such attempts were intensified in the 1980s and carried forward into the early 1990s. Many investigations were originally done on quantum dots that were embedded in glass matrices or suspended in colloidal solutions. Such crystallites were fabricated with very small radii (1–100 nm). Even today, the growth of semiconductor quantum dots inside a glass matrix is the most generally used technique followed by the synthesis of semiconductor crystallites in liquid solvent. Direct bandgap semiconductors were used in the preparation of quantum dots until early 2000. Mostly, III–IV and II–IV compounds were used, but CuCl and CuBr were also studied. The quantum-confined, group II–VI nanocrystallites, embedded in the glass matrix have great potential for optical device applications due to their nonlinear optical properties and ultrafast response time [5]. A slow response is associated with high nonlinearity and a compromise has to be reached between the two. The majority of quantum dot systems are based on direct bandgap semiconductors, but indirect bandgap quantum dots have also been studied. It has been observed that these quantum dots have intensity responses many orders of magnitude better than indirect bandgap bulk materials. The carbon quantum dots are also reported to be the fascinating class of carbon nanoparticles with a size around 10 nm. Their unique properties like sustainable raw material, low toxicity, chemical inertness, biocompatibility, tunable photoluminescence, cost-effectiveness, eco-friendly behaviors, etc., have attracted many research groups worldwide [6]. Most of the research activities in the low-dimensional semiconductor area have been carried out on the confinement studies using photoluminescence, X-ray scattering, Raman scattering, absorption spectroscopy, etc. These experiments are usually done on commercially available colored glass filters and semiconductor doped glasses prepared in the laboratory using melt quenching techniques. Such quantum dots act as good material systems in which optical properties can be tailored by changing particle size and composition. Sometimes semiconductors are coated with Ag and Au to obtain enhanced nonlinear effects. The basic technological goal is to control the optoelectronic and vibrational properties to tailor the material for optical, electro-optical, and nonlinear optical applications.

The low-dimensional structures that are fabricated under size restrictions to a few nanometers are finding applications in modern technological advances. The excitations in these structures undergo a quantum confinement effect due to finite restrictions in motion along the confinement axis and infinite motion in other directions. The number of confinement directions decides whether the material structure will act as quantum well, wire, or dot. The zero-dimensional systems, however, are named as quantum *cubes, parallelepipeds, cylinders, spheres (dot),* etc. It is useful to calculate energy levels in such structures with infinite barrier height to understand their use in different applications. In this chapter, we will derive the energy expressions for quasi-particles under the assumptions of the infinite energy barrier for such quantum structures.

10.2 SOLVING SCHRODINGER'S WAVE EQUATION

10.2.1 THREE-DIMENSIONAL CRYSTAL (BULK CRYSTAL)

A solid consists of a large number of atoms packed in close proximity to one another. The interatomic distances in the solids are so small that each atom is exposed to the sufficiently strong electromagnetic field of neighboring atoms. Therefore, atoms in a crystal cannot be considered as independent of each other and such interactions among the atoms influence the valence electron behavior. The potential energy of valence electrons in solids is due to mutual Coulomb's interactions of these valence electrons with positive ions located at sites a, $2a$, $3a$, ... Na, where N is a large number. This situation can be approximated by considering electrons in the crystal moving in a region of periodically varying potentials as shown in Figure 10.1.

This periodic potential is defined by $V(x+a) = V(x)$. The wavefunctions that satisfy Schrodinger's equation with this potential are of the form $\psi(x) = e^{jkx}u(x)$, where $u(x)$ has periodicity of lattice

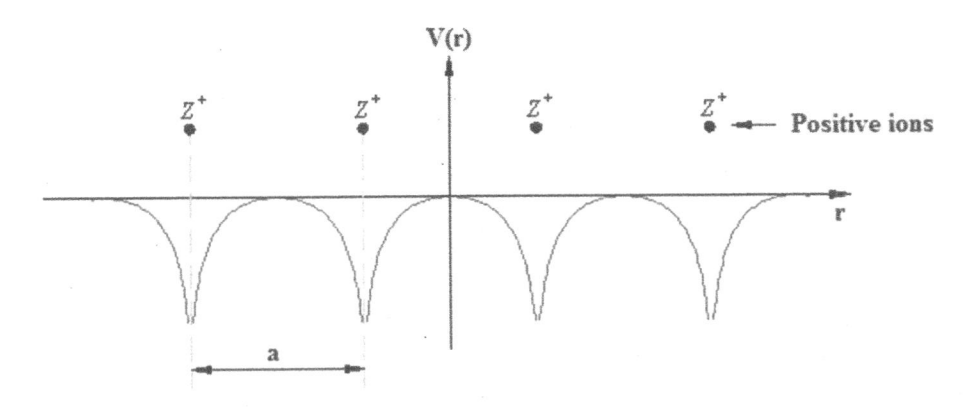

FIGURE 10.1 Electron energy under a periodic array of positive ions.

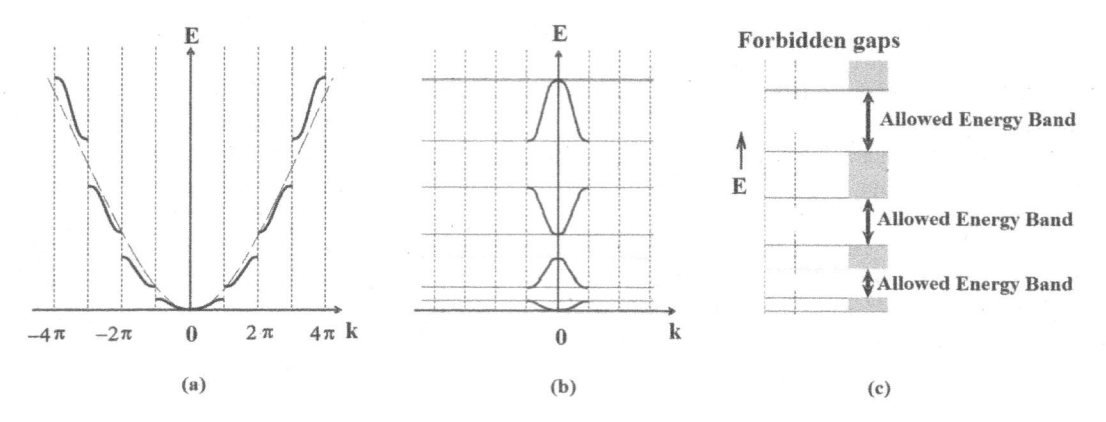

FIGURE 10.2 Dispersion law of a particle in a one-dimensional periodic potential. (a) Corresponds to the free particle, $E = \hbar^2 k^2/2m$ function (dashed curve) and solids (dark curve), (b) reduced dispersion curve, and (c) energy bands for solids. ($\hbar$ is reduced plank constant.)

$u(x + a) = u(x)$[Bloch's theorem]. The simplified model (the Kronig-Penney model) showed that the energies of the electrons under the effect of such periodic potential in the crystal could be illustrated by the dispersion curve presented in Figure 10.2. The extended and reduced dispersion law of particles in a one-dimensional array of periodic potential and the corresponding spatial energy bands are shown in Figure 10.2.

The dashed curve in Figure 10.2(a) corresponds to the $E = \hbar^2 k^2/2m$ function that represents the kinetic energy of the free particle. The dispersion curve is at the points of discontinuities at $k = \pi n/a$, where n is an integer. The value of k corresponds to standing waves that cannot propagate due to multiple reflections at periodic boundaries. The energy spectrum, therefore, breaks into bands that are separated by forbidden gaps as the k values differing by $\pi n/a$ appear equivalent to the translation symmetry of space. The reduced dispersion curve is shown in Figure 10.2(b). It results from Figure 10.2(a) by a shift of several branches equal to $2\pi/a$.

10.2.2 Two-Dimensional Crystal (Quantum Well)

If the size of a three-dimensional crystal is reduced to nanometers in one of the three dimensions, a two-dimensional structure would result called a quantum well. In this case, the motion of the particle will be restricted in one dimension while it is free to move in the other two dimensions. Let us consider a two-dimensional structure, with potential walls of infinite depth, and assume that l is the size along the confinement direction. Consider a particle of mass μ and energy E inside this potential.

Mathematically, this potential well is defined by $V(x)$ and illustrated in Figure 10.3.

$$V(x) = \begin{array}{l} 0, \, 0 \leq x \leq l \\ \infty, \, otherwise \end{array}$$

Schrodinger's equation, within the well, can be written as

$$\nabla^2 \psi + \frac{2\mu}{\hbar^2} E \psi = 0 \qquad (10.1)$$

$$\frac{\partial^2 \psi}{\partial x^2} + \frac{\partial^2 \psi}{\partial y^2} + \frac{\partial^2 \psi}{\partial z^2} + \frac{2\mu}{\hbar^2} E \psi = 0$$

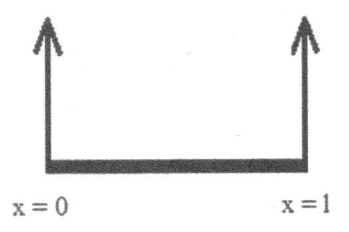

$$x = 0 \qquad\qquad x = 1$$

FIGURE 10.3 Potential energy of particle in quantum well.

Let $\psi(x,y,z) = X(x)Y(y)Z(z)$

$$Y(y)\,Z(z)\frac{d^2X(x)}{dx^2} + X(x)Z(z)\frac{d^2Y(y)}{dy^2} + X(x)\,Y(y)\frac{d^2Z(z)}{dz^2} + \frac{2\mu}{\hbar^2}E\,X(x)Y(y)\,Z(z) = 0$$

$$\frac{1}{X(x)}\frac{d^2X(x)}{dx^2} + \frac{1}{Y(y)}\frac{d^2Y(y)}{dy^2} + \frac{1}{Z(z)}\frac{d^2Z(z)}{dz^2} = -\frac{2\mu}{\hbar^2}E \tag{10.2}$$

In Equation (10.2), the first term is the function of x variable alone, the second term is the function of the y variable alone, and the third term is the function of the z variable alone; therefore, each term is required to be set equal to a constant. Hence we may write,

$$\frac{1}{X(x)}\frac{d^2X(x)}{dx^2} = -k_x^2 \tag{10.3}$$

$$\frac{1}{Y(y)}\frac{d^2Y(y)}{dy^2} = -k_y^2 \tag{10.4}$$

$$\frac{1}{Z(z)}\frac{d^2Z(z)}{dz^2} = -k_z^2 \tag{10.5}$$

So that,

$$k_x^2 + k_y^2 + k_z^2 = \frac{2\mu E}{\hbar^2} \tag{10.6}$$

The confinement is only along the x-axis, so that we have solutions for Equation (10.3) as

$$X(x) = A\,\sin(k_x x) + B\,\cos(k_x x)$$

Since $\psi(x,y,z) = X(x)Y(y)Z(z)$ has to vanish at all points on the surface $x = 0$ and $x = 1$, we must have $B = 0$ and

$$k_x = \frac{n\pi}{1},\ n = 1,\,2,\,3,\ldots$$

$$E_{x,\,n} = \frac{n^2\pi^2\hbar^2}{2\mu l^2},\ n = 1,\,2,\,3,\ldots \tag{10.7}$$

The dispersion relation can be given by

$$E(k) = \frac{\hbar^2}{2\mu} k_x^2 + \frac{\hbar^2}{2\mu} k_y^2 + \frac{\hbar^2}{2\mu} k_z^2$$

$$E(k) = E_{x,n} + \frac{\hbar^2}{2\mu}\left(k_y^2 + k_z^2\right) \tag{10.8}$$

The energy of the particle inside the quantum well is described by infinite motion in y- and z-directions and, finite motion along the confinement direction, i.e., along the x-axis. Hence, the wave function is given by

$$\psi(x,y,z) = C \sin\left(\frac{n\pi}{l}\right) e^{\pm j k_y y} e^{\pm j k_z z} \tag{10.9}$$

10.2.3 ONE-DIMENSIONAL CRYSTAL (QUANTUM WIRE)

If the size of a three-dimensional crystal is reduced to nanometers in two of the three dimensions, a one-dimensional structure would result called a quantum wire as shown in Figure 10.4. In this case, the motion of the particle will be restricted in two directions. A quantum wire may be approximated by a particle (of a certain mass) confined inside an infinitely deep cylinder of radius a. Because the potential is symmetric, the potential depends only on the magnitude of the distance from the origin, i.e., position (radial) vector, $\vec{r}$, i.e.

$$V(r) = \begin{array}{l} 0,\, r \leq a \\ \infty,\, otherwise \end{array}$$

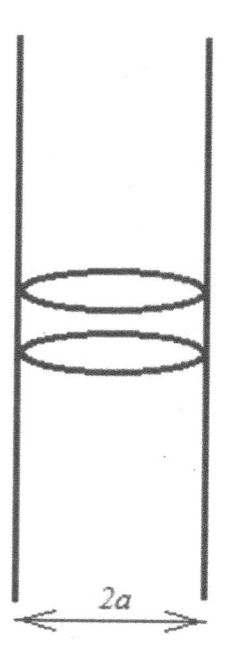

FIGURE 10.4 Potential energy of particle in quantum wire.

Schrodinger's equation in cylindrical coordinates could be written as

$$\frac{1}{r}\frac{\partial}{\partial r}\left(r\frac{\partial}{\partial r}\psi\right)+\frac{1}{r^2}\frac{\partial^2}{\partial\theta^2}\psi+\frac{\partial^2}{\partial z^2}\psi+\frac{2\mu}{\hbar^2}\left[E-V(r)\right]\psi=0 \tag{10.10}$$

$$\text{Let, } \psi\ (r,\ \theta,\ z)=R\ (r)\ \Theta(\theta)\ Z(z) \tag{10.11}$$

Substituting Equation (10.11) into Equation (10.10)

$$\frac{\Theta Z}{r}\frac{d}{dr}\left(r\frac{dR}{dr}\right)+\frac{R Z}{r^2}\frac{d^2\Theta}{d\theta^2}+R\ \Theta\frac{d^2Z}{dz^2}+\frac{2\mu}{\hbar^2}\left[E-V(r)\right]R\ \Theta\ Z=0$$

$$\frac{1}{R}\frac{d^2R}{dr^2}+\frac{1}{R r}\frac{d\ R}{dr}+\frac{1}{\Theta r^2}\frac{d^2\Theta}{d\theta^2}+\frac{1}{Z}\frac{d^2Z}{dz^2}+\frac{2\mu}{\hbar^2}\left[E-V(r)\right]=0$$

Multiplying by r^2 gives,

$$\frac{r^2}{R}\frac{d^2R}{dr^2}+\frac{r}{R}\frac{d\ R}{dr}+\frac{2\mu r^2}{\hbar^2}\left[E-V(r)\right]+\frac{r^2}{Z}\frac{d^2Z}{dz^2}=-\frac{1}{\Theta}\frac{d^2\Theta}{d\theta^2}=+m^2 \tag{10.12}$$

The right side of Equation (10.12) is the function of variables r and z, whereas the left side is a function of θ only; therefore, both sides should be equal to constant m^2, which gives two equations, the first equation is

$$\frac{d^2\Theta}{d\theta^2}+m^2\Theta=0$$

That has the solutions as

$$\Theta\ (\theta)=Ae^{\pm jm\theta}=\frac{1}{\sqrt{2\pi}}e^{\pm jm\theta}\text{ where } m=0,\ \pm1,\ \pm2,\ \pm3,\dots \tag{10.13}$$

The second equation is

$$\frac{r^2}{R}\frac{d^2R}{dr^2}+\frac{r}{R}\frac{d\ R}{dr}+\frac{2\mu r^2}{\hbar^2}\left[E-V(r)\right]+\frac{r^2}{Z}\frac{d^2Z}{dz^2}=+m^2$$

Or

$$\frac{1}{R}\frac{d^2R}{dr^2}+\frac{1}{rR}\frac{d\ R}{dr}+\frac{2\mu}{\hbar^2}\left[E-V(r)\right]+\frac{m^2}{r^2}=-\frac{1}{Z}\frac{d^2Z}{dz^2}=+k_z^2$$

Again the left two terms are equated to a constant because the first term depends on r and the second term depends on z only. The first equation may be written as

$$\frac{d^2Z}{dz^2}+k_z^2z=0$$

whose solutions are

$$Z\ (z)=Be^{\pm jk_z z} \tag{10.14}$$

The second equation is given as

$$\frac{d^2R}{dr^2} + \frac{1}{r}\frac{d\,R}{dr} + \frac{2\mu}{\hbar^2}\left[E - V(r) - \frac{m^2\hbar^2}{2\mu r^2} - \frac{k_z^2\hbar^2}{2\mu}\right]R = 0 \tag{10.15}$$

Because $V(\vec{r}) = 0$, for $r \leq a$, the equation is rewritten, and after multiplication of r^2 and putting in $E_1 = E - \frac{k_z^2\hbar^2}{2\mu}$, we get

$$r^2\frac{d^2R}{dr^2} + r\frac{d\,R}{dr} + \left[\frac{2\mu E_1}{\hbar^2}r^2 - m^2\right]R = 0$$

Substituting $t^2 = \frac{2\mu E_1}{\hbar^2}$, we get

$$r^2\frac{d^2R}{dr^2} + r\frac{d\,R}{dr} + \left[t^2r^2 - m^2\right]R = 0 \tag{10.16}$$

Substituting $s = t\,r$, we get

$$s^2\frac{d^2R}{ds^2} + s\frac{d\,R}{ds} + \left[s^2 - m^2\right]R = 0 \tag{10.17}$$

Equation (10.16) is the standard Bessel differential equation and particular solutions are called Bessel functions of the mth order. Therefore Equation (10.17) has solutions given by

$$R\,(r) = C\,J_m\,(tr)$$

Applying the boundary conditions that at $r = a$, $R\,(r) = 0$, we get $J_m\,(ta) = 0$. This implies
    $ta = \alpha_{nm} = n$th root of Bessel's Function
    Hence,

$$R(r) = C\,J_m\left(\alpha_{nm}\frac{r}{a}\right) \tag{10.18}$$

Normalization gives,

$$R(r) = \frac{\sqrt{2}}{aJ_{m+1}\left(\alpha_{nm}\right)}\,J_m\left(\alpha_{nm}\frac{r}{a}\right) \tag{10.19}$$

Since $t = \frac{\alpha_{nm}}{a}$, or $t^2 = \frac{2\mu E_1}{\hbar^2} = \left(\frac{\alpha_{nm}}{a}\right)^2$ so

$$E_{nm} = \frac{\hbar^2}{2\mu}\left(k_z^2 + \frac{\alpha_{nm}^2}{a^2}\right) \tag{10.20}$$

and the corresponding wave function can be given by

$$\psi\,(r,\,\theta,\,z) = K\,J_m\left(\alpha_{nm}\frac{r}{a}\right)e^{\pm jk_z z}\,e^{\pm jmP} \tag{10.21}$$

10.2.4   ZERO-DIMENSIONAL CRYSTAL (QUANTUM CYLINDER)

If the size of the crystal is reduced in all possible dimensions, we get a quasi-zero-dimensional crystal. If we reduce the length of the quantum wire to a finite value, we will get one of such structure.

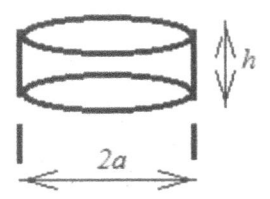

FIGURE 10.5 Potential energy of particle in quantum cylinder.

The structure of this type is called a quantum cylinder (Figure 10.5). If we restrict the height of the infinitely long cylinder to a height h, such that

$$V(z) = \begin{array}{l} 0, \, 0 \leq z \leq h \\ \infty, \, otherwise \end{array}$$

Applying these boundary conditions to Equation (10.14), we get

$$k_z = \frac{l\,\pi}{h}, \text{ where } l = 0, 1, 2\ 3, \ldots \tag{10.22}$$

and,

$$Z(z) = \sqrt{\frac{2}{h}} \sin\left(\frac{l\,\pi}{h}\,z\right) \tag{10.23}$$

Substituting these results into Equations (10.20) and (10.21), we can obtain the expression for quantized energy and wavefunctions, respectively, as

$$E_{nlm} = \frac{\hbar^2}{2\mu} \left\{ \left(\frac{l\,\pi}{h}\right)^2 + \left(\frac{\alpha_{nm}}{a}\right)^2 \right\} \tag{10.24}$$

$$\psi_{nlm}(r, \theta, z) = \left[\frac{\sqrt{2}}{a J_{m+1}(\alpha_{nm})} J_m\left(\alpha_{nm}\frac{r}{a}\right) \right] \left[\sqrt{\frac{2}{h}} \sin\left(\frac{l\,\pi}{h}\,z\right) \right] \left[\frac{1}{\sqrt{2\pi}} e^{\pm jm\theta} \right] \tag{10.25}$$

Defined by three quantum numbers, viz. n, l, and m.

10.2.5 DENSITY OF STATES

The density of states for electrons and holes in low-dimensional structures may be expressed in following general form:

$$g(E) \propto E^{\left(\frac{d}{2}-1\right)}, \text{ where } d \equiv \text{Dimensionality}, d = 1, 2, \text{ or } 3 \tag{10.26}$$

For measuring electron energy, a reference is taken from the bottom of the conduction band, and for holes it is measured from the top of the valence band. In a three-dimensional system, the density of state is the square root function of energy and for $d = 2$ and $d = 1$, a number of discrete sub-bands appears because of quantum confinement effects. The density of states in these cases obeys Equation (10.26) within every sub-band, as depicted in Figure 10.6. For $d = 0$ *i.e.* zero dimensional structure the density of states is characterized by a discrete δ – function appearing at the allowed discrete energies.

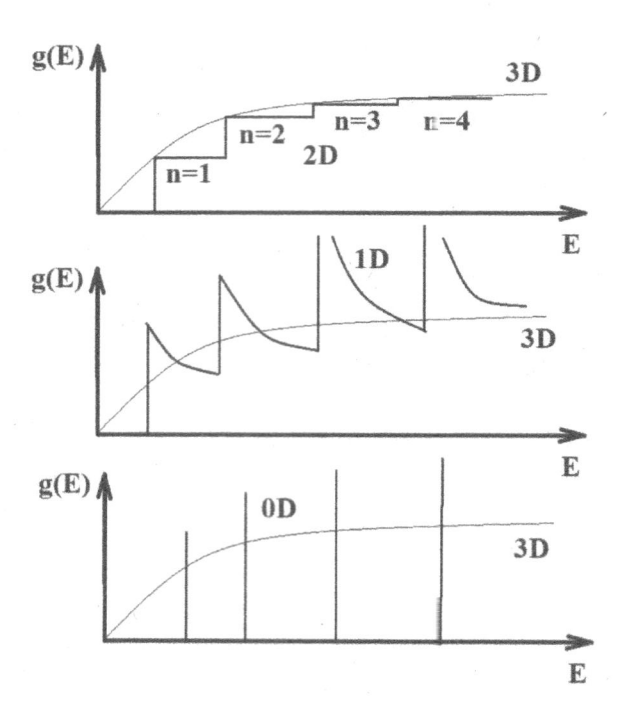

FIGURE 10.6   Density of states of electrons for various dimensionalities.

10.3   IDEAL NANOCRYSTALS

An ideal nanocrystal can be considered as an ideal crystal with a spherical shape, referred to as a quantum dot. Generally, nanocrystal and quantum dots would be used interchangeably. A nanocrystal has dimensions comparable to the lattice constant of the crystal. The potential is spherically symmetric, therefore, the potential depends only on the magnitude of the distance from a fixed point, i.e., only on the magnitude of the position vector, $\vec{r}$. Hence,

$$V(\vec{r}) = V(r)$$

10.3.1   PARTICLE IN QUANTUM DOT

A particle in nanocrystal quantum dot may be described by that particle, with certain effective mass μ, in an infinitely deep spherical well. The motion of the particle in a quantum dot is restricted in all three dimensions, which is called three-dimensional confinement. The three-dimensional Schrodinger's wave equation is given by

$$\nabla^2 \psi + \frac{2\mu}{\hbar^2}\{E - V(r)\}\psi = 0 \tag{10.27}$$

In the spherical coordinate, $\phi = \psi(r, \theta, \phi)$,

$$\nabla^2 \psi = \frac{1}{r^2}\frac{\partial}{\partial r}\left(r^2\frac{\partial \psi}{\partial r}\right) + \frac{1}{r^2}\left\{\frac{1}{\sin\theta}\frac{\partial}{\partial \theta}\left(\sin\theta\frac{\partial \psi}{\partial \theta}\right) + \frac{1}{(\sin\theta)^2}\frac{\partial^2 \psi}{\partial \phi^2}\right\} \tag{10.28}$$

In quantum mechanics, the total angular momentum is associated with an operator given by

$$L^2 = -\hbar^2 \left[\frac{1}{\sin\theta} \frac{\partial}{\partial\theta} \left(\sin\theta \frac{\partial}{\partial\theta} \right) + \frac{1}{(\sin\theta)^2} \frac{\partial^2}{\partial\phi^2} \right] \tag{10.29}$$

Substituting Equation (10.29) into Equation (10.28), we get

$$\nabla^2\psi = \frac{1}{r^2} \frac{\partial}{\partial r} \left(r^2 \frac{\partial\psi}{\partial r} \right) - \frac{L^2\psi}{\hbar^2 r^2} \tag{10.30}$$

Substituting Equation (10.30) into Equation (10.27), we get

$$\frac{1}{r^2} \frac{\partial}{\partial r} \left(r^2 \frac{\partial\psi}{\partial r} \right) - \frac{L^2\psi}{\hbar^2 r^2} + \frac{2\mu}{\hbar^2} \{E - V(r)\}\psi = 0 \tag{10.31}$$

In this case, the wave function may be written as separable functions of $r, \theta,$ and ϕ

$$\psi(r,\theta,\phi) = R(r)\Theta(\theta)\phi(\phi) \tag{10.32}$$

$$\text{Let } \phi(r,\theta,\phi) = R(r)\, Y(\theta,\phi) \tag{10.33}$$

Substituting Equation (10.33) into Equation (10.31), we get

$$\frac{Y}{r^2} \frac{d}{dr} \left(r^2 \frac{dR}{dr} \right) - \frac{L^2 R\, Y}{\hbar^2 r^2} + \frac{2\mu R\, Y}{\hbar^2} \{E - V(r)\} = 0$$

Dividing by $R\, Y$, we get

$$\frac{1}{Rr^2} \frac{d}{dr} \left(r^2 \frac{dR}{dr} \right) - \frac{L^2}{\hbar^2 r^2} + \frac{2\mu}{\hbar^2} \{E - V(r)\} = 0$$

Multiplying by r^2

$$\frac{1}{R} \frac{d}{dr} \left(r^2 \frac{dR}{dr} \right) + \frac{2\mu\, r^2}{\hbar^2} \{E - V(r)\} = \frac{L^2}{\hbar^2} = \lambda \text{ (constant)} \tag{10.34}$$

Each term in Equation (10.34) is equated to a constant λ because the first term on the left side is a function of the position vector's magnitude, r, only while the second function is a function of (θ,ϕ). Equation (10.34) defines two independent equations, first for the radial part and the other for the angular parts, i.e.

$$\frac{1}{R} \frac{d}{dr} \left(r^2 \frac{dR}{dr} \right) + \frac{2\mu r^2}{\hbar^2} \{E - V(r)\} = \lambda \tag{10.35}$$

and

$$\frac{L^2}{\hbar^2} = \frac{1}{\hbar^2} \frac{1}{Y(\Theta,\phi)} L^2\, Y(\Theta,\phi) = \lambda \tag{10.36}$$

Or

$$L^2 Y(\Theta,\phi) = \lambda\, \hbar^2 Y(\Theta,\phi) \tag{10.37}$$

Equation (10.37) is the eigenvalue equation for the operator, L^2, hence L^2 has solutions for only certain values of $\lambda \, \hbar^2$. If $\lambda \, \hbar^2$ is not equal to one of these discrete values, then there exists no single valued continuous function $Y(\theta,\phi)$ that satisfies eigenvalue Equation (10.37).

Substituting value of L^2 into Equation 10.37,

$$\hbar^2\left[\frac{1}{\sin\theta}\frac{\partial}{\partial\theta}\left(\sin\theta\frac{\partial\,Y(\theta,\phi)}{\partial\theta}\right)+\frac{1}{(\sin\theta)^2}\frac{\partial^2Y(\theta,\phi)}{\partial\phi^2}\right]+\lambda\,\hbar^2Y(\theta,\phi)=0 \qquad (10.38)$$

Substituting $Y(\theta,\phi)=\Theta(\theta)\phi(\phi)$ and then multiplying by $\frac{(\sin\theta)^2}{\Theta(\theta)\varphi(\phi)}$, we get

$$\frac{\sin\theta}{\Theta(\theta)}\left\{\frac{d}{d\theta}\left(\sin\theta\frac{d\Theta(\theta)}{d\theta}\right)+\lambda\,\Theta(\theta)\right\}=-\frac{1}{\phi(\phi)}\frac{d^2\phi(\phi)}{d\phi^2}=m^2 \qquad (10.39)$$

Solving for $\phi(\phi)=A\,e^{jm\phi}$ for the wave function to be unique single valued, we must have $\phi(\phi+2\pi)=\phi(\phi)$ or, $e^{jm\phi}=1$, this implies $m=0,\pm1,\pm2,\pm3,\dots$ and normalization gives,

$$\phi(\phi)=\frac{1}{\sqrt{2\pi}}\,e^{jm\phi}\quad\text{where }m=0,\pm1,\pm2,\pm3,\dots \qquad (10.40)$$

The θ dependent part of Equation 10.39 can be written as

$$\sin\theta\frac{d}{d\theta}\left(\sin\theta\frac{d\Theta(\theta)}{d\theta}\right)\{\lambda\,\sin\theta-m^2\}\Theta(\theta)=0 \qquad (10.41)$$

Substituting $x=\cos\theta$, and taking $m=0$, Equation 10.41 may rewritten as,

$$\frac{d}{dx}(1-x^2)\frac{d}{dx}\Theta(x)+\lambda\,\Theta(x)=0 \qquad (10.42)$$

This is the Legendre differential equation. To solve this equation, let $\Theta(x)=\sum_{k=0}^{\infty}a_kx^k$, this implies,

$$\sum_{k=0}^{\infty}x^k\left[(k+1)(k+2)a_{k+2}-\{k(k+1)-\lambda\}a_k\right]=0$$

$$a_{k+2}=\frac{k\,(k+1)-\lambda}{(k+1)(k+2)}\,a_k \qquad (10.43)$$

To have a finite value of the wave function, this series must terminate. If the series terminates for $k=l$, then $a_l\neq0$ and $a_k=0$ for $k>l$, therefore $a_{l+2}=C$. Putting this into Equation (10.43) we get

$$\lambda=l(l+1)\text{ Where }l=0,\,1,\,2,\,3,\dots$$

Also $L^2=\lambda\,\hbar^2$, therefore angular momentum can take discrete values. Hence, the total angular momentum is quantized and given by $L=\sqrt{l(l+1)}\,\hbar$ and integer l is, therefore, called the orbital quantum number.

If $m \neq 0$ then Equation (10.41), can be written as,

$$\frac{d}{dx}(1-x^2)\frac{d}{dx}\Theta(x)+\left\{\lambda-\frac{m^2}{(1-x^2)}\right\}\Theta(x)=0$$

That has the solutions as

$$\Theta(x)=P_l^m(x)=(1-x^2)^{|m|/2}\frac{d^{|m|}}{dx^{|m|}}P_l(x)$$

where $P_l(x)$ is the Legendre polynomial, and for $|m|>l$, $P_l^m(x)$ has to vanish. Thus $|m|$ is restricted to values less than or equal to l, i.e., $-l \leq m \leq l$.

The complete solution for the nonradial dependent part of the wave is given by

$$Y_l^m(\theta, \phi)=\Theta(\theta)\phi(\phi)=C\ P_l^m(\cos\ \theta)e^{jm\phi} \tag{10.44}$$

where C is the arbitrary constant and $Y_l^m(\theta, \phi)$ represents spherical harmonics. Now let us solve the radial part of Equation (10.34),

$$\frac{1}{R}\frac{d}{dr}\left(r^2\frac{dR}{dr}\right)+\frac{2\mu\ r^2}{\hbar^2}\{E-V(r)\}=l(l+1)=\lambda$$

or,

$$\frac{d^2R}{d^2r}+\frac{2}{r}\frac{dR}{dr}+\frac{2\mu}{\hbar^2}\left\{E-V(r)-\frac{l(l+1)\hbar^2}{2\mu r^2}\right\}R=0$$

Putting, $u(r)=r\ R(r)$, we get

$$\frac{d^2u}{d^2r}+\frac{2\mu}{\hbar^2}\left\{E-V(r)-\frac{l(l+1)\hbar^2}{2\mu\ r^2}\right\}u=0$$

For an infinitely deep potential well we can write,

$$V(r)=\begin{matrix}0, r\leq a\\ \infty, r>a\end{matrix}$$

In the region, $r \leq a$, we have

$$\frac{d^2u}{dr^2}+\left\{k^2-\frac{l(l+1)}{r^2}\right\}u=0 \tag{10.45}$$

$$\text{where,}\ k=\frac{\sqrt{2\mu E}}{\hbar^2} \tag{10.46}$$

and it has the solutions of the form, $R(r)=A\ j_l(kr)+B\ n_l(kr)$, where

$$j_l(kr)=\sqrt{\frac{\pi}{2r}}J_{l+\frac{1}{2}}(r)=\text{Spherical Bessel function}$$

and

$$n_l(kr) = \sqrt{\frac{\pi}{2r}} N_{l+\frac{1}{2}}(r) = \text{Spherical Neumann function}$$

However, we need to reject the Neumann function as it diverts at the origin, therefore

$$R(r) = A \, j_l(kr)$$

Now applying the boundary condition that $\psi(r = a) = 0$, we have $A \, j_l(ka) = 0$. If κ_{nl} represents the nth root of the spherical Bessel function, then $\kappa_{nl} = ka$. Substituting this into Equation (10.46), we get

$$E = \frac{\hbar^2 \kappa_{nl}^2}{2\mu \, a^2} \tag{10.47}$$

$n = 1, 2, 3, \ldots$ and is called the principal quantum number.

Thus, wave function is associated with the particle in deep quantum dot and can be written as,

$$\psi_{n \, l \, m}(r, \theta, \phi) = K \, j_l\left(\kappa_{nl} \frac{r}{a}\right) Y_l^m(\theta, \phi) \tag{10.48}$$

where K is the arbitrary constant.

10.3.2 Inferences and Discussions

1. Equation (10.47) indicates that the boundary condition $\phi(r = a) = 0$ limits the allowed energies to be at a certain discrete set of values (or quantization) of energy.
2. The ground state energy, i.e., the minimum zero-point energy that corresponds to $n = 1$ and $l = 0$, is

$$E = \frac{\hbar^2 \kappa_{10}^2}{2\mu \, a^2} = \frac{\pi^2 \hbar^2}{2\mu a^2}$$

Hence, a particle confined in any finite sphere will have positive minimum zero-point energy.

3. According to *Heisenberg's uncertainty principle*, $\Delta p \, \Delta x \geq \hbar/2$, if the particle has to be restricted in the spherical well then the uncertainty in position will be $\Delta x = a$. Hence,

$$\Delta p \geq \hbar/2a \text{ or } p_{min} = \hbar/2a$$

Thus, the minimum uncertainty in energy will be

$$\Delta E_{min} = \frac{p_{min}^2}{2\mu} = \frac{\hbar^2}{8\mu a^2}$$

4. The energy Equation (10.47) may be considered as a good approximation if and only if V_0 is large enough for a spherical well with potential V_0, i.e.

$$V_0 \gg \Delta E_{min} = \frac{\hbar^2}{8\mu a^2} \cong 9.5 \text{ meV } \left(\text{for } \mu = \mu_0 \text{ and } a_0 = 10 \text{ nm}\right)$$

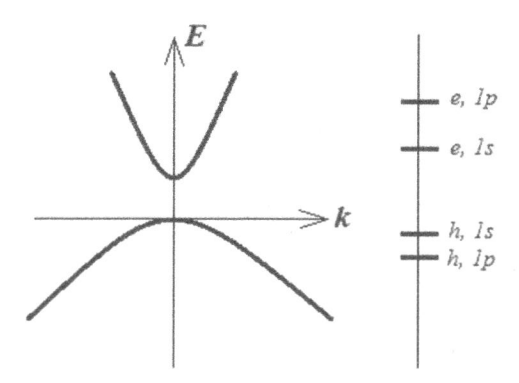

FIGURE 10.7 Energy spectrum for free particle (left) and confined (right) in an ideal quantum dot.

Hence for all practical purposes where the quantum dots are grown in an insulator matrix (e.g., glass with $V_0 = 4$ eV), an infinitely deep spherical potential well is a good approximation. If $V_0 < E_{min}$, no state exists in the well at all.

5. For given values of n and l we have $(2l+1)$ states, which have the same energy because, for the same n and l, we can have $(2l+1)$ values of m. Therefore, in a quantum dot we have $(2l+1)$ fold degeneracy.

6. Figure 10.7 shows the schematic plot of the single-particle energy spectrum in a bulk semiconductor (left) and single-particle energy of the electron (e) and hole (h) in a small quantum dot (right). It is usual for the n, l − electrons or hole eigenstates to be denoted as $1s$, $1p$, $1d$, etc., corresponding to $l = 0, 1, 2, ...$, respectively. The point is that a $1p$ state in case of spherical confinement potential is possible in contrast to the well-known case of Coulomb's potential case.

7. The kinetic energy of a quantum mechanical particle takes discrete values given by

$$\text{For electron: } E_{e, \, nlm} = E_g + \frac{\hbar^2 \kappa_{nl}^2}{2 \, m_e a^2} \tag{10.49}$$

$$\text{For hole: } E_{h, \, nlm} = \frac{\hbar^2 \kappa_{nl}^2}{2 \, m_h a^2} \tag{10.50}$$

The energy of single-particle electron-hole pair (EHP) has energy given by,

$$E = E_{e, nlm} - E_{h, nlm} = E_g + \frac{\hbar^2 \kappa_{nl}^2}{2 \, m_e a^2} + \frac{\hbar^2 \kappa_{nl}^2}{2 \, m_h a^2} \tag{10.51}$$

The lowest confined state of single-particle EHP has energy given by,

$$E_{min} = E_g + \frac{\hbar^2 \pi^2}{2 \, m_e a^2} + \frac{\hbar^2 \pi^2}{2 \, m_h a^2}$$

Hence, there is an energy increase with respect to the bulk semiconductor bandgap, given by

$$\Delta E = \frac{\hbar^2 \pi^2}{2 \mu a^2} \tag{10.52}$$

where μ is reduced *electron-hole* mass, given by $\mu^{-1} = m_e^{-1} + m_h^{-1}$.

10.4   QUANTUM CONFINEMENT REGIMES

The previous description is based on assumptions that electrons and holes are noninteracting, because these are the only elementary particles that correspond to the so-called single-particle presentation. However, the electron and hole are charged particles that can interact with each other via Coulomb's potential, thereby forming an extra quasi-particle. Such quasi-particles correspond to a bound state similar to hydrogen; hence, these EHPs are denoted as an exciton. The exciton Rydberg energy may be introduced as,

$$Ry = \frac{\hbar^2}{2\mu a_B} = \frac{\mu e^4}{m_0 \; \epsilon^2} \times 13.6 \text{ eV} \tag{10.53}$$

where Bohr radius is $a_B = \frac{\epsilon \hbar^2}{\mu e^2}$.

The reduced electron-hole mass (μ) is smaller than the electron mass (m_0), whereas the dielectric constant of the background semiconductor material (ϵ) is many times larger than that of the vacuum. Therefore, the exciton Bohr radius is larger, exciton Rydberg energy is significantly larger, and the exciton Rydberg energy is significantly smaller than these relevant values for the hydrogen atom. The absolute values of the Bohr radius (a_B) for the common semiconductor range in the internal are 0.1–10 nm and the exciton Rydberg energy takes the approximate values of 1–100 meV. The energy shift in terms of these quantities may be given by Equation (10.52), which that may be written as

$$\Delta E = Ry \left[\frac{\pi a_B}{a} \right]^2 \tag{10.54}$$

Based on the pioneering investigations for the quantum confinement in semiconductor quantum dots, Efros and Efros introduced three quantum confinement regimes depending on the ratio of quantum dot radius to the Bohr radius of electrons, holes, and EHPs [7].

10.4.1   Weak Confinement Regime

The weak confinement regime refers to the cases when the quantum dot radius (a) is only a few times larger than the exciton Bohr radius (a_B). Relatively small confinement effects are found in this regime. In the weak confinement regime, we get a large number of states, but these states are separated by small gaps. Also, transitions between these states are not forbidden by momentum conservation law. Every level has a homogeneous width with additional inhomogeneous broadening due to the size distribution of different quantum dots. So the energy levels overlap, which leads to an effectively continuous spectrum.

10.4.2   Intermediate Confinement Regime

In the cases for which the effective mass of the holes is larger than that of the electrons (i.e., $m_h / m_e \ll 1$), we have situations where the radius of the quantum dot is small relative to the electron Bohr radius, but they are still big compared with the Bohr radius of the hole, so that $a_h < a < a_e$. This corresponds to the intermediate confinement regime.

10.4.3   Strong Confinement Regime

The strong confinement regime refers to the cases in which the quantum dot radius (a) is very small compared with the exciton Bohr radius (a_B). For such cases the confined electrons and holes observe no bound state like the hydrogen exciton. In such cases it is a good approximation to ignore Coulomb's interactions completely as the movement of charge carriers is quantized in all spatial directions. The elaborated numerical calculations, however, never justify such negligence for

Coulomb's interactions. If we refer to Equation (10.54), it is clear that strong confinement ($a \ll a_B$) induces a larger energy shift.

The electron and hole exhibit higher kinetic energy compared with Ry because of strong confinement. The selection rules that allow the optical transitions at the same principal and orbital quantum numbers and couple electron and hole states are due to energy and momentum conservation laws. The absorption spectrum is, therefore, a set of these discrete bands with peak energies at,

$$E_{nl} = E_g + \frac{\hbar^2 \kappa_{nl}^2}{2\mu \, a^2} \tag{10.55}$$

Due to these reasons, the quantum dots under strong confinement limits are also referred to as artificial atoms or hyper-atoms. However, quantum dots exhibit a discrete optical spectrum that is controlled by size dispersion (i.e., by the number of atoms used to make quantum dots). This is in contrast to an atom that has a discrete spectrum controlled by the number of nucleons. Under strong confinement conditions, the discrete confinement levels are distant enough to observe the discrete spectrum. This situation is in contrast to weak confinement conditions in which we observe continuous spectrum due to overlapping energy levels.

10.5 CONCLUSIONS

This chapter derived the basic results that are elementary for the study of quantum confinement effects in the low-dimensional nanostructures. These results have been obtained using Schrodinger's wave equation in different coordinate systems as they deemed fit for the situation. The mathematical background is essential for the proper understanding of the beginner in this emerging field of study. The results are discussed keeping the basic need for optical studies of quantum nanostructures that may be made under the glass matrices for optical applications.

REFERENCES

1. W. Xiao, Y. Feng, P. Dong and J. Huang, "A Mini Review on Carbon Quantum Dots: Preparation, Properties, and Electrocatalytic Application," *Frontiers in Chemistry*, vol. 7, p. 671, 2019.
2. S. Lüker and D. E. Reiter, "A Review on Optical Excitation of Semiconductor Quantum Dots Under the Influence of Phonons," *Semiconductor Science and Technology*, vol. 34, p. 063002, 2019.
3. A. Kumar, S. Bhusan and P. K. Tiwari, "A Threshold Voltage Model of Silicon-Nanotube-Based Ultrathin Double Gate-All-Around (DGAA) MOSFETs Incorporating Quantum Confinement Effects," *IEEE Transactions on Nanotechnology*, vol. 15, no. 5, p. 868–875, 2017.
4. P. Vimala and N. R. Nithin Kumar, "Analytical Quantum Model for Germanium Channel Gate-All-Around (GAA) MOSFET," *Journal of Nano Research*, vol. 59, p. 137–148, 2019.
5. A. J. Nozik and O. I. Micic, "Colloidal Quantum Dots of III-V Semiconductors," *MRS Bulletin*, vol. 23, no. 2, p. 24–30, 1998.
6. U. A. Rani, L. Y. Ng, C. Y. Ng and E. Mahmoudi, "A Review of Carbon Quantum Dots and Their Applications in Wastewater Treatment," *Advances in Colloid and Interface Science*, vol. 278, p. 102124, 2020.
7. A. L. Efros and A. L. Efros, "Interband Absorption of Light in A Semiconductor Sphere," *Soviet Physics Semiconductor*, vol. 16, no. 7, p. 772–775, 1982.

FURTHER READINGS

Gopanenko S. V. (1998). *Optical Properties of Semiconductor Nanocrystals Illustrated Edition*. Cambridge University Press, Cambridge, UK. https://doi.org/10.1017/CBO9780511524141

Koch S. W. et al. (1993). *Semiconductor Quantum Dots*. World Scientific Publishing, Toh Tuck Link, Singapore.

Yükselici H. et al. (2013). Optical Studies of Semiconductor Quantum Dots. In: Ünlü H., Horing N. (eds), *Low Dimensional Semiconductor Structures. NanoScience and Technology*. Springer, Berlin, Heidelberg. https://doi.org/10.1007/978-3-642-28424-3_6

11 Simulation of Reconfigurable FET Circuits Using Sentaurus TCAD Tool

Remya Jayachandran, Rama S. Komaragiri, and K. J. Dhanaraj

CONTENTS

11.1 INTRODUCTION

Technology computer-aided design (TCAD) tools simplify the microscopic level analysis of various field-effect transistors (FETs) found in the literature [1–5]. TCAD tools enable the implementation of new device architectures and the study of its device characteristics. Beyond Moore's era, in the literature, multigate transistors, carbon nanotubes, FETs based on tunneling phenomenon, single-electron devices, and reconfigurable devices are proposed as the potential replacement for the conventional complementary metal oxide semiconductor (CMOS) devices [6–13]. Among these, the reconfigurable device is an emerging area. By changing the polarity of the gate bias, the functionality of a FET can be altered from an n-type or a p-type transistor [6], thus the name reconfigurable field-effect transistor (RFET). Implementation of analog and digital circuits using RFET can provide insight into the device's reconfigurable nature and its circuit applications. Many variants of RFET devices (triple-gate RFET, double-gate RFET, and single-gate RFET [SG-RFET]) are reported in the literature. This chapter introduces the Sentaurus TCAD [14] tool flow. Later, the device characteristic simulations and circuit-level simulations using the SG-RFET device are explored. The Synopsis Sentaurus structure editor and Sentaurus Device (sdevice) simulator are used to create the device structure and simulate the electrical characteristics of an SG-RFET. The challenges associated with device physics are analyzed including the convergence issues faced in the simulations. Various physics-based models used to simulate the device characteristics are also briefly discussed.

    This chapter presents insight into the circuit designed using the multigate device structures such as the gate-all-around FETs (GAA-FETs) and SG-RFETs. A calibrated simulation environment is used to explore new device architectures with different materials. The simulations also provide an insight into proposing new device architectures to alleviate the existing challenges. Instead of the examples given in the Sentaurus TCAD tool, the practical cases presented in this chapter are used to understand and analyze the simulation results of the physics and electrical properties of the SG-RFET device. An inverter circuit implemented using SG-RFET devices is studied for its voltage transfer characteristics, transient response, and step response. An amplifier circuit designed using the SG-RFET inverter is also demonstrated using Sentaurus TCAD. This chapter also addresses other universal logic gates, NAND and NOR, implemented using reconfigurable transistors in the TCAD. The challenges faced in implementing the logic circuits using new device architectures are explained. The simulation results highlight the reconfigurable nature of analog and digital circuits by using the SG-RFET device. The electrical characteristics are analyzed using the sdevice, and the circuits designed using SG-RFET are analyzed using mixed-mode simulation in the Sentaurus TCAD. Transport properties are modeled using the physics models, namely the drift-diffusion model, density-gradient (DG) model, mobility model, channel quantization models, and band-to-band tunneling and recombination models.

The device characteristics are also studied for different gate and channel lengths. The electron and hole tunneling of the SG-RFET device is characterized, and the tunneling current is enhanced by including the required physics models. The current components of the SG-RFET device are thermionic emission current and tunneling current. Unity gain analog buffer amplifier using the SG-RFET device is also demonstrated in this chapter to analyze the feasibility of using SG-RFET devices in analog circuits.

11.2 EMERGING NEW DEVICE ARCHITECTURES

The semiconductor industry has remarkably followed Moore's law [1] for the last 50 years. Integrated circuits are fabricated on bulk silicon wafers. As the device dimensions are scaled down, the bulk planar transistors are replaced by silicon-on-insulator (SOI) substrates. SOI technology enhanced the current driving capability of a transistor by reducing the parasitic capacitance. The geometrical scaling of transistors in conventional planar structures has reached its limit due to the pronounced short channel effects (SCEs) [3], power consumption, and fabrication complexity. Hence, an alternative transistor geometry, new materials, channel engineering, strain engineering, and gate engineering are necessary to improve the performance of the FET.

Scaling the transistors demanded control on the electric fields in the channel, which resulted in multigate structures such as double-gate, triple-gate, Ω-gate, FinFET transistors, quadruple-gate, and surround-gate transistors [2–5]. As the number of gate terminals in a MOSFET increases, the degree of gate controllability increases. FinFET transistors are currently the most used transistors in the semiconductor industry, giving more volume than a planar gate for the same planar area. From the structure point of view, Tunnel FET (TFET) is very similar to MOSFET except that the source and drain regions of a TFET are doped differently. This device is a promising candidate for low power applications. Beyond 7 nm, FinFET transistors show degraded performance. To further optimize the channel geometry, research for new device architectures led to a GAA devices in which the gate electrode wraps around the channel. RFETs, carbon nanotubes, and nanowires are emerging nonclassical device architectures [12].

Device reconfigurability promises to comprehend more complex systems with a lower device count. Exciting concepts have been proposed to understand such a device-level reconfiguration [15–26]. Triple-gate RFET, double-gate RFET, and SG-RFET are the existing RFETs that use surround-gate/GAA architecture. Figure 11.1 shows different RFET configurations. Among these, the SG-RFET is a simple and efficient device that can be configured into an n-type or p-type by applying a suitable bias to the control gate terminal [9, 10].

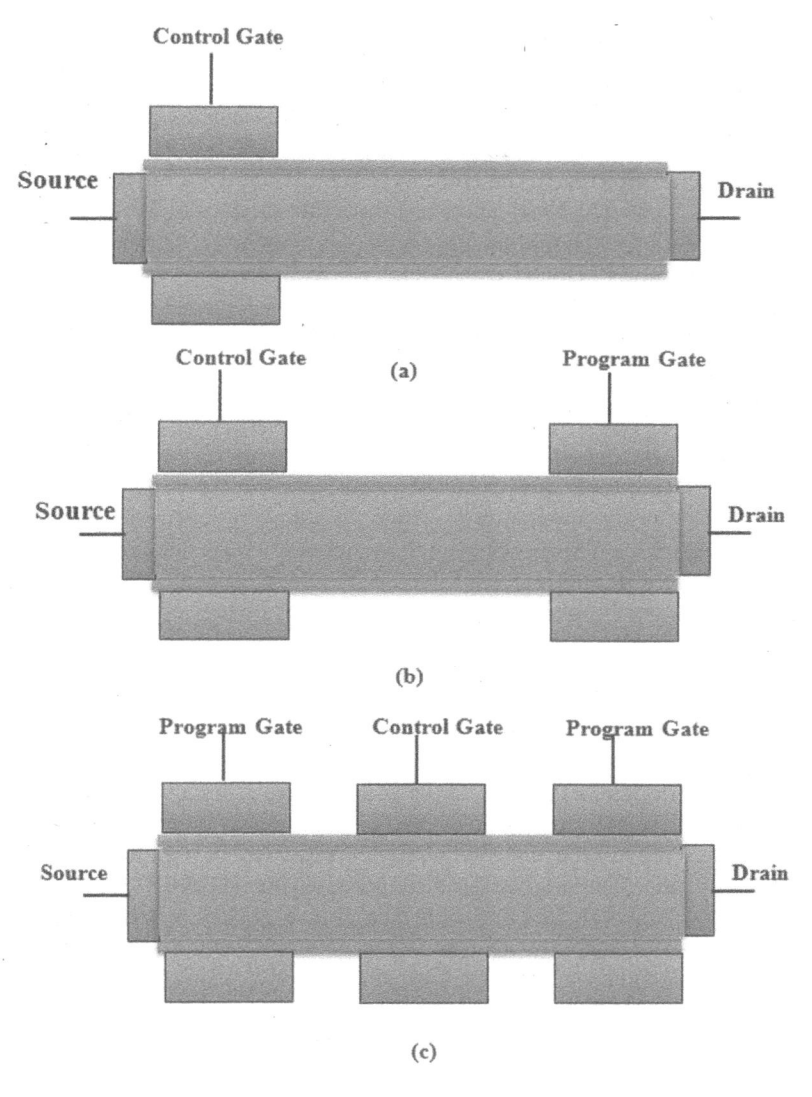

FIGURE 11.1 Reconfigurable FET structures: (a) Single-gate RFET (SG-RFET), (b) double-gate RFET (DG-RFET), and (c) triple-gate RFET (TG-RFET).

11.3 RECONFIGURABLE FETs

Depending on the gate bias, RFETs provide n- and p-type FET characteristics as selected simply by an electric signal. The working of these devices mainly depends on the unique properties of nanoscale Schottky junctions in nanowires. The main advantage of RFETs is that these devices do not require doping. The most challenging part of multigate or three-dimensional (3D) structure fabrication is to attain accurate and reproducible doping distribution control [11] and doping efficiency [12]. Various reconfigurable devices using single or multiple gates with new device architectures were recently reported [4]. A Schottky junction is formed at the S/D (semiconductor interface region), and the gate potential enables band bending at the interface. The band bends so that the predominant charge carriers are injected into the valence or conduction band that contributes the current flow.

11.4 IMPACT OF ELECTRON AND HOLE TUNNELING IN RFET

The main components of drain current in RFET devices are thermionic emission current and tunneling current [15–26]. Thermionic emission current dominates at small gate voltages and tunneling current dominates at large gate voltages. As the electron and hole barrier voltages are different, the current-voltage characteristics for n-type and p-type RFETs are not similar. The drain currents of n-type and p-type RFETs with Si-NiSi$_2$ junctions (barrier for electrons ~0.66 eV, for holes ~0.46 eV) are not symmetric. Complementary circuit operation requirements using conventional n- and p-type MOSFETs can be attained using an RFET device.

Moreover, the drive currents for n-type and p-type configurations vary by approximately a factor of 10, and the threshold voltages and the subthreshold slopes are not the same. The main reason is the difference in barrier height for electrons and holes. Symmetrical current-voltage characteristics are achievable by implementing S/D contacts that exactly align their Fermi level with the channel's mid-gap energy. However, due to differences in the injection efficiency and mobilities of electrons and holes, some mismatch can still exist.

Device simulations (sdevice) were executed to validate the theoretical analysis. A 3D drift-diffusion model is used in the simulation in which the tunneling currents are obtained using the Wentzel-Kramers-Brillouin (WKB) approximation model [15, 16]. The I_D-V_G characteristics display the behavior for both p- and n-type FETs. The result of the simulations also reproduces the characteristic kink in the subthreshold region, as expected. In the subthreshold region, flat band conditions are attained at the source electrode. This interprets that the barrier height is equal to the natural Schottky barrier. The tunneling transmission dominates the ON-state conduction of the RFET through the barrier, and the thermionic emission current dominates the subthreshold region. By enhancing the tunneling probability in RFET, the ON current can be increased. In this chapter, a high-performance SG-RFET device structure is analyzed.

In SG-RFET, the gate is placed close to the source region. On applying a drain voltage (V_D) with gate open ($V_G = 0$ V), no current flows as the tunneling of charge carriers at the semiconductor-metal interface are not feasible. In SG-RFET, there is a barrier for charge carriers under an equilibrium condition. The energy band diagram of the SG-RFET device under different bias conditions is explained in Ref. [16]. On applying a positive (negative) gate voltage, the energy band at the source region bends downward (upward), which reduces the tunneling barrier at the source-semiconductor junction. Due to the band bending, the charge carriers tunnel into the channel from the source terminal resulting in carrier conduction. The current flowing through an SG-RFET is mainly due to two components: thermionic emission current and quantum mechanical tunneling current. In the subthreshold region of operation, the dominant current is the thermionic emission current. As gate voltage exceeds the threshold voltage, tunneling current dominates, and the device enters into the ON state. Supply voltages greater than 1.5 V are required to achieve a better reconfigurable characteristic with a significant I_{ON}/I_{OFF} ratio. In this work, a supply voltage of 2 V is used, as the device's reconfigurable nature is dominant at higher supply voltages.

The required working region of an SG-RFET for logic applications is around the threshold voltage. The simulations of the SG-RFET indicates that quantum mechanical tunneling dominates in this region. To enhance the drive current, control the injection of electrons and holes through the barrier by applying compressive strain to the channel [16]. The compressive strain lowers the effective electron masses m_n^* and enhances electron mobility for <110> oriented channels. On the other hand, hole effective mass, m_p^* is increased. The tunneling probability through the barrier can be varied by tuning the m_n^* to m_p^* ratio with the application of strain. The strained channel SG-RFET performance can be analyzed in Sentaurus TCAD by changing the parameters m_n^* and m_p^* of the silicon material in the channel region. The parameter file can be modified, and the new values for the device parameter can be included in the workbench of TCAD.

11.5 SIMULATIONS OF SG-RFET USING SENTAURUS TCAD

Sentaurus Workbench, through a graphical user interface, provides a framework to design, organize, and automatically execute TCAD device simulation. Using simulations and visualization tools, Sentaurus Workbench allows automatic execution of fully parameterized projects through the graphical user interface. The sdevice simulator, through numerical computations, can simulate the electrical characteristics of one or more devices. Based on a set of physical device equations that describe the carrier distribution and conduction mechanisms in the device, the simulator computes charge distribution, electric field distribution, voltages, and terminal currents. The device's structure is discretized by forming a nonuniform "grid" (or "mesh") of nodes. Using the Sentaurus device editor (sde), a user can create one-, two-, and three-dimensional structures of a device and define various regions using the TCAD material library. Alternately, a user can execute a program to create a device structure through simulations.

The 3D device structure can be created in sde, and properties like charge distribution, depletion region, and band diagram can be visualized after executing the sdevice file for electrical characterization. Figure 11.2 presents the electrically characterized 3D MOSFET structure in the TCAD tool.

The parameter of interest can be selected to visualize its variation in the structure. The sdevice simulator computes the electrical behavior of a single device or a circuit containing few devices. Terminal currents, voltages, and charges at each node or region are computed based on a set of physical device equations that describes the carrier distribution and conduction mechanisms.

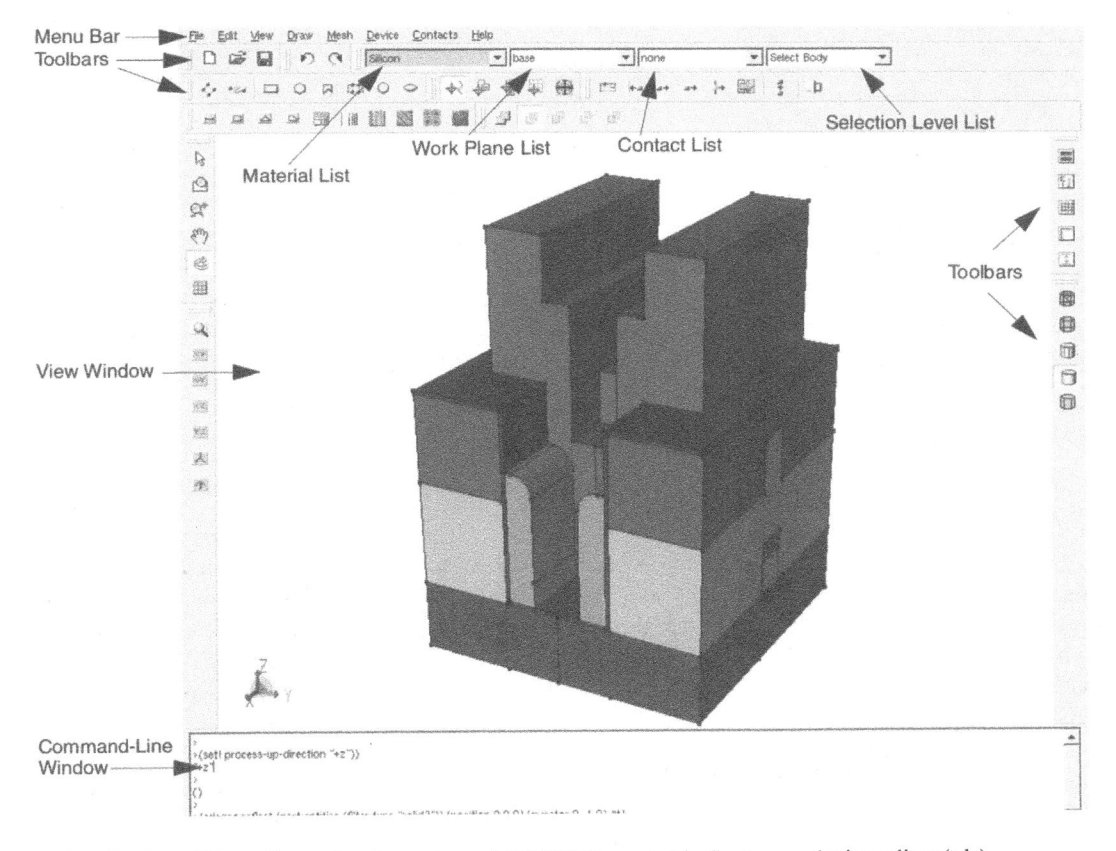

FIGURE 11.2 Three-dimensional structure of MOSFET created in Sentaurus device editor (sde).

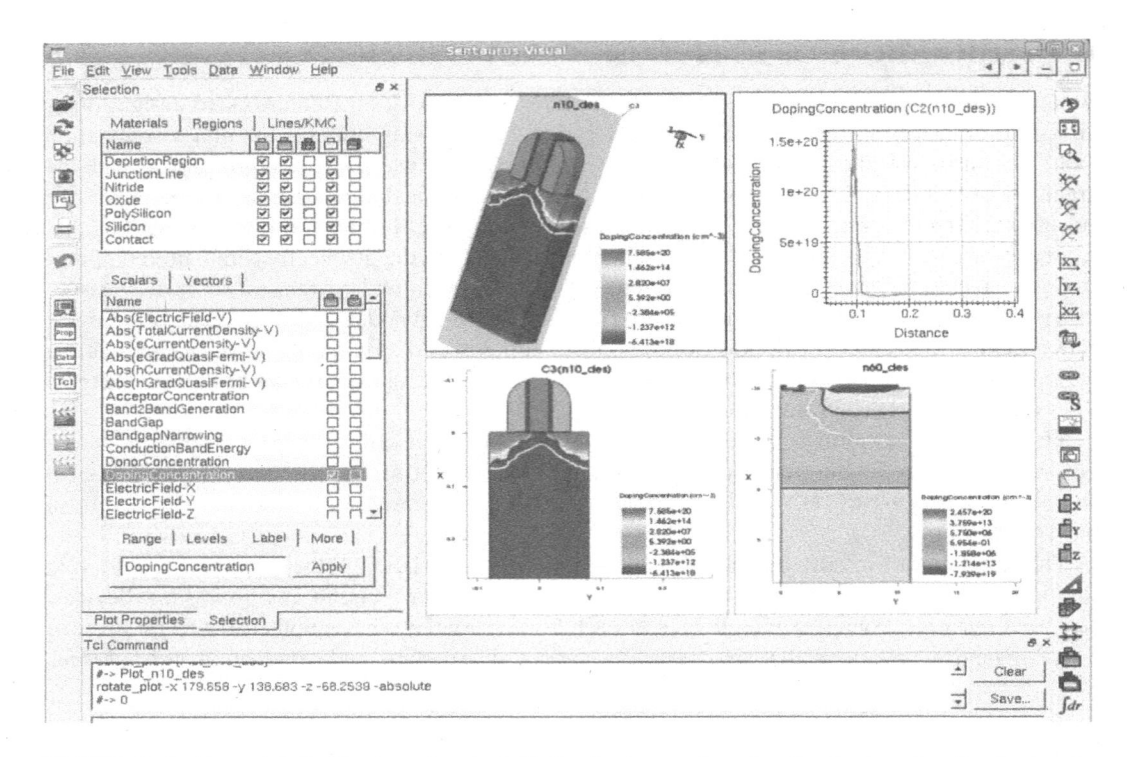

FIGURE 11.3 Potential distribution, electric field distribution, and carrier distribution visualized using the Sentaurus visual tool (svisual) in TCAD Sentaurus.

Figures 11.3–11.5 represent the electrically characterized structures in the sdevice. The carrier distribution, potential distribution, electric field, and band diagram of the structure can be visualized in the svisual tool. Using sdevice, the electrical characterization of the SG-RFET device is possible. Figure 11.6 shows the 3D structure of SG-RFET created in sde of the Sentaurus TCAD tool. The device dimension used for the SG-RFET device is the same as that in Ref. [16]. It is required to select the physics models as per the device and the biasing conditions. The 3D structure of an SG-RFET is created using the sde environment in Sentaurus TCAD. The electrical characterization of the SG-RFET structure is performed in the sdevice.

11.6   MIXED-MODE SIMULATIONS FOR CIRCUIT SIMULATION USING TCAD

In Sentaurus TCAD, the sde is used to create the device structure and sdevice for direct-current (DC) and alternating-current (AC) device simulations. In sdevice, mobility degradation mechanisms due to doping, interface effects, and electric fields (i.e., high-field saturation and normal-field degradation), the doping-dependent mobility model, Lombardi mobility degradation models at interfaces, and Shockley-Read-Hall recombination are included in the simulations. The band-to-band tunneling model and nonlocal tunneling model are used to simulate the tunneling effects. The DG model is used in the device physics section to capture the quantum effects.

The electrical characterized SG-RFET device for a different gate and channel length is presented in Figure 11.7. Figure 11.8 shows the comparison of the I_D-V_D characteristics of 2D and 3D SG-RFET structures. The SG-RFET performs well for gate length, L_G = 50nm, and device length L_T 220 nm. We have used the SG-RFET device structure and parameters used in the [13] in this chapter.

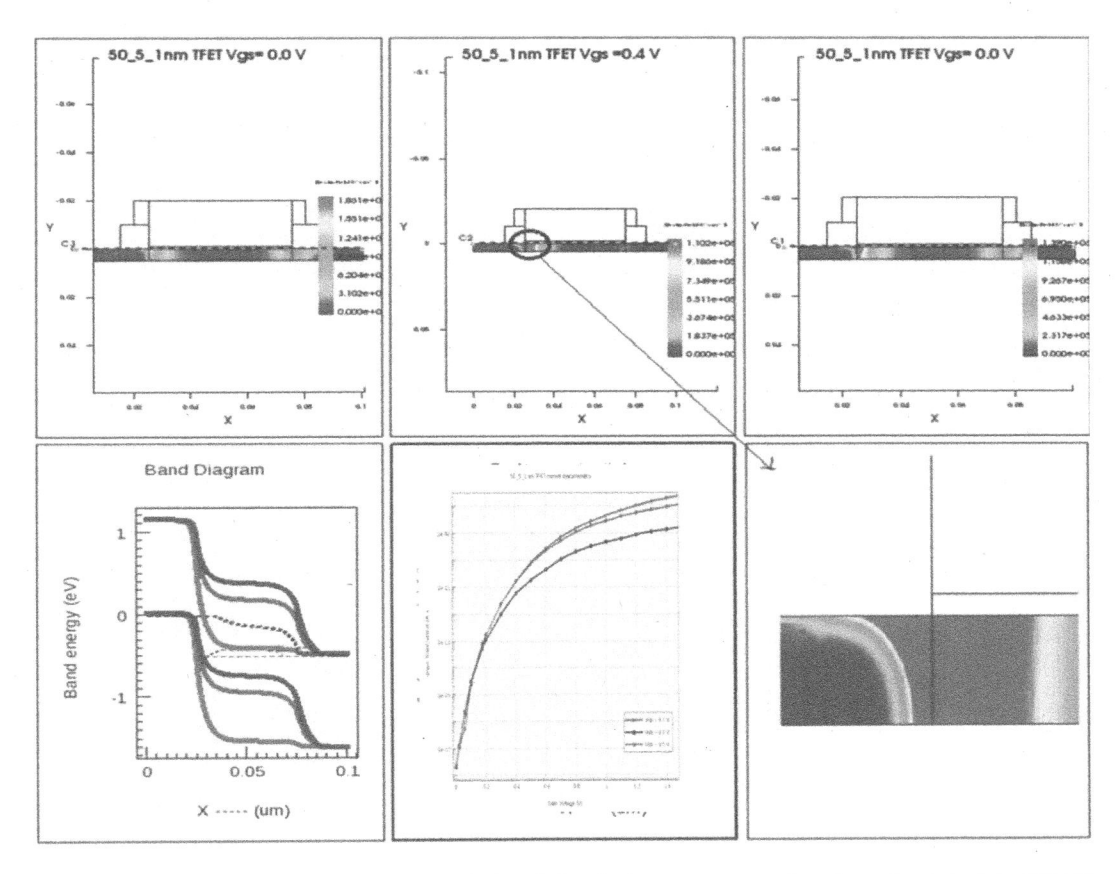

FIGURE 11.4 Energy band diagram analysis of the device structure using the svisual tool in TCAD Sentaurus.

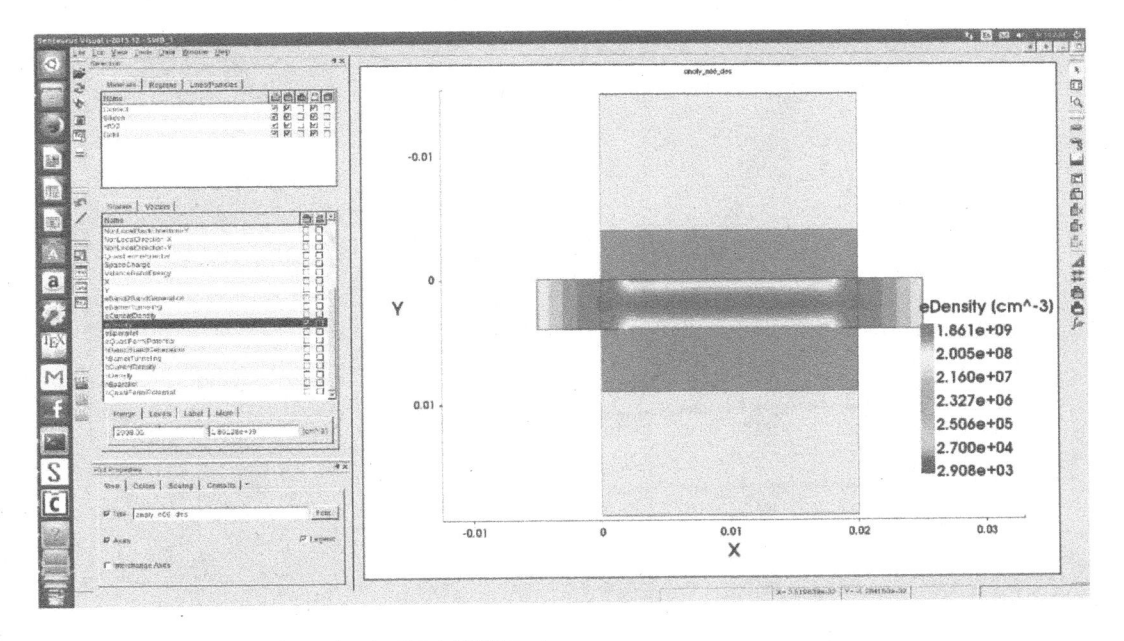

FIGURE 11.5 Volume inversion in GAA-FET device.

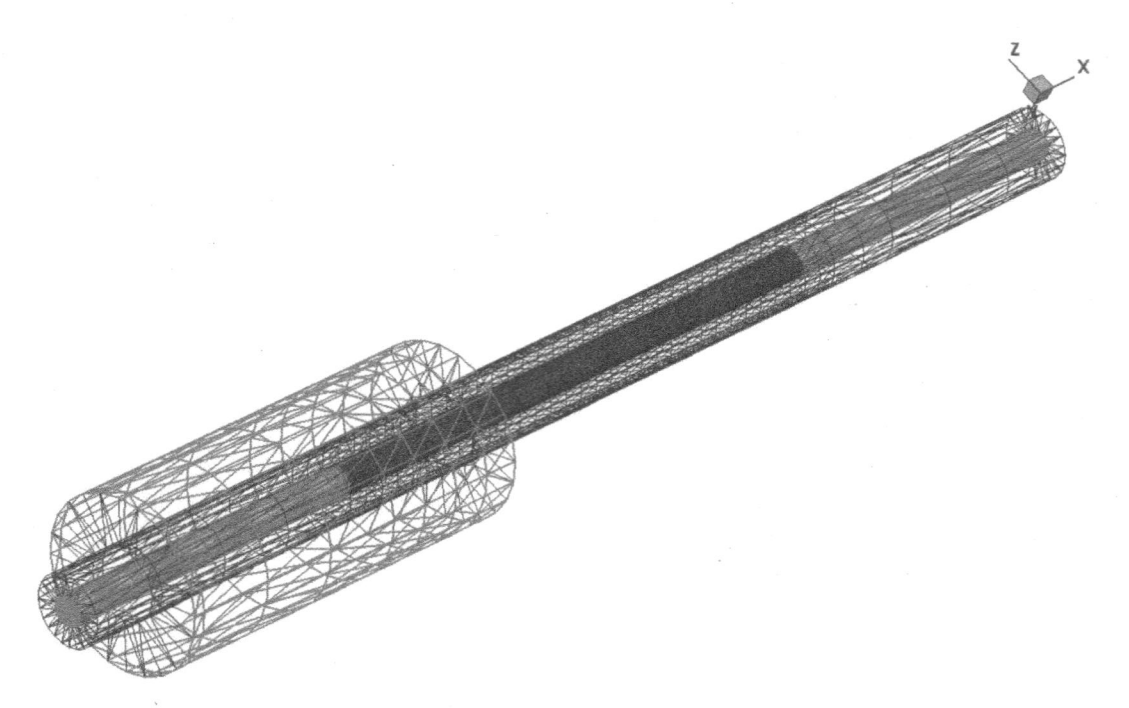

FIGURE 11.6 Three-dimensional structure of the SG-RFET device created in the sde module of TCAD Sentaurus.

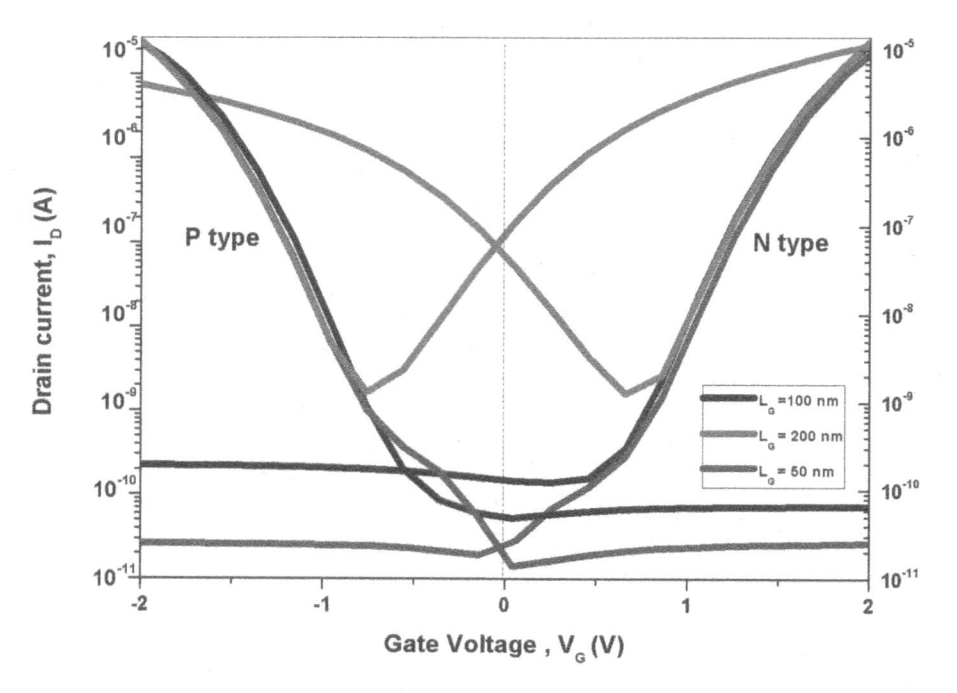

FIGURE 11.7 I–V characteristics of SG-RFET device with gate length $L_G = 50$ nm, 100 nm, 200 nm.

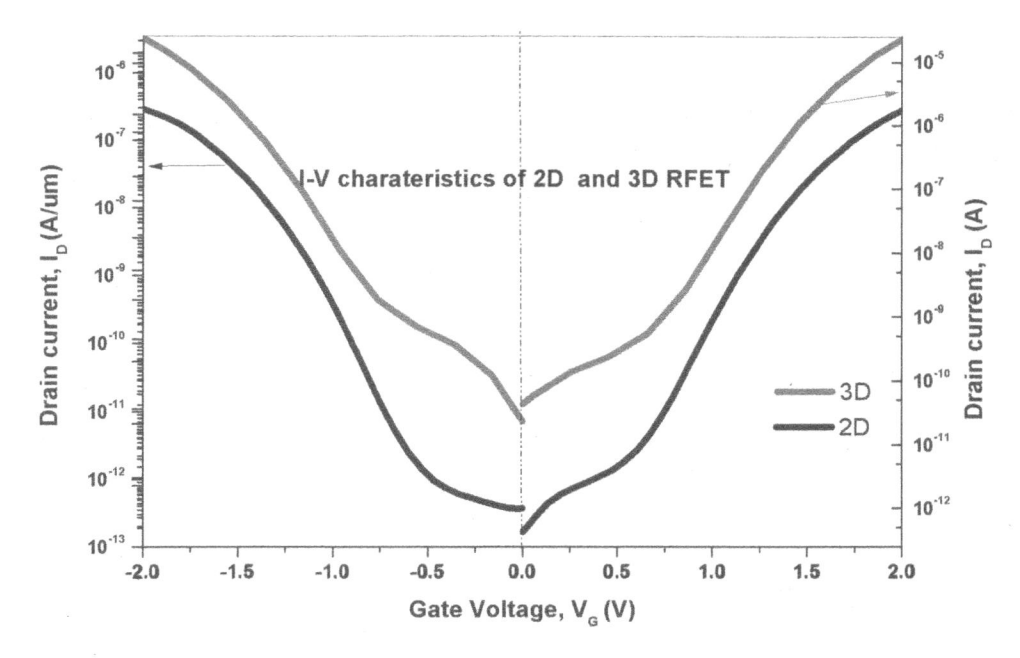

FIGURE 11.8 Comparison of I-V characteristics of a 3D SG-RFET and 2D SG-RFET device.

11.6.1 INVERTER/BUFFER OPERATION USING SG-RFET

The reconfigurable nature of the device lessens the complexity of the digital circuits. The inverter circuit is implemented using the SG-RFET structure, as shown in Figure 11.9(a) [16]. The inverter circuit is reconfigured as two p-type or n-type SG-RFETs connected in parallel, as shown in Figure 11.9(b) [16]. The reconfigurable nature is useful for implementing programmable logic arrays (PLAs). The static inverter characteristics at different supply voltage are shown in Figure 11.10(a). A high DC gain of −120 (V/V) and a switching threshold (V_M) of 1.04 V (~V_{DD}/2) at V_{DD} = 2 V is obtained for the SG-RFET inverter, whereas for the nanowire FET inverter DC gain of −105 (V/V), and V_M of 1.04 V for channel diameter ratio for p-type to n-type is 2.5:1. Figure 11.10(b) represents

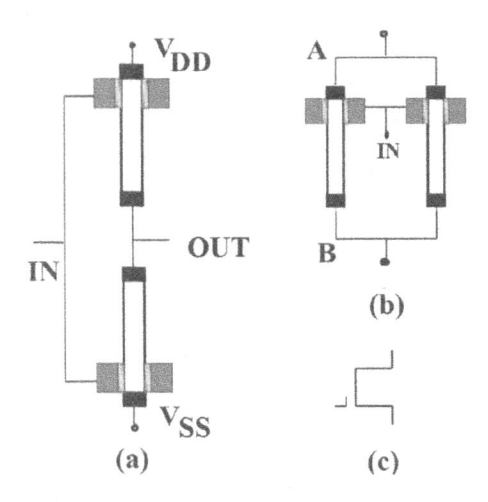

FIGURE 11.9 SG-RFET reconfigurable circuit: (a) inverter using SG-RFET, (b) reconfigured inverter circuit to operate as parallel transistors (n-type and p-type), and (c) SGRFET symbol.

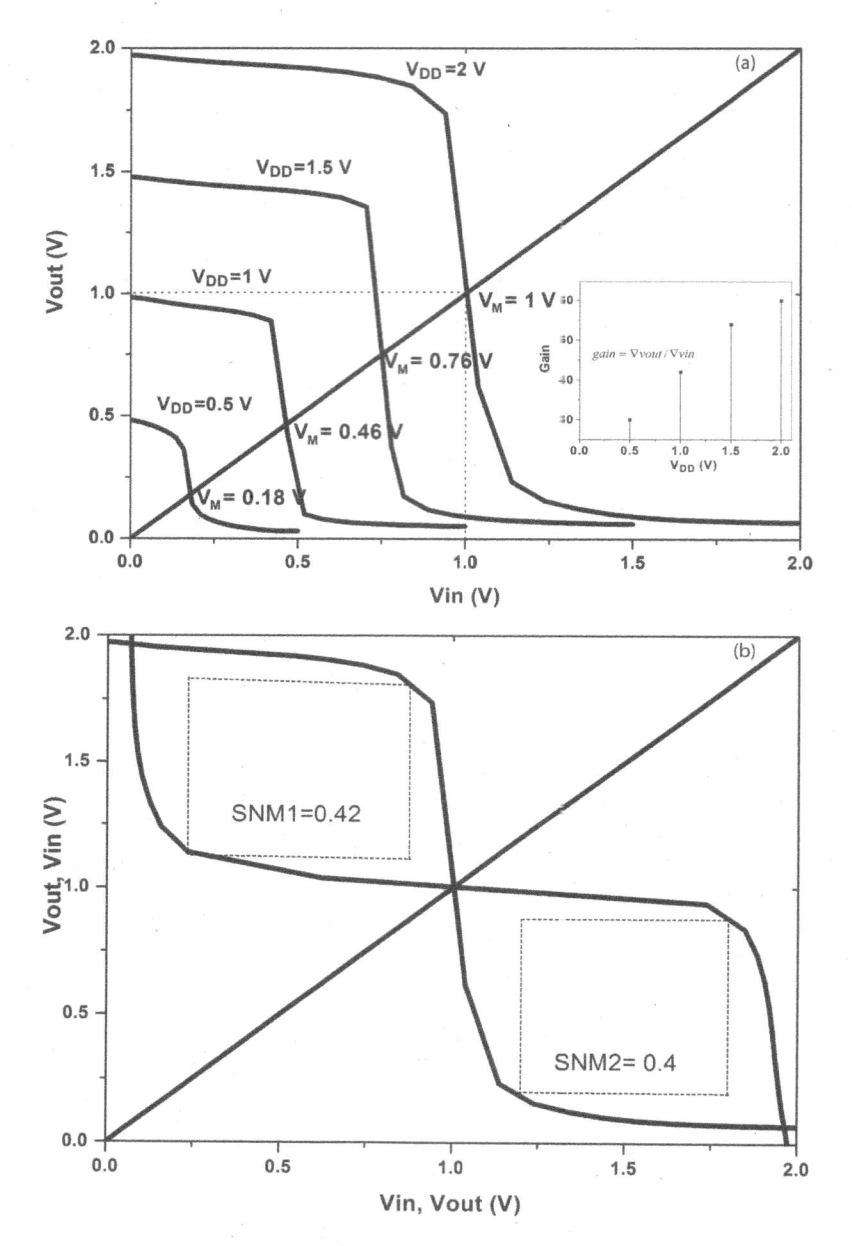

FIGURE 11.10 (a) Voltage transfer characteristics of an SG-RFET inverter at different V_{DD}) (inset graph represents the gain of the SG-RFET inverter at different V_{DD}). (b) Butterfly graph of SG-RFET inverter at $V_{DD} = 2$ V.

the butterfly plot of the inverter circuit using SG-RFET. The static noise margin is determined from the graph. As the supply voltage is reduced below 0.5 V, the gain of the SG-RFET inverter reduces to −1 as in conventional CMOS inverters.

A square pulse train of 2 V amplitude and frequency 5 MHz is applied to the inverter, and the simulation result is depicted in Figure 11.11. The value of high to low (τ_{PHL}) and low to high (τ_{PLH}) transition delays extracted from the dynamic characteristics of an inverter are 19 ps and 21 ps, respectively.

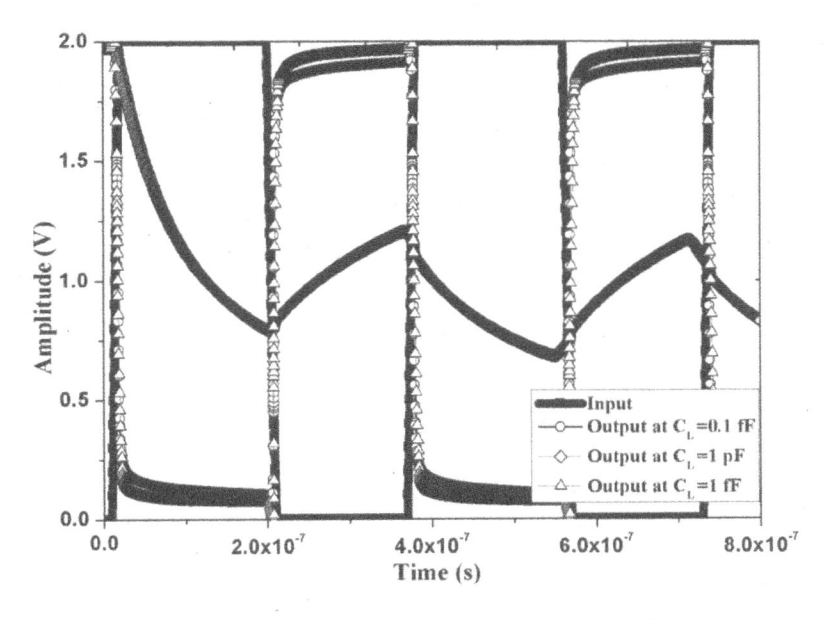

FIGURE 11.11   Input and output waveforms of the SG-RFET inverter with different capacitive loads.

The τ_{pHL} and τ_{pLH} are estimated at the crossing points of $V_{DD}/2$. The static power dissipation of the inverter designed using nanowire FET is greater compared with the inverter designed as an SG-RFET due to the low static current of an SG-RFET. Fabrication steps to develop an RFET device to possess symmetrical n-type and p-type characteristics are presented in Ref. [27–28]. The fabrication procedure can be simulated in the TCAD tool in the sprocess. Figure 11.12(a and b) shows the static characteristics at different supply voltages ranging from 0.5 to 2 V and 0 to 0.5 V, respectively.

The SG-RFET inverter exhibits a gain of −120 V/V at the switching threshold voltage. Figure 11.13 depicts the input and output response of an inverter circuit biased to operate as an amplifier with an input of 1 mV peak to peak. The static random access memory (SRAM) implemented using RFET is reported in Ref. [29], which depicts the TCAD simulations to implement SRAM using RFET structure. The RFET device used in biosensor application is reported in Ref. [30] and shows the feasibility of using an RFET device in many applications.

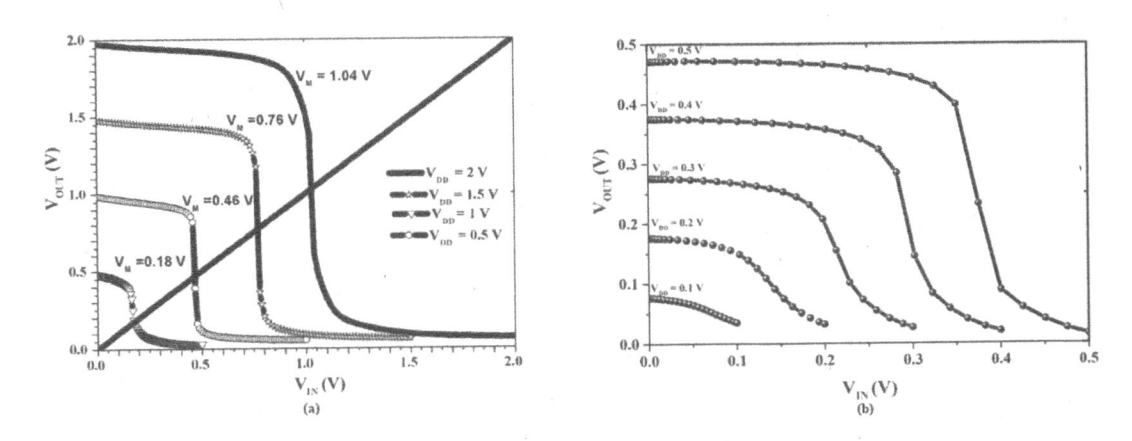

FIGURE 11.12   SG-RFET inverter VTC curves: (a) $V_{DD} \geq 0.5$ V and (b) $V_{DD} < 0.5$ V.

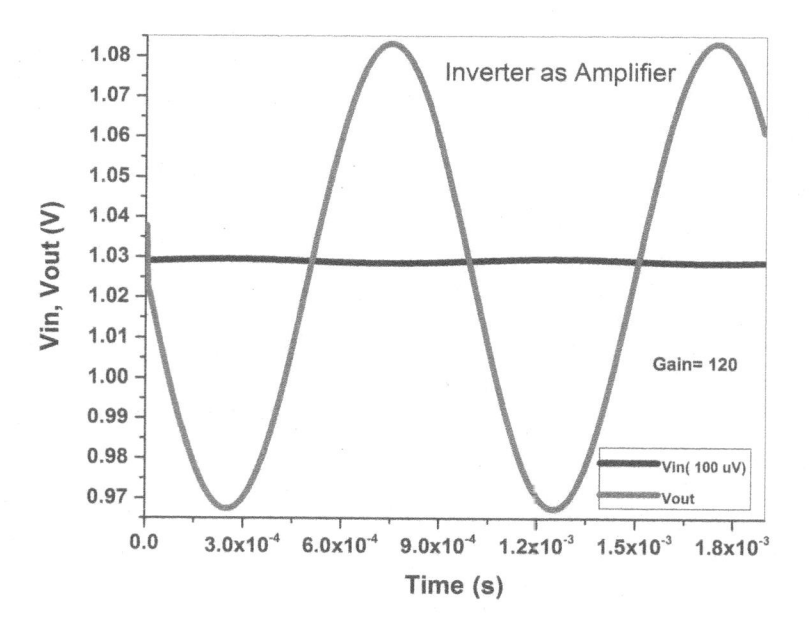

FIGURE 11.13   Input and output waveforms of SG-RFET inverter biased to work as an amplifier.

These results are favorable for the CMOS inverters implemented using nanowire transistors and existing reconfigurable transistors in the literature [28]. The drive current of the SG-RFET can be increased by connecting an array of parallel nanowires to form a single SG-RFET device.

11.6.2   SG-RFET Ring Oscillator

Three-stage and five-stage ring oscillators using SG-RFET inverters are simulated in the sdevice. The command file to simulate a ring oscillator circuit is created using the SG-RFET device structure file. From the simulation results, the operating frequencies (*fosc*) of the three-stage and five-stage ring oscillators are found to be 1 GHz and 0.5 GHz, respectively. The output waveforms of a three-stage ring oscillator and five-stage ring oscillator designed using SG-RFET are shown in Figure 11.14(a and b), respectively. The delay/stages of the three-stage ring oscillator and five-stage ring oscillator are 0.17 ns and 0.2 ns, respectively

11.6.3   Bram Nauta's Transconductor-Operational Transconductance Amplifier (OTA)

The ideal transconductance element for realizing a Gm-C filter would have infinite bandwidth and infinite DC gain. The main building block is an operational transconductance amplifier (OTA) using inverters, as shown in Figure 11.15 [31]. The whole circuit acts as a differential amplifier with an input and output common mode at $V_{DD}/2$. An amplified signal can be obtained at the output terminals when applying a sine wave at the input. This OTA can be used for implementing Gm-C filters by connecting the capacitor at the output. Nauta's OTA circuit is realized using 2D SG-RFET. The common-mode output voltage is 1.04 V at DC condition, which is almost equal to $V_{DD}/2$.

11.6.4   Buffer Amplifier Using SG-RFET OTA

An OTA buffer circuit capable of driving capacitive loads is used in LCD column drivers. Unity gain single OTA buffer circuit driving resistive load is analyzed in detail in Ref. [32]. A buffer amplifier is a circuit with high input resistance and low output resistance that enables maximum

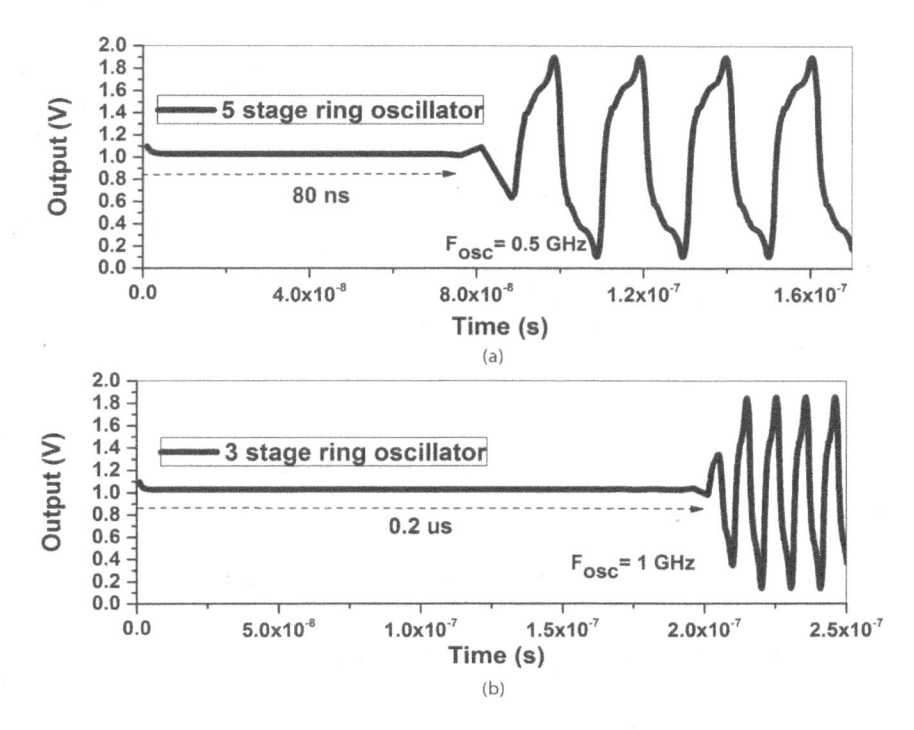

FIGURE 11.14 SG-RFET oscillator output of (a) five-stage ring oscillator and (b) three-stage ring oscillator.

power transfer. CMOS buffer circuits capable of driving capacitive loads and resistive loads are available in the literature. We have implemented a simple unity gain buffer amplifier using an OTA to drive low-resistance load. The buffer amplifier design is demonstrated using an SG-RFET OTA in the Sentaurus TCAD tool. The detailed study of the OTA buffer amplifier using MOSFET is presented in Ref. [32]. The gain variation depicts the low-resistance load driving capability of the SG-RFET unity gain buffer amplifier similar to the MOSFET OTA buffer amplifier. The gain is reduced as in the conventional unity gain buffer circuit with a decrease in load.

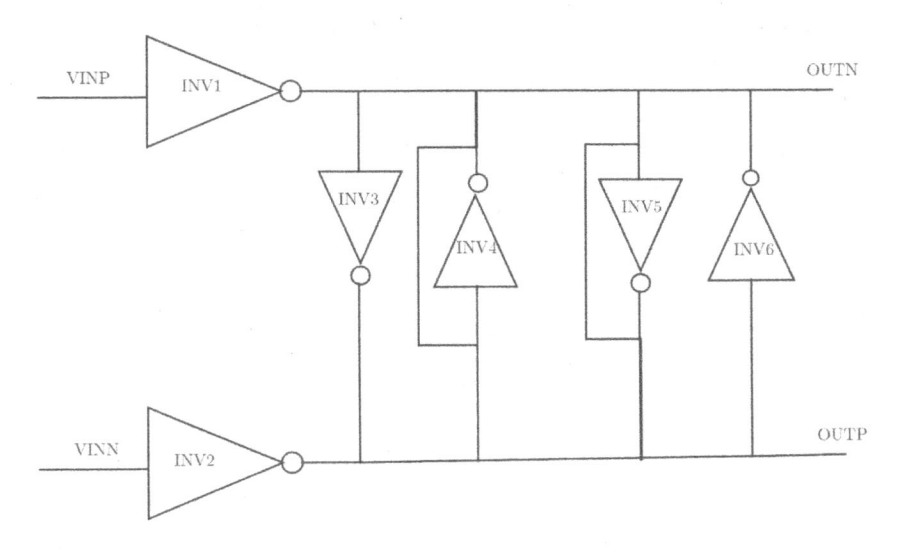

FIGURE 11.15 Bram Nauta's transconductor-OTA.

TABLE 11.1

Unity Gain Buffer Amplifier Using Different OTA

	Unity Gain Amplifier - Single OTA			
Parameters	LM13700 IC	GAA-FET OTA	SG-RFET OTA	Strained Silicon SG-RFET OTA
Supply voltage	±15 V	2	2	2
g_m (mS)	9.6	1	0.79	0.9
R_L (kΩ)	10	1	1	1
Bandwidth (MHz)	0.88	5800	560	780
Gain (V/V)	0.99	0.99	0.95	0.95

The load driving capability is increased by increasing the g_m of the OTA. The best way to increase the g_m is by using the transconductance-multiplier technique. A unity gain buffer amplifier circuit using OTA with improved load driving capability and wide bandwidth is proposed [32]. The output impedance of the buffer circuit is inversely proportional to the number of OTA blocks (N). A two-stage buffer amplifier circuit using OTA is realized using an SG-RFET OTA. The OTA circuit implemented using SG-RFET is characterized in the Sentaurus TCAD tool to study the performance of the buffer circuit. Using sdevice, electrical characterization of the two-stage unity gain OTA buffer circuit is created with resistive load. This configuration can drive low resistance loads as the output impedance of the configuration decreases with the increase in the number of stages N. The circuit parameters used in the simulation are supply voltage, $V_{DD} = 2$ V; output load resistance, $R_L = 1$ kΩ; input sinusoidal signal amplitude, $V_{pp} = 0.5$ V; and frequency 1 kHz. The output obtained has a peak-to-peak amplitude of 0.495 V from the simulation results, resulting in a gain of 0.99 V/V. As the R_L reduces, the buffer circuit's gain reduces, as expected. The hardware of the two-stage OTA buffer circuit is implemented using OTA LM13700 IC and the performance is verified.

The simulation results of buffer amplifiers implemented using different OTAs are given in Tables 11.1 and 11.2. The performance of the SG-RFET OTA buffer amplifier is compared with the GAA-FET and CMOS OTAs. The circuit can provide wide bandwidth with large resistance load driving capability. The performance of the two-stage OTA buffer circuit using an SG-RFET OTA is discussed in the previous section. The simulations are performed with different load resistance values ($R_L = 1$ kΩ, 5 kΩ, and 10 kΩ). The simulation results provide an insight into the application of the simple SG-RFET device in analog circuits. The gain reduces with a decrease in R_L. The SG-RFET buffer circuit can operate in a frequency range from DC to 1 GHz with a gain of 0.95 V/V at R_L 800 Ω and $V_{DD} = 2$ V.

TABLE 11.2

Unity Gain Buffer Amplifier Using Different OTA

	Unity Gain Amplifier - Two OTA			
Parameters	LM13700 IC	GAA-FET OTA	SG-RFET OTA	Strained Silicon SG-RFET OTA
Supply voltage	±15 V	2	2	2
g_m (mS)	9.6	1	0.79	0.9
R_L (kΩ)	10	1	1	1
Bandwidth (MHz)	0.81	5000	510	720
Gain (V/V)	0.99	0.99	0.99	0.99

In this chapter, a buffer amplifier circuit for driving low-resistance load using an SG-RFET device is discussed. In Ref. [33], a simple variable gain buffer amplifier using OTA is proposed that can drive a resistive load. Unity gain buffer amplifier using CMOS OTA is reported in Ref. [32] is simulated in the TCAD tool using SG-RFET OTA and GAA-FET OTA. Unity gain buffer circuit is implemented using the strained silicon SG-RFET device and compared with the LM13700 OTA IC, SG-RFET OTA, and GAA-FET OTAs. The simulation results show that the SG-RFET device can be used for analog circuit implementation and digital circuits. With scaling, OFF current increases in an RFET device. To overcome this and to set symmetrical characteristics in RFET, new materials have to be used instead of the silicon channel and the gate electrode.

11.7   CONCLUSION

The device characteristics of SG-RFETs are studied with different gates and channel lengths to explore the possibilities in the TCAD tool. The electron and hole tunneling of the SG-RFET device is studied by including the required physics models. Various analog and digital circuits are implemented and characterized using the SG-RFET device. A unity gain analog buffer amplifier driving low-resistance load is demonstrated using the SG-RFET OTA for the first time, which provides insight into the application of SG-RFETs in analog circuits.

REFERENCES

1. Colinge, Jean-Pierre, ed. *FinFETs and Other Multi-Gate Transistors*. Vol. 73. New York: Springer, 2008.
2. Colinge, Jean-Pierre, ed. "The SOI MOSFET: From Single Gate to Multi-Gate." In *FinFETs and Other Multi-Gate Transistors*. pp. 1–48. Boston, MA: Springer, 2008.
3. Ning, Tak H., and Yuan Taur. *Fundamentals of Modern VLSI Devices*. pp. 270–271. Cambridge, UK: Cambridge University Press, 1998.
4. Taur, Y., and T. H. Ning. *Fundamentals of Modern VLSI Devices*. Cambridge: Cambridge University Press, 2013.
5. Rai, Shubham, Jens Trommer, Michael Raitza, Thomas Mikolajick, Walter M. Weber, and Akash Kumar. "Designing Efficient Circuits Based on Runtime-Reconfigurable Field-Effect Transistors." *IEEE Transactions on Very Large Scale Integration (VLSI) Systems* 27, no. 3 (2018): 560–572.
6. Yao, Yan, Yabin Sun, Xiaojin Li, Yanling Shi, and Ziyu Liu. "Novel Reconfigurable Field-Effect Transistor with Asymmetric Spacer Engineering at Drain Side." *IEEE Transactions on Electron Devices* 67, no. 2 (2020): 751–757.
7. Weber, W. M., A. Heinzig, J. Trommer, D. Martin, M. Grube, and T. Mikolajick. "Reconfigurable Nanowire Electronics–A Review." *Solid-State Electronics* 102 (2014): 12–24.
8. Weber, Walter M., André Heinzig, Jens Trommer, Matthias Grube, Franz Kreupl, and Thomas Mikolajick. "Reconfigurable Nanowire Electronics-Enabling A Single CMOS Circuit Technology." *IEEE Transactions on Nanotechnology* 13, no. 6 (2014): 1020–1028.
9. Heinzig, André, Stefan Slesazeck, Franz Kreupl, Thomas Mikolajick, and Walter M. Weber. "Reconfigurable Silicon Nanowire Transistors." *Nano Letters* 12, no. 1 (2012): 119–124.
10. Weber, Walter M., Lutz Geelhaar, Luca Lamagna, Marco Fanciulli, Franz Kreupl, Eugen Unger, Henning Riechert, Giuseppe Scarpa, and Paolo Lugli. "Tuning the Polarity of Si-Nanowire Transistors Without the Use of Doping." In *2008 8th IEEE Conference on Nanotechnology*, pp. 580–581. IEEE, 2008.
11. Wessely, F., T. Krauss, and U. Schwalke. "CMOS Without Doping: Multi-Gate Silicon-Nanowire Field-Effect-Transistors." *Solid-State Electronics* 70 (2012): 33–38.
12. Sacchetto, Davide, Yusuf Leblebici, and Giovanni De Micheli. "Ambipolar Gate-Controllable SiNW FETs for Configurable Logic Circuits with Improved Expressive Capability." *IEEE Electron Device Letters* 33, no. 2 (2011): 143–145.
13. Darbandy, Ghader, Martin Claus, and Michael Schröter. "High-Performance Reconfigurable Si Nanowire Field-Effect Transistor Based on Simplified Device Design." *IEEE Transactions on Nanotechnology* 15, no. 2 (2016): 289–294.

14. "Synopsys TCAD Sentaurus device user's manual version 2012.06," 2017. https://www.synopsys.com/silicon/tcad.html

15. Ranjith, R., Remya Jayachandran, K. J. Suja, and Rama S. Komaragiri. "Two Dimensional Analytical Model for A Reconfigurable Field Effect Transistor." *Superlattices and Microstructures* 114 (2018): 62–74.

16. Jayachandran, Remya, Rama S. Komaragiri, and P. C. Subramaniam. "Reconfigurable Circuits Based on Single Gate Reconfigurable Field-Effect Transistors." In *2020 IEEE International Conference on Electronics, Computing and Communication Technologies (CONECCT)*, pp. 1–5. IEEE, 2020.

17. Jayachandran, Remya, Rama S. Komaragiri, and P. C. Subramaniam. "Study of Circuits Based on SOI-vertical Gate-All-Around FET." In *2018 15th IEEE India Council International Conference (INDICON)*, pp. 1–6. IEEE, 2018.

18. Lin, Yu-Ming, Joerg Appenzeller, Joachim Knoch, and Phaedon Avouris. "High-Performance Carbon Nanotube Field-Effect Transistor with Tunable Polarities." *IEEE Transactions on Nanotechnology* 4, no. 5 (2005): 481–489.

19. Weber, Walter M., Jens Trommer, Matthias Grube, André Heinzig, Markus König, and Thomas Mikolajick. "Reconfigurable Silicon Nanowire Devices and Circuits: Opportunities and Challenges." In *2014 Design, Automation & Test in Europe Conference & Exhibition (DATE)*, pp. 1–6. IEEE, 2014.

20. Trommer, Jens, André Heinzig, Tim Baldauf, Stefan Slesazeck, Thomas Mikolajick, and Walter M. Weber. "Functionality-Enhanced Logic Gate Design Enabled by Symmetrical Reconfigurable Silicon Nanowire Transistors." *IEEE Transactions on Nanotechnology* 14, no. 4 (2015): 689–698.

21. Weber, Walter M., André Heinzig, Jens Trommer, Matthias Grube, Franz Kreupl, and Thomas Mikolajick. "Reconfigurable Nanowire Electronics-Enabling A Single CMOS Circuit Technology." *IEEE Transactions on Nanotechnology* 13, no. 6 (2014): 1020–1028.

22. De Marchi, Michele, Davide Sacchetto, Stefano Frache, Jian Zhang, P-E. Gaillardon, Yusuf Leblebici, and Giovanni De Micheli. "Polarity Control in Double-Gate, Gate-All-Around Vertically Stacked Silicon Nanowire FETs." In *2012 International Electron Devices Meeting*, pp. 8–4. IEEE, 2012.

23. Lin, Yu-Ming, Joerg Appenzeller, Joachim Knoch, and Phaedon Avouris. "High-Performance Carbon Nanotube Field-Effect Transistor with Tunable Polarities." *IEEE Transactions on Nanotechnology* 4, no. 5 (2005): 481–489.

24. De Marchi, Michele, Jian Zhang, Stefano Frache, Davide Sacchetto, Pierre-Emmanuel Gaillardon, Yusuf Leblebici, and Giovanni De Micheli. "Configurable Logic Gates Using Polarity-Controlled Silicon Nanowire Gate-All-Around FETs." *IEEE Electron Device Letters* 35, no. 8 (2014): 880–882.

25. Wang, Dunwei, Bonnie A. Sheriff, and James R. Heath. "Complementary Symmetry Silicon Nanowire Logic: Power-Efficient Inverters with Gain." *Small* 2, no. 10 (2006): 1153–1158.

26. Gaillardon, Pierre-Emmanuel, Luca Amaru, Jian Zhang, and Giovanni De Micheli. "Advanced System on a Chip Design Based on Controllable-Polarity FETs." In *2014 Design, Automation & Test in Europe Conference & Exhibition (DATE)*, pp. 1–6. IEEE, 2014.

27. Simon, Maik, Boshen Liang, Dustin Fischer, Martin Knaut, Alexander Tahn, Thomas Mikolajick, and Walter M. Weber. "Top-Down Fabricated Reconfigurable FET with two Symmetric and High-Current on-States." *IEEE Electron Device Letters* 41, no. 7 (2020): 1110–1113.

28. Wang, D. W., B. A. Sheriff, and J. R. Heath, "Complimentary Symmetry Silicon Nanowire Logic: Power-Efficient Inverters with Gain." *Small* 2, no. 10, (2006) 1153–1158,

29. Justeena, A. Nisha, and R. Srinivasan, "Reconfigurable FET-Based SRAM and Its Single Event Upset Performance Analysis Using TCAD Simulations." *Microelectronics Journal* 101 (2020): 104815.

30. Saha, Priyanka, Dinesh Kumar Dash, and Subir Kumar Sarkar. "Nanowire Reconfigurable FET as Biosensor: Based on Dielectric Modulation Approach." *Solid-State Electronics* 161 (2019): 107637.

31. Nauta, Bram. "A CMOS Transconductance-C Filter Technique for Very High Frequencies." *IEEE Journal of Solid-State Circuits* 27, no. 2 (1992): 142–153.

32. Jayachandran, Remya, P. C. Subramaniam, and K. J. Dhanaraj. "A Novel Tunable Gain CMOS Buffer Amplifier for Large Resistive Loads." *Integration –The VLSI Journal* 7 (March 2021): 1–12.

33. Jayachandran, Remya, R. Ranjith, Rama S Komaragiri, and P. C. Subramaniam. "High-Performance Reconfigurable FET for a Simple Variable Gain Buffer Amplifier Design." *International Journal of Electronics* 2021 (Published Online). https://doi.org/10.1080/00207217.2021.1908618

12 NEGF Method for Design and Simulation Analysis of Nanoscale MOS Devices

Chhaya Verma and Jeetendra Singh

CONTENTS

12.1 INTRODUCTION

The non-equilibrium Green's function (NEGF) method primarily originated in the 1960s from the informative works of Martin and Schwinger to examine the carrier interaction only in the channel [1]. Because of the dominant wavelike behavior of electrons in mesoscopic physics, the quantum transport theory approach is utilized at the interface of the channel along with the scattering effect in the channel, which combines the effect of the dynamic [2]. The transportation of electrons is caused by two effects, i.e., force and entropy [3]. Even in the absence of an external force, the electrons are in motion and it is not possible to predict their exact location; one can only tell where the probability of finding the electron is higher or lower. The trajectory in which the electron moves is also complex and is represented in the wave function [4]. The flow of electrons is affected by electron-electron interaction or electron-phonon interaction. These interactions are random and depend on effects like internal energy and spatial arraignment of sub-particles [5]. With the application of an external force, the motion of the electron is distributed and the vibration of lattice atoms increases, which results in higher scattering [6, 7]. At the macroscopic level, these internal effects are small and, hence, ignored, but at the mesoscopic regimes, these effects are vital and need to be considered for an accurate analysis of electron flow [8]. The NEGF method is advanced and carefully considers all the dynamic and scattering processes and provides a reliable result [9]. The following are the main features of NEGF [10–14]:

1. The Green function is a function of two coordinates, i.e., space and time, and using this function in quantum transport, time-dependent expectation values such as current, densities, energy, and so forth, can be calculated.

DOI: 10.1201/9781003126393-12

2. NEGF is applicable to both extended and finite systems.
3. Diagrammatical analysis of dissipative processes and memory effect in quantum transport can be done using the NEGF method.
4. NEGF can handle strong and complicated perturbative external fields as electron-electron interactions are included in terms of infinite summations.

This chapter starts with the modified Ohm's law for ballistic transports followed by a brief explanation of diffusive and ballistic transport. Next, two transport theories are discussed followed by thermodynamics and the Schrödinger wave equation. Then, an overview of the many-body system and the NEGF method is discussed.

12.2 MODIFIED OHM'S LAW FOR BALLISTIC TRANSPORT

The eminent physicist and mathematician Georg Simon Ohm (1787–1854) defined a relationship between current and voltage, and this relationship is known as Ohm's law [15]. The law states that the potential difference V across any material is directly proportional to the current I flowing through it under constant temperature and pressure [16]. The generalized relation is given as

$$V = IR \tag{12.1}$$

The constant of proportionality R is the resistance of the material and for macroscopic and microscopic size, it is defined as $R = \rho \frac{L}{A}$. The ability of a material to oppose the current flow is termed as resistivity (ρ), which is independent of the geometry property of the material. Figure 12.1 depicts how R linearly varies with L and is inversely related to cross-sectional area A of the current-carrying conductor. It is a common understanding that reducing the length reduces resistance and decreasing the cross-sectional area increases resistance.

In a field-effect transistor (FET), the flow of electrons can be visualized easily by considering it as a resistor that has two contacts and a channel. The electrons flow from source contact to drain contact via the channel. The channel is like a controlling breeze that is operated by a gate terminal. From the standard Ohm's law, it can be stated that if the size of the FET is reduced to nanoscale then its resistance will decrease significantly. Theoretically, if the length L is reduced to zero then resistance R approaches zero, but practically there is still some resistance present in the interface referred to as interface resistance [17].

In a long resistor, electron flow from one end to the other suffers with an inelastic collision in the path, as shown in Figure 12.2, and such a transport is termed as diffusive transport. The resistor is called an inelastic resistor as energy is dissipated. Figure 12.2 shows only the source channel and drain of the metal oxide semiconductor field-effect transistor (MOSFET) consisting of a long channel.

When a short resistor whose length of the channel is almost equal to the mean free path (λ), the electron travels unscattered without dissipating energy like the elastic resistor, as depicted in Figure 12.3, and such transport is termed as ballistic transport. The mean free path, λ, is the path in

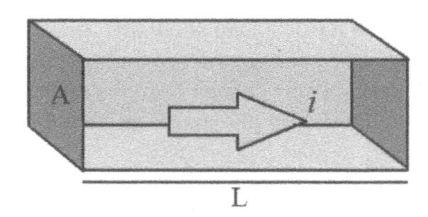

FIGURE 12.1 Current carrying conductor.

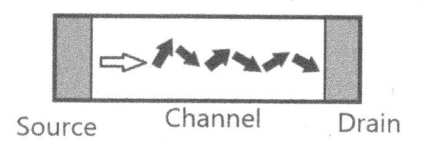

FIGURE 12.2 Diffusive means of transport in the MOSFET system of long channel length.

which the electron travels ballistically without any hindrance through the channel, and it is generally less than 1 μm, but its value also depends on temperature and material type [18].

Depending on the length of the conductor, it can be classified as ballistic or diffusive. If the length $L \ll \lambda$, the electron travels without being scattered and it is a ballistic conductor. If $L \gg \lambda$, diffusive transport of electrons occurs and it is a diffusive conductor [17]. For diffusive transport, standard Ohm's law is adequate, but in ballistic transport a modification in Ohm's law is made as expressed [19]

$$R = \frac{\rho}{A}(L + \lambda) \tag{12.2}$$

where resistivity ρ is expressed as

$$\rho = \frac{h}{q^2} \frac{A}{M\lambda} \tag{12.3}$$

where h/q^2 is a fundamental constant with Plank constant (h) and electron charge q, and M denotes the effective number of channels present for conduction (channel is not referring to the physical channel but to the parallel mean free path), and λ is the mean free path. So, the resistance for a ballistic conductor can be given by

$$R_B = \frac{h}{q^2} \frac{1}{M} \tag{12.4}$$

The short resistors are referred to as elastic resistors when there is no energy dissipation during the electron's ballistic transportation. The long resistor can also be represented as a series combination of the elastic resistor. Such a division of the long resistor into small parts of length shorter than L_{in}, where L_{in} is an average length traveled by an electron before being scattered, helps in obtaining linear conductance [17, 18]. It must be noted that the elastic resistor model can be used only for short length ($<L_{in}$) and the voltage drop across this length should be less thermal voltage (V_T).

12.3 BALLISTIC AND DIFFUSIVE TRANSPORT

The ballistic conductance [20] can be given by

$$G_B = \frac{q^2}{h} M \tag{12.5}$$

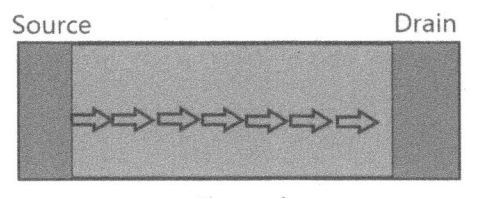

FIGURE 12.3 Ballistic means of transport in the MOSFET system of short channel length.

The conductance G of a long resistor of length L can be represented as ballistic conductance G_B and mean free path λ, as

$$G = \frac{G_B \lambda}{L + \lambda} \tag{12.6}$$

The long channel conductance G can also be rewritten in terms of travel time t and density of states $D(E)$, since the number of channels $M(E)$ is proportional to the density of states $D(E)$. Therefore, it can be represented as

$$G = \frac{q^2}{2} \frac{D}{t} \tag{12.7}$$

The density of states and travel time depends on channel dimensions for a long resistor [21]; hence, two transport regimes are broadly divided based on the travel time of electron:

1. *Diffusive regime:* Travel time $t \sim L^2$
2. *Ballistic regime:* Travel time $t \sim L$

Since the density of the state is proportional to $A*L$, it replaces length L in terms of travel time t. Ballistic conductance can be written as Equation (12.8)

$$G_B \sim \frac{q^2}{2} \frac{AL}{L} \tag{12.8}$$

or

$$G_B \sim \frac{q^2}{2} A \tag{12.9}$$

Therefore, it can be concluded that ballistic conductance does not depend on the length and has non-Ohmic behavior, which is also noticed in short conductors. Whereas diffusive conductor G depends on channel length L, as can be noticed from Equation (12.10), which shows ohmic behavior,

$$G \sim \frac{q^2}{2} \frac{D}{L} \tag{12.10}$$

In the ballistic transport, electrons move in a straight trajectory, and the travel time t can be assumed as

$$\text{Travel time} = \text{Distance traveled/mean velocity of the electron} \tag{12.11}$$

$$t_B = \frac{L}{\bar{u}} \tag{12.12}$$

where $\bar{u} = \langle |v_z| \rangle$ represents the mean velocity of electrons in the Z-direction. Practically even in short resistors, few electrons also receive scatter randomly under the influence of phonons and other defects [18]. Thus, the travel time can be rewritten as

$$t = \frac{L}{\bar{u}} + \frac{L^2}{2\bar{D}} \tag{12.13}$$

Here the travel time has been represented as a summation of ballistic limit $\frac{L}{\bar{u}}$ and diffusive limit $\frac{L^2}{2\bar{D}}$ from the theory of random motion [22], where $\bar{D}$ is the diffusion constant given by $\bar{D} = \langle v_z^2 \tau \rangle$, and τ is the mean free time [17, 18]. An average, which is represented by the symbol $\langle \ \rangle$, provides the different numerical factor over the angular variation of velocities depending on the conductor dimensionality.

Rewriting the travel time as

$$t = t_B \left(1 + \frac{L\bar{u}}{2D} \right)$$

(12.14)

Hence, the mean free path will become

$$\lambda = \frac{2\bar{D}}{\bar{u}}$$

(12.15)

12.4 DRIFT AND TRANSPORT THEORIES

The flow of electrons in any conductor is classically described by drift and diffusion phenomena. The drift current is caused by an electric field that occurs due to an electrostatic potential gradient. The driving electrochemical potential (μ) across the conductor is a combination of two different potentials, i.e., chemical potential ($\mu - U$) and electrostatic potential (U) [3], as depicted in Figure 12.4.

Current density is given as $J = J_{Drift} + J_{Diffusion}$, which is called the drift-diffusion equation [23]. Considering drift in the calculation of current in such a nanoscale regime is complex and requires deep knowledge of quantum mechanics. Therefore, to ease the evaluation, the effects of drift are ignored. The main cause of current flow was considered only due to the difference in the electrochemical potential, neglecting the effect of drift current, resulting in an inaccurate evaluation of current. To resolve this issue and obtain an accurate value of the current, the concept of transport theory was introduced. In these transport theories, two driving factors, force and entropy, are considered as the main cause of electron flow. The transport theory explains the flow of electrons in terms of two different processes: an elastic transfer which is force driven, and heat generation leading to randomization which is entropy driven [3]. The two transport theories that are used to study the transport theory are as follows

1. Semi-classical transport theory
2. Quantum transport theory

The semi-classical transport theory explains the transport of electrons by a combination of Newtonian mechanics for force-driven transport and microscopic thermodynamics for entropy-driven transport using the Boltzmann transport equation (BTE) [24, 25]. In a semi-classical approach, particle behavior of the electron is considered and quantum input is used and both forces drive the energy-momentum relation and entropy-driven scattering operator.

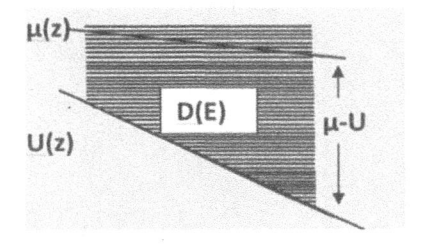

FIGURE 12.4 Electrostatic potential U/q, electrochemical μ/q, and varying density of state $D(\mu, E)$.

12.4.1 BOLTZMANN TRANSPORT EQUATION (BTE)

In the BTE, the force-driven factor is explained using a distribution function that gives information about the occupancy of state in terms of momentum and velocity in any particular direction at any instant, and the entropy-driven factor is represented by a scattering operator [25, 26].

$$\frac{\partial f}{\partial t} + v_z \frac{\partial f}{\partial z} + F_z \frac{\partial f}{\partial p} = S_{op} f \tag{12.16}$$

where f is the distribution function and F_z is the force experienced by the electrons in the z-direction at time t, such that electrons gain a momentum P_z and start to flow with the velocity of v_z. While moving in the z-direction, these electrons scatter along with different velocity states, which are represented by scattering operator S_{op}. If there is no heat generation entropy effect then the right set is equal to zero (i.e., without scattering).

12.4.2 QUANTUM TRANSPORT

In mesoelectronics and nanoelectronics, the wavelike behavior of electrons is more dominant; hence, force-driven factors cannot be explained using classical dynamics. In this transport the Schrödinger wave equation is used to represent force-driven factors, and thermodynamics is used for entropy-driven factors [27]. The method that combines the quantum mechanics force-driven equation with the entropy-driven factor using many-body perturbation theory (MBPT) is the NEGF method [28].

12.5 THERMODYNAMICS

A system like a molecule that consists of a large number of discrete particles whose energies and physical state constantly change with time needs a large number of variables to describe the system's state. It does this by using microscopic thermodynamics. A macrostate represents the collective behavior of a system, whereas a microstate describes the entire system in terms of physical quantities of discrete particles. The total number of a microstate is a function of energy E, several particles N, and volume V and is denoted by $W(N, V, E)$ [29]. The thermodynamic properties of any given system can be determined by function $W(N, V, E)$. Figure 12.5 shows the energy, particles, and volume exchange of two systems that are brought into contact.

The total entropy of the combined system is given by

$$S(E,V,N) = S_1\left(E_1,V_1,N_1\right) + S_2\left(E_2,V_2,N_2\right) \tag{12.17}$$

If macrostates A_1 and A_2 have $\Omega_1(N_1, V_1, E_1)$ and $\Omega_2(N_2, V_2, E_2)$ microstates, respectively, then Ω is a statistical quantity [29]. The macrostate of the total system when these two systems A_1 and A_2 are brought in contact is given by

$$\Omega(N,V,E) = \Omega_1\left(N_1,V_1,E_1\right)\Omega_2\left(N_2,V_2,E_2\right) \tag{12.18}$$

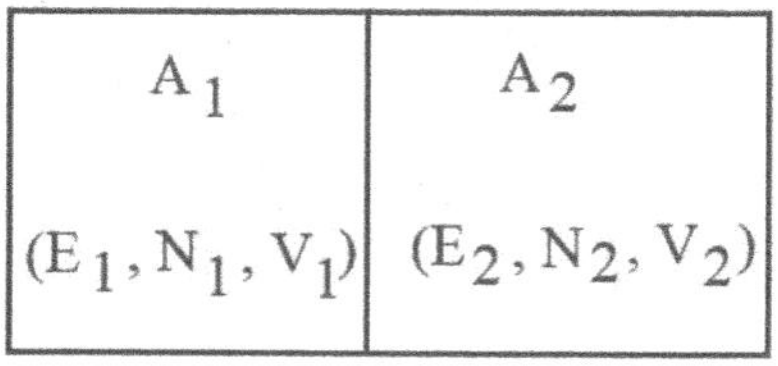

A_1	A_2
(E_1, N_1, V_1)	(E_2, N_2, V_2)

FIGURE 12.5 Energy, particles, and volume exchange of two systems that are brought into contact.

A statistical function is a set of similar macrostates that have different microstates. For any physical system, there exists a relationship between thermodynamics extensive quantity entropy S, and statistical quantity Ω is given by

$$S = k_B \ln \Omega \tag{12.19}$$

Equation (12.19) explains the randomness of microscopic particles in small systems, where k_B is Boltzmann's constant, which acts as a gateway between microscopic and macroscopic physics.

12.6 SCHRÖDINGER WAVE EQUATION

The wave behavior and discrete energy level of particle-like atoms and molecules are well explained in quantum mechanics. Schrödinger's wave equation explains both wave-particle duality and quantization of energy levels for dynamic particles [30]. The flow of current depends on the energy of the channel and the density of state $D(E)$ available. In Schrödinger's wave function, the physical density of the particle is described using wave function $\psi(x, t)$ and is given by $|\psi(x, t)|^2$ [29]. The spatial mass density is denoted by $m|\psi(x, t)|^2$, and charge density is denoted by $-e\,|\psi(x, t)|^2$. At the mesoscale and nanoscale the wave property of the particle is more dominant, and the interpretation of this wave function implies that wave property prevails over particle property. The Schrödinger wave equation is given as

$$i\hbar \frac{\partial}{\partial t} \psi(x,t) = -\frac{\hbar}{2m} \nabla^2 \psi(x,t) + V(x)\psi(x,t) \tag{12.20}$$

where $\hbar = \frac{h}{2\pi}$ is the reduced Plank constant, $V(x)$ is potential energy, and $\psi(x, t)$ denotes the wave function. Schrödinger's wave equation can be derived from the conservative force of classical mechanics by replacing the dynamical variables with the operators. For a non-relativistic particle in motion, the conservative force $F(x) = -\nabla V(x)$ can be described in Newton's equation [31]

$$m \frac{d^2 x(t)}{dt^2} = -\nabla V(x,t) \tag{12.21}$$

The wave function can be described with the help of energy, momentum of the particles; total energy is the addition of potential energy $V(x)$ and kinetic energy $\frac{P^2}{2m}$, and from the energy-momentum relation, total energy can be written as

$$E = \frac{P^2}{2m} + V(x) \tag{12.22}$$

From the de Broglie wavelength [32] for wavelike behavior of an electron,

$$\lambda = \frac{h}{P} \tag{12.23}$$

From the Compton effect, the energy of the photon,

$$E = hf \tag{12.24}$$

Assuming a monochromatic plane wave of frequency f, wavelength λ, and direction of motion $\hat{k}$, the wave function can be described as

$$\psi(x,t) = Ae^{2\pi i\left(\frac{\hat{k}}{\lambda} - ft\right)} \tag{12.25}$$

Substituting the values of wavelength and frequency in the wave Equation (12.25) we get

$$\psi(x,t) = Ae^{i\left(\frac{Px}{\hbar} - \frac{P2}{2m\hbar}t\right)} \tag{12.26}$$

Partially differentiating Equation (12.26) with respect to time t, the obtained result is

$$i\hbar \frac{\partial \psi(x,t)}{\partial t} = \frac{P^2}{2m}\psi(x,t) \tag{12.27}$$

Partially differentiating Equation (12.26) with respect to position x, the obtained result is

$$\psi(x,t) = \frac{-\hbar^2}{P^2}\nabla^2\psi(x,t) \tag{12.28}$$

Substituting the value of $\psi(x,t)$ from Equation (12.28) into Equation (12.27), Equation (12.27) will become

$$i\hbar \frac{\partial \psi(x,t)}{\partial t} = \frac{P^2}{2m}\psi(x,t) = -\frac{\hbar}{2m}\nabla^2\psi(x,t) \tag{12.29}$$

Generalizing the Equation (12.29) in terms of total energy, Schrödinger's wave equation is obtained

$$i\hbar \frac{\partial}{\partial t}\psi(x,t) = -\frac{\hbar}{2m}\nabla^2\psi(x,t) + V(x)\psi(x,t) \tag{12.30}$$

Equation (12.30) is the time-dependent Schrödinger wave equation. This equation is also written as

$$E\psi = \hat{H}\psi \tag{12.31}$$

where $\hat{H}$ is the Hamiltonian operator given as

$$\hat{H} = -\frac{\hbar^2}{2m}\nabla^2 + V(r) \tag{12.32}$$

12.7   MANY-BODY SYSTEMS

The characteristic study of microscopic systems, which include multiple active particles, is dealt with by quantum many-body theory [29]. Inaccurately specifying such systems, the interparticle potential must be considered while determining the many-body Schrödinger equation.

12.7.1   PRIMARY QUANTIZATION

In quantum mechanics, it cannot be stated that a certain wave function represents the motion of a particular particle, i.e., only by knowing the total state of all particles. If only total states of the particles are known then the wave function of an N particle system is given as

$$\psi\left(x_1, x_2, ..., x_N\right) = \left\langle x_1, x_2, ..., x_N \middle| \psi \right\rangle$$

where the location of the ith particle is denoted by x_i. This location is determined by spatial coordinate r_i and spin coordinate σ_i. In this quantization method, the dynamics variables are replaced by specified operators, whereas time-dependent Schrödinger equations are set for motion expressions of single-body systems with N particles [33]. Here, the wave function relates only to the state of a system with fixed particles in an instant but neglects the effect of the force field, especially the electromagnetic field and special relativity. For an N-body state, the Hamiltonian operator is expressed as a combination of summation kinetic energy, and background energy with the summation of the interaction potential [34]

$$\hat{H} = \hat{H}_0 + \hat{H}_{int} \tag{12.33}$$

Since this quantization neglects the force field, the solution of the Hamiltonian becomes difficult for a system with interaction potential.

12.7.2 SECONDARY QUANTIZATION

The second quantization describes the many-body systems by quantizing the fields using a basis that describes the number of particles occupying each state. The many-body systems are described by the second quantization scheme, in which fields are quantized after assuming that each state is occupied and described by several particles [35]. Using this quantization concept, a one-body state is articulated in terms of occupation number. For any order, a complete one-body basis is given as $|v_i\rangle$ and for an N-body system, the basis states can be represented as $|n_{vi}, n_{v2}, ...\rangle$ in the occupation number representation. This notation means n_{vi} particles in the state v_i with $\sum_i n_{vi} = N$ [29, 33]. To draw up statistical mechanics in the form of the grand-canonical ensemble (a thermodynamic ensemble that describes a system in contact with a heat and particle bath, where chemical potential and the temperature are defined while the particle number and energy of the particles are not specified), one should treat each state with a distinct number of particles. Therefore, Fock space F is used to define the direct addition of all N-body Hilbert spaces [36].

$$f = H^0 \oplus H^1 \oplus H^2..... \tag{12.34}$$

where H represents Hilbert space.

12.8 NON-EQUILIBRIUM GREEN FUNCTION (NEGF)

The NEGF method combines quantum mechanics specified by the Schrödinger equation and entropy, which are described using the MBPT [37, 38]. An earlier NEGF method only focused on the interactions occurring throughout the channel, which did not include the entropy-driven processes. Due to inaccurate results from semi-classical transport, the new quantum transport theory, which modified the NEGF method by including entropy-driven processes, was started. The NEGF method uses two types of inputs. The first type is Hamiltonian $[\hat{H}]$ and it describes the dynamics of an elastic channel since the eigenvalue of the Hamiltonian matrix denotes the allowed energy level in the channel [3]. The second type is self-energy $[\Sigma]$, which is used to describe the interconnection to the contacts. The word "contacts" here denotes all kinds of entropy-driven processes. In Figure 12.6, the quantum transport model of the MOSFET is represented, where Σ_1 and Σ_2 are physical contacts, whereas Σ_0 denotes an entropy-driven process in the channel only. The NEGF method is an esoteric and advanced tool.

In mesoscopic physics, the current through the elastic resistors can evaluate accurately by using the NEGF method. The scattering theory can be combined with the Schrödinger wave equation that effectively adds the inflow and outflow terms as given in Equation (12.35) with open boundary conditions.

$$E(\psi) = [H]\{\psi\} + [\Sigma]\{\psi\} + \{s\} \tag{12.35}$$

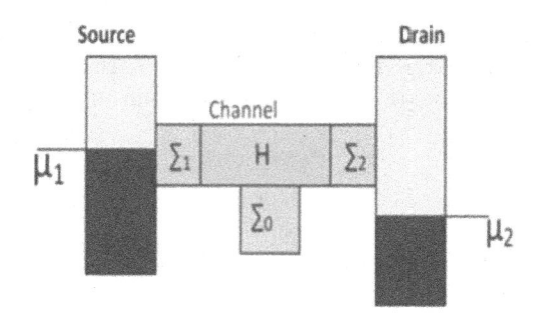

FIGURE 12.6   Quantum transport model of the MOSFET.

where $[\Sigma]\{\psi\}$ denotes the outflow term and $\{s\}$ is the inflow term

The NEGF method very diligently handles all kinds of interactions in both non-dissipative and dissipative channels and gives reliable results.

12.9   CONCLUSION

The quantum transport theory approach at the mesoscopic level is well explained here by quantum mechanics, which uses the force-driven transport of electrons. In this regard, initially modified Ohm's law for ballistic transport is discussed, which concludes that ballistic conductance does not depend on the length of the resistor. Further ballistic and diffusive transport is mentioned, which reveals that electron flow in a short resistor (length $< L_{in}$) is like a bullet in a straight trajectory. The flow of electrons can be analyzed by two transport theories, i.e., semi-classical transport theory, which uses BTE, and quantum transport theory, which uses the NEGF method. A brief overview of thermodynamics is given followed by the Schrödinger's wave equation, which is derived using the classical approach of energy momentum. Since a macrosystem consists of a large number of particles whose microstates are different irrespective of similar physical traits, the study of many-body systems is important to analyze such systems. Finally, the NEGF method is discussed briefly. The NEGF method is very advanced and to study it in greater detail a good knowledge of advanced quantum mechanics, thermodynamics for microscopic particles, and advanced engineering mathematics is needed.

REFERENCES

1. Kadanoff, L.P. and Martin, P.C., 1961. Theory of many-particle systems. II. Superconductivity. *Physical Review, 124*(3), p.670.
2. Martinez, A., Bescond, M., Barker, J.R., Svizhenko, A., Anantram, M.P., Millar, C. and Asenov, A., 2007. A self-consistent full 3-D real-space NEGF simulator for studying nonperturbative effects in nano-MOSFETs. *IEEE Transactions on Electron Devices, 54*(9), p.2213–2222.
3. Datta, S., 2018. *Lessons from nanoelectronics: A new perspective on transport*, Singapore: World Scientific.
4. Vaníček, J. and Heller, E.J., 2003. Uniform semiclassical wave function for coherent two-dimensional electron flow. *Physical Review E, 67*(1), p.016211.
5. Dirac, P.A.M., 1930. A theory of electrons and protons. *Proceedings of the Royal Society of London. Series A, Containing Papers of a Mathematical and Physical Character, 126*(801), p.360–365.
6. Peng, L.M., 1999. Electron atomic scattering factors and scattering potentials of crystals. *Micron, 30*(6), p.625–648.
7. Levi, A.C. and Suhl, H., 1979. Quantum theory of atom-surface scattering: Debye-Waller factor. *Surface Science, 88*(1), p.221–254.
8. Datta, S., 1997. *Electronic transport in mesoscopic systems*. Cambridge, UK: Cambridge University Press.

9. Papior, N., Lorente, N., Frederiksen, T., García, A. and Brandbyge, M., 2017. Improvements on non-equilibrium and transport Green function techniques: The next-generation transiesta. *Computer Physics Communications, 212*, p.8–24.

10. van Leeuwen, R. and Dahlen, N.E., 2005 An Introduction to Nonequilibrium Green Functions, Theoretical Chemistry Materials Science Centre Rijksuniversiteit Groningen. https://tddft.org/TDDFT2008/lectures/RL_LN.pdf

11. Perfetto, E., Uimonen, A.M., van Leeuwen, R. and Stefanucci, G., 2016, March. Time-resolved photoabsorption in finite systems: a first-principles NEGF approach. *Journal of Physics: Conference Series, 696*, p.012004.

12. Zeng, L., He, Y., Povolotskyi, M., Liu, X., Klimeck, G. and Kubis, T., 2013. Low rank approximation method for efficient Green's function calculation of dissipative quantum transport. *Journal of Applied Physics, 113*(21), p.213–707.

13. Datta, S., 2000. Nanoscale device modeling: the Green's function method. *Superlattices and Microstructures, 28*(4), p.253–278.

14. Anantram, M.P., Lundstrom, M.S. and Nikonov, D.E., 2008. Modeling of nanoscale devices. *Proceedings of the IEEE, 96*(9), p.1511–1550.

15. Gee, B., 1969. Georg Simon Ohm 1789-1854. *Physics Education, 4*(2), p.106.

16. Gupta, M.S., 1980. Georg Simon Ohm and Ohm's Law. *IEEE Transactions on Education, 23*(3), p.156–162.

17. Datta, S., 2005. *Quantum transport: atom to transistor.* Cambridge, UK: Cambridge University Press.

18. Areshkin, D., Gunlycke, D. and White, C., 2007. Ballistic transport in graphene nanostrips in the presence of disorder: importance of edge effects. *Nano Letters, 7*(1), p.204–210.

19. Joshi, A.A. and Majumdar, A., 1993. Transient ballistic and diffusive phonon heat transport in thin films. *Journal of Applied Physics, 74*(1), p.31–39.

20. Chklovskii, D.B., Matveev, K.A. and Shklovskii, B.I., 1993. Ballistic conductance of interacting electrons in the quantum Hall regime. *Physical Review B, 47*(19), p.12605.

21. Haug, H. and Jauho, A.P., 2008. *Quantum kinetics in transport and optics of semiconductors* (Vol. 2). Berlin: Springer.

22. Santos, E., 1969. A Lagrangian formulation of the theory of random motion. *Il Nuovo Cimento B (1965-1970), 59*(1), p.65–81.

23. Hänsch, W., 2012. *The drift diffusion equation and its applications in MOSFET modeling.* Cham, Switzerland: Springer Science & Business Media.

24. Bonitz, M., 2016. *Quantum kinetic theory* (Vol. 412). Berlin: Springer.

25. Rammer, J. and Smith, H., 1986. Quantum field-theoretical methods in transport theory of metals. *Reviews of Modern Physics, 58*(2), p.323.

26. Goldsman, N., Henrickson, L. and Frey, J., 1991. A physics-based analytical/numerical solution to the Boltzmann transport equation for use in device simulation. *Solid-State Electronics, 34*(4), p.389–396.

27. Jacoboni, C., 2010. *Theory of electron transport in semiconductors: a pathway from elementary physics to nonequilibrium Green functions* (Vol. 165). Berlin: Springer.

28. Camsari, K.Y., Chowdhury, S. and Datta, S., 2020. The non-equilibrium green function (NEGF) method. *arXiv Preprint arXiv, 2008*, p.01275.

29. Pourfath, M., 2014. *The non-equilibrium Green's function method for nanoscale device simulation.* Vienna: Springer.

30. Dong, S.H. and Garcia-Ravelo, J., 2007. Exact solutions of the s-wave Schrödinger equation with Manning–Rosen potential. *Physica Scripta, 75*(3), p.307.

31. Dick, R., 2012. *Advanced quantum mechanics.* New York: Springer.

32. De Broglie, L., 1924. Recherches sur la théorie des quanta. Doctoral dissertation, Masson, Paris.

33. Fetter, A.L. and Walecka, J.D., 2012. *Quantum theory of many-particle systems.* Minneola, NY: Dover Publications.

34. Mahan, G.D., 2013. *Many-particle physics.* New Delhi: Springer.

35. Mattuck, R.D., 1992. *A guide to Feynman diagrams in the many-body problem.* Minneola, NY: Dover Publications.

36. Chandler, C. and Gibson, A.G., 1977. N-body quantum scattering theory in two Hilbert spaces. I. The basic equations. *Journal of Mathematical Physics, 18*(12), p.2336–2347.

37. Kadanoff, L.P. andBaym, G., 2018. *Quantum statistical mechanics.* Boca Raton, FL: CRC Press.

38. Datta, S., 2000. Nanoscale device modeling: the Green's function method. *Superlattices and Microstructures, 28*(4), p.253–278.

13 Performance Investigation of a Novel Si/Ge Heterojunction Asymmetric Double-Gate DLTFET for Low-Power Analog/RF and IoT Applications

Suruchi Sharma, Rikmantra Basu, and Baljit Kaur

CONTENTS

13.1 INTRODUCTION

For the last five decades, the incessant scaling of metal oxide semiconductor field-effect transistor (MOSFET) technology nodes has enhanced functioning of complementary MOS (CMOS) transistors in the sense of speed, active current capability, footprint area, and improved electrical and radiofrequency (RF) specificities. But then again, it instigates short channel effects (SCEs) leading to large-power consumption by integrated circuits. These restrictions necessitate inquiry of novel structures pertaining to contemporary conduction devices and upgraded system designs with the aim of MOSFET succession and to subdue its subthreshold swing (SS) constraint (SS > 60 mV/dec). Consequently, several approaches have been stated in the literature such as work-function (WF) engineering (Raad et al. 2016), gate-dielectric engineering (Duan et al. 2018; Patel et al. 2019), spacer engineering (Raushan et al. 2018), bandgap engineering (Yoon et al. 2018), material engineering (Kumar et al. 2019), pocket engineering (Ashita et al. 2019), and so on, to deal with the issues mentioned earlier. Among them, the charge-plasma (CP) concept-based dopingless tunnel FET (DLTFET) (Kumar and Janardhanan 2013) has arisen as an energy-efficient expedient for low-power (LP) applications. The DLTFET provides immunity against process variations, random dopant fluctuations (RDFs), and compatibility with classical CMOS fabrication process flow. In the CP concept, electron and hole plasmas are formed on each side of the intrinsic semiconductor body by choosing the appropriate WF of the two gates and layer thickness of the semiconductor body. The CP requires that the WF of the gate should be different from that of semiconductor material, and the length of the semiconductor body should be lower than the Debye length, i.e., $L_D = \sqrt{\left(\frac{\varepsilon_{Sem} V_T}{qN}\right)}$, where ε_{Sem} represents the dielectric constant of the semiconductor material, V_T is the thermal voltage, q is the electronic charge, and N is the carrier concentration of the semiconductor body. This CP concept (Hueting et al. 2008; Rajasekharan et al. 2010; Kumar and Nadda 2012) was first introduced in a DLTFET by Kumar and Janardhanan (2013), wherein the p+ and n+ sections are made in the intrinsic semiconductor body by selecting suitable WF for the source and drain metal electrodes, respectively. Then, a gate-engineered DLTFET (GEDL-TFET) was proposed by Bashir et al. (2015) in which dual material was used for the top gate to enhance tunneling. Subsequently, there was significant enhancement in ON current (53 times) and I_{ON}/I_{OFF} ratio (~68 times). However, the realization of multioxide thickness and different WFs are the fabrication challenges associated with this device. Then, a drain WF-engineered DLTFET was proposed by Raad et al. (2016), wherein the drain electrode is separated into two different segments of different WF; the electrode with a larger WF is kept near the channel to constrain hole tunneling at the drain/channel interface, thus, suppressing ambipolar conduction and enhancing RF performance. Additionally, for enhancement of I_{ON}, a low bandgap material (InAs) has been exploited for the source region (Yadav et al. 2017), and a metallic film is positioned in the dielectric layer amid the gate and source electrode (Raad et al. 2017) to create an abrupt tunneling junction.

After that, Yadav et al. (2018) proposed a novel metal strip heterogate dielectric-engineered dual metal gate DLTFET, wherein a metal strip is inserted within the source/channel junction to enhance band-to-band tunneling (BTBT) with a drop in bandgap among the conduction band (CB) and valence band (VB). Furthermore, Raushan et al. (2018) proposed an Si-DLTFET with an oversized back gate to suppress ambipolarity issues. Then, Duan et al. (2018) proposed a GEDL-TFET in which $In_{0.75}Ga_{0.25}N$ is employed as a semiconductor body and gate engineering is done to enhance device performance. Furthermore, an $In_{0.53}Ga_{0.47}As/In_{0.52}Al_{0.48}As$ heterojunction (HJ) DLTFET (Liu et al. 2019) suppresses both I_{OFF} and the ambipolar current attributable to larger bandgap and higher hole effective mass, with an HfO_2/SiO_2 heterogate dielectric, in which HfO_2 provides the greater electric field for obtaining higher I_{ON}, and SiO_2 can avoid increasing the ambipolar current. A heteromaterial (HT) channel-based dual-gate DLTFET was proposed by Kumar and Raman (2020) that can be optimized according to the physical conditions for sensing applications.

From the TFET application point of view, the transfer characteristic, as well as its reliability, ought to be guaranteed. The reliability concerns in the TFET are very significant because of the establishment of an interface trap charge (ITC) across the semiconductor-dielectric interface (Pala et al. 2012; Wang et al. 2013). The presence of ITC is investigated by trapping mobile ionic and immobile charges across the semiconductor-oxide interface. These ITCs are created throughout the fabrication procedures by process-induced (Poindexter 1989), stress-induced (Trabzon and Awadelkarim 1998), and radiation-induced damage (Lho and Kim 2005), which deteriorates the device performance in the context of reliability and lifetime. Generally, there are two types of ITC, i.e., donor (PITC – positive ITC) and acceptor (NITC – negative ITC). The donor and acceptor ITCs lie among the VB and the CB, and they are formed primarily because of the unsaturated fourth bond identified at the semiconductor-dielectric interface. Without interface trap charge (WITC) is $N_f = 0$ cm^{-2}. There have been plenty of published reports concerning the influence of ITC on TFET (Venkatesh et al. 2017; Chandan et al. 2018; Gupta et al. 2018; Kumar and Raman 2019). Nevertheless, there is a need to evaluate the impact of ITC, as heterojunction asymmetric double-gate DLTFETs (HJ-ADG-DLTFETs) possess Si-HfO$_2$ and Ge-HfO$_2$ interfaces, and this issue has not been addressed much to date. Generally, the high-κ dielectric (Robertson 2004; Huang et al. 2010) leads to degradation of the interface surface, but due to proper annealing and layer deposition techniques, interface quality can be improved. Therefore, in this chapter, interface trap density (N_f) is taken as 10^{12} cm^{-2} (Sarkar et al. 2015). There is a need to evaluate the influence of temperature on HJ-ADG-DLTFET as the bandgap of the semiconductor material depends on the temperature (Varshni 1967) and the increased chip density results in increased heat dissipation, which leads to the increment in operating temperature of the chip. Consequently, increased operating temperature affects the on-chip operation and reliability of the device, and the suitability of the device to function properly in extreme conditions such as high- and low-operating temperatures ought to be investigated. Therefore, it is imperative to inspect the device's performance at a broad array of temperatures (Madan and Chaujar 2018). In this regard, an HJ-ADG-DLTFET (Sharma and Kaur 2020) was proposed by the authors that utilizes a low bandgap material for its source (i.e., Ge), which raises the BTBT rate and, hence, I_{ON}, while maintaining high I_{ON}/I_{OFF}. The reliability peculiarities of HJ-ADG-DLTFET are examined for different ITC densities and polarities in the context of the analog/RF parameters. Furthermore, process, voltage, and temperature (PVT) disparities affect switching speed and I_{OFF} associated with the transistor (Xiang et al. 2019). To overcome the PVT variability issue in the nanometer regime, the device must be aware of variations.

This chapter starts by discussing the need and fundamentals of TFET and DLTFET. This is followed by a description of published reports in the literature to overcome the limitations of conventional DLTFET. After that, the proposed device, i.e., HJ-ADG-DLTFET, is discussed in the context of device design, simulation setup, and fabrication feasibility of the device. Finally, HJ-ADG-DLTFET is analyzed comparatively on a simulation basis with ADG-DLTFET, in the context of analog/RF, and linear parameters. Additionally, reliability characteristics are also compared for both devices in terms of different ITC density, polarity, and PVT variations.

13.2 DEVICE DESIGN, SIMULATION SETUP, AND FABRICATION FEASIBILITY

The schematic device views of the ADG-DLTFET (Raushan et al. 2018) and the HJ-ADG-DLTFET (Sharma and Kaur 2020) are illustrated in Figure 13.1(a and b), correspondingly. The authors have employed Ge (low bandgap and high mobility material relative to Si) for the source region in HJ-ADG-DLTFET to intensify I_{ON}, but Ge material as a source also leads to increments in I_{OFF}. Therefore, to suppress the enhanced leakage current associated with the high mobility of charge carriers of the Ge-source region, we have used 4.6 eV as the drain electrode WF in HJ-ADG-DLTFET and high-κ gate dielectric. Subsequently, HJ-ADG-DLTFET exhibits high I_{ON} while maintaining a high I_{ON}/I_{OFF} ratio. Furthermore, the WF value has been chosen pragmatically by referring to the literature for the range of values that is practical (Uddin Shaikh and Loan 2019) to achieve better results for the device.

An equivalent oxide thickness (EOT) of 0.37 nm of high-κ (HfO$_2$) is made separating the source electrode and semiconductor body (1) to avoid silicide formation and direct tunneling of carriers through the gate (Robertson 2004), (2) to create large hole plasma, and (3) to enhance BTBT due

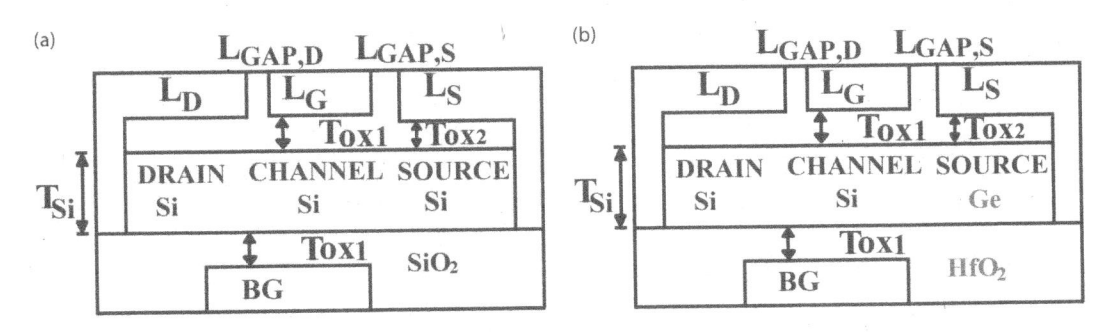

FIGURE 13.1 Schematic illustration of (a) ADG-DLTFET and (b) HJ-ADG-DLTFET.

TABLE 13.1

Physical Parameters of the Device Used in the Simulation

Parameters	Unit	ADG-DLTFET	HJ-ADG-DLTFET
Channel length (L_G)	nm	20	20
Silicon thickness (T_{si})	nm	5	5
Drain length (L_D)	nm	20	20
Source length (L_S)	nm	20	20
Gate-oxide material	-	SiO_2	HfO_2
Gate-oxide thickness (T_{OX1})	nm	5	5
Length of asymmetric back gate	nm	40	40
Spacer length for gate and drain ($L_{GAP,D}$)	nm	5	5
Spacer length for gate and source ($L_{GAP,S}$)	nm	5	5
Source work function	eV	5.93	5.93
Drain work function	eV	4.4	4.6
Gate work function	eV	4.5	4.5
Oxide thickness at Source (T_{OX2})	nm	0.37	0.37

to enhanced electrical coupling across the gate and tunneling junction by virtue of increased gate capacitance (Boucart and Ionescu 2007). The physical and simulation parameters are displayed in Table 13.1.

This chapter utilizes the Silvaco ATLAS for the simulation of conventional ADG-DLTFET and HJ-ADG-DLTFET (Silvaco 2016). Here, the nonlocal BTBT (BBT.NONLOCAL) model has been taken into consideration for the evaluation of BTBT in the horizontal direction as it estimates the BTBT rates by the energy band diagram. The FERMI and NI.FERMI models represent the integration of Fermi-Dirac characteristics so that the induced charge carriers do not surpass the limitation of the effective density of states of the semiconductor material. The Wentzel-Kramer-Brillouin (WKB) method is used for solving the Poisson and continuity equations. Additionally, the Lombardi mobility model (CVT) and concentration-dependent Shockley-Read-Hall (CONSRH) (Shockley and Read 1952) recombination models have been used to account for mobility variation and leakage current when other parameters such as voltage and temperature are constant. The Hansch model (HANSCHQM) is used to account for quantum confinement in the HJ-ADG-DLTFET. The TEMP statement is used to analyze the temperature variations in device physics. The calibration of device physical simulation models has been done with already published reports. The simulations have been done at a 1-MHz input frequency for gate voltage (V_G) = 1 V and drain voltage (V_D) = 1.

The proposed HJ-ADG-DLTFET can be mass produced under the state-of-the-art process technology discussed in the following section. Initially, high-quality intrinsic Ge film can be grown on an intrinsic Si substrate via molecular beam epitaxy (MBE) (Nakatsuka and Zaima 2015), ultrahigh vacuum chemical vapor deposition (UHVCVD), (Wang and Tsui 2015), or direct current (DC) magnetron sputtering (Steglich et al. 2013). After that, a bottom gate (BG) can be formed independently before the front gate (FG) formation (Widiez et al. 2005). Thus, reactive ion etching (RIE) is used to etch the epitaxy and then a high-κ gate dielectric layer (HfO_2) is deposited by using the Atomic Layer Deposition (ALD) technique (Johnson et al. 2014). Then, patterning and wet etching of the bottom side gate oxide in Hf are done. After that, the patterning of the BG region is done using lithography, and finally, BG can be formed via the metallization process. The same set of procedures is repeated to form the FG. Additionally, both BG and FG can be aligned using the same alignment marks. Then, patterning and etching of the FG oxide is done to form drain and source electrode regions. Subsequently, the process is completed by metallization of the source and drain electrodes.

13.3 RESULTS AND DISCUSSIONS

13.3.1 DC AND ANALOG/RF ANALYSIS

In the OFF state, a large barrier exists among the VB of the source and the CB of the channel area as illustrated in Figure 13.2(b), resulting in inhibition of BTBT. In this matter, the previously mentioned barrier has been abridged to a certain degree for HJ-ADG-DLTFET by adopting a small bandgap material in the source section; nonetheless, the electrons cannot tunnel through the source-channel junction. However, the application of positive V_G makes the source-channel junction reverse biased. Consequently, tunneling occurs as depicted in Figure 13.2(a), hence, the drain current (I_D) increases.

The I_D-V_G curve of both devices is shown in Figure 13.3(a). HJ-ADG-DLTFET illustrates an advancement in I_{ON} by one order with regard to ADG-DLTFET because of the adoption of a small bandgap semiconductor material for the source. For $V_D = 1$ V and $V_G = 1$ V, I_{ON} ~1.5 × 10⁻⁵A/μm, I_{OFF} ~1.5 × 10⁻¹⁵A/μm, and I_{ON}/I_{OFF} ~10¹⁰ has been attained. Figure 13.3(b) illustrates the g_m of both devices with regard to the V_G. For analog applications, transconductance appears to be a critical device criterion that alters V_G into I_D and regulates device gain and is formulated in Equation 13.1. Figure 13.3(b) shows that the g_m of the HJ-ADG-DLTFET is higher than the conventional one as HJ-ADG-DLTFET exhibits higher I_D. The improved g_m validates superior sensitivity for the conversion of V_G to I_D, ameliorated linearity, and high-frequency specificities.

The device efficiency (DE) is a momentous criterion that determines the efficacy to translate I_D into g_m for analog/RF applications and is given by Equation 13.2. Figure 13.3(c) illustrates better DE for HJ-ADG-DLTFET compared with ADG-DLTFET. Furthermore, the transconductance frequency product (TFP) is a considerable design criterion that is defined as the product of the DE and f_T and is given by Equation 13.3. In Figure 13.3(c) HJ-ADG-DLTFET illustrates high TFP as it rises linearly and achieves a highest value in the saturation condition and drops at higher V_G due to mobility conditions.

For TFET devices, C_{gd} dominates at high frequencies because of the formation of a feedback track among output and input, resulting in parasitic fluctuations and signal aberration. The formula for C_{gd} is given in Equation 13.4. As illustrated in Figure 13.4(a), HJ-ADG-DLTFET structures have somewhat more C_{gd} than ADG-DLTFET at higher V_G due to a rise in quantity of conduction states and capacitive coupling across the drain-channel interface.

$$g_m = \frac{\partial I_{DS}}{\partial V_{GS}} \tag{13.1}$$

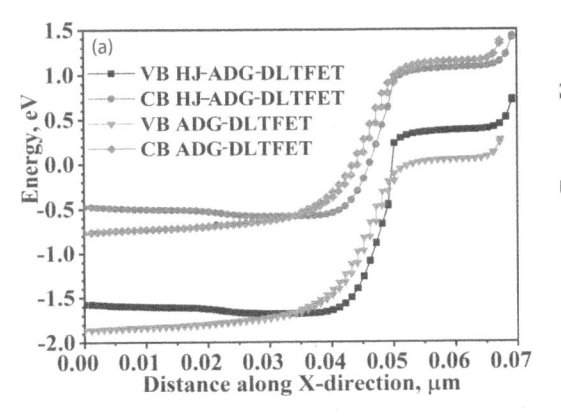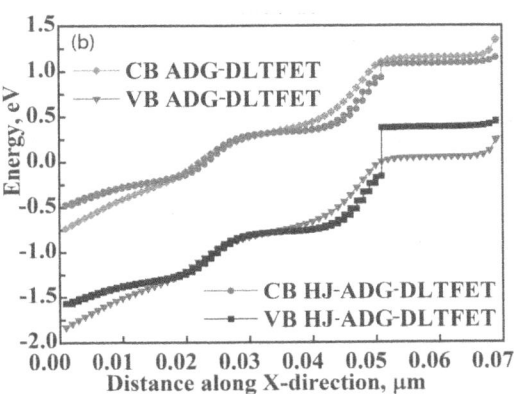

FIGURE 13.2 Energy band diagram of ADG-DLTFET and HJ-ADG-DLTFET: (a) ON state ($V_G = V_D = 1$ V) and (b) OFF state ($V_G = 0$ V, $V_D = 1$ V).

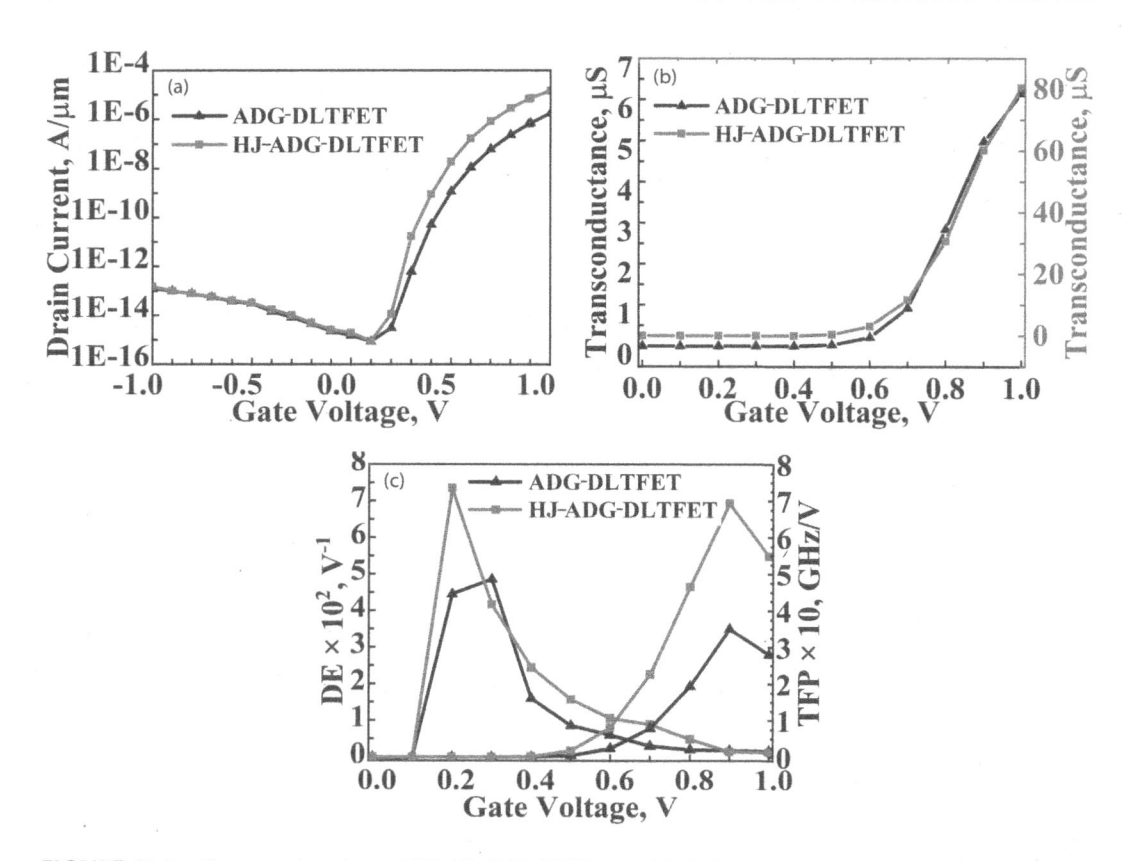

FIGURE 13.3 Comparative plots of HJ-ADG-DLTFET and ADG-DLTFET for (a) transfer characteristics, (b) transconductance, and (c) DE and TFP.

$$DE = \frac{g_m}{I_D} \tag{13.2}$$

$$TFP = \frac{g_m}{I_{ds}} \times f_T \tag{13.3}$$

$$C_{gd} = \frac{\partial Q_G}{\partial V_D} \tag{13.4}$$

$$f_T = \frac{g_m}{2\pi\left(C_{gd} + C_{gs}\right)} \tag{13.5}$$

$$GBP = \frac{g_m}{2\pi C_{gd}} \tag{13.6}$$

The f_T and gain-bandwidth product (GBP) are significant criteria for RF applications and pronounced as that frequency, whereas the short circuit current gain approaches toward unity and is expressed by Equation 13.5. Figure 13.4(b) specifies the change in f_T with regard to V_G. It is detected that f_T increases at higher V_G as g_m increases at high V_G. Furthermore, Figure 13.4(c) illustrates that the HJ-ADG-DLTFET shows higher GBP as GBP depends on g_m and C_{gd}. It is given by Equation 13.6.

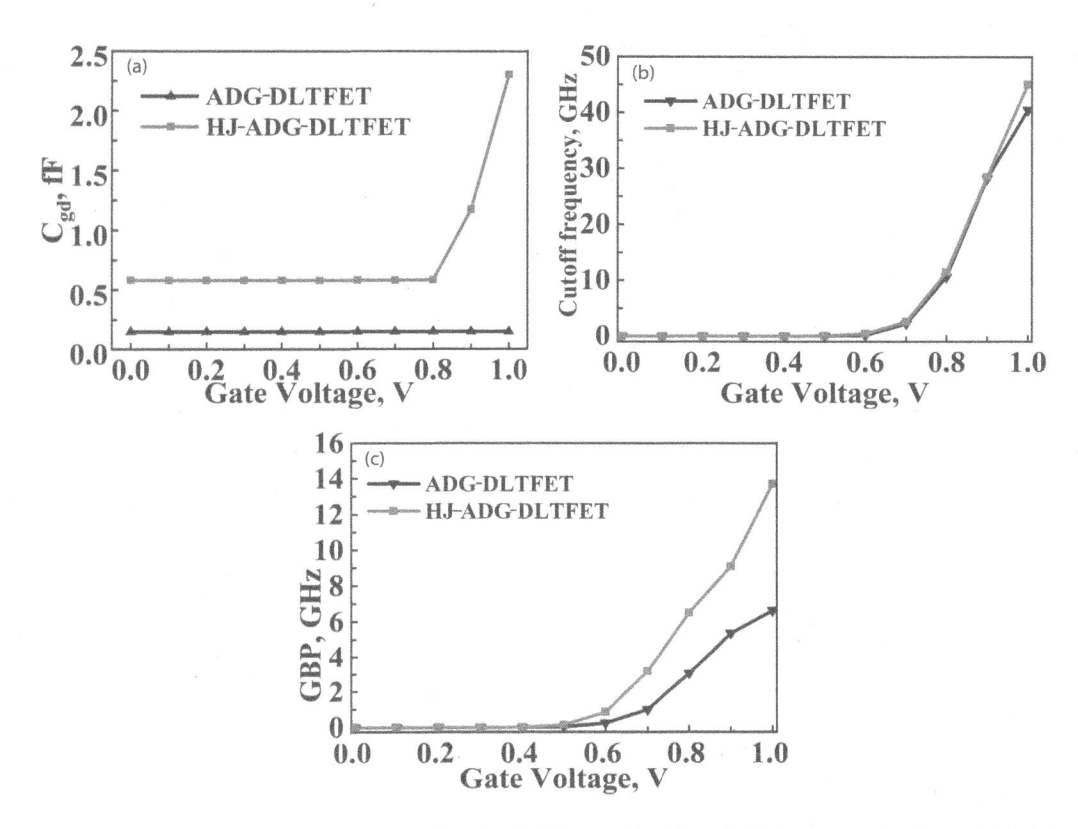

FIGURE 13.4 Comparative plots of HJ-ADG-DLTFET and ADG-DLTFET for (a) C_{gd}, (b) f_T, and (c) GBP.

13.3.2 LINEARITY ANALYSIS

Linearity parameters are crucial for investigating the device performance for RF integrated circuit (RFIC) and wireless communication systems. If a device does not maintain linearity, then non-linearity across the output may distort the desired signal. For a device to be linear, g_m should be constant over the given input voltage. However, for TFET, g_m depends on the input voltage (i.e., V_G), which reflects the nonlinear behavior of the TFET. The g_m determines gain, and the high point of the g_m curve gives the optimum bias points at which maximum gain can be achieved. In this regard, the second-order transconductance coefficient (g_{m2}) and the third-order transconductance coefficient (g_{m3}) play a substantial role in determining the linearity of a device. The general formula for the nth-order coefficient of transconductance is given in Equation 13.7. Figure 13.5(a and b) demonstrates the g_{m2} and g_{m3} of the ADG-DLTFET and HJ-ADG-DLTFET with regard to V_G. It is inferred from Figure 13.5(b) that the peak of g_{m3} of HJ-ADG-DLTFET is obtained at a low V_G gate bias compared with the conventional DLTFET. However, g_{m3} determines lower bounds on the aberration and serves as the primary controlling criteria.

$$g_{mn} = \frac{1}{n!} \frac{\partial^n I_{ds}}{\partial V_{gs}^n} \quad n = 1, 2, 3 \ldots \tag{13.7}$$

$$IIP_3 = \frac{2}{3} \times \frac{g_{m1}}{g_{m3} \times R_S}; \tag{13.8}$$

where $R_S = 50 \, \Omega$ for most RF applications.

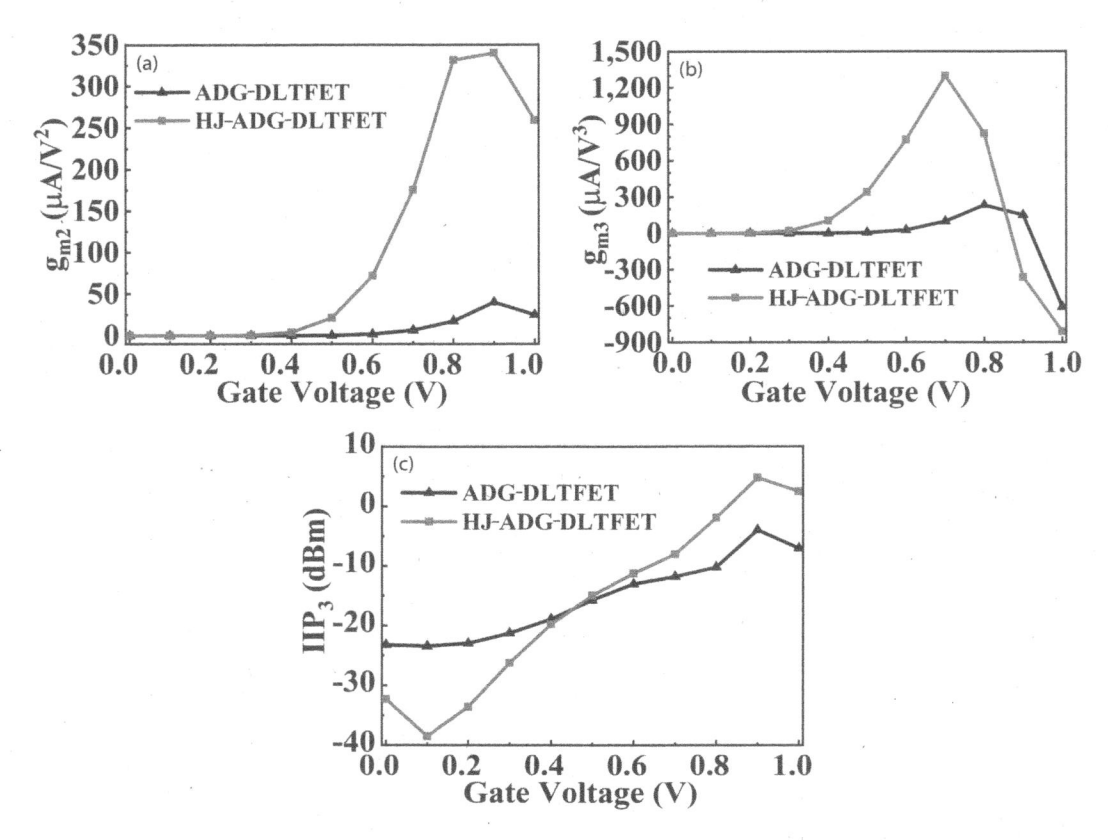

FIGURE 13.5 Comparative plots of HJ-ADG-DLTFET and ADG-DLTFET for (a) g_{m2}, (b) g_{m3}, and (c) IIP_3.

Additionally, IIP_3 is an exclusive quantity that can appear as a criterion for comparison of linearity and is stated in Equation 13.8. The low value of IIP_3 specifies the nonlinearity of the system as it imitates the capability of the gate to regulate the carrier flow across the channel. Figure 5(c) illustrates the effect of IIP_3 and shows that HJ-ADG-DLTFET obtains higher values than ADG-DLTFET. The HJ-ADG-DLTFET attains high peak values at lower gate voltages, signifying that it can achieve high linearity at low gate voltages.

Figure 13.6(a and b) represents the VIP_2 and VIP_3 of both devices concerning V_G. VIP_2 symbolizes the extrapolated input voltage, whereas first- and second-order harmonic voltages are equivalent. VIP_3 corresponds to the extrapolated input voltage, whereas first- and third-order harmonic voltages are equivalent and are stated in Equations 13.9 and 13.10, respectively. The g_{m3} is negated out through the highest point of the VIP_3 (Madan and Chaujar 2016); hence, its value should be high. Looking at Figure 13.6(b), a peak point is perceived at low V_G, which correlates to a moderate inversion region. It can be observed that the amplitude of VIP_2 and VIP_3 is supposed to be the highest to obtain the desired output signal free of any kind of distortion. Furthermore, IMD_3 is a figure of merit (FOM) to inspect the distortion performance as it characterizes extrapolated intermodulation current, whereas first- and third-order intermodulation harmonic currents are equal. IMD_3 instigates via nonlinearity indicated by the I_D-V_G curve and higher-order derivatives of g_m and prompts signal aberration. Figure 13.6(c) explains the disparity of IMD_3 with V_G, and it can be perceived that HJ-ADG-DLTFET suffers the issue of intermodulation distortion. Thus, more techniques have to be explored so that the IMD_3 of the proposed device can be minimized. However, the HJ-ADG-DLTFET possesses input power (IIP_3) higher than amplitude distortion (IMD_3) corresponding to

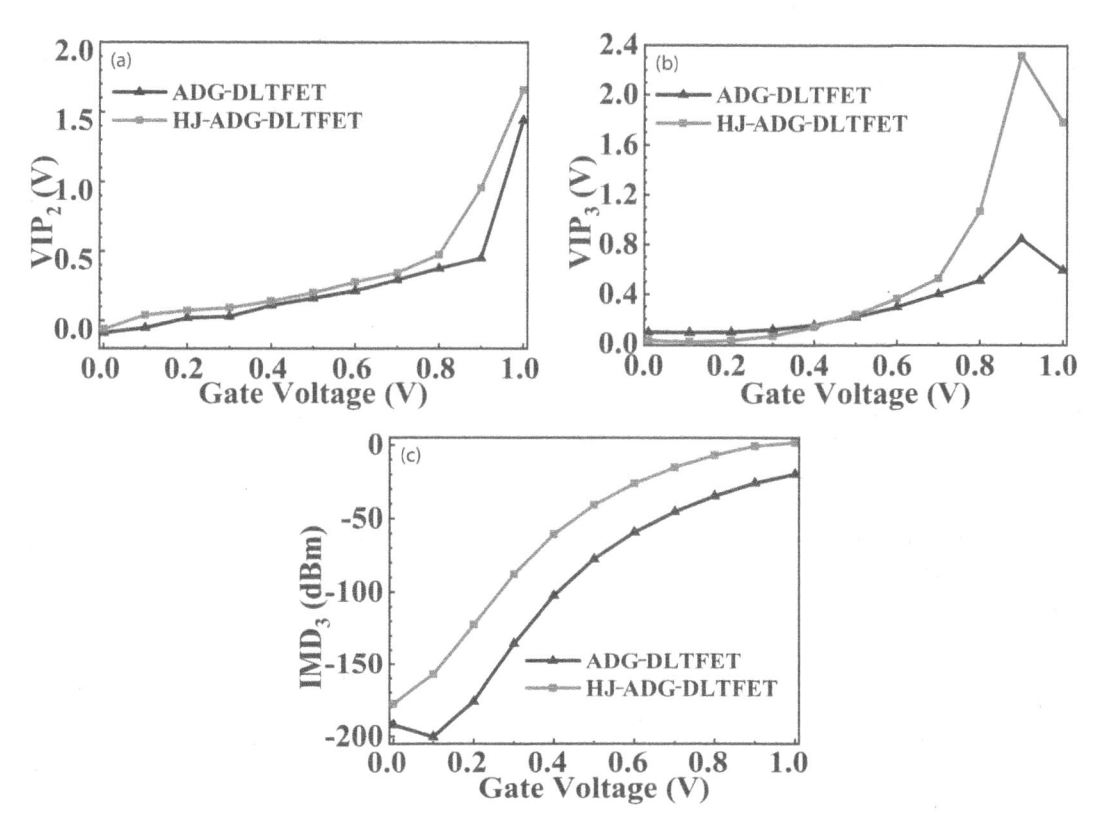

FIGURE 13.6 Comparative plots of HJ-ADG-DLTFET and ADG-DLTFET for (a) VIP_2, (b) VIP_3, and (c) IMD_3.

enhanced device performance in terms of better linearity. Therefore, the increment in IMD_3 can be traded off.

$$VIP_2 = 4 \times \frac{g_{m1}}{g_{m2}} \tag{13.9}$$

$$VIP_3 = \sqrt{24 \times \frac{g_{m1}}{g_{m3}}} \tag{13.10}$$

$$IMD_3 = \left(\frac{9}{2} \times \left(VIP_3 \right)^2 \times g_{m3} \right)^2 \times R_s; \tag{13.11}$$

where $R_S = 50\ \Omega$ for most RF applications.

13.3.3 ITC Analysis

In this section, the impact of ITC has been demonstrated by investigating the effect of ITC density on an Si/Ge HJ-ADG-DLTFET. After that, comparative analysis between an Si/Ge HJ-ADG-DLTFET and ADG-DLTFET for different ITC polarity in the context of the I_D-V_G curve, g_m, f_T, and DE is carried out. The variation in these performance parameters represents the degree of the device immunity toward ITC.

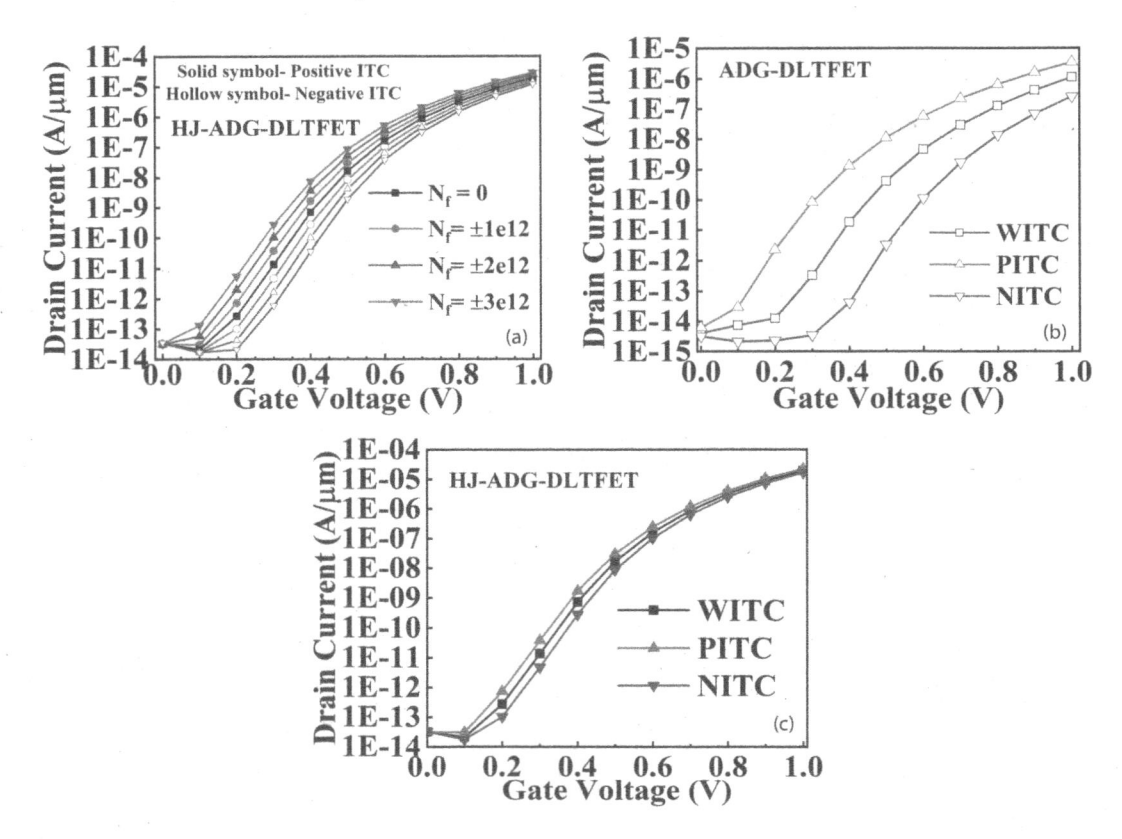

FIGURE 13.7   Effect on I_D-V_G considering ITC density for (a) HJ-ADG-DLTFET, and ITC polarity for (b) ADG-DLTFET and (c) HJ-ADG-DLTFET. (PITC represents positive ITC and NITC represents negative ITC.)

Figure 13.7(a) illustrates the influence of different ITC densities on the I_D-V_G curve of HJ-ADG-DLTFET concerning the applied V_G. The results demonstrate that the increment (decrement) in the PITC and NITC enhances (reduces) the flat-band voltage, and, consequently, improves (worsens) the band bending of the CB and the VB, following the lowering (increasing) of the tunneling barrier width. Consequently, the I_D increases (decreases) with PITC (NITC). The outcome of different types of ITCs on the I_D-V_G curve is illustrated in Figure 13.7(b and c) for ADG-DLTFET and HJ-ADG-DLTFET, correspondingly, on the logarithmic scale. A more significant quantity of electrons is accumulated across the semiconductor-oxide interface due to the presence of PITC, causing an increment in I_{ON}. Figure 13.7(b and c) shows that the variance in the I_D concerning the V_G is negligible for HJ-ADG-DLTFET with regard to ADG-DLTFET, demonstrating the insusceptibility of HJ-ADG-DLTFET toward different types of ITCs. Figure 13.8(a) illustrates the influence of different ITC densities on the transconductance (g_{m1}) of HJ-ADG-DLTFET in accordance with the applied V_G, at room temperature. It is apparent from Figure 13.9(a) that when PITC (NITC) is introduced at the semiconductor-dielectric, nonlocal BTBT increases (decreases) and transconductance increases (decreases). The effect of ITC on g_m is demonstrated in Figure 13.8(b and c) for ADG-DLTFET and HJ-ADG-DLTFET, correspondingly, and it is evident that when PITC (NITC) is introduced at the semiconductor-dielectric, nonlocal BTBT increases (decreases), and g_m increases (decreases) for ADG-DLTFET as well as HJ-ADG-DLTFET. However, the impact of ITC on g_m for HJ-ADG-DLTFET is insignificant because of the introduction of low bandgap material for the source region, an asymmetric

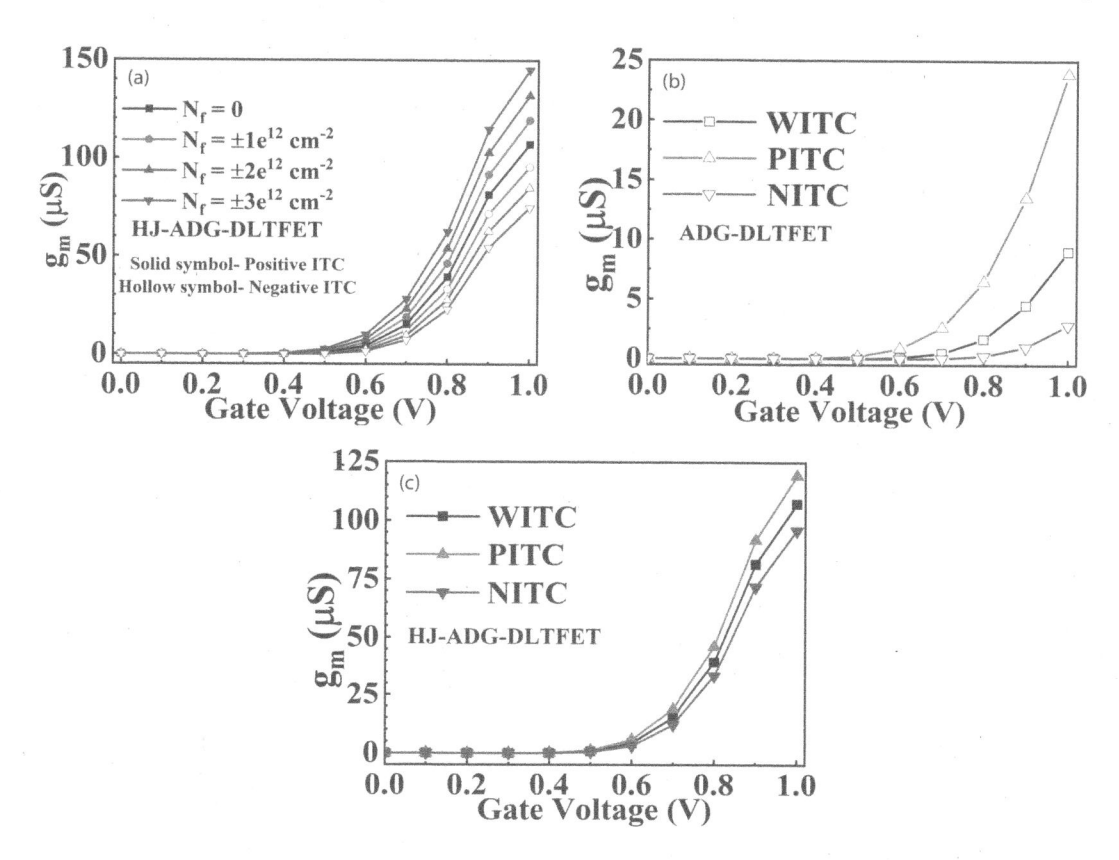

FIGURE 13.8 Effect on g_m with regard to V_G considering ITC density for (a) HJ-ADG-DLTFET, and ITC polarity for (b) ADG-DLTFET and (c) HJ-ADG-DLTFET.

double-gate structure, and a high-κ dielectric. Figure 13.9(a) illustrates the effect of different ITC densities on the f_T of HJ-ADG-DLTFET, at room temperature. Figure 13.9(a) shows that, as the V_G increases, f_T increases are due to the increment in g_m. Additionally, f_T increases (decreases) with PITC (NITC). The effect of ITC on f_T is shown in Figure 13.9(b and c) for ADG-DLTFET and HJ-ADG-DLTFET, individually, and it is illustrated that, for both devices, as the V_G rises, f_T rises due to an increase in g_m. Also, f_T increases (decreases) with PITC (NITC). However, variation in f_T due to the ITCs is less in HJ-ADG-DLTFET, unlike ADG-DLTFET.

Figure 13.10(a) illustrates the effect of different ITC densities on the DE of HJ-ADG-DLTFET in response to the applied gate voltage, at room temperature. Figure 13.10(a) shows that DE achieves a maximum at a small gate bias and then reduces as the gate bias is further increased due to small variations of I_D in the saturation region. The DE increases (decreases) due to the impact of PITC (NITC) due to more extensive variation in I_D compared with the transconductance with PITC (NITC). The effect of ITC on DE is shown in Figure 13.10(b and c) for ADG-DLTFET and HJ-ADG-DLTFET, correspondingly, outlining that HJ-ADG-DLTFET displays better DE as it relies on g_m and I_D.

13.3.4 PVT ANALYSIS

13.3.4.1 Gate Length Variations

In this section, the effect of variance in gate length is evaluated as gate length is an important process parameter. For HJ-ADG-DLTFET, FG length (L_g) is altered in the range of 10 to 30 nm with an

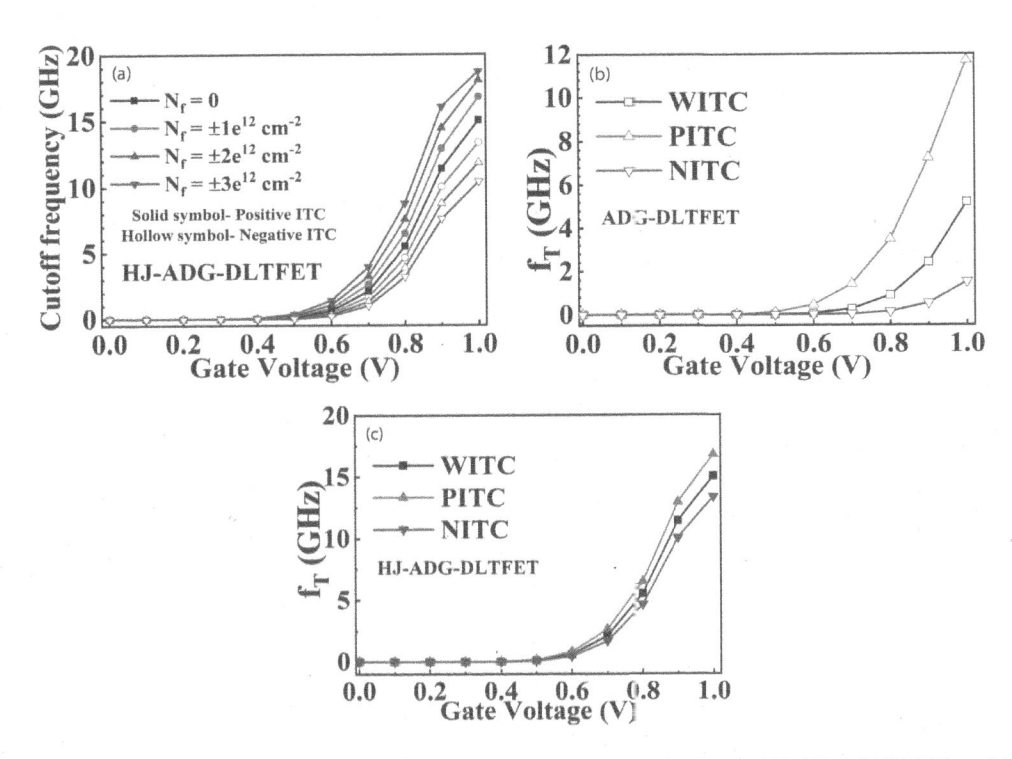

FIGURE 13.9 Effect on f_T with regard to V_G considering ITC density for (a) HJ-ADG-DLTFET, and ITC polarity for (b) ADG-DLTFET and (c) HJ-ADG-DLTFET.

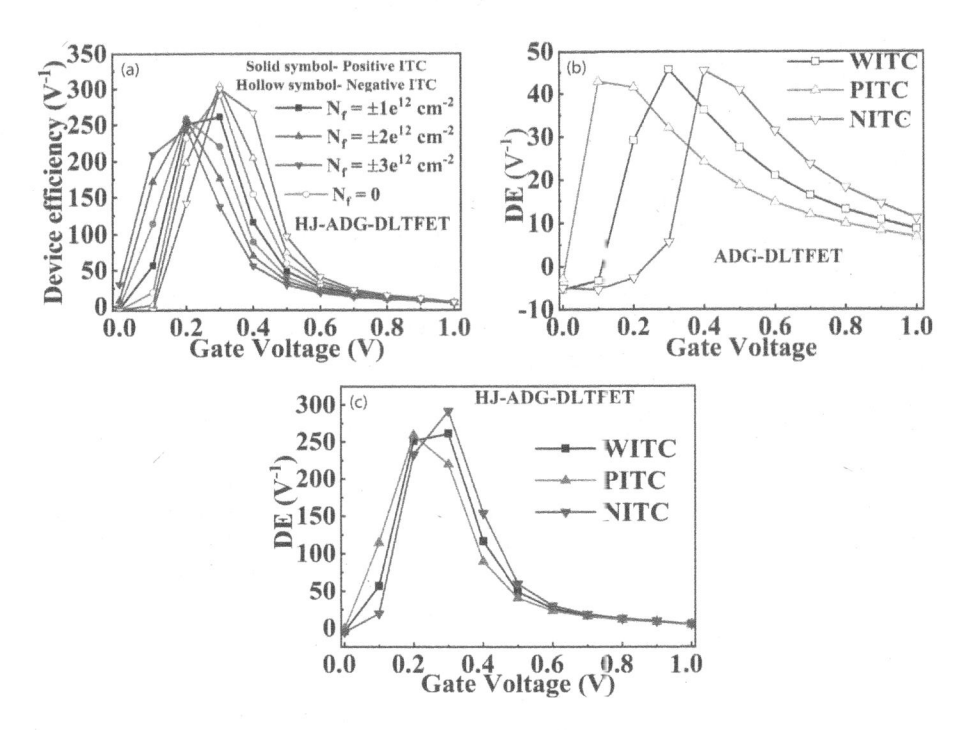

FIGURE 13.10 Effect on device efficiency (DE) with regard to V_G considering ITC density for (a) HJ-ADG-DLTFET, and ITC polarity for (b) ADG-DLTFET and (c) HJ-ADC-DLTFET.

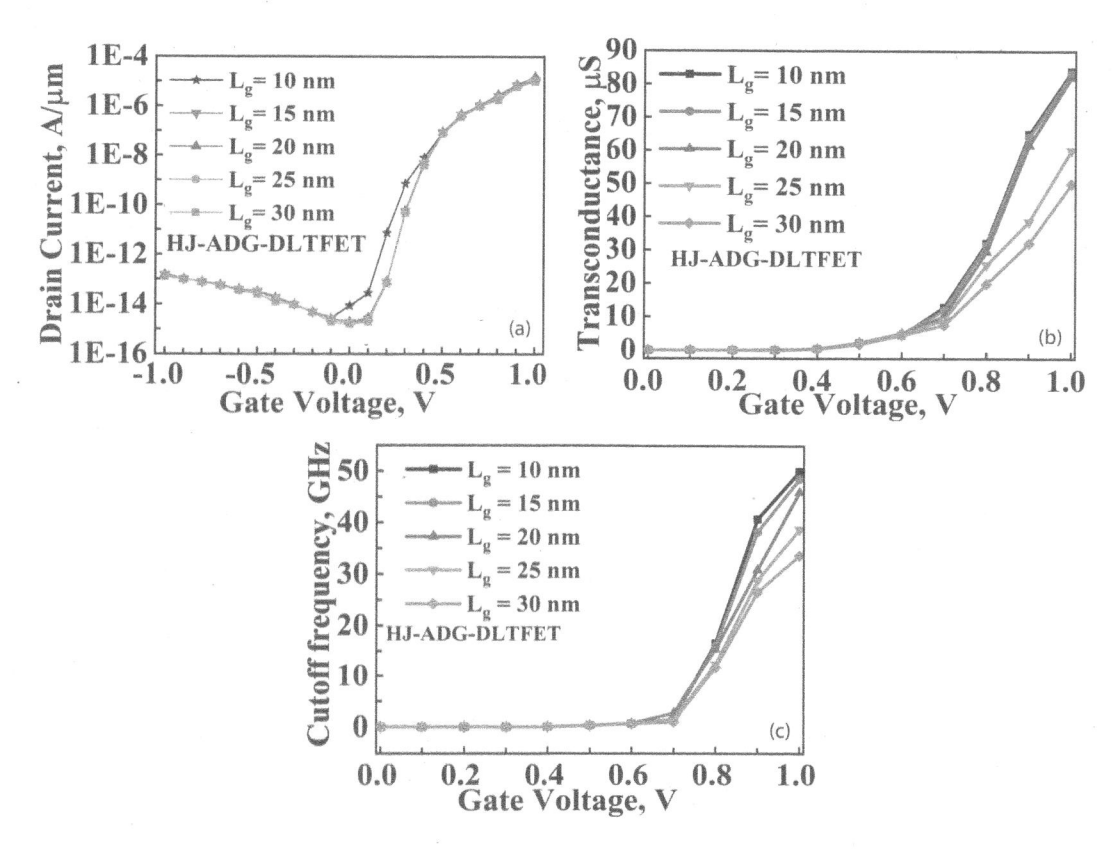

FIGURE 13.11 Effect of gate length variations for HJ-ADG-DLTFET on (a) transfer characteristics, (b) g_m, and (c) f_T.

interval of 5 nm, and BG length is reserved as $L_g + 20$ nm. Figure 13.11(a) shows the outcome of L_g variations on the I_D-V_G curve. It is noted that I_D manifests trivial change with regard to V_G because I_D is more precarious toward source-channel resistance. Therefore, it is resolved that L_g variations have an insignificant consequence on the I_D-V_G curve of HJ-ADG-DLTFET, with ensuing insusceptibility to ambipolarity, and enhanced scalability. Figure 13.11(b) denotes the difference of the g_m of HJ-ADG-DLTFET with regard to V_G for different L_g. As V_G rises, charge carriers in the source rise, with ensuing boosted BTBT. It is noted in Figure 13.11(b) that greater g_m has been attained for lower L_g, although negligible deviation is found when the L_g is reduced below 20 nm and an alteration in I_D also remains irrelevant for $L_g \leq 20$ nm. The presentation of f_T with regard to V_G at different L_g has been revealed with the help of Figure 13.11(c). As L_g has been augmented along with V_G, f_T also has been reduced because of the increment in g_m and C_{gd}.

13.3.4.2 Drain Voltage Variations

This section discusses drain voltage (V_D) variations from 0.4 V to 1 V with a 0.2-V step size for HJ-ADG-DLTFET. Figure 13.12(a) demonstrates the I_D-V_G characteristics of the HJ-ADG-DLTFET from which the I_D has shown a tiny increment with regard to V_G when V_D has been raised. When V_G is negative, the arrangement of bands happens at the drain-channel junction, followed by BTBT. Subsequently, an ambipolar current flows in the device largely because of the holes that govern the I_{OFF}. Thus, when V_D increases for a given V_G, V_{GD} declines, leading to curtailment in electron concentration (EC) and a rise in electron concentration (EC), followed by a rise of ambipolar behavior. Moreover, the maximum I_D has been attained at $V_D = 1$ V amid the previously mentioned V_D. As a

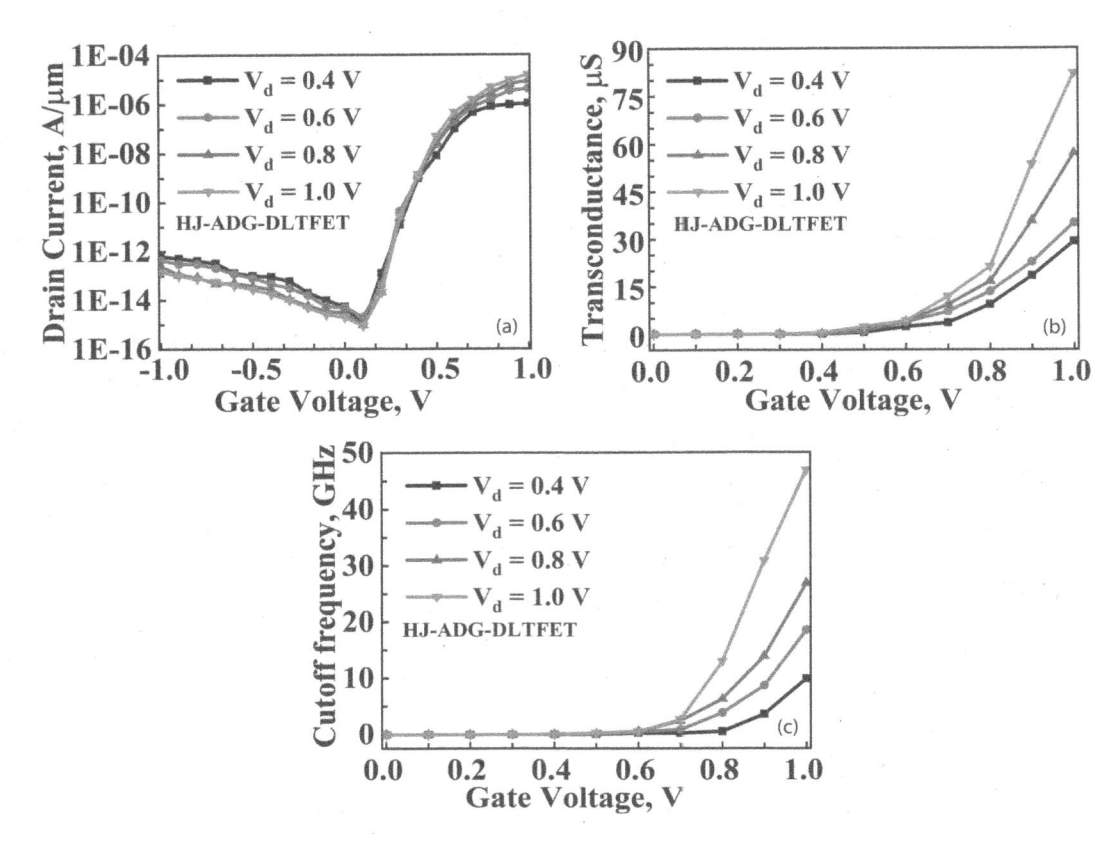

FIGURE 13.12 Effect of drain voltage variations for HJ-ADG-DLTFET on (a) transfer characteristics, (b) g_m, and (c) f_T.

slight increment occurs in I_D with regard to V_G when the V_D is diversified across 0.4 V to 1 V, there is a trivial rise in g_m, as shown in Figure 13.12(b). The presentation of f_T with regard to V_G at different V_D is shown in Figure 13.12(c). It is noted that as V_D has been raised together with V_G, f_T also has been enhanced because of augmentation in g_m and decreasing C_{gd}.

13.3.4.3 Temperature Analysis

In this segment, the impact of temperature is investigated for HJ-ADG-DLTFET at different ITC densities for several parameters of the device. For TFET, BTBT depends on temperature. Additionally, the energy bandgap of semiconductor materials is also affected by temperature and is defined as $E_g(T) = E_g(0) - \frac{\alpha_E T^2}{T + \beta_E}$, where $E_g(0)$ represents the energy bandgap at zero kelvin, α_E and β_E represent the material parameters, and T represents the absolute temperature in kelvin. The rise in the temperature results in the decrement of the bandgap of the semiconductor. Consequently, the smaller bandgap results in decreasing the tunneling barrier height, which enhances the BTBT rate. Thus, the improvement in the BTBT rate results in a massive drive current of the device. Figure 13.13(a) illustrates the deviation of I_D in response to the applied V_G, for the temperature range 200–500 K. It is apparent from Figure 13.13(a) that temperature susceptibility on the I_D-V_G characteristics relies on the applied bias, as a result of the different carrier transport schemes in the OFF and ON states. The I_{ON} increases exponentially with rising temperature because Shockley-Read-Hall (SRH) recombination has an exponential temperature dependence at low V_G. However, at high V_G, BTBT dominates over SRH and results in small increments in the I_D. Furthermore, I_{OFF} degrades more significantly at different temperatures along with SS, whereas I_{ON} is enhanced, resulting in

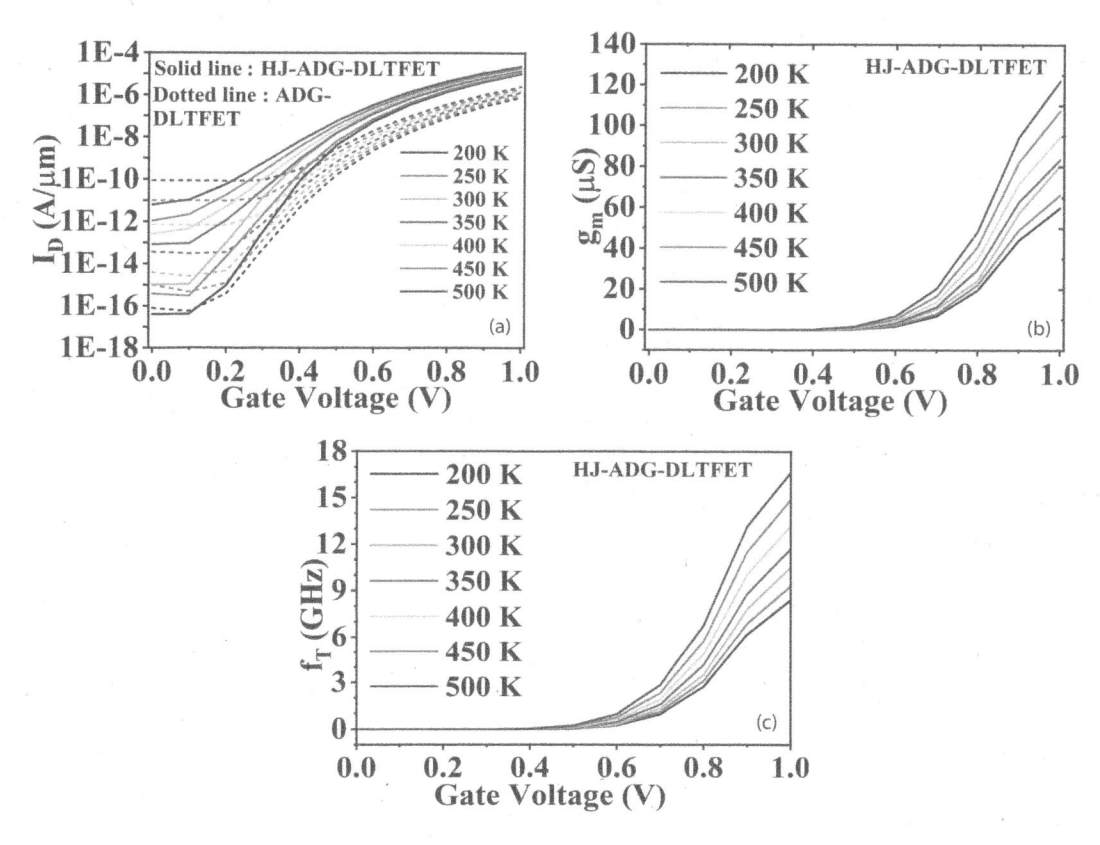

FIGURE 13.13 Effect of temperature variations for HJ-ADG-DLTFET on (a) transfer characteristics, (b) g_m, and (c) f_T.

reduction of the threshold voltage. Figure 13.13(b and c) illustrates the deviation in g_m and f_T with regard to V_G, for the temperature range 200–500 K, respectively. As HJ-ADG-DLTFET illustrates the increment of the I_D with increasing temperature, g_m also shows an increment in the temperature. The HJ-ADG-DLTFET illustrates the positive temperature coefficient of the I_D. Consequently, f_T increases with temperature as a result of the increment in the I_{ON} and g_m.

13.4 CONCLUSION

This section concludes the performance assessment of HJ-ADG-DLTFET in terms of DC, analog/RF, and linearity parameters, relative to ADG-DLTFET through the adoption of low bandgap material (Ge) in the source and HfO$_2$ as the high-κ gate dielectric. The Ge source diminishes tunneling barrier width across the source-channel junction, leading to better BTBT and intensified I_{ON}. The high-κ dielectric reduces the I_{OFF} associated with the adoption of the low bandgap source material. Thus, $I_{ON} \sim 15\ \mu A/\mu m$, $I_{OFF} \sim 10^{-15}\ A/\mu m$, and $I_{ON}/I_{OFF} \sim 10^{10}$ has been achieved for $V_D = 1$ V and $V_G = 1$ V. Furthermore, this chapter includes the impact of ITC density and polarity on HJ-ADG-DLTFET via simulation because the examination of device behavior in the presence of ITC is crucial as ITCs always exist in the real device and, hence, deteriorate device characteristics. The occurrence of PITC (NITC) enhances (reduces) the I_D of HJ-ADG-DLTFET, which consequently results in the increment (decrement) of device performance. Moreover, PVT variations are also investigated for HJ-ADG-DLTFET to predict device behavior in harsh environmental conditions. Thus, for a gate length of 20 nm and drain voltage of 1 V, HJ-ADG-DLTFET demonstrates improved DC and analog/RF attributes. Additionally, HJ-ADG-DLTFET appears to have a positive temperature coefficient for I_D, and, hence, for g_m and f_T for complete gate bias range. Thus, the investigation helps to optimize several device parameters during fabrication so that the device can behave according to the results obtained from the simulation.

REFERENCES

Ashita, Sajad A. Loan, and Mohammad Rafat. 2019. "Insights into the Impact of Pocket and Source Elevation in Vertical Gate Elevated Source Tunnel FET Structures." *IEEE Transactions on Electron Devices* 66 (1): 752–758. doi:10.1109/TED.2018.2878010.

Bashir, Faisal, Sajad A. Loan, M. Rafat, Abdul Rehman M. Alamoud, and Shuja A. Abbasi. 2015. "A High Performance Gate Engineered Charge Plasma Based Tunnel Field Effect Transistor." *Journal of Computational Electronics* 14 (2): 477–485. doi:10.1007/s10825-015-0665-5.

Boucart, Kathy, and Adrian Mihai Ionescu. 2007. "Double-Gate Tunnel FET with High-κ Gate Dielectric." *IEEE Transactions on Electron Devices* 54 (7): 1725–1733. doi:10.1109/TED.2007.899389.

Chandan, Bandi Venkata, Kaushal Nigam, Dheeraj Sharma, and Sunil Pandey. 2018. "Impact of Interface Trap Charges on Dopingless Tunnel FET for Enhancement of Linearity Characteristics." *Applied Physics A: Materials Science and Processing* 124 (7): 503. doi:10.1007/s00339-018-1923-8.

Duan, Xiaoling, Jincheng Zhang, Shulong Wang, Yao Li, Shengrui Xu, and Yue Hao. 2018. "A High-Performance Gate Engineered InGaN Dopingless Tunnel FET." *IEEE Transactions on Electron Devices* 65 (3): 1223–1229. doi:10.1109/TED.2018.2796848.

Gupta, Sarthak, Dheeraj Sharma, Deepak Soni, Shivendra Yadav, Mohd. Aslam, Dharmendra Singh Yadav, Kaushal Nigam, and Neeraj Sharma. 2018. "Examination of the Impingement of Interface Trap Charges on Heterogeneous Gate Dielectric Dual Material Control Gate Tunnel Field Effect Transistor for the Refinement of Device Reliability." *Micro & Nano Letters* 13 (8): 1192–1196. doi:10.1049/mnl.2017.0869.

Huang, A.P., Z.C. Yang, and Paul K. Chu. 2010. "Hafnium-Based High-k Gate Dielectrics." In *Advances in Solid State Circuit Technologies*. InTechOpen.

Hueting, Raymond J.E., Bijoy Rajasekharan, Cora Salm, and Jurriaan Schmitz. 2008. "The Charge Plasma P-N Diode." *IEEE Electron Device Letters* 29 (12): 1367–1369. doi:10.1109/LED.2008.2006864.

Johnson, Richard W., Adam Hultqvist, and Stacey F. Bent. 2014. "A Brief Review of Atomic Layer Deposition: From Fundamentals to Applications." *Materials Today* 17 (5): 236–246. doi:10.1016/j.mattod.2014.04.026.

Kumar, M. Jagadesh, and Sindhu Janardhanan. 2013. "Doping-Less Tunnel Field Effect Transistor: Design and Investigation." *IEEE Transactions on Electron Devices* 60 (10): 3285–3290. doi:10.1109/TED.2013.2276888.

Kumar, M. Jagadesh, and Kanika Nadda. 2012. "Bipolar Charge-Plasma Transistor: A Novel Three Terminal Device." *IEEE Transactions on Electron Devices* 59 (4): 962–967. doi:10.1109/TED.2012.2184763.

Kumar, Naveen, and Ashish Raman. 2019. "Performance Assessment of the Charge-Plasma-Based Cylindrical GAA Vertical Nanowire TFET With Impact of Interface Trap Charges." *IEEE Transactions on Electron Devices* 66 (10): 4453–4460. doi:10.1109/ted.2019.2935342.

Kumar, Naveen, and Ashish Raman. 2020. "Low Voltage Charge-Plasma Based Dopingless Tunnel Field Effect Transistor: Analysis and Optimization." *Microsystem Technologies* 26 (4): 1343–50. doi:10.1007/s00542-019-04666-y.

Kumar, Satyendra, Km. Sucheta Singh, Kaushal Nigam, Vinay Anand Tikkiwal, and Bandi Venkata Chandan. 2019. "Dual-Material Dual-Oxide Double-Gate TFET for Improvement in DC Characteristics, Analog/RF and Linearity Performance." *Applied Physics A: Materials Science and Processing* 125 (5): 353. doi:10.1007/s00339-019-2650-5.

Lho, Young Hwan, and Ki Yup Kim. 2005. "Radiation Effects on the Power MOSFET for Space Applications." *ETRI Journal* 27 (4): 449–452. doi:10.4218/etrij.05.0205.0031.

Liu, Hu, Lin An Yang, Zhi Jin, and Yue Hao. 2019. "An In0.53ga0.47as/In0.52al0.48as Heterojunction Dopingless Tunnel FET with a Heterogate Dielectric for High Performance." *IEEE Transactions on Electron Devices* 66 (7): 3229–3235. doi:10.1109/TED.2019.2916975.

Madan, Jaya, and Rishu Chaujar. 2016. "Interfacial Charge Analysis of Heterogeneous Gate Dielectric-Gate All Around-Tunnel FET for Improved Device Reliability." *IEEE Transactions on Device and Materials Reliability* 16 (2): 227–234. doi:10.1109/TDMR.2016.2564448.

Madan, Jaya, and Rishu Chaujar. 2018. "Temperature Associated Reliability Issues of Heterogeneous Gate Dielectric-Gate All Around-Tunnel FET." *IEEE Transactions on Nanotechnology* 17 (1): 41–48. doi:10.1109/TNANO.2017.2650209.

Nakatsuka, Osamu, and Shigeaki Zaima. 2015. "Heteroepitaxial Growth of Si, Si1−xGex-, and Ge-Based Alloy." In *Handbook of Crystal Growth*, Vol. 3, pp. 1301–1318. Elsevier. doi:10.1016/B978-0-444-63304-0.00032-9.

Pala, M.G., D. Esseni, and F. Conzatti. 2012. "Impact of Interface Traps on the IV Curves of InAs Tunnel-FETs and MOSFETs: A Full Quantum Study." *2012 International Electron Devices Meeting*, San Francisco, CA, 135–38. doi:10.1109/IEDM.2012.6478992.

Patel, Jyoti, Dheeraj Sharma, Shivendra Yadav, Alemienla Lemtur, and Priyanka Suman. 2019. "Performance Improvement of Nano Wire TFET by Hetero-Dielectric and Hetero-Material: At Device and Circuit Level." *Microelectronics Journal* 85 (January): 72–82. doi:10.1016/j.mejo.2019.02.004.

Poindexter, E H. 1989. "MOS Interface States: Overview and Physicochemical Perspective." *Semiconductor Science and Technology* 4 (12): 961. doi:10.1088/0268-1242/4/12/001.

Raad, Bhagwan Ram, Dheeraj Sharma, Pravin Kondekar, Kaushal Nigam, and Dharmendra Singh Yadav. 2016. "Drain Work Function Engineered Doping-Less Charge Plasma TFET for Ambipolar Suppression and RF Performance Improvement: A Proposal, Design, and Investigation." *IEEE Transactions on Electron Devices* 63 (10): 3950–3957. doi:10.1109/TED.2016.2600621.

Raad, Bhagwan Ram, Sukeshni Tirkey, Dheeraj Sharma, and Pravin Kondekar. 2017. "A New Design Approach of Dopingless Tunnel FET for Enhancement of Device Characteristics." *IEEE Transactions on Electron Devices* 64 (4): 1830–1836. doi:10.1109/TED.2017.2672640.

Rajasekharan, Bijoy, Raymond J.E. Hueting, Cora Salm, Tom Van Hemert, Rob A.M. Wolters, and Jurriaan Schmitz. 2010. "Fabrication and Characterization of the Charge-Plasma Diode." *IEEE Electron Device Letters* 31 (6): 528–530. doi:10.1109/LED.2010.2045731.

Raushan, Mohd Adil, Naushad Alam, Mohd Waseem Akram, and Mohd Jawaid Siddiqui. 2018. "Impact of Asymmetric Dual-k Spacers on Tunnel Field Effect Transistors." *Journal of Computational Electronics* 17 (2): 756–765. doi:10.1007/s10825-018-1129-5.

Raushan, Mohd Adil, Naushad Alam, and Mohammad Jawaid Siddiqui. 2018. "Dopingless Tunnel Field-Effect Transistor with Oversized Back Gate: Proposal and Investigation." *IEEE Transactions on Electron Devices* 65 (10): 4701–4708. doi:10.1109/TED.2018.2861943.

Robertson, J. 2004. "High Dielectric Constant Oxides." *EPJ Applied Physics* 28 (3): 265–291. doi:10.1051/epjap:2004206.

Sarkar, Deblina, Xuejun Xie, Wei Liu, Wei Cao, Jiahao Kang, Yongji Gong, Stephan Kraemer, Pulickel M. Ajayan, and Kaustav Banerjee. 2015. "A Subthermionic Tunnel Field-Effect Transistor with an Atomically Thin Channel." *Nature* 526 (7571): 91–95. doi:10.1038/nature15387.

Sharma, Suruchi, and Baljit Kaur. 2020. "Performance Investigation of Asymmetric Double-Gate Doping Less Tunnel FET with Si/Ge Heterojunction." *IET Circuits, Devices and Systems* 14 (5): 695–701. doi:10.1049/iet-cds.2019.0290.

Shockley, W., and W.T. Read. 1952. "Statistics of the Recombinations of Holes and Electrons." *Physical Review* 87 (5): 835–842. doi:10.1103/PhysRev.87.835.

Silvaco. 2016. *ATLAS User's Manual*, Silvaco, Santa Clara, CA.

Steglich, Martin, Christian Patzig, Lutz Berthold, Frank Schrempel, Kevin Füchsel, Thomas Höche, Ernst Bernhard Kley, and Andreas Tünnermann. 2013. "Heteroepitaxial Ge-on-Si by DC Magnetron Sputtering." *AIP Advances* 3 (7): 072108. doi:10.1063/1.4813841.

Trabzon, L., and O.O.O. Awadelkarim. 1998. "Damage to N-MOSFETs From Electrical Stress Relationship to Processing Damage and Impact on Device Reliability." *Microelectronics Reliability* 38 (4): 651–657. doi:10.1016/S0026-2714(97)00194-7.

Uddin Shaikh, Mohd Rizwan, and Sajad A. Loan. 2019. "Drain-Engineered TFET With Fully Suppressed Ambipolarity for High-Frequency Application." *IEEE Transactions on Electron Devices* 66 (4): 1628–1634. doi:10.1109/TED.2019.2896674.

Varshni, Y.P. 1967. "Temperature Dependence of the Energy Gap in Semiconductors." *Physica* 34 (1): 149–154. doi:10.1016/0031-8914(67)90062-6.

Venkatesh, Pulimamidi, Kaushal Nigam, Sunil Pandey, Dheeraj Sharma, and Pravin N. Kondekar. 2017. "Impact of Interface Trap Charges on Performance of Electrically Doped Tunnel FET With Heterogeneous Gate Dielectric." *IEEE Transactions on Device and Materials Reliability* 17 (1): 245–252. doi:10.1109/TDMR.2017.2653620.

Wang, Pei Yu, and Bing Yue Tsui. 2015. "Experimental Demonstration of P-Channel Germanium Epitaxial Tunnel Layer (ETL) Tunnel FET with High Tunneling Current and High ON/OFF Ratio." *IEEE Electron Device Letters* 36 (12): 1264–1266. doi:10.1109/LED.2015.2487563.

Wang, Runsheng, Xiaobo Jiang, Tao Yu, Jiewen Fan, Jiang Chen, David Z. Pan, and Ru Huang. 2013. "Investigations on Line-Edge Roughness (LER) and Line-Width Roughness (LWR) in Nanoscale CMOS Technology: Part II-Experimental Results and Impacts on Device Variability." *IEEE Transactions on Electron Devices* 60 (11): 3676–3682. doi:10.1109/TED.2013.2283517.

Widiez, Julie, Jérôme Lolivier, Maud Vinet, Thierry Poiroux, and Bernard Previtali, Frédéric Daugé, Mireille Mouis, and Simon Deleonibus. 2005. "Experimental Evaluation of Gate Architecture Influence on DG SOI MOSFETs Performance." *IEEE Transactions on Electron Devices* 52 (8): 1772–1779. doi:10.1109/ TED.2005.851824.

Xiang, Yang, Anne S. Verhulst, Dmitry Yakimets, Bertrand Parvais, Anda Mocuta, and Guido Groeseneken. 2019. "Process-Induced Power-Performance Variability in Sub-5-Nm III-V Tunnel FETs." *IEEE Transactions on Electron Devices* 66 (6): 2802–2808. doi:10.1109/TED.2019.2909217.

Yadav, Dharmendra Singh, Dheeraj Sharma, Ashish Kumar, Deepak Rathor, Rahul Agrawal, Sukeshni Tirkey, Bhagwan Ram Raad, and Varun Bajaj. 2017. "Performance Investigation of Hetero Material (InAs/Si)-Based Charge Plasma TFET." *Micro & Nano Letters* 12 (6): 358–363. doi:10.1049/mnl.2016.0688.

Yadav, Dharmendra Singh, Abhishek Verma, Dheeraj Sharma, and Neeraj Sharma. 2018. "Study of Metal Strip Insertion and Its Optimization in Doping Less TFET." *Superlattices and Microstructures* 122 (March): 577–586. doi:10.1016/j.spmi.2018.06.046.

Yoon, Jun Sik, Kihyun Kim, M. Meyyappan, and Chang Ki Baek. 2018. "Bandgap Engineering and Strain Effects of Core-Shell Tunneling Field-Effect Transistors." *IEEE Transactions on Electron Devices* 65 (1): 277–281. doi:10.1109/TED.2017.2767628.

14 Synthesis of Graphene Nanocomposites Toward the Enhancement of Energy Storage Performance for Supercapacitors

Monojit Mondal, Avik Sett, Dipak Kumar Goswami, and Tarun Kanti Bhattacharyya

CONTENTS

DOI: 10.1201/9781003126393-14

14.1   INTRODUCTION

The increasing energy impulse in concurrence with the degrading environment and depletion of natural resources like fossil fuels is the genesis for developing energy storage devices. Conversion of energy has become a potential issue in both applied and fundamental research in the technology domain. Electrochemical supercapacitors (SCs) are a promising device for energy storage as they can exceed the power density of a rechargeable battery and enhance energy density equated to conventional capacitors. SCs are categorized into two broad segments: electric double-layer capacitors (EDLCs) and pseudocapacitors. They depend on the mechanism of energy storage and material of electrode. The EDLCs accumulate electrical energy through physical adsorption methods, i.e., gathering electrostatic charge on the electrolyte-electrode interface. In pseudocapacitors, the charge occurrence's net storage is controlled by the faradic methods ascribed by the reversible and fast redox reaction at the electrode's interfacing surface. Today, for scientific upgrading, material science researchers are committed to fabricating nanomaterial-based active material for SC electrodes. The as-fabricated nano morphologies can notably ameliorate the execution as an electrode by creating fewer ion and elect transport pathways by furnishing a huge active contact area interface with the electrolytes. The porous structure with the higher active surface area appreciably improves the ion diffusion rate, enhancing the electrode's operation [1, 2]. So, to admit the previously mentioned intimidating challenges, modern scientists and researchers have focused on a novel class of materials regarding the pseudocapacitive and EDLC storage response. All of the procedures have the capability of great amounts of energy storage. Their physicochemical attributes, rated capacity, specific capacitance, and long cycle stability can contribute to producing effectual energy solutions in forthcoming real-time applications.

Graphene oxide (GO) or reduced graphene oxide (rGO) is a favorable two-dimensional (2D) material for energy storage that has been vastly used in electrochemical systems. Its enormously high active specific surface area, enhanced electrical properties, thermal energy, and most significant mechanical properties contribute largely to various applications. The 2D graphene is the root of different carbon allotropes. The fabrication of a high-quality graphene sheet is an overlong process as it can easily get agglomerated or restacked. The low gain of single-layer graphene becomes costly. Graphene is synthesized using well-known methods like microwave exfoliation, mechanical exfoliation, laser ablation process, chemical derivation method, and chemical vapor decomposition (CVD) process. The graphene electrode may be fabricated into free-standing carbon dot (0D), fiber (1D), flexible thin-film (2D), and foam (3D). The graphene electrode is sometimes the perfect option for flexible SCs to attribute less diffusion resistance to ion transport, high flexibility, high mechanical properties, and increased active surface area to ion movement [3, 4].

A few layers of graphene are rearranged stack wise for the van der Waals force, which intercepts the transportation of the movement of ions in electrolytes. The problem of stacking or interlayer spacing can be neutralized by incrementing using a spacer, which generates more ion storage and transport within the process of charging. The short diffusion path of the 2D graphene-based electrode creates a propitious material for in-plane SC devices for high cycle stability and ultrahigh-power density. The properties of low micropore volume, polymer-binder free material, and good connection among the active materials and collector of current make this SC electrode material prominent. Heteroatom-doped graphene is a prominent future time material for SCs, strengthening cyclic stability and electrochemical performance. It is crucial to fabricate water-soluble graphite or graphene typically. Graphene's composite with metal oxide and the conductive polymer is ameliorated with the low electrical conductivity and unstable structures of polymers of graphene. Graphene has gained attention in large energy applications for its distinctive chemical and physical features, along with its extraordinary mechanical properties, excellent electrical conductivity, and a large active specific surface area. Also, graphene shows lithium-ion storage capacity and electric double-layer (EDL) capacitance to work as an active material. The incorporation of graphene components also attributes some new perspectives to the polymer composites, such as transparency, self-healing, flexibility, etc. There are different graphene and conventional research products in graphene-based green energy applications being executed today, like lithium-ion batteries (LIBs), fuel cells, solar cells, and dye-sensitized solar cells (DSSCs); SCs; photo-electrochemical water-splitting cells; and in sensing platforms [5–9].

In the present study, the graphene properties that lead to energy storage performance enhancement have been analyzed. Different synthesis techniques for graphene fabrication, along with its nanocomposites, have been discussed [10]. The study then focuses on the energy storage device's performance when graphene-modified electrodes are introduced. The effect of the doping heteroatom in the graphene matrix has been demonstrated. The EDLC formation is delineated in Figure 14.1. The graphene-conductive polymer composite, a newly emerging research area, has adequately improved electrochemical performance and has been discussed. The benefits of attaching graphene with porous metal oxide (binary, ternary, etc.) nano counterparts, with improved surface area toward enhancing EDLC and pseudocapacitive storage response, are discussed. Finally, the challenges to be addressed and prospects of graphene-based EDLC devices are analyzed and elucidated [11, 12].

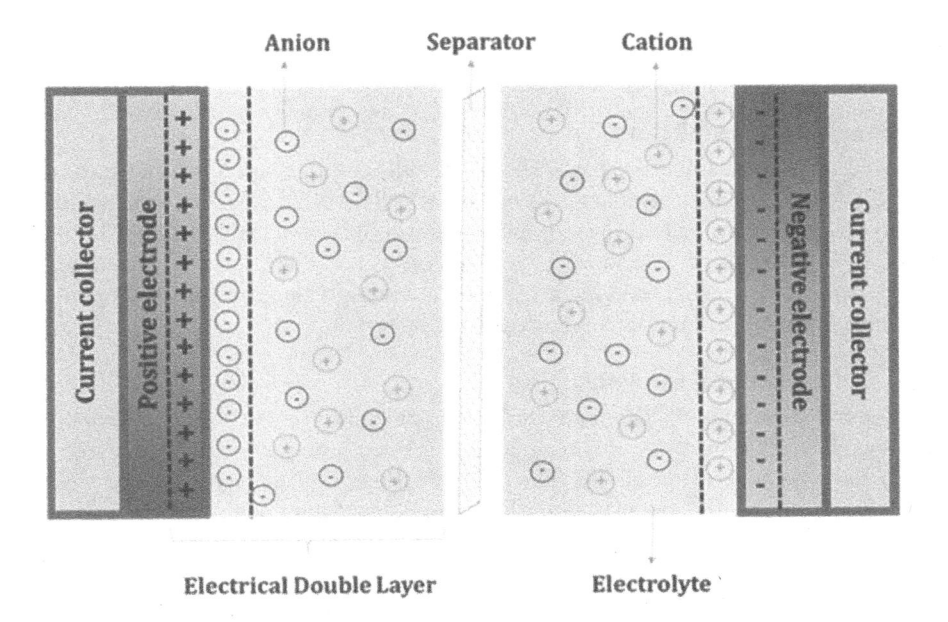

FIGURE 14.1 Storage process of a charge in an electrical double-layer capacitance capacitor. (Reprinted from Gowrisankar, A., T. Saravanakumar, and T. Selvaraju. "Graphene-Based Composite Materials for Flexible Supercapacitors." In Nanostructured, Functional, and Flexible Materials for Energy Conversion and Storage Systems, pp. 345–372. Elsevier, Amsterdam, 2020 with permission from Elsevier.)

The main advantages of using SCs are their high-power density, fast charging-discharging procedure, long cycle stability, working ability in a broad range of temperatures, exceptional safety standards, and increased rate capabilities, which make these devices highly anticipated for implementation as devices of energy storage. This is a static and passive energy-storage system, and it can deliver and store energy at comparatively higher rates. SCs have good long cycling stability and operational safety, delineating them as excellent energy-storage platforms with substantial prospective. They are perfect when a rapid charge is desirable to fill an immediate power necessity, wherever batteries are preferred to deliver long-standing energy. In applications requiring many speedy charge-discharge cycles rather than lasting compact storage of energy, rapid storage of energy, or burst-mode delivery of power, this SC platform is one of the most acceptable methods [13–16]. In a constant voltage current response operation system like the semiconductor device area, the SC power source platform is a novel and prominent pathway to maintain device operation.

14.2 SYNTHESIS OF GRAPHENE NANOMATERIALS AND NANOCOMPOSITES

Researchers worldwide explore various techniques to synthesize graphene, one of the fascinating emerging materials with many novel applications. The graphene preparation may be classified into two major clusters: (1) bottom-up and (2) top-down methodologies. The van der Waals forces acting between graphene's thin layers allow the layers' stacking to form a rigid structure, commonly known as graphite. Top-down approaches can separate the bulk graphite's thin layers if the detaching forces applied are more significant than the existing van der Waals forces. However, during separation, there are possibilities that the exfoliated or the separated layers become agglomerated again. The possibility of film damage by introducing certain defects during micromechanical cleaving also is possible. Hence, proper care and expertise are required to prevent film damage and agglomeration of the layers, which may sufficiently impact device performance. The bottom-up approach also requires extensive care during the graphitic phase formation. The formation of graphitic layers and their related nano counterparts depends significantly on the growth process and demands superior temperature control and pressure during synthesis.

14.2.1 TOP-DOWN TECHNIQUES

14.2.1.1 Mechanical Exfoliation

When cleaved micromechanically by adhesive tape, bulk graphite structures move to realize thin graphene layers; this was first demonstrated by Geim et al. in 2004 [17]. With this technique, they successfully isolated single to multilayer graphene and could comfortably transfer it to a 300-nm-thick layer of silicon dioxide developed over a silicon wafer. They successfully investigated the number of graphene layers by observing color contrast through an optical microscope that worked on the mismatch principle in refractive indices of the materials.

14.2.1.2 Electrochemical Exfoliation

The electrochemical exfoliation technique facilitated graphene layers to be available in floating conditions in the electrolyte medium; this technique employed highly crystalline graphite electrodes as a source and sacrificial material [18]. The freely floating graphene layers can be single layer to multilayer depending on the electrochemical conditions and parameters. These graphene layers tend to become agglomerated due to nonsupportive electrolytic surface tension. The prevention of agglomeration can be done by employing a surfactant (a hydrophilic group) in the electrolytic solution. This, in turn, may attach the surfactant group to the suspended network of graphene, hence, altering the intrinsic property of the film. Counter groups are also added in an optimized amount to preserve the property of graphene layers. Sonication of the resultant graphene network attached with surfactant and counter groups in dimethylformamide (DMF) [19] and water facilitates detachment of the chemical entities' graphene layer.

14.2.1.3   Exfoliation of Graphite Intercalated Compounds (GICs)

A very different graphene synthesis domain is graphite intercalated compounds (GICs) created by pressure-assisted techniques and solvothermal methods. The residing intercalated elements between the graphite layers produce gas molecules, enhancing the interlayer spacing and readily exfoliating graphite. Graphene's exfoliation in water and ethanol is conducted by sonication of GICs with lithium ions and alkali metals [20]. When intercalated with N-methyl pyrrolidine (NMP) and ethanol, graphite powders produce one-layer to few-layer graphene after sonication in DMF solution. The surface tension of NMP is 40 mJ/m^2, which allows the facile exfoliation of graphene sheets. The size and concentration of the graphene layer rely on the type of solvent. Solvents with higher boiling points support the non-agglomeration of graphene layers, whereas the accumulation of layers occurs using low boiling point solvents [21]. The non-agglomerated graphene layers from the high boiling point solvents impose complexity during coating and transferring graphene from solution to devices. The density of graphene layers increases with sonication time. Still, the long duration of sonication reduces the size of the graphene flake alongside the incorporation of defects in the matrix, which will be sorted out by introducing a surfactant with a high concentration, though it can make an impact on the attribution of graphene.

14.2.1.4   Chemical Synthesis of Graphene Oxide and Reduced Graphene Oxide

The fabrication of rGO and its composites has been a critical research area for many applications. The advantages of using rGO are (1) ease of synthesis; (2) achievement of single layers to multilayers; and (3) the various nanocomposites for improving its thermal, mechanical, electrical, and optical properties. The graphene layers in suspension can easily be transferred into many different substrates for device applications in energy storage, bio applications, solar cells, electrochemical sensors, field-effect transistor (FET)-based devices, and many more. Brodie et al. first developed graphite oxide by heating a blend of graphite powder, fuming nitrate acid, and potassium chlorate at 60°C in four consecutive cycles. The weight ratio of C:O:H was found to be 61.04:37.11:1.85 [22]. A subsequent reduction in the number of oxidation steps by adding sulfuric acid was demonstrated by Staudenmaier et al. A century later, Hummers and Offerman made a breakthrough by synthesizing graphite oxide using potassium permanganate, sodium nitrate, and sulfuric acid as an oxidizing agent. This method offered better oxidation compared with the previously mentioned methods. Various research groups have modeled the GO structure to study the properties related to these functional groups. Hofmann and Holst postulated a structure in which the entire basal plane consists of an epoxy group having sp^3 hybridization with the C=C backbone matrix [23]. The model developed by Leaf-Klinowski showed that groups containing oxygen like hydroxyl and 1,2-epoxide groups are attached to graphene's basal matrix. The oxide form of graphite is called graphene oxide. The numerous oxygen-containing groups present, like carbonyl, carboxylic, hydroxyl, and epoxide, control the exfoliated graphene sheet's thickness and defects on different degrees of reduction.

The reduction methods of GO are essential to minimize the defect sites to be device compatible. Various reduction techniques and different reducing agents have been developed to remove the functional groups of oxygen. It is very significant to note that no reduction technique can nullify the defects or remove the oxygen-containing sites as a whole. Hence, the reduction of GO has its maximum limit, and one of the most frequently used reducing agents is sodium borohydride (NaBH$_4$). This reducing agent shows selective reduction toward aldehyde and ketone groups, but it is less effective in reducing ester and amide groups. The concentration of reducing NaBH$_4$ in the reacting medium varies according to the graphene sheets. Numerous research groups have used halo acids as reducing agents, such as HCl, HI, and HBr, due to their specific reactivity to intercalated hydroxyl and epoxide groups [24]. One of the most potential reducing agents is hydrazine hydrate (N$_2$H$_4$), which several research groups used to reduce GO. Hydrazine readily forms hydrazone with the anchored carbonyl group and reacts with the oxygen and hydroxyl groups to form nitrogen and water.

14.2.1.5 Laser-Induced Graphene Synthesis

Apart from graphene's ablation and modification, laser irradiation on specific polymeric materials can also be applied. This laser irradiation enables direct induction of graphene formation due to the carbonization process. The very recently developed CC_2 laser [25] induces graphene from a polyimide substrate at optimally ambient conditions. Due to the sharp rise in localized temperature (>2500°C) at the PI's surface, carbonization occurs at the surface. Due to this localized heating, C=O, C-O, and N-C bonds in the PI network are broken, forming graphitic structures from the aromatic counterparts. The color of the PI film after laser treatment becomes black from light orange. A distinct 2D peak centered around 2700 cm^{-1} is analyzed in the Raman spectrum, confirming the aromatic compound's transformation to graphene. The quality of laser-induced graphene is directly related to the polymer source and its structural features. Apart from PI, several other materials such as cloth, paper, wood, lignin, and carbon nanodots are used successfully to fabricate graphene structures. These unique advantages allow laser-induced graphene in flexible electronics such as SCs and sensors.

14.2.2 BOTTOM-UP TECHNIQUES

14.2.2.1 High-Temperature Sublimation

A high-temperature sublimation technique that preferentially separates carbon atoms from the silicon carbide framework (SiC) has shown great potential in the electronic device, as a silicon substrate is considered to grow graphene for fabrication purposes. Silicon sublimates from the top stacking of the crystalline silicon carbide surface at a temperature of approximately 1500°C. This leaves behind the atomic carbon on the underlying SiC layer. It requires ultrahigh vacuum conditions with argon and a disilane-controlled atmosphere. Graphene grown on carbon-rich SiC leads to multilayer graphene, whereas graphene growth on Si-rich SiC limits bilayer graphene formation [26].

14.2.2.2 Chemical Vapor Deposition (CVD)

One of the critical parameters for the synthesis of graphene through CVD is very high temperatures. Unlike the sublimation process, carbon atom formation occurs when the reactant gases decompose at high temperatures. This technique is renowned for mainly producing monolayer graphene with the highest degree of crystallinity. Two mechanisms can categorize this growth technique of graphene:

1. *Surface catalysis:* This is a self-limiting process where monolayer graphene is obtained on transition metals. It decomposes carbon-containing groups to the transition metals, pacifying the surface at high temperatures.
2. *Diffusion:* In this process, the cooling rate, temperature, and solubility of carbon in a metal substrate play a vital role in synthesizing high-quality graphene.

Many reports that elucidate graphene's synthesis on various metal substrates controlled three important parameters during deposition: pressure, temperature, and carbon exposure. Metal substrate selection and its reactivity to carbon atoms at high temperatures play a significant role during deposition and determine the defects induced on the synthesized graphene layer. Copper and nickel as a growth substrate for graphene have drawn the most consideration from the research community [27], because copper and nickel support proper segregation and catalysis mechanism, respectively, during graphene synthesis. The monocrystalline metal substrate is used to obtain a high-quality graphene layer; however, a monocrystalline metal substrate's employment enhances the material's cost and subsequent techniques.

14.3 PROPERTIES OF GRAPHENE-ENABLING ENHANCEMENT OF ENERGY STORAGE PERFORMANCE

The 2D graphene materials are the potential candidates for future-generation energy storage, such as SC devices, because of their distinctive internal properties. Different graphene materials are used for SC electrodes like 3D, 2D, 1D yarn, and 0D dots. First, a 2D material, i.e., graphene, contains a single carbon atom sheet oriented in a honeycomb system lattice. Graphene's attributes appear for the long-distance π bond conjugation with higher carrier mobility and good thermal conductivity [28]. Graphite lattice crystals are oriented in the highly oriented pyrolytic graphite (HOPG) form, which binds the graphene together. There are four electrons, one π, and three σ and a bond for graphene in the carbon atom outer shell. The sp^2 hybridized π bond provides a strong atomic binding force for atoms in its neighbors. The π bond creates a Pz orbital that enhances electrons' free movement in a half-field band, delineating metallic characteristics. Also, graphene shows many important properties like better electrical charge mobility (230,000 cm^2/Vs), best specific surface area (2600 m^2/g), highest strength (130 GPa), and thermal conductivity (3000 W/mK) [29].

14.3.1 ELECTRICAL PROPERTIES

Graphene is a zero-energy bandgap semiconductive material because of its 2D honeycomb lattice structure carbon. It has an excellent charge carrier's density of holes and electrons [30]. These charge carriers move freely and very rapidly through the graphene. It has higher conductivity when in a monolayer, which has less defect density in the crystal lattice, leading to its superior quality. Those charge carriers' mean free paths are hindered by defect scattering, impacting ion mobility, and lowered electrical conductivity. Also, inherent construction, including traps of surface charge, substrate ripples, beneath substrate interaction in measurement time, and the phonons at interfacial surface, are the controlling factors that modify graphene's electrical conductivity.

14.3.2 MECHANICAL PROPERTIES

Due to the variation of the interatomic gap, incorporating exterior stress to the rearrangement of electrical charges happens within any crystal [31]. So, pointedly the mobility of electrons is modified, and bandgap is introduced into the electronic structure. It has been frequently observed in graphene material that the modulus of elasticity and strength varies according to dimension [32]. In graphene, monolayers (2D) have the ultimate ~130-GPa tensile strength and 1.02-TPa stiffness, which is nearer to 3D graphite. Graphene shows approximately 1.3% high tensile strain in tension compared with compression.

14.3.3 THERMAL PROPERTIES

Graphene has C–C strong covalent bonds and phonon scattering features, displaying its high thermal conductivity. Moreover, graphene's monolayer has generally exhibited ~5000 W/mK thermal conductivity at 25°C (room temperature [RT]). It is proved that single-layer graphene shows a maximum of ~5300 W/mK thermal conductivity, as analyzed with the benefit of confocal micro-Raman spectroscopy, which equates to the suspension of graphene flake with a conductivity up to ~4800 W/mK.

14.3.4 NEXT-GENERATION BATTERIES

There are many potential features of graphene including its excellent electrical conductivity, fragile flexible nature, chemical stability, and great active surface area-to-volume ratio (SA/V); the ability to make composites by conducting polymers and transition metal oxide (TMO) nanoparticles that enhance theoretical capacities [33]; and the ability to make an advanced impact on energy storage applications, including the Li-O$_2$, Li-ion batteries, Na-ion, and Li-S. Numerous SC materials for electrodes are implemented for LIBs deepened on graphene nanomaterials, enhancing the specific capacitance [34].

14.3.5 SUPERCAPACITORS

The improvement of electrolyte solution and electrode material improves the whole SC performance. Therefore, the SC's electrode materials should possess high electrical conductivity, good stability, suitable pore sizes, high surface area, highest volumetric power, and energy densities [35]. The different types of nanoscale carbon materials show the previously mentioned features that enhance the electrochemical response. The SC devices store the charges electrostatically on the electrolyte surface, i.e., the electrolyte ions are adsorbed into the electrode's active surface area. So, the active materials containing high active surface area contribute more ion storage in energy storage applications. Graphene is described as a prominent electrode material for SCs. The graphene nanosheet's active surface area plays an important part in the SC's energy storage performance, but the main drawbacks are an agglomeration of nanosheets and, after that, restacking for van der Waals forces in between nearby graphene nanosheets. This reduces the active surface area, and capacitance is decremented. Therefore, incorporating different metal oxides with the graphene sheets and fabricating a suitable composite can prevent aggregation and enhance the capacitance value. The positive interface between graphene sheets and nanoparticles can help advance an SC charge storage system. Researchers found that the restacking by the face-to-face method is hindered in the graphene nanosheets because of its curved nature, which maintains the large porosity (2–25 nm). Also, the rapid charging-discharging possibilities with no deficiency of capacitance and energy density leads to preparing a prominent candidate compared with the present storage. Graphene is reported frequently as a potential SC material for electrodes. It is fabricated with different structures like free-standing graphene in 0D morphologies, in the form of fiber as a 1D formation, as thin films sheets like in a 2D formation, and in composite network formation or foamlike formation with metal hydroxides-oxide nanomaterials or with any 3D formation with conductive polymers.

14.4 GRAPHENE-MODIFIED ELECTRODES FOR SUPERCAPACITOR APPLICATIONS

14.4.1 0D GRAPHENE POWDERS AND DOTS

Graphene powders and dots are generally fabricated using the oxidation and graphene reduction process, which contains rapid ion transfer and good power density with a higher electrical conductivity and surface area value [36]. Furthermore, the composite is prepared with metal oxides and conductive polymers. This works as hybrid electrodes with the response of EDLC and pseudocapacitance; hence, capacitance value is enhanced. This powder is used as a filler material in the electrode fabrication process and helps to increase conductivity.

14.4.2 1D GRAPHENE-BASED FIBERS AND YARNS

These graphenes are potential candidates for recent research on SC wearable and portable electronic devices because they are lightweight, have better electrical conductivity, have high flexibility, and are small volume. The CVD process synthesizes the carbon nanotubes (CNTs) as hybrid electrode materials on the 2D graphene sheets enhancing the properties and stability [37].

14.4.3 2D GRAPHENE-BASED SHEETS AND FILMS

Thin films are generally worked to fabricate flexible SCs because they have tunable thickness, are lightweight, have high electrical conductivity, have high energy and power densities, and there is a huge active surface area for ion transport and mechanical stability. The main fabrication methods are interfacial self-assembly, Langmuir- Blodgett, vacuum filtration, deposition by layer-by-layer, and spin coating [38]. The interplanar van der Waals forces and π-π interaction attractive forces cause graphene films to restock, reducing the effective surface area, and hindering the diffusion of ions of electrolytes in between. The applicable incorporation is used to solve this problem, like a conducting polymer, metal oxide nanoparticles, metal ions, and 1D CNTs.

14.4.4 3D GRAPHENE-BASED AEROGELS AND FOAMS

These carbons are suitable for higher SC performance, best energy, and power densities for the continuous meso-, micro-, and macro-interconnected pores that lead to the vast active surface area for fast ion transport. The graphene aerogels are a unique class of very lightweight, porous carbon materials. This depicts higher SA/V and strength-to-weight ratios [39]. This aerogel formation corroborated that the ion transport pathway is reduced between the bulk electrode material and electrolyte and provides multidirectional ion or electron movement pathways for storage. So, the graphene aerogel is very prominently used as a binder-free and high-performing SC application.

14.5 EFFECT OF DOPING HETEROATOMS IN GRAPHENE MATRIX FOR IMPROVEMENT IN ENERGY STORAGE PERFORMANCE

The new graphene electrode material is stored physically by the ions of electrolytes accumulated on the surface, which is how EDLC is formed. To enhance the capacitance, graphene-functionalized and graphene-doped composites have been fabricated that incorporate extra pseudocapacitance [40]. Electrolyte ions access both sides of the surfaces of the graphene nanosheet to store charge. The experimental capacity value is higher than the theoretical value because the defect is introduced into the plane in the experiment, and they can access more amounts of ion storage-active sites [41]. Due to the van der Waals interaction, the sheets are restacked and the surface area is not accessed, so charge storage is hampered. A high amount of electric conductivity can make graphene electrochemical performance better, but it is not easy to maintain the vast active surface area with higher graphene conductivity in the same synthesis. Incorporating metal or metal oxides and decorating with them provides different active charge sites and leads to incrementing the specific capacitance with energy density. Thus, it is very challenging and crucial to implement any defect or impurities on the graphene sheet in synthesis time. The heteroatom-doped graphitic composite can efficiently tailor graphene's morphologies, with pristine electrical and thermal features. Doped graphene exhibits higher electrical, mechanical, and physicochemical properties, helping the electrochemical response. N, B, S, P, and O are the several types of dopants that have been used for the fabrication of composite.

14.5.1 NITROGEN (N)-DOPED GRAPHENE (NG)

Electron-rich N atom doping leads to improved conductivity, reduced graphene sheets agglomeration, created holes, and improved electrolyte ion migration. Also, nitrogen atoms are covalently joined with a carbon network of graphene and provide more stability. The CVD and hydrothermal, plasma treatment, pyrolysis, ball-milling, supercritical, and electrochemical methods are the primary synthesis methods for nitrogen doping. The highest doping percentage is obtained from the CVD process, hydrothermal process, and ball-milling method. Like nitriles and aliphatic amines, the moieties modulate the change-of-growth kinetics and help fabricate highly ordered graphene in the carbon nanostructure. Mohana Reddy et al. and Li et al. showed that the liquid precursor-based CVD technique can grow the nitrogen-doped graphene (NG) layer [42, 43] directly on the current copper collectors, and it exhibits very high-power density capacitors. The two reasons for the low doping rate are fewer defects in pristine graphene and the high annealing C-N bond-breaking temperature. Li et al. researched it., GO is annealed with NH_3 ambience at 500°C, and it is reduced to rGO, and NG is obtained with 5 at. % of N doping. At 500°C, the highest N doping level of 10.1 at. % was obtained. Also, N plasma is a process in which NG can be prepared, carbon is kept in the N plasma atmosphere, and N atoms partially replace the carbon atoms. At the time of capacitive behavior analysis, the surface redox occurs in the functional groups of NG composite, exhibiting the pseudocapacitive response. The pyrolysis method is carried out using the nitrogen-containing precursor. Then those materials are pyrolyzed at high temperatures in the atmosphere of inert gas [44]. It was recently reported that there are many new precursors like the shell of soybean, cocoon, resin (4-aminobenzoyl), okara, polypyrrole (PPy), chitosan, polyaniline (PANI), glycine, urea and glucose, polyurethane acrylate, asphalt/dicyandiamide, and graphitic-C_3N_4 [45–47]. The graphitic C–C mechanochemical cracking leads to the fabrication of N_2 accommodated active carbon and edge-nitrogenated graphene nanosheets. The electrochemical and electrical features have potential for energy storage from glycine ammonia, ethyl ammonium nitrate, $(NH_4)_2SO_4/NH_3 \cdot H_2O$, and ammonium nitrate [48].

14.5.2   BORON (B)-DOPED GRAPHENE (BG)

As neighboring element boron is highly acquiescent to graphene doping. First, the boron-carbon bond length is higher than carbon-carbon bonding in pristine conditions; the slightly altered lattice parameters with less persuasive strain energy make more detailed reaction kinetics [49]. Second, higher conductivity was analyzed and obtained for graphitic B doping, delivering many holes to the graphene valence band. It is doped by gas-solid reactions, thermal decomposition, microwave plasma, direct solid-state reaction, and reflux and hydrothermal processes. The highest amount of doping is obtained hydrothermally. Sathe et al., at 600°C GO, reduced borane tetrahydrofuran (BH_3-THF) material thermally and then reacted using the reflux method, and then boron-doped graphene (BG) was obtained [50]. The B-doping amount is modulated in carbon crystal by variation in composition, temperature, and gaseous atmosphere. The solid-state reaction is completed using H_3BO_3 or B_2O_3 and GO, and after complete grinding, it is heated at a high temperature of approximately 900–1200°C in argon ambience. Lee's group applied microwave plasma for the fabrication of BG by trimethyl boron. The transport properties and the bandgap are tuned by the boron graphene composite synthesized by reactive microwave plasma. Solvothermal and hydrothermal methods have been vastly implemented to fabricate graphene composite with surface functionalization and unique morphology. $NABH_4$ was used as a source of B-dopant atoms and reductants in the GO hydrothermal process with polydiallyldimethylammonium chloride [51].

14.5.3   SULFUR (S)-DOPED GRAPHENE (SG)

Sulfur also shows polarization effects that modulate the charge transfer. The S doping in graphene was done in two steps: a change of the small bandgap of the pristine graphene and incorporation of defect sites with S-S bond rapture. The bond length of carbon-sulfur is ~25% higher than the carbon-carbon bond [52]. Also, Denis et al. assumed that Si-doped graphene is the potential for O reduction reaction (ORR). The spin density in the graphene is probably enhanced, and obtained by S doping is the main reason for those properties. Similarly, it is also doped by hydrothermal, thermal, exfoliation, ball-milling, and ion-exchange methods and remarkably high doping percentage obtained by the ion-exchange method. The high-temperature GO reaction with sulfur-containing precursors has corroborated the sulfur functionalized graphene, which holds sulfur atoms by substituting carbon atoms [53]. Poh et al. fabricated sulfur-doped graphene using thermal exfoliation of GO in SO_2, CS_2, and H_2S gases [54].

14.5.4 PHOSPHORUS (P)-DOPED GRAPHENE (PG)

The strong hybridization between C2p and P3p orbitals changes the carbon hybridization from sp^2 to sp^3, and phosphorus can create a pyramidal configuration of three carbon atoms bonding. This doping is executed by pyrolysis, post-thermal annealing, and a liquid process, and pyrolysis creates the highest yield. The C atom's electronegativity is higher than that of the P atom [55]; so, the polarity between the carbon-phosphorus bond and the C–N bond is the opposite. Moreover, phosphorus has a different orbital and high electron-donating tendency, so compared with nitrogen doping, the doping of carbon leads to fundamentally different properties. The pyrolysis process of modified H_2PO_4 accomplished the phosphorus-doped graphene (PG) preparation-alginate under inert ambience at 900°C. PG is also synthesized using a post-thermal heating synthesis method in which a phosphorus-containing precursor and graphene are used in closed environments [56]. Wen et al. prepared PG using graphene and phosphoric acid mixture annealing. In reality, graphene delineates less activity for inert π electrons [57].

On the other hand, π electrons are activated by heteroatoms like S, N, P, and B doping. Creating new charge store active sites extensively increases graphene activity in different prospects with regulating chemical properties [58]. Moreover, with different electronegativity, the co-doping by two or three elements can modulate the specific electron distribution; it exhibits very definite properties. The heteroatom-doped graphene composite exhibits positive enhancement in ORR applications. The dependency on heteroatom doping of different atoms and its comparison is a new attribution used to further research NG catalysis.

14.5.5 DOPED GRAPHENE COMPOSITE FOR ELECTROCHEMICAL ENERGY STORAGE

Pristine graphenes are chemically inert; thus, pristine graphene nanosheets are doped by heteroatoms like P, N, S, and B, which enhance the charge transport in neighboring carbon atoms and increment the performance of EDLCs. They also enhance faradaic reactions between electrolytes and graphene. N-doped graphene can modulate the electronic configuration by incorporating updated energy levels in the conduction band lower section of sp^2-bonded carbon atoms. These are corroborated by three standard carbon lattice bonding configurations, including quaternary (graphitic) N, pyridinic N, and pyrrolic N. Large interfacial capacitance is motivated to increase charge carrier density. NG exhibits capacitive properties that are crucial factors in designing the electrode material for SCs. Both pyrrolic N and pyridinic N depict high pseudocapacitance, because those are active electrochemically in alkali aqueous solution; moreover, graphitic nitrogen enriches the electrical conductivity of the graphene. The pyridinic N basal-plane of NG shows high cation K^+ binding energy that enhances the larger number of ions from electrolyte accumulations on the electrode's surface area. The BG shows a higher conductivity and an excellent specific capacitance in the necessary electrolyte. Still, the energy storage is not fully matched with excellent value for storing low active surfaces [59]. Conway [60] suggested that a carbon SC with high surface area and conductivity should exhibit a few hundred specific capacitances of Fg^{-1}. So, B doping may be incorporated redox reactions, which are responsible for high capacitance [61].

Interestingly, high B content-doped graphene exhibited high performances compared with pristine graphene. About 4 at. % of boron-doped composites depict a capacitance increment up to 86% and increased volumetric energy density 5–10 times more than the pristine graphene. The heteroatom doping enhances the Li diffusion and storage for graphene-based electrodes in the SC energy storage capacity. This can also introduce huge defects in the graphene surface, which helps create a disoriented carbon configuration that improves the lithium-ion intercalation features. In the lithium-sulfur battery field, the doping of heteroatoms manipulates the distribution of electronics and generates functional groups of the surface. The heteroatom-doped graphene has been fabricated and introduced in electrocatalysts for metal air batteries' air cathodes [62]. Presently, different carbon materials, like carbon fibers [63], sphere-like carbons [64], graphene [65], and CNTs [66] were found to work in the sodium ions intercalation/de-intercalation into/from the host system of the sodium-ion battery.

14.6 CONDUCTING POLYMER INCORPORATION IN GRAPHENE MATRIX AS HIGH-ENERGY SUPERCAPACITOR ELECTRODE MATERIAL

Metal oxides or sulfides and conducting polymers (CPs) are the utmost potential electroactive materials for configuring pseudocapacitors. CPs have distinctive attributions like higher theoretical specific capacitance, good physicochemical properties, high electrical conductivity, quick large-scale production, and lower cost [67]. The charge or discharge methods are corroborated with the de-doping and doping methods of CPs in the case of pseudocapacitor application. Two kinds of doping variations occur, i.e., reduction (n-doping) and oxidation (p-doping), and the increment of the total polymer conductivity is comparable to the level of metal [68]. In the polymers, p-doping, which is controlled by the polymer's (P) partial oxidation, and the intercalation of respective counter anions (A^-) occur to maintain the system charge neutrality, as illustrated in the following equations. In contrast, the polymer partial reduction can modulate the n-doping of polymers, and counter-cations (M^+) are inserted to withstand the charge neutrality,

$$P_m - xe^- + xA^- \quad \leftrightarrow \quad P_m^{x+}A_x^- \tag{14.1}$$

$$P_m + xe^- + xM^+ \quad \leftrightarrow \quad P_m^{x-}M_x^+ \tag{14.2}$$

where m is the degree of polymerization and x is the charge's transferred number. Only using these redox materials, the low power density and poor stability in capacitors are enhanced. Both redox-active materials and carbonaceous materials can be worked in the fabrication as a hybrid SC to conquer the downside of pseudocapacitors and EDCLs. Generally, carbon works as a framework to sustain electroactive components, and the composites work as electrodes for hybrid SCs. Hybrid SCs depict higher power and energy density with comparatively high long stability. CPs, like PPy, PANI, and polythiophene (PT), and their respective derivatives, are potential elements for applications of charge storage for high electrical conductivity, low cost, and high specific capacitance, mainly relating to the pseudo-faradaic application. PPy shows 620 Fg^{-1} [69], PANI shows 2000 Fg^{-1}, and PT shows 485 Fg^{-1} [70], depending on the doping level. Still, the substantial degradation mechanically (e.g., shrinking and swelling) and irrevocable structure changes in the time of the charge-discharge processes are decaying stability rapidly. The present analysis concludes that graphene may modulate the molecular orientation of CP chains and their conformation, which leads to a positive collective effect on the composite for achieving higher stability and good electrical conductivity rather than the single material itself. The advantages of composite materials are considerable: structural variation, right effective surface area (to increment the charge storage), short ion transfer pathway, small particles, uniform structure, and controlled morphology [71].

Several essential characteristics granted for the fabrication of nanocomposites of graphene-CP include (1) polymer-graphene surface compatibility, (2) the appropriate process of synthesis, (3) surface features of the nanocomposite (e.g., hydrophilicity, surface roughness), and (4) controlling morphology and thickness of the polymer film. There are two critical tactical methods for the fabrication of graphene-CP nanocomposites: (1) with the intercalation of the CP in between graphene nanosheets for making 2D materials and (2) deployment of graphene nanosheets as a system matrix for the fabrication of nanocomposites with a different structure for the 3D network [72]. This is synthesized by incorporating at the time of GO reduction. Polymer or metal with micro or nano morphologies is used for the layer-by-layer stacking at the monomers' in situ polymerization on the rGO surface. Tanaka et al. proposed assembly of graphene/CP with electrolyte interfaces containing three distinct layers: graphene and CP sheets are covalently interconnected at the bounded layer, electrostatic or ionic forces, and interfacial region, which interacts on the surface of the

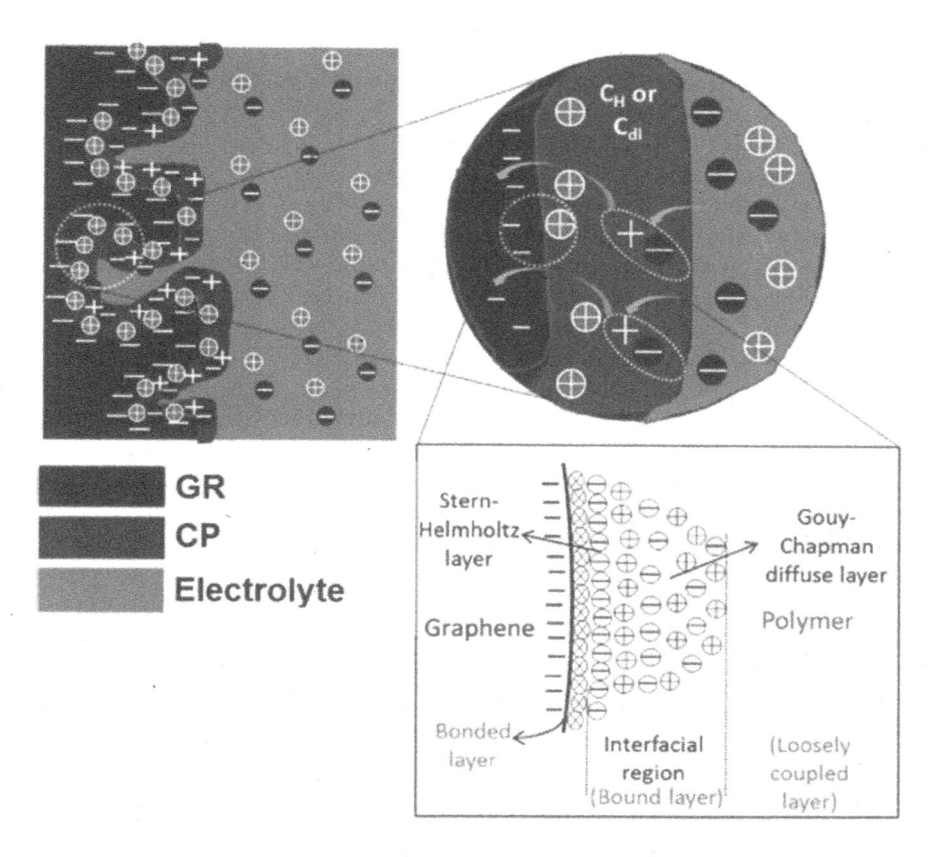

FIGURE 14.2 Graphene-polymer composite and electrolyte interfaces in between. (Reprinted from Shen, Fei, Dmitry Pankratov, and Qijin Chi. (2017). "Graphene-conducting polymer nanocomposites for enhancing electrochemical capacitive energy storage." *Current Opinion in Electrochemistry* 4(1): 133–144 with permission from Elsevier.)

nanomaterials, and the externally joined layer with improved movement and significant CP chain changes [73]. The graphene-CP composite and its charge storage EDLC formation are depicted in Figure 14.2. The interfacial layer structure variation at the time of polarization of charge or discharge must happen several times and play a crucial role in charge-storing performance. Both pseudocapacitance and double-layer capacitance are developed as active surface phenomena that control the graphene-CP composites and the stability of their devices. Liman et al. reported that, the graphene sheets are evenly covered by CP, followed by the parallel stacking. It is designated as a series connection of parallel capacitor-resistance components [74].

14.6.1 Different Conducting Polymers

PVDF is a strongly non-reactive fluoropolymer with thermal stability, better mechanical strength, good aging resistance, and high chemical resistance. PVDF is extensively used as a binder for graphene-based SCs to attach graphene nanosheets firmly to the current collector and produce mechanical strength to maintain the electrode characteristics. First, graphene and PVDF 10–20 wt% are combined to fabricate the graphene SC working electrode. To make the mixture uniform and homogeneous, organic solvents like DMF or NMP are added to make a slurry. The composite slurry is coated over the working electrode. A few conductive carbons like acetylene black and carbon black (CB) are mixed into the mixture to nullify PVDF's negative impact on conductivity [75].

A few coating methods like drop cast, bar coating, and doctor blade coating can be used. The conductive carbon nanoparticles work as a filler material to graphene nanosheets, improve the long cycle stability of the SCs, and withstand the good electrode conductivity. Ninety-one percent capacitance retention with a specific capacitance of 175 Fg^{-1} is obtained from the rGO-PVDF-CB composites.

Also, PTFE has generally been used as a polymer binder for SCs, and it is an alternative category of fluoropolymer. PTFE's highest response relies on different factors, including electrolytes and active electrode materials. Tsay et al. showed that the 5 wt% PTFE carbon composite SCs exhibited the highest energy density and the specific capacitance using the Na_2SO_4 electrolyte. Less PTFE binder creates adhesion problems with the active electrode materials, and the value of specific capacitance and the retention of capacitance are also poorly impacted over long cycles. Five percent of PVDF binders are optimal for electrode use [76]. Zhu et al. analyzed that the chemically altered graphene and PTFE composite used as the SC electrode's active material and measured it in organic and aqueous electrolytes [77].

PANI is an extensively analyzed CP. It is a perspective electrode material because of its electroactivity, higher conductivity, good stability, and specific capacitance. A proton is required for PANI to conduct and sufficiently charge and discharge. Thus, an acidic solvent, an aprotic solvent, or aprotic ionic liquid is mandatory for PANI to work in an SC [78]. The established fabricated procedures are electrochemical polymerization and oxidative polymerization. Composites of graphene and PANI may be fabricated by the in situ polymerization of graphene suspension and monomer of aniline in an acidic solution. In the process of oxidative polymerization, ammonium peroxydisulfate (APS; $(NH_4)_2S_2O_8$) was used as the oxidizing agent in 1 M aqueous acidic media of HCl. The GO-PANI composite with the varying GO concentrations corroborated the different composite structures and influenced the SCS's electrochemical response. Moreover, the retention of capacitance in long cycle stability analysis is prominently incremented with a higher concentration of GO in the composites.

The GO nanosheets can expiate for volumetric change of PANI in the time of charge and discharge cycles. Furthermore, GO is insulating in nature, and its nanosheets contribute significantly less specific capacitance in the composite. The composite's total capacitance is exhibited for the pseudocapacitance of PANI nanofibers. The high specific capacitance can reduce the ESR of the SC electrode for an increase in conductivity. Wang et al. implemented a three-step fabrication process by reducing or doping-de-doping in the in situ polymerization to fabricate rGO and PANI composites. PPy is also a useful CP material in SC applications. The fabrications of the polymer are comfortable and have better thermal stability and conductivity. The agglomerated formation of the dense film of PPy is not acceptable for SCs because it reduces the active surface area and hinders the electrolyte ion transportation. These particulate morphologies can collapse because of the stress generated due to volumetric change at the time of charge and discharge cycles; in general, pure PPy film exhibits inadequate extended cycle response as SCs compared with PANI films [79].

However, to reduce the agglomeration effect of graphene nanosheets, a normal in situ polymerization process can fabricate graphene-PEDOT composites. First, ethylenedioxythiophene (EDOT) and PSS and monomers are combined in an aqueous medium like DI water [80] and HCl, and the final solution is placed with sonication and stirring. Next, iron (III) sulfate [$Fe_2(SO_4)_3$] and sodium persulfate ($Na_2S_2O_8$) or iron (III) chloride ($FeCl_3$) [81] and ammonium peroxydisulfate [$(NH_4)_2S_2O_8$)] oxidant is mixed to start the process of polymerization. Lastly, as synthesizing composites of graphene-PEDOT are obtained. The composite film of rGO-PEDOT creates a planar, curved film of 200-nm thickness [82]. The graphene-PEDOT conductivity was enhanced and almost doubled than that of pristine PE DOT with the incorporation of graphene.

Moreover, the mechanical strength was also enhanced sixfold using the same process. The graphene creates a charge propagation and percolation pathway, enhancing PEDOT's charge transport behavior. This polymer composite also improved the long cycle stability performance and specific capacitance in supercapacitors application. Graphene plays an essential role in the composites: (1) graphene or rGO contains conductivity compared with PEDOT, and PSS and production of the polymer composite enhance its conductivity; (2) heterogeneous structure is formed with PEDOT, which strongly decreases the structural destruction due to volumetric change at the time of the charge-discharge cycles; and (3) graphene contributes to significant enhancement of a specific capacitance by arranging a huge active surface area, which leads to electrolyte ion interaction and the redox reactions [83].

14.7 GRAPHENE SHEET COMPOSITE WITH METAL OXIDE NANOPARTICLES AND ITS BEHAVIOR AS SUPERCAPACITOR MATERIAL

Two-dimensional graphene, with its excellent attributes like optical and electrical properties, mechanical strength, and high conductivities, is one of the potential materials for creating energy with renewable sources. The graphene composites with similar configuration materials like layered TMOs are prospective combinations to enhance their conducting, mechanical, and optoelectronic features. Moreover, the transition metal chalcogenide materials gained prominence for features like conducting and electrical characteristics and enormous optical and atomically thin layers. The layered materials mainly involve a 2D plane, and are thin atomically for higher conductivity ion or charge transport. The transition metal dichalcogenides (TMDs) have become the prominent materials in energy harvesting technologies because of their bandgap, thickness at the atomic level, optoelectronics properties, and coupling of spin-orbit.

14.7.1 TRANSITION METAL OXIDES AND SULFIDES WITH GRAPHENE COMPOSITES

Molybdenum disulfide (MoS_2) is one of the vastly used and analyzed layered TMDs. It shows outstanding electrical, chemical, mechanical, and optical features. MoS_2 is a potential candidate as a transport material in solar cell photocatalysts or a catalyst in hydrogen evolution reaction (HER) as active material or electrodes in lithium batteries. The bulk MoS_2 has an indirect bandgap of 1.2 eV. The bandgap modifies after being exfoliated to a monolayer to 1.9 eV. The rGO/MoS_2 composite is generated without any binder, delineating a better 585.4 m^2g^{-1} active specific area, which is a planar-like structure. The as-fabricated composite contains mesopores (2.5–9 nm) and micropores (1–2 nm) and provides a sizeable active energy storage area. Also, nanospheres of MoS_2 are created on the active surface of rGO and the rGO sheets to create structure layer by layer with interactions of van der Waals between -OH, -COOH, and -S of GO functional groups. The rGO device exhibits specific capacitance at discharge as low as 90 Fg^{-1} with 67% columbic efficiency. After MoS_2 nanospheres incorporate into rGO nanosheets, the planar MoS_2-rGO device delineates an improved 323 Fg^{-1} specific capacitance at 0.2 Ag^{-1} current density and a columbic efficiency of 72.6% is obtained.

A semiconductor type 2H-WS_2 shows a 1.3-eV indirect bandgap, becoming a direct bandgap after the exfoliation of monolayer WS_2. These WS_2 monolayers are restacked by van der Waals interactions that create a thin 2D layer. Because of the exceptional structures, $MoSe_2$ and $NiSe_2$ have shown features like electrochemistry, storage, and HER catalytic activity [84] in the research. An electrochemically viable metal oxide is tungsten oxide, and with its great potential applications include gas detectors, energy storage devices, field-emission devices, and as a photocatalyst. Due to their low-charge movement resistance and significant surface area, nanostructured tungsten oxides (WO_3) as components of electrodes are depicted as having enhanced electrochemical efficiency. This material is a potential part of the SC electrode and different energy storage devices for de-intercalation and the intercalation of electrons and protons into the oxide [85]. The composite material of oxide with CPs, rGOs, and various carbon fibers are prepared to enhance the tungsten oxide's features and performance. The device's specific capacitance was boosted by the metal oxides combined on the active surface of graphene. The composites depicted the reactivity of ions and redox properties.

It is necessary to fabricate composites of SnO_2 with graphene to enhance those features as graphene has good electrical conductivity and a large surface area [86]. Pulse microwave deposition methods, hydrothermal synthesis, and one-step synthesis have been employed to prepare those hybrid electrode composites that modulate the SC response. Li et al. fabricated SnO_2 on graphene sheets with urea and hydrochloric acid helped by graphite oxide reduction with $SnCl_2$. The nano-composite of graphene–CuO has been prepared using a hydrothermal method with an ammonium solution for SC application. Accordingly, those composites exhibit enhanced capacitance of 906 Fg^{-1} at 1 Ag^{-1} and 946 Fg^{-1} at 10 mVs^{-1}. The long cycle stability is as high as 89% in 5000 cycles at 5 Ag^{-1} densities of current. Three-dimensional GO nanosheet makes a composite with iron oxide nanoparticles, and this hybrid material is fabricated by a one-pot microwave method. The value of specific capacitance of iron oxide GO 3D hybrid composite is 317 Fg^{-1} and 455 Fg^{-1} at 27 mVs^{-1} scan rates [87]. The ZnO-rGO is fabricated when the hydrothermal synthesis method exhibits a total value of specific capacitance of 156 Fg^{-1} at a 5 mVs^{-1} scan rate. The composite electrode manifests a good cycling ability of 90% after long 2000 cycles [88]. The RuO_2/Gr hydrous composite with varying Ru loads shows that the specific capacity and corresponding integrated area is enhanced up to 55.8% at the 10 mVs^{-1} scanning rate. The retention of capacitance is 80.5% of that composite at a high current density of 20 Ag^{-1}. The GO composite's V_2O_5 and powders are synthesized by mixing in a 1:1 ratio before GO and adding the V_2O_5 dissolute in distilled water. The rGO/V_2O_5 composites show specific capacitance of 186 and 290 Fg^{-1} in 10 and 5 Ag^{-1} scanning rates.

Generally, the metal oxide-based SC demonstrates a higher amount of power and energy densities. The TMOs like NiO, RuO_2, Mo_2O_3, Co_3O_4, MnO_2, and V_2O_5 acquire variable oxidation states, delineating higher capacitance performances and enhancing rich redox reactions in SCs. A planar-type and stretchable graphene/d-MnO_2 composite nanostructure composite has been fabricated in the prominent implementations of portable and wearable devices of electronics. The 2D nanostructure of graphene/d-MnO_2 is synthesized by the technique of vacuum filtration, which helps control film thickness. Two working electrodes are filled with PVA/H_3PO_4 polymer gel electrolytes in between. In a 0.2-Ag^{-1} density of the current, these flexible devices have delineated a higher specific capacitance of 267 Fg^{-1} and illustrated better long cycle stability and higher retention of the capacitance at 92% after 7000 cycles.

14.8 AREAS OF FOCUS AND IMPROVEMENT CORNERS

Metal-organic frameworks (MOFs) are crystalline materials containing metal ions coordinating in organic bridging ligands. This material has several features like high porosity, structure tenability, and large surface area [89]. With these potential attributions, MOFs have become a prominent material in the field of SCs [90], batteries [91], etc. MOFs show weak responses in the field of conductivity and have less operational stability, and they are preventing the performance in the electrochemistry area. The idea of generating MOF-based composites with other materials is encouraged to nullify the previously stated difficulties, which may be ideal replacements instead of using only MOFs. The composite is fabricated traditionally. Presently, incorporating nanomaterials that are graphene-based into different nanomaterials can help electrical conductivity developments and stability. The composites of graphene/MOF-based nanomaterials can incorporate the improvements and entirely alleviate the discrete component's defects, i.e., the stability, electrical conductivity, selectivity, and template effects all are improved and enhanced [92]. The aromatic sp^2 domains and ionic groups participate in bonding interactions in these graphene-based materials, creating structural or functional nodes and enhancing MOF activity. Pyridine and carboxylate are in the graphene-based composite groups and work as essential roles in accumulation, controlling the MOF growth, and incrementing the coordination bonding by exhibiting a more favorable structure. Carbonized ZIF-9 may be researched as a superior electrocatalyst in the oxygen evolution reaction (OER) and in the ORR [93]. Carbon composite-derived Co_3O_4/nanoporous from MOFs used as SC electrodes depicts a superior performance. MOF/graphene-based materials become potential materials for better battery electrode material. In recent studies, LIBs are the most effective power delivery system. In the configuration of LIBs, the electrode materials are more crucial constituents and bring more attention to the research. Graphene/MOF-based materials lead to their position in this case. The solvothermal process has been a widely used synthesis method in the application of batteries [94]. Jin and coworkers synthesized rGO/Fe-MOF using the experiment's solvothermal method, and the electrodes had higher results. In general, the columbic efficiency is 33.3%, and the value is low for a number of irreversible reactions in the Fe-MOF first cycle. When the composite with graphene is fabricated, it is enhanced up to 43%, and the charge storing capacity and long cycle stability are improved [95].

Sodium ion batteries' have gained more attention because of low prices, abundant resources, and high system safety. Moreover, graphene's composites with MOF-based materials are used as potential candidates for SIB anode material. Jin et al. synthesized Cd(L) MOF ($[Cd(L)(H_2O)]n\cdot 2nH_2O$)/rGO (L = 5 amino isophthalic acid) and Co(L) MOF ($[Co(L) (H_2O)]n\cdot 2nH_2O$)/rGO using the direct mixing method to analyze properties. The long cycle stability, rate capacity of Cd(L) MOF, Co(L) MOF, and its hybrid are investigated. Similarly, as before, the composite with graphene shows a significantly better performance in energy storage. The large surface area and specific porous property are the MOF's controlling merits in the composites, and the composites of the same material exhibited a tremendous super capacitive study response. The highly conductive graphene can also be corroborated by the MOF/graphene-based materials in the porous structure, accelerating phase transfer reactions and reducing diffusion time significantly, resulting in excellent cycling stability to enhance conductivity. Moreover, the graphene layer works as MOF passivation, basically for the graphene layer small lattice constant; this can also increase the MOF porous structure's electrochemical stability. Furthermore, graphene composites stop electrode agglomeration and control the volume contraction or expansion in the time charge and discharge analysis. The incorporation of graphene improves the stability and electrochemical performance, and MOF's disadvantages are hidden [96]. Saraf et al. synthesized the rGO/Cu-MOF composite using GCE to fabricate supercapacitor electrodes. This can enhance the discharging time. Moreover, the rGO/Cu-MOF/GCE electrode reduces the electrodes' voltage drop, basically minimizing internal resistance and higher conductivity. The composite Ni-MOFs@GO is another composite with flower-like thin tailored morphologies. The active MOF centers were provided to deposit the Ni-MOF on GO's nanosheets, which can assist the thorough exfoliation of GO nanosheets. These composite formations are helped to lower the transfer resistance and increase ion transportation. The Ni-MOFs@GO exhibits specific capacitance at $\approx$2192.4 Fg^{-1} at 1 Ag^{-1} current density [97, 99].

14.9 CONCLUSION

Graphene is a well-known 2D material with excellent attributes that exhibit several potential applications like SC devices. Graphene works as a perfect electrode material to enhance power and energy densities, specific capacitance, and more prominent energy storage devices. In pristine graphene, the energy storage response is sometimes limited, and the answer is not high. The electrical conductivity for fast ion transportation, stability at the time of charging and discharging, restacking of the layer due to van der Waals attraction force, and the reduction of active surface area are responsible for minimizing its charge storage capability. This is achieved by making a graphene composite with metal oxide or hydroxide, conductive polymers, or elemental doping to improve the SC's overall response. Creating a defect state over the nanosheets, incorporating more active charge storage areas, and enhancing conductivity with mechanical stability leads the graphene nanocomposite to be a potential energy storage study. Accordingly, future real-time research should move on to the genesis of the free-standing, nanocomposite, flexible, and multifunctional graphene composite-based SCs to realize the combinational effect of EDL capacitance and high-performance pseudocapacitance. This research to contribute to the worldwide demand of prominent energy storage systems. On the other hand, MOF is one of the potential materials for graphene composites, which improves the features, leading to improved energy storage. Graphene-based materials in variant forms like 0D, 1D, 2D, and 3D are excellent electrochemical energy storage materials.

The properties and refabricating of those materials have to be ameliorated for realizing the real-time application. The development of graphene materials for energy storage electrodes depends on the morphological intricacy: 3D (hydrogels and graphene-based foams), 2D (graphene-based nanocomposites thin films), 1D (fiber and yarn type structures), and 0D (free-standing graphene dots and particles) have been incorporated into this chapter to analyze graphene's role on charge storage. Many different fabrication methods can help assemble free-standing graphene SCs, which work as a current collector. The future aspects should be attained as follows. (1) The graphene-based morphologies are designed rationally to work as electrode materials. The difference in morphologies depicts specific mechanical, physical, and chemical attributes, creating an effect in energy storage. The graphene-based electrode fabrication's primary contribution is the porous interconnected structure, modified for huge active surface area, fast ion movement paths, and avoiding the system's collapse. (2) The graphene nanocomposites contain pseudocapacitive materials along with its EDLC charge storage features, i.e., graphene-metal oxides or hydroxides, graphene-conductive polymers, metal-doped graphene, and MOF-graphene composites. They are propitious to achieve the enhanced and improved higher energy and power density. Moreover, the nanocomposite morphologies and interfacial interaction between pseudocapacitive materials and graphene have to be explored to increase the total interfacial faradic processes. (3) The development of flexible electronics in the real-time application needs a foldable and flexible device of energy storage. The mechanical flexibility of graphene-based composite materials in primary research focused on the genesis of energy storage devices like an SC. (4) In the present time, the flexible composite of graphene and nanomaterials for SCs, energy devices (Li-ion batteries), and smart electronic devices (i.e., photovoltaic cells, nano generators) were analyzed and discussed. Therefore, the graphene-based and composite SCs' integration with these devices is the potential future and real-time challenge with considerable values. For an SC for a precise application that needs high power density or high energy density or both, appropriate electrode materials and a proper electrolyte must be preferred. The blend of graphene-based materials and graphene materials as electrode materials corroborates exceptional performance in SCs. Graphene-based materials can be attained by a different fabrication method in a more controllable manner. The improved and requisite graphene attributes with their composite for intelligent capacitive behavior can be obtained by the suitable configuration of graphene sheets and the interlinked nanoscale channels.

Moreover, these studies consist of different types of modified electrode configurations in an energy storage operation. The effect of doping, composite with metal oxide, CP, MOF, and flexible graphene, and graphene composite electrode fabrication is concise. The technical improvement's main novelty is the enhanced fabrication of graphene, with its attributes precisely distinct and attuned to applying electrode material in the devices of energy storage. This would be attained through defining processing parameters to tailor-made graphene with size, active specific surface area, and their respective electrical features is a new concern with its potential composites.

ACKNOWLEDGMENT

The authors want to thank the DST NNetra project for its extensive support.

REFERENCES

1. Lai, Enping, Xinxia Yue, Wan'E. Ning, Jiwei Huang, Xinlong Ling, and Haitao Lin. (2019). "Three-dimensional graphene-based composite hydrogel materials for flexible supercapacitor electrodes." *Frontiers in Chemistry* 7: 660.
2. Cossutta, Matteo, Viliam Vretenar, Teresa A. Centeno, Peter Kotrusz, Jon McKechnie, and Stephen J. Pickering. (2020). "A comparative life cycle assessment of graphene and activated carbon in a supercapacitor application." *Journal of Cleaner Production* 242: 118468.

3. Pottathara, Yasir Beeran, Hanuma Reddy Tiyyagura, Zakiah Ahmad, and Kishor Kumar Sadasivuni. (2020). "Graphene based aerogels: fundamentals and applications as supercapacitors." *Journal of Energy Storage* 30: 101549.

4. Lokhande, A. C., I. A. Qattan, Chandrakant D. Lokhande, and Shashikant P. Patole. (2020). "Holey graphene: an emerging versatile material." *Journal of Materials Chemistry A* 8(3): 918–977.

5. Elessawy, Noha A., J. El Nady, W. Wazeer, and A. B. Kashyout. (2019). "Development of high-performance supercapacitor based on a novel controllable green synthesis for 3D nitrogen doped graphene." *Scientific Reports* 9(1): 1–10.

6. Li, Qi, Michael Horn, Yinong Wang, Jennifer MacLeod, Nunzio Motta, and Jinzhang Liu. (2019). "A review of supercapacitors based on graphene and redox-active organic materials." *Materials* 12(5): 703.

7. Sett, Avik, Kunal Biswas, Santanab Majumder, Arkaprava Datta, and Tarun Kanti Bhattacharyya. (2021). "Graphene and Its Nanocomposites Based Humidity Sensors: Recent Trends and Challenges." [Online First], IntechOpen. doi:https://www.intechopen.com/online-first/76765https://www.intechopen.com/online-first/76765

8. Sett, Avik, and Tarun Kanti Bhattacharyya. (2020). "Functionalized gold nanoparticles decorated reduced graphene oxide sheets for efficient detection of mercury." *IEEE Sensors Journal* 20(11): 5712–5719.

9. Sett, Avik, Santanab Majumder, and Tarun Kanti Bhattacharyya. (2021). "Flexible room temperature ammonia gas sensor based on low-temperature tuning of functional groups in graphene." *IEEE Transactions on Electron Devices* 68(7): 3181–3188.

10. Wang, Bo, Tingting Ruan, Yong Chen, Fan Jin, Li Peng, Yu Zhou, Dianlong Wang, and Shixue Dou. (2020). "Graphene-based composites for electrochemical energy storage." *Energy Storage Materials* 24: 22–51.

11. Novoselov, K. S., A. K. Geim, S. V. Morozov, D. Jiang, Y. Zhang,, S. V. Dubonos, et al. (2004). "Electric field effect in atomically thin carbon films." *Science* 306(5696): 666–669.

12. Wang, G., B. Wang, J. Park, Y. Wang, B. Sun, and J. Yao. (2009). "Highly efficient and large-scale synthesis of graphene by electrolytic exfoliation." *Carbon* 47(14): 3242–3246.

13. Mondal, Monojit, Dipak Kumar Goswami, and Tarun Kanti Bhattacharyya. (2021). "Microwave synthesized manganese vanadium oxide: high performing electrode material for energy storage." *Materials Today: Proceedings* 50 (2022): 74–80.

14. Mondal, Monojit, Dipak Kumar Goswami, and Tarun Kanti Bhattacharyya. (2021). "Lignocellulose based bio-waste materials derived activated porous carbon as superior electrode materials for high-performance supercapacitor." *Journal of Energy Storage* 34: 102229.

15. Mondal, Monojit, Dipak Kumar Goswami, and Tarun Kanti Bhattacharyya. "Solvent dependent fabrication of Manganese Vanadium Oxide as cathode material for high performing supercapacitor." In *2020 4th International Conference on Electronics, Materials Engineering & Nano-Technology (IEMENTech)*, pp. 1–6. IEEE, 2020.

16. Mondal, M., B. Das, P. Howli, N. S. Das, and K. K. Chattopadhyay. (2018). "Porosity-tuned NiO nanoflakes: Effect of calcination temperature for high performing supercapacitor application." *Journal of Electroanalytical Chemistry* 813: 116–126.

17. Geim, Andre K., and Konstantin S. Novoselov. (2010). "The rise of graphene." *Nanoscience and Technology: A Collection of Reviews from Nature Journals*, 11–19.

18. Mondal, Monojit, Arkaprava Datta, and Tarun Kanti Bhattacharyya. (2021). "IPMC based flexible platform: a boon to the alternative energy solution." [Online First]. IntechOpen. doi:https://www.intechopen.com/online-first/78105

19. Wang, J., K. K. Manga, Q. Bao, and K. P. Loh. (2011). "High-yield synthesis of few-layer graphene flakes through electrochemical expansion of graphite in propylene carbonate electrolyte." *Journal of the American Chemical Society* 133(23): 8888–8891.

20. Huang, H., Y. Xia, X. Tao, J. Du, J. Fang, Y. Gan, and W. Zhang. (2012). "Highly efficient electrolytic exfoliation of graphite into graphene sheets based on Li ions intercalation–expansion–microexplosion mechanism." *Journal of Materials Chemistry* 22(21): 10452–10456.

21. Khan, U., A. O'Neill, M. Lotya, S. De, and J. N. Coleman. (2010). "High-concentration solvent exfoliation of graphene." *Small* 6(7): 864–871.

22. Brodie, Benjamin Collins. (1859). "XIII. On the atomic weight of graphite." *Philosophical transactions of the Royal Society of London* 149: 249–259.

23. Hofmann, Ulrich, and Rudolf Holst. "Über die Säurenatur und die Methylierung von Graphitoxyd." *Berichte der deutschen chemischen Gesellschaft* (A and B Series) 72, no. 4 (1939): 754–771.

24. Muszynski, R., B. Seger, and P. V. Kamat. (2008). "Decorating graphene sheets with gold nanoparticles." *The Journal of Physical Chemistry C* 112(14): 5263–5266.

25. Choi, W., I. Lahiri, R. Seelaboyina, and Y. S. Kang. (2010). "Synthesis of graphene and its applications: a review." *Critical Reviews in Solid State and Materials Sciences* 35(1): 52–71.

26. Avouris, P. (2010). "Graphene: electronic and photonic properties and devices." *Nano Letters* 10(11): 4285–4294.

27. Chang, Haixin, and Hongkai Wu. (2013). "Graphene-based nanocomposites: preparation, functionalization, and energy and environmental applications." *Energy & Environmental Science* 6(12): 3483–3507.

28. Gilje, Scott, Song Han, Minsheng Wang, Kang L. Wang, and Richard B. Kaner. (2007). "A chemical route to graphene for device applications." *Nano Letters* 7(11): 3394–3398.

29. Meyer, Jannik C., Andre K. Geim, Mikhail I. Katsnelson, Konstantin S. Novoselov, Tim J. Booth, and Siegmar Roth. (2007). "The structure of suspended graphene sheets." *Nature* 446(7131): 60–63.

30. Makarova, Tatiana L., Bertil Sundqvist, Peter Scharff, M. E. Gaevski, Eva Olsson, V. A. Davydov, A. V. Rakhmanina, and L. S. Kashevarova. (2001). "Electrical properties of two-dimensional fullerene matrices." *Carbon* 39(14): 2203–2209.

31. Liu, Chang, Feng Li, Lai-Peng Ma, and Hui-Ming Cheng. (2010). "Advanced materials for energy storage." *Advanced Materials* 22(8): E28–E62.

32. Chou, Shu-Lei, Jia-Zhao Wang, Sau-Yen Chew, Hua-Kun Liu, and Shi-Xue Dou. (2008). "Electrodeposition of MnO2 nanowires on carbon nanotube paper as free-standing, flexible electrode for supercapacitors." *Electrochemistry Communications* 10(11): 1724–1727.

33. Akhavan, O., M. Choobtashani, and E. Ghaderi. (2012). "Protein degradation and RNA efflux of viruses photocatalyzed by graphene–tungsten oxide composite under visible light irradiation." *The Journal of Physical Chemistry C* 116(17): 9653–9659.

34. Wu, Songping, Rui Xu, Mingjia Lu, Rongyun Ge, James Iocozza, Cuiping Han, Beibei Jiang, and Zhiqun Lin. "Graphene-containing nanomaterials for lithium-ion batteries." *Advanced Energy Materials* 5, no. 21 (2015): 1500400..

35. Tung, Vincent C., Jaemyung Kim, Laura J. Cote, and Jiaxing Huang. (2011). "Sticky interconnect for solution-processed tandem solar cells." *Journal of the American Chemical Society* 133(24): 9262–9265.

36. Cao, Xiehong, Yumeng Shi, Wenhui Shi, Gang Lu, Xiao Huang, Qingyu Yan, Qichun Zhang, and Hua Zhang. (2011). "Preparation of novel 3D graphene networks for supercapacitor applications." *Small* 7(22): 3163–3168.

37. Gómez-Navarro, Cristina, Marko Burghard, and Klaus Kern. (2008). "Elastic properties of chemically derived single graphene sheets." *Nano Letters* 8(7): 2045–2049.

38. Emtsev, Konstantin V., Aaron Bostwick, Karsten Horn, Johannes Jobst, Gary L. Kellogg, Lothar Ley, Jessica L. McChesney, et al. (2009). "Towards wafer-size graphene layers by atmospheric pressure graphitization of silicon carbide." *Nature Materials* 8(3): 203–207.

39. Ambrosi, Adriano, Chun Kiang Chua, Alessandra Bonanni, and Martin Pumera. (2014). "Electrochemistry of graphene and related materials." *Chemical Reviews* 114(14): 7150–7188.

40. Meyer, Jannik C., C. Kisielowski, R. Erni, Marta D. Rossell, M. F. Crommie, and A. Zettl. (2008). "Direct imaging of lattice atoms and topological defects in graphene membranes." *Nano Letters* 8(11): 3582–3586.

41. Wang, Hua, Hongbin Feng, and Jinghong Li. (2014). "Graphene and graphene-like layered transition metal dichalcogenides in energy conversion and storage." *Small* 10(11): 2165–2181.

42. Reddy, Arava Leela Mohana, Anchal Srivastava, Sanketh R. Gowda, Hemtej Gullapalli, Madan Dubey, and Pulickel M. Ajayan. (2010). "Synthesis of nitrogen-doped graphene films for lithium battery application." *ACS Nano* 4(11): 6337–6342.

43. Li, Xiaolin, Hailiang Wang, Joshua T. Robinson, Hernan Sanchez, Georgi Diankov, and Hongjie Dai. (2009). "Simultaneous nitrogen doping and reduction of graphene oxide." *Journal of the American Chemical Society* 131(43): 15939–15944.

44. Peng, Yue, and Junhua Li. (2013). "Ammonia adsorption on graphene and graphene oxide: a first-principles study." *Frontiers of Environmental Science & Engineering* 7(3): 403–411.

45. He, Chunyong, Zesheng Li, Maolin Cai, Mei Cai, Jian-Qiang Wang, Zhiqun Tian, Xin Zhang, and Pei Kang Shen. (2013). "A strategy for mass production of self-assembled nitrogen-doped graphene as catalytic materials." *Journal of Materials Chemistry A* 1(4): 1401–1406.

46. Gao, Zhiyong, Lan Wang, Jiuli Chang, Xiao Liu, Dapeng Wu, Fang Xu, Yuming Guo, and Kai Jiang. (2016). "Nitrogen doped porous graphene as counter electrode for efficient dye sensitized solar cell." *Electrochimica Acta* 188: 441–449.

47. Shi, Penghui, Ruijing Su, Fengzhi Wan, Mincong Zhu, Dengxin Li, and Shihong Xu. (2012). "Co_3O_4 nanocrystals on graphene oxide as a synergistic catalyst for degradation of Orange II in water by advanced oxidation technology based on sulfate radicals." *Applied Catalysis B: Environmental* 123: 265–272.

48. Sheng, Zhen-Huan, Hong-Li Gao, Wen-Jing Bao, Feng-Bin Wang, and Xing-Hua Xia. (2012). "Synthesis of boron doped graphene for oxygen reduction reaction in fuel cells." *Journal of Materials Chemistry* 22(2): 390–395.

49. Denis, Pablo A. (2010). "Band gap opening of monolayer and bilayer graphene doped with aluminium, silicon, phosphorus, and sulfur." *Chemical Physics Letters* 492(4–6): 251–257.

50. Sathe, Bhaskar R., Xiaoxin Zou, and Tewodros Asefa. (2014). "Metal-free B-doped graphene with efficient electrocatalytic activity for hydrogen evolution reaction." *Catalysis Science & Technology* 4(7): 2023–2030.

51. Wang, Zegao, Pingjian Li, Yuanfu Chen, Jiarui He, Wanli Zhang, Oliver G. Schmidt, and Yanrong Li. (2014). "Pure thiophene–sulfur doped reduced graphene oxide: synthesis, structure, and electrical properties." *Nanoscale* 6(13): 7281–7287.

52. Zhang, Chenzhen, Nasir Mahmood, Han Yin, Fei Liu, and Yanglong Hou. (2013). "Synthesis of phosphorus-doped graphene and its multifunctional applications for oxygen reduction reaction and lithium ion batteries." *Advanced Materials* 25(35): 4932–4937.

53. Denis, Pablo A., and C. Pereyra Huelmo. "Structural characterization and chemical reactivity of dual doped graphene." Carbon 87 (2015): 106–115.

54. Poh, Hwee Ling, Petr Simek, Zdenek Sofer, and Martin Pumera. (2013). "Sulfur-doped graphene via thermal exfoliation of graphite oxide in H2S, SO2, or CS2 gas." *ACS nano* 7(6): 5262–5272.

55. Kong, Xiang-Kai, Chang-Le Chen, and Qian-Wang Chen. (2014). "Doped graphene for metal-free catalysis." *Chemical Society Reviews* 43(8): 2841–2857.

56. Bhandary, Sumanta, and Biplab Sanyal. (2012). "Graphene-boron nitride composite: a material with advanced functionalities." [Online First], Intech Open. doi:https://www.intechopen.com/chapters/38414

57. Wen, Yangyang, Bei Wang, Congcong Huang, Lianzhou Wang, and Denisa Hulicova-Jurcakova. (2015). "Synthesis of phosphorus-doped graphene and its wide potential window in aqueous supercapacitors." *Chemistry–A European Journal* 21(1): 80–85.

58. Borowiec, Joanna, and Jingdong Zhang. (2015). "Hydrothermal synthesis of boron-doped graphene for electrochemical sensing of guanine." *Journal of The Electrochemical Society* 162(12): B332.

59. Song, Jiangxuan, Terrence Xu, Mikhail L. Gordin, Pengyu Zhu, Dongping Lv, Ying-Bing Jiang, Yongsheng Chen, Yuhua Duan, and Donghai Wang. (2014). "Nitrogen-doped mesoporous carbon promoted chemical adsorption of sulfur and fabrication of high-areal-capacity sulfur cathode with exceptional cycling stability for lithium-sulfur batteries." *Advanced Functional Materials* 24(9): 1243–1250.

60. Yoo, Jung Joon, Kaushik Balakrishnan, Jingsong Huang, Vincent Meunier, Bobby G. Sumpter, Anchal Srivastava, Michelle Conway et al. (2011). "Ultrathin planar graphene supercapacitors." *Nano letters* 11(4): 1423–1427.

61. Sapkota, Prabal, and Honggon Kim. (2009). "Zinc–air fuel cell, a potential candidate for alternative energy." *Journal of Industrial and Engineering Chemistry* 15(4): 445–450.

62. Yan, Yang, Ya-Xia Yin, Yu-Guo Guo, and Li-Jun Wan. (2014). "A sandwich-like hierarchically porous carbon/graphene composite as a high-performance anode material for sodium-ion batteries." *Advanced Energy Materials* 4(8): 1301584.

63. Liu, Ying, Feifei Fan, Jiangwei Wang, Yang Liu, Hailong Chen, Katherine L. Jungjohann, Yunhua Xu, et al. (2014). "In situ transmission electron microscopy study of electrochemical sodiation and potassiation of carbon nanofibers." *Nano Letters* 14(6): 3445–3452.

64. Cao, Yuliang, Lifen Xiao, Maria L. Sushko, Wei Wang, Birgit Schwenzer, Jie Xiao, Zimin Nie, Laxmikant V. Saraf, Zhengguo Yang, and Jun Liu. (2012). "Sodium ion insertion in hollow carbon nanowires for battery applications." *Nano Letters* 12(7): 3783–3787.

65. Zhang, Jiao, Chuanqi Li, Zhikun Peng, Yushan Liu, Jianmin Zhang, Zhongyi Liu, and Dan Li. (2017). "3D free-standing nitrogen-doped reduced graphene oxide aerogel as anode material for sodium ion batteries with enhanced sodium storage." *Scientific Reports* 7(1): 1–7.

66. Meng, Qiufeng, Kefeng Cai, Yuanxun Chen, and Lidong Chen. (2017). "Research progress on conducting polymer based supercapacitor electrode materials." *Nano Energy* 36: 268–285.

67. MacDiarmid, Alan Graham, R. J. Mammone, R. B. Kaner, and Lord Porter. (1985). "The concept of 'doping' of conducting polymers: the role of reduction potentials." *Philosophical Transactions of the Royal Society of London. Series A, Mathematical and Physical Sciences* 314(1528): 3–15.

68. Zhao, Junhong, Jinping Wu, Bing Li, Weimin Du, Qingli Huang, Mingbo Zheng, Huaiguo Xue, and Huan Pang. (2016). "Facile synthesis of polypyrrole nanowires for high-performance supercapacitor electrode materials." *Progress in Natural Science: Materials International* 26(3): 237–242.

69. Kim, Byung Chul, Jin-Yong Hong, Gordon G. Wallace, and Ho Seok Park. (2015). "Recent progress in flexible electrochemical capacitors: electrode materials, device configuration, and functions." *Advanced Energy Materials* 5(22): 1500959.

70. Zhang, Jintao, and X. S. Zhao. (2012). "On the configuration of supercapacitors for maximizing electrochemical performance." *ChemSusChem* 5(5): 818–841.

71. Wang, Meng, Xidong Duan, Yuxi Xu, and Xiangfeng Duan. (2016). "Functional three-dimensional graphene/polymer composites." *ACS Nano* 10(8): 7231–7247.

72. Yan, Jun, Tong Wei, Bo Shao, Zhuangjun Fan, Weizhong Qian, Milin Zhang, and Fei Wei. (2010). "Preparation of a graphene nanosheet/polyaniline composite with high specific capacitance." *Carbon* 48(2): 487–493.

73. Fugetsu, Bunshi, Eiichi Sano, Hongwen Yu, Kenichiro Mori, and Tomo Tanaka. (2010). "Graphene oxide as dyestuffs for the creation of electrically conductive fabrics." *Carbon* 48(12): 3340–3345.

74. Liman, Md Luthfar Rahman, M. Tauhidul Islam, and Md Milon Hossain. (2021). "Mapping the progress in flexible electrodes for wearable electronic textiles: materials, durability, and applications." *Advanced Electronic Materials*, 2100578.

75. Zhu, Zhentao, Shuihua Tang, Jiawei Yuan, Xiaolong Qin, Yuxiao Deng, Renjie Qu, and Geir Martin Haarberg. (2016). "Effects of various binders on supercapacitor performances." *International Journal of Electrochemical Science* 11(10): 8270–8279.

76. Tsay, Keh-Chyun, Lei Zhang, and Jiujun Zhang. (2012). 'Effects of electrode layer composition/thickness and electrolyte concentration on both specific capacitance and energy density of supercapacitor." *Electrochimica Acta* 60: 428–436.

77. Zhu, Xianjun, Yanwu Zhu, Shanthi Murali, Meryl D. Stoller, and Rodney S. Ruoff. (2011). "Nanostructured reduced graphene oxide/Fe2O3 composite as a high-performance anode material for lithium ion batteries." *ACS nano* 5(4): 3333–3338.

78. Peng, Chuang, Jun Jin, and George Z. Chen. (2007). "A comparative study on electrochemical co-deposition and capacitance of composite films of conducting polymers and carbon nanotubes." *Electrochimica Acta* 53(2): 525–537.

79. Wang, Yan, Xinming Wu, Wenzhi Zhang, and Shuo Huang. (2015). "Facile synthesis of Ni/PANI/RGO composites and their excellent electromagnetic wave absorption properties." *Synthetic Metals* 210: 165–170.

80. Alvi, Farah, Manoj K. Ram, Punya A. Basnayaka, Elias Stefanakos, Yogi Goswami, and Ashok Kumar. (2011). "Graphene–polyethylenedioxythiophene conducting polymer nanocomposite based supercapacitor." *Electrochimica Acta* 56(25): 9406–9412.

81. Xu, Kongli, Guangming Chen, and Dong Qiu. (2013). "Convenient construction of poly (3, 4-ethylenedioxythiophene)–graphene pie-like structure with enhanced thermoelectric performance." *Journal of Materials Chemistry A* 1(40): 12395–12399.

82. Zhao, Zhiheng, Georgia F. Richardson, Qingshi Meng, Shenmin Zhu, Hsu-Chiang Kuan, and Jun Ma. (2015). "PEDOT-based composites as electrode materials for supercapacitors." *Nanotechnology* 27(4): 042001.

83. Reddy, B. Narsimha, Melepurath Deepa, Amish G. Joshi, and Avanish Kumar Srivastava. (2011). "Poly (3, 4-ethylenedioxypyrrole) enwrapped by reduced graphene oxide: how conduction behavior at nanolevel leads to increased electrochemical activity." *The Journal of Physical Chemistry C* 115(37): 18354–18365.

84. Ferrari, Victoria C., Ivaldete S. Dupim, Vinicius Sousa, and Flavio L. Souza. (2018). "Photoactive multilayer WO3 electrode synthesized via dip-coating." *Ceramics International* 44(18): 22983–22990.

85. Zhao, Tao, Yuan Ren, Guangyou Jia, Yuye Zhao, Yuchi Fan, Jianping Yang, Xin Zhang, Wan Jiang, Lianjun Wang, and Wei Luo. (2019). "Facile synthesis of mesoporous WO₃@ graphene aerogel nanocomposites for low-temperature acetone sensing." *Chinese Chemical Letters* 30(12): 2032–2038.

86. Berger, C., E. H. Conrad, and W. A. de Heer. "The electronic band structure of graphene." In *Physics of Solid Surfaces*, pp. 674–682. Springer, Berlin, Heidelberg, 2018.

87. Yan, Fangfang, Jie Ding, Yushan Liu, Zeng Wang, Qiang Cai, and Jianmin Zhang. (2015). "Fabrication of magnetic irregular hexagonal-Fe3O4 sheets/reduced graphene oxide composite for supercapacitors." *Synthetic Metals* 209: 473–479.

88. Wang, Lu, Yuzhen Han, Xiao Feng, Junwen Zhou, Pengfei Qi, and Bo Wang. (2016). "Metal–organic frameworks for energy storage: batteries and supercapacitors." *Coordination Chemistry Reviews* 307: 361–381.

89. Wei, Tao, Mi Zhang, Ping Wu, Yu-Jia Tang, Shun-Li Li, Feng-Cui Shen, Xiao-Li Wang, Xin-Ping Zhou, and Ya-Qian Lan. (2017). "POM-based metal-organic framework/reduced graphene oxide nanocomposites with hybrid behavior of battery-supercapacitor for superior lithium storage." *Nano Energy* 34: 205–214.

90. Morozan, Adina, and Frédéric Jaouen. (2012). "Metal organic frameworks for electrochemical applications." Energy & Environmental Science 5(11): 9269–9290.

91. Yang, Zhiwang, Xueqing Xu, Xixi Liang, Cheng Lei, Yuli Wei, Peiqi He, Bolin Lv, Hengchang Ma, and Ziqiang Lei. (2016). "MIL-53 (Fe)-graphene nanocomposites: efficient visible-light photocatalysts for the selective oxidation of alcohols." *Applied Catalysis B: Environmental* 198: 112–123.

92. Yang, Qihao, Qiang Xu, and Hai-Long Jiang. (2017). "Metal–organic frameworks meet metal nanoparticles: synergistic effect for enhanced catalysis." *Chemical Society Reviews* 46(15): 4774–4808.

93. Young, Christine, Jie Wang, Jeonghun Kim, Yoshiyuki Sugahara, Joel Henzie, and Yusuke Yamauchi. (2018). "Controlled chemical vapor deposition for synthesis of nanowire arrays of metal–organic frameworks and their thermal conversion to carbon/metal oxide hybrid materials." *Chemistry of Materials* 30(10): 3379–3386.

94. Zhang, Chuanhui, Weiqiang Hu, Heng Jiang, Jeng-Kuei Chang, Mingsen Zheng, Qi-Hui Wu, and Quanfeng Dong. (2017). "Electrochemical performance of MIL-53 (Fe)@ rGO as an organic anode material for Li-ion batteries." *Electrochimica Acta* 246: 528–535.

95. Bareschino, Piero, Giuseppe Diglio, Francesco Pepe, Giovanni Angrisani, Carlo Roselli, and Maurizio Sasso. (2017). "Numerical study of a MIL101 metal organic framework based desiccant cooling system for air conditioning applications." *Applied Thermal Engineering* 124: 641–651.

96. Jin, Guo-Xia, Xue Niu, Jia Wang, Jian-Ping Ma, Tong-Liang Hu, and Yu-Bin Dong. (2018). "APPT-Cd MOF: acetylene adsorption mechanism and its highly efficient acetylene/ethylene separation at room temperature." *Chemistry of Materials* 30(21): 7433–7437.

97. Saraf, Mohit, Richa Rajak, and Shaikh M. Mobin. (2016). "A fascinating multitasking Cu-MOF/rGO hybrid for high performance supercapacitors and highly sensitive and selective electrochemical nitrite sensors." *Journal of Materials Chemistry A* 4(42): 16432–16445.

98. Lin, Kun-Yi Andrew, Fu-Kong Hsu, and Wei-Der Lee. (2015). "Magnetic cobalt–graphene nanocomposite derived from self-assembly of MOFs with graphene oxide as an activator for peroxymonosulfate." *Journal of Materials Chemistry A* 3(18): 9480–9490.

99. Björk, Jonas, Felix Hanke, Carlos-Andres Palma, Paolo Samori, Marco Cecchini, and Mats Persson. (2010). "Adsorption of aromatic and anti-aromatic systems on graphene through π- π stacking." *The Journal of Physical Chemistry Letters* 1(23): 3407–3412.

15 Design and Analysis of Dopingless Charge-Plasma-Based Ring Architecture of Tunnel Field-Effect Transistor for Low-Power Application

Ashok Kumar Gupta, Ashish Raman, Naveen Kumar, Deep Shekhar, and Prateek Kumar

CONTENTS

15.1 INTRODUCTION

Semiconductor industries are changing continuously and a great deal of research work goes on every year [1–4]. In 1974, the first transistor was introduced by the Brattain, Shockley, and Bardeen [5, 6]. Transistors are three-terminal devices that use the third terminal as a controller that controls the resistance between the first and the second terminal. In the 1970s bipolar junction transistors (BJTs) are more popular and used for the high-frequency applications [7, 8], but they are too bulky so metal oxide semiconductor field-effect transistors (MOSFETs) were introduced and are

DOI: 10.1201/9781003126393-15

still popular in the semiconductor industry. MOSFETs provide higher drain current and are smaller in size [9, 10]. Further scaling the device below the submicron region (below the 10-nm scale) elicits large leakage current, short channel effects (SCEs), and random dopant fluctuations (RDFs) [11, 12]. As technologies are continuously scaled down to create a high-speed MOSFET device by reducing the channel length, the power consumption has become a major problem in electronics devices. When the number of transistors increases in unit area, the leakage or thermal power affects the device's battery life, which is undesirable in electronic equipment [13, 14]. MOSFETs show leakage current in the OFF state, so power dissipation is higher in MOSFETs. The static power dissipation is defined as the product of supply voltage (V_{DD}) and leakage current (I_{OFF}) [15].

$$P_{Static} = V_{DD} \times I_{OFF} \tag{15.1}$$

MOSFETs also work as a switch, which means it changes they transition the state from OFF to ON and vice versa. MOSFETs also show dynamic power consumption, which depends on the number of times switching (F) can take place and load capacitance (C_L). The dynamic power consumption has a quadratic relationship with a supply voltage (V), as shown in Equation (15.2) [16].

$$P_{Dynamic} = V^2{}_{DD} \times C_L \times F \tag{15.2}$$

MOSFETs show higher static power consumption even in the OFF state. The dynamic power consumption with scaled-down technology is shown in Figure 15.1 [17]. This higher power consumption provides a heating problem so the performance of the operating device degrades. Another problem with MOSFETs is its subthreshold slope (SS); MOSFETs have an SS of 60 mV/dec (ideal value), so further degradation of the SS is not possible. This means the drain current (I_{DS}) of the MOSFETs only increases 10-fold when the input voltage is increased by 63 mV [18, 19]. To reduce these problems, different structures and different physics have been introduced. The decrease in performance of the MOSFET device occurs, due to the occurrence of SCEs and second-order effects, as the technology is scaled down. This results in a high OFF-state current in the device, which highly affects the OFF-state condition. The SS of the MOSFET device is approximately

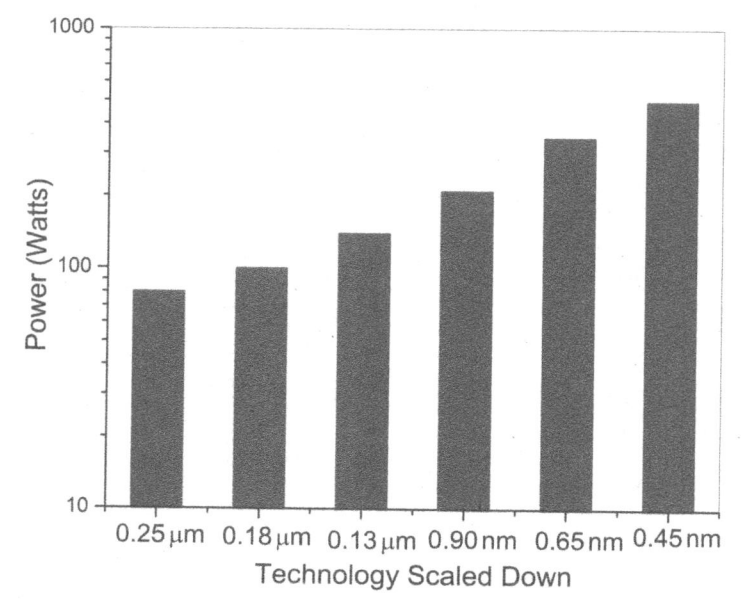

FIGURE 15.1 Dynamic power consumption increased when technology scaled down (From [17].)

60 mV/dec at room temperature, which is highly dependent on the thermal voltage. The ON-current to OFF-current ratio is quite a bit less in this device due to the low ON current. The drain current depends on the drift and diffusion mechanism of this device due to which reduction in SS is impossible [20]. Therefore, we are moving toward different devices in which a different current conduction mechanism is used for the transport of majority and minority carriers from source to drain. These devices are FinFET [21], carbon nanotube field-effect transistor (CNTFET) [22], high electron mobility transfer (HEMT) [23], nanowires [24], quantum dot [25], tunnel field-effect transistor (TFET) [26], and dopingless TFET (DLTFET) [27]. The TFETs is a more promising device with these effects. TFETs provide lower leakage current, negligible SCEs and RDFs, and low-power consumption. TFETs have a much better SS ($\ll$60 mV/dec) compared with MOSFETs. Due to the lower SS, switching activity is improved. The drawback of TFETs is the lower ON-state current (I_{ON}). To reduce this drawback, different architectures have been described, such as dual-gate TFET (DG-TFET), FinFET structure (Fin-TFET), nanotube TFET, and gate-all-around structure (GAA-TFET) [28–32].

15.1.1 TFET Overview

TFET is a more promising device compared with leakage current, subthreshold swing, lower power consumption, higher I_{ON}/I_{OFF} current ratio, and negligible SCEs. TFETs are similar to MOSFETs, but the operating physics of both devices is different. The basic principle of MOSFETs is thermal emission, but the operating principle of TFET is band-to-band tunneling (BTBT) [26].

15.1.1.1 Device Structure

The basic structure of a TFET is much similar to the basic structure of MOSFETs. The basic structures of both MOSFETs and TFETs are shown in Figure 15.2.

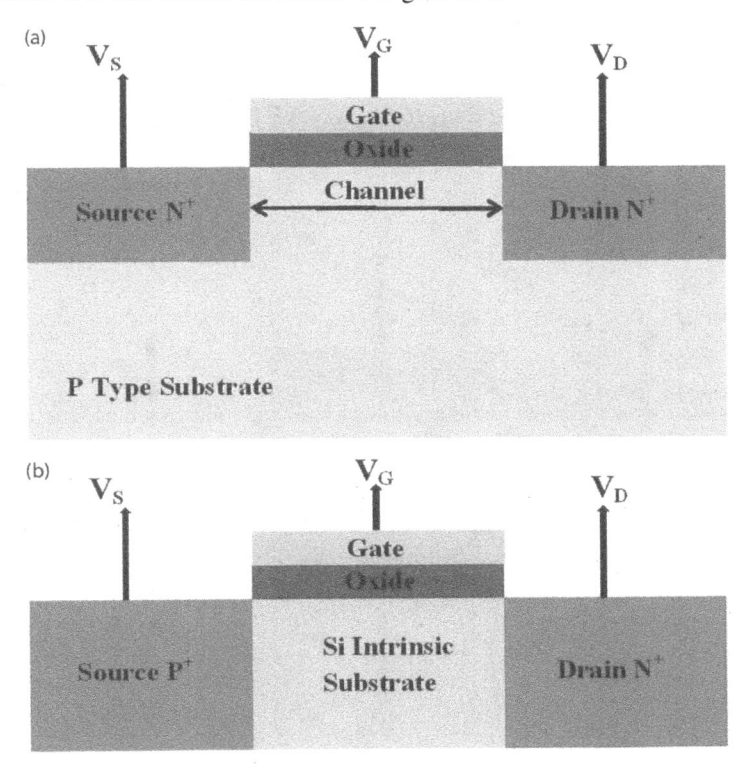

FIGURE 15.2 (a) Basic structure of N-type MOSFET. (b) Basic structure of N-type conventional TFET.

In the TFET structure, both the source and drain regions are different. The source is P-type and the drain region is N-type. TFET is an asymmetrical device. In the P-type TFET device, the source is N-type and the drain is a doped P-type. The channel region is an intrinsic or lower doped p-type or n-type. The operating mechanism of the TFET is a BTBT mechanism. The fabrication of the TFET device is similar to that of MOSFET [26].

15.1.1.2 Different Structure of TFET

The disadvantage of the conventional TFET is that it has lower ON current (I_{ON}) and because of this the I_{ON}/I_{OFF} current ratio is reduced. The structure of the conventional TFET is already described in Figure 15.2(b).

To improve the ON-state current (I_{ON}) of the conventional TFET, a great deal of modification is done on both the structure and the material. There are different methods to boost the ON current (I_{ON}) of the device. Some of the techniques are described here. The following changes in the architecture of the conventional TFET enhance the ON current (I_{ON}): decreasing the gate oxide thickness, increasing the dielectric constant (high-k) of the gate oxide, decreasing the body thickness of the device, and reducing the smaller bandgap material so that BTBT will increase and the ON current of (I_{ON}) the device will increase. There are different novel architectures that boost the ON current (I_{ON}), such as DG-TFET [33], dual-gate material TFET (DM-DG-TFET) [34], heterojunction TFET (HJ-TFET) [35], III–V compound-based HJ-TFET (III–V HJ-TFET) [36], graphene ribbon-based TFET (GNR-TFET) [37], and GAA-TFET [38]. Figure 15.3(a) examines the architecture of the DG-TFET device, which improves the control of drain current and boosts the ON current. Figure 15.3(b) describes the DM-DG-TFET, which also improves the ON current. Figure 15.3(c) describes the HJ-TFET, where high-k material is used to improve the device performance; ferroelectric material is also used to improve device performance. Figure 15.3(d) describes the group III–V material-based TFET device. Using III–V bandgap material improves the tunneling rate, SS, and ON current of the device. Figure 15.3(e) describes the GNR-based TFET structure, wherein the channel region carbon alloy graphene is used. The GNR-based TFET device improves the ON current in the range of ~10^{-3} A (mA), which also is used for low-power application. Figure 15.3(f) examines the GAA-TFET, which improves the gate controllability and ON current of the device.

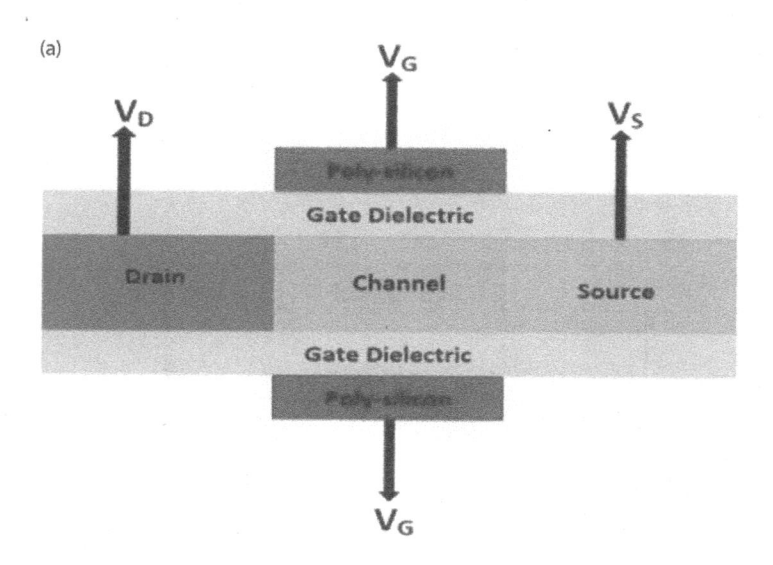

FIGURE 15.3 Different architecture is used to boost the ON current: (a) dual-gate TFET (DG-TFET).

(Continued)

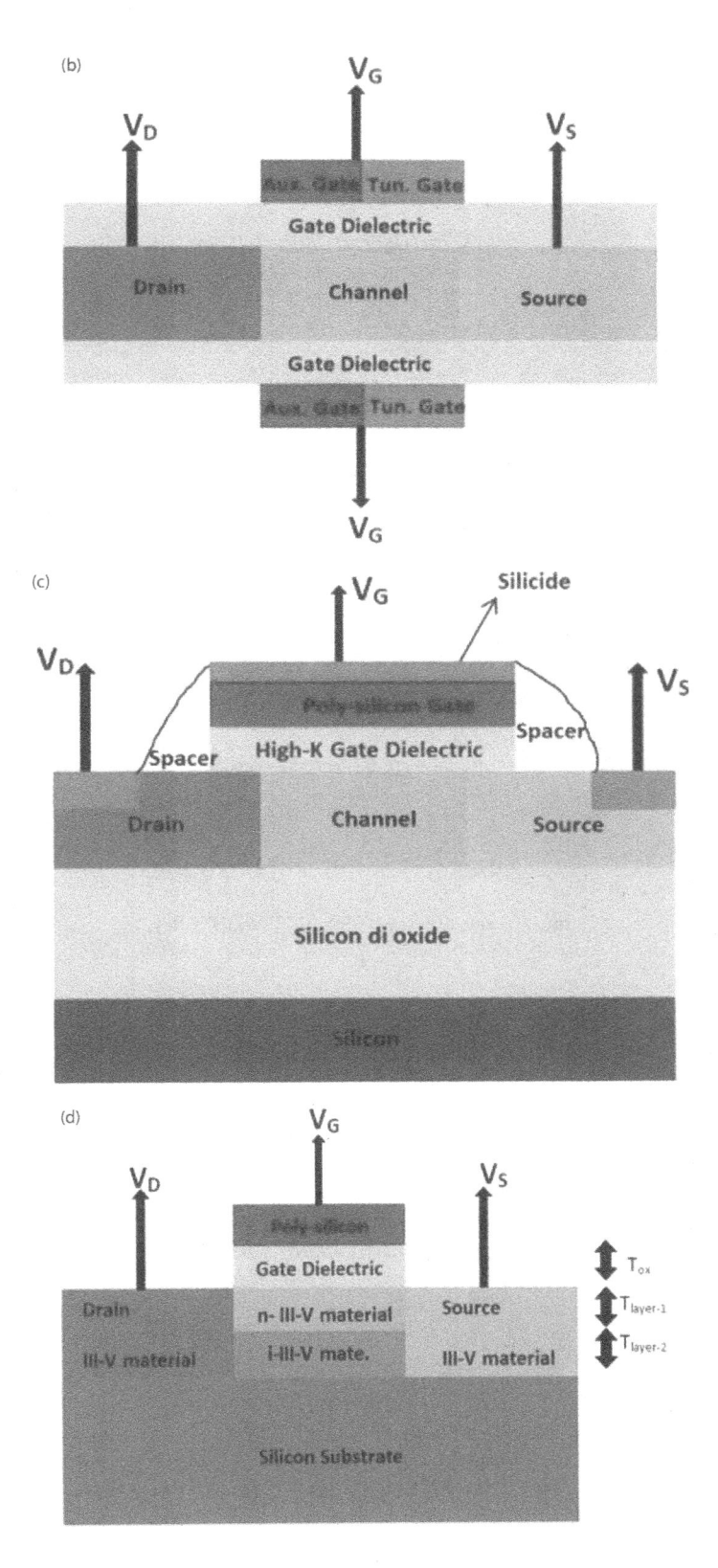

FIGURE 15.3 (*Continued*) Different architecture is used to boost the ON current: (b) dual-material DG-TFET (DM-DG-TFET), (c) heterojunction TFET (HJ-TFET), (d) III–V HJ-TFET.

(Continued)

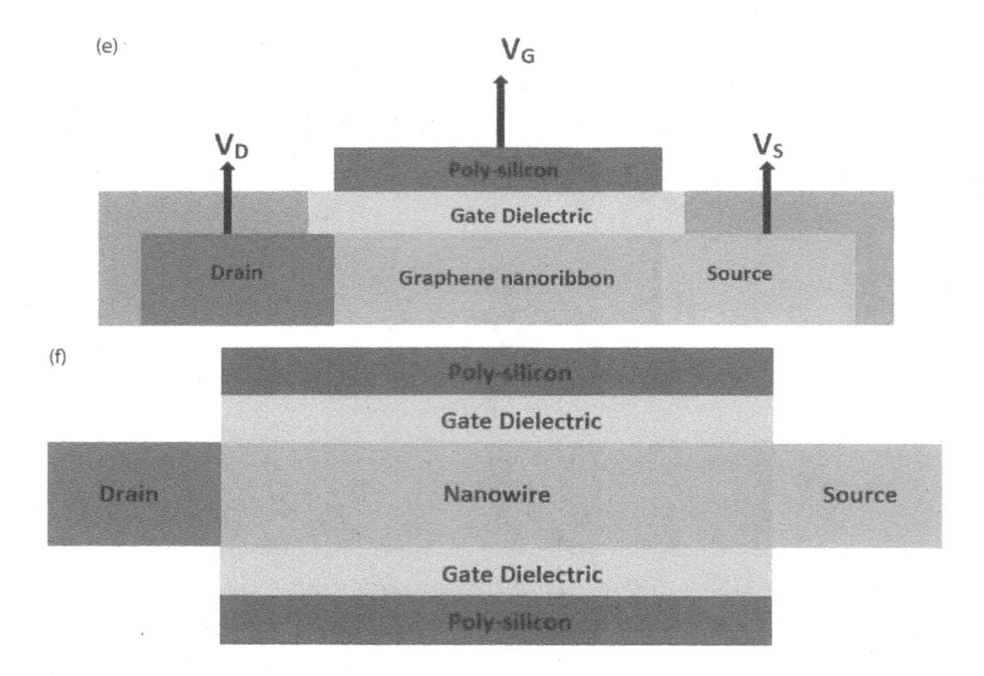

FIGURE 15.3 (*Continued*)   Different architecture is used to boost the ON current: (e) graphene nanoribbon (GNR)-based TFET, and (f) gate-all-around (GAA)-based TFET.

15.1.1.3   Advantage of TFET

TFET has the following advantages over the conventional MOSFET [26]:

- Steep SS
- Lower leakage or OFF current
- Higher I_{ON}/I_{OFF} current ratio
- Minimum SCEs

15.2   RING ARCHITECTURE

Discussion of the methodology, structure specification, model specification, and numerical calculation will be covered. The ATLAS simulator Silvaco technology computer-aided device (TCAD) tool is used to obtain the results of the ring architecture device. The two- and three-dimensional structures are also examined.

15.2.1   2D DEVICE STRUCTURE

The two-dimensional (2D) structure of proposed ring architecture is examined in Figure 15.4(a), which describes only a one-fourth view of the architecture. The source region is containing outer side and drain region is containing the inner region of the proposed ring architecture. If the drain electrode is kept at the center of the ring architecture and moves the source region 360 degrees, a complete three-dimensional (3D) ring architecture of the proposed device obtained. Figure 15.4(b) describes the full view structure of the proposed 2D ring architecture. The source region is 10 nm, the drain region is 10 nm, and the length of the channel region is 20 nm.

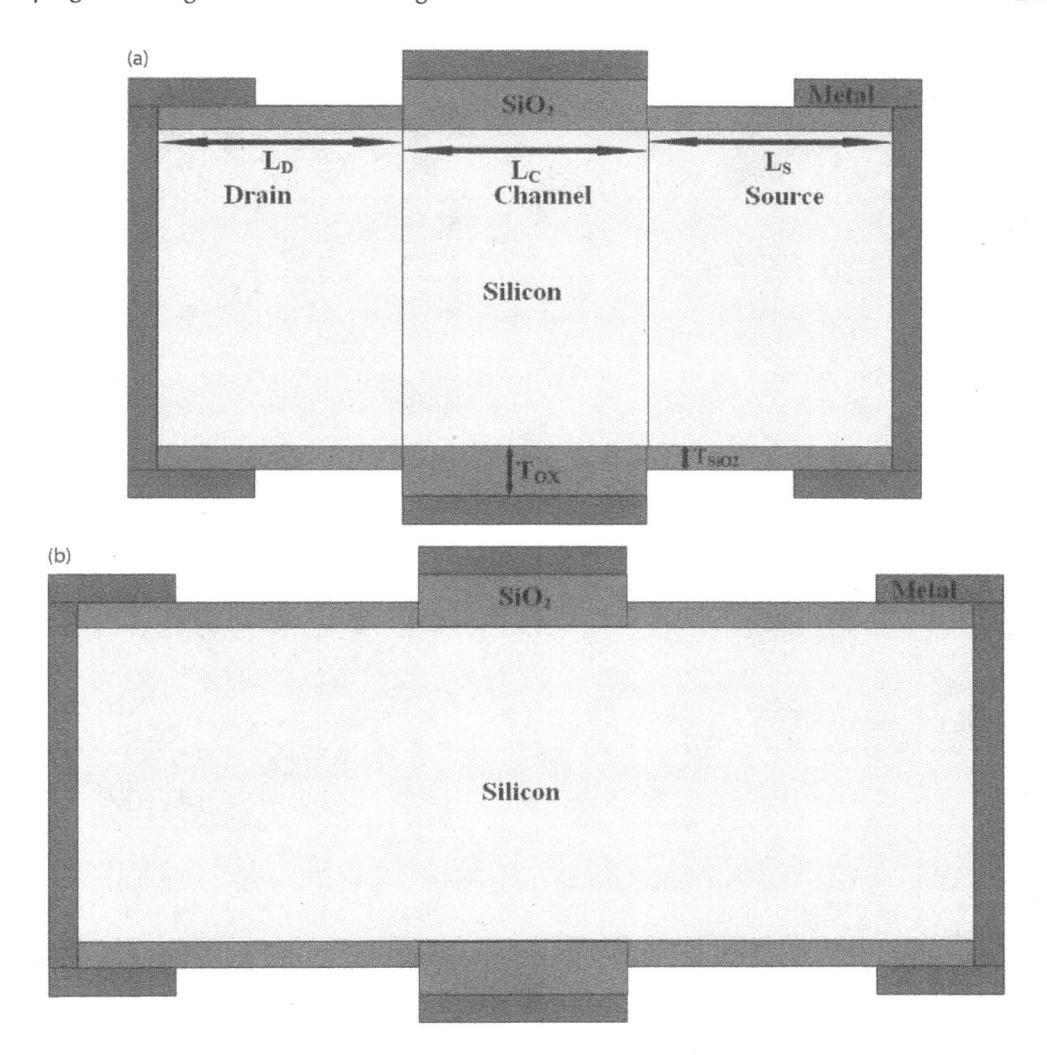

FIGURE 15.4 (a) Two-dimensional structure of ring architecture (one-fourth view). (b) Two-dimensional structure of ring architecture (full).

The doping of the device is intrinsic. The width of the gate oxide is 2 nm. The source/drain spacer length is 5 nm. The total width of the silicon is 10 nm. The proposed device is operating at the charge-plasma (CP) technique.

15.2.2 3D DEVICE STRUCTURE

The meshing in the proposed device is called cylindrical meshing. Rotation of the 2D structure (one-fourth view) 360 degrees describes the actual 3D structure, where drain metal works as centered. The various 3D structures of ring architecture are shown in Figure 15.5 with various cross-sectional views. Figure 15.5(a) describes the 90-degree angle view with the drain side at the inner portion and source side at the outer section. Figure 15.5(b and c) shows the 180-degree angle views and 270-degree angle views, respectively. Figure 15.5(d) illustrates the full 3D structure of the proposed ring-TFET architecture The Device specifications are shown in Table 15.1.

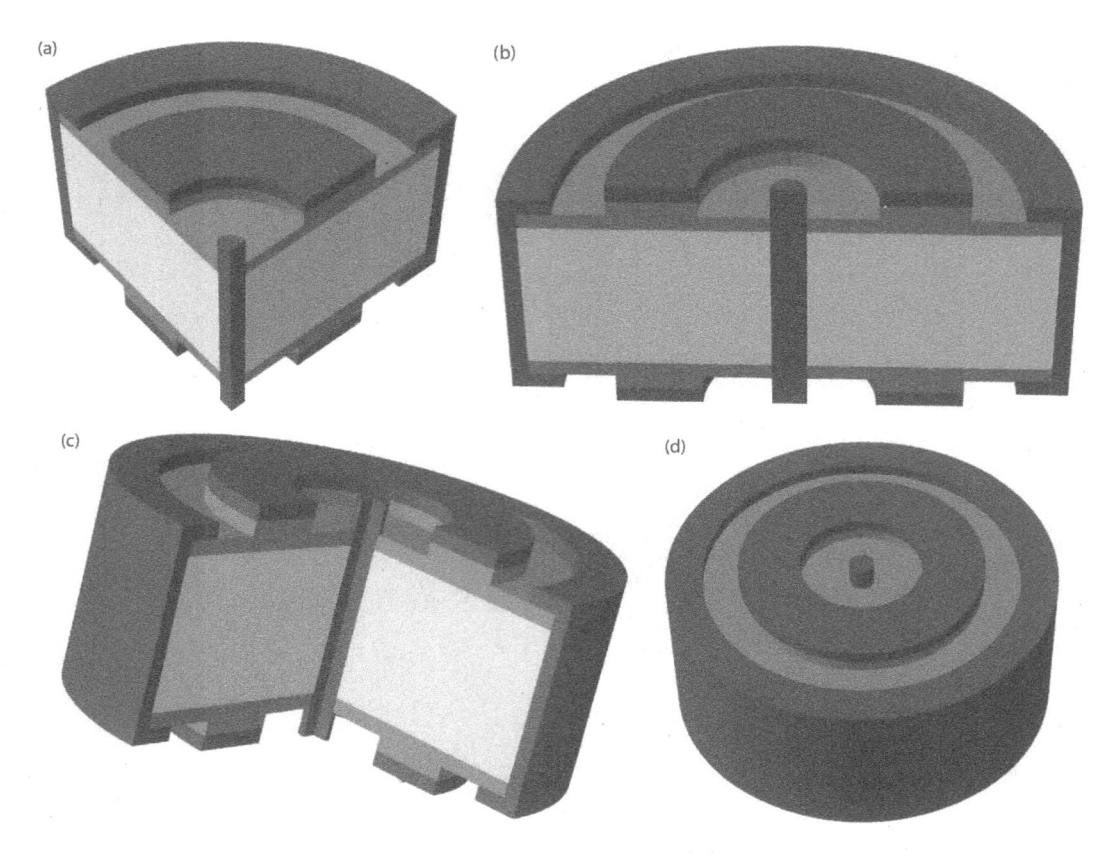

FIGURE 15.5 Three-dimensional architecture of the proposed ring-TFET: (a) Cross-sectional view with 90-degree angle. (b) Cross-sectional view with 180-degree angle (c) Cross-sectional view with 270-degree angle. (d) Cross-sectional view with 360-degree angle.

15.2.3 RING TFET DEVICE SPECIFICATION WITH SYMBOL AND UNITS

TABLE 15.1

Device Specification of Proposed Ring Architecture TFET with Proper Symbols and Units

Parameter	Symbol	Dimension
Drain length	L_D	10 nm
Channel length	L_G	20 nm
Source length	L_S	10 nm
Gate oxide	T_{OX}	2 nm
Device (silicon) thickness	T_{Si}	10 nm
Source/drain spacer length	L_{SS}/L_{DD}	5 nm
Source/drain oxide thickness	$T_{CP\text{-}OX}$	1 nm
Doping concentration	N_D	Intrinsic
Gate work function	ϕ_{MG}	4.2 eV
Source work function	ϕ_{MS}	5.93 eV
Drain work function	ϕ_{MD}	4.2 eV

15.3 DEVICE SIMULATION WITH ATLAS

In this section, discussion of the methodology, structure specifications, model specifications, and numerical calculations are covered. To obtain the results of the ring architecture device, the ATLAS simulator Silvaco TCAD tool was used [39].

15.3.1 METHODOLOGY

For designing electronics devices such as transistors, there are now many 2D and 3D simulation tools available, which provide all facilities required for virtual simulation and fabrication of a device, and they become very popular. One such tool is Silvaco TCAD [39], which is generally based on the finite-element methods (FEMs), where device fabrication and its parameters (electrical characteristics) are assessed by a physical mechanism rather than electrical processing in the device. TCAD provides a virtual fabrication and characterization facility without using an actual fabrication process for the actual behavior of the device. Thus, both static and dynamic solutions are provided by these TCAD simulators for the electrical characterization. The ATLAS simulator is a tool designed for the physical realization of two dimensions as well as a 3D virtual prototype of semiconductor devices. ATLAS is outlined in a manner such that it can be utilized with the Virtual Wafer Fabrication (VWF) Interactive Tools. The VWF Interactive Tools include Maskviews, Editor, DeckBuild, TonyPlot, TonyPlot 3D, and Deceit, which is shown in Figure 15.6.

Now the designing of any architecture using ATLAS Silvaco TCAD tools is analyzed. The designing portion includes three parts: (1) INPUT statements, (2) the MODELS, NUMERICAL Methods, and ANALYSIS Portion, and (3) OUTPUT files. INPUT statements include meshing, which is used to create the node points for the calculation. There are two types of meshing, including the ATLAS TCAD tool. These include normal mesh and quantum tunneling mesh (QT mesh), which is defined in TFETs where tunneling occurs. Region formation is used to define the region portion that works as a source, drain, channel, or insulator. After defining the region, material selection and doping is done, and at last the electrode is produced. The designing of a device and its characteristics is illustrated in Figure 15.7.

15.3.2 DEVICE/STRUCTURE SPECIFICATION

Device specification or structure specification describes the meshing, region formation, material selection, doping, electrode selection, and QT region formation. In meshing, different nodes are created and regions are formed. In region formation, the source, drain, oxide, and channel regions are described. In material selection, different materials such as silicon and others are defined. In electrode selection, different types of electrode are described, like source, drain, and channel electrodes. The QT region describes the tunneling region of the device. In the tunneling region most of the BTBT occurs. The operating phenomenon of the TFET is tunneling, so QT meshing is essential.

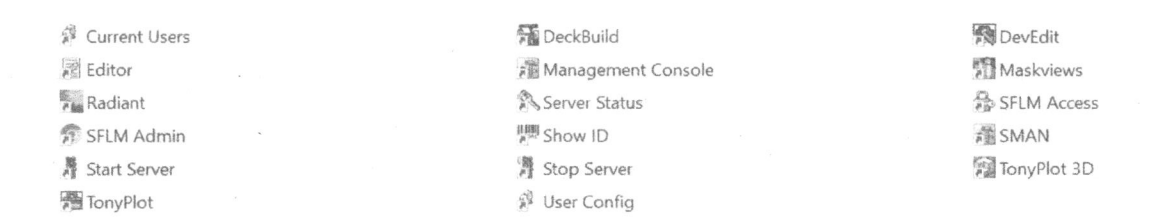

FIGURE 15.6 Shortcuts for tools available in the Silvaco ATLAS TCAD (From [39].)

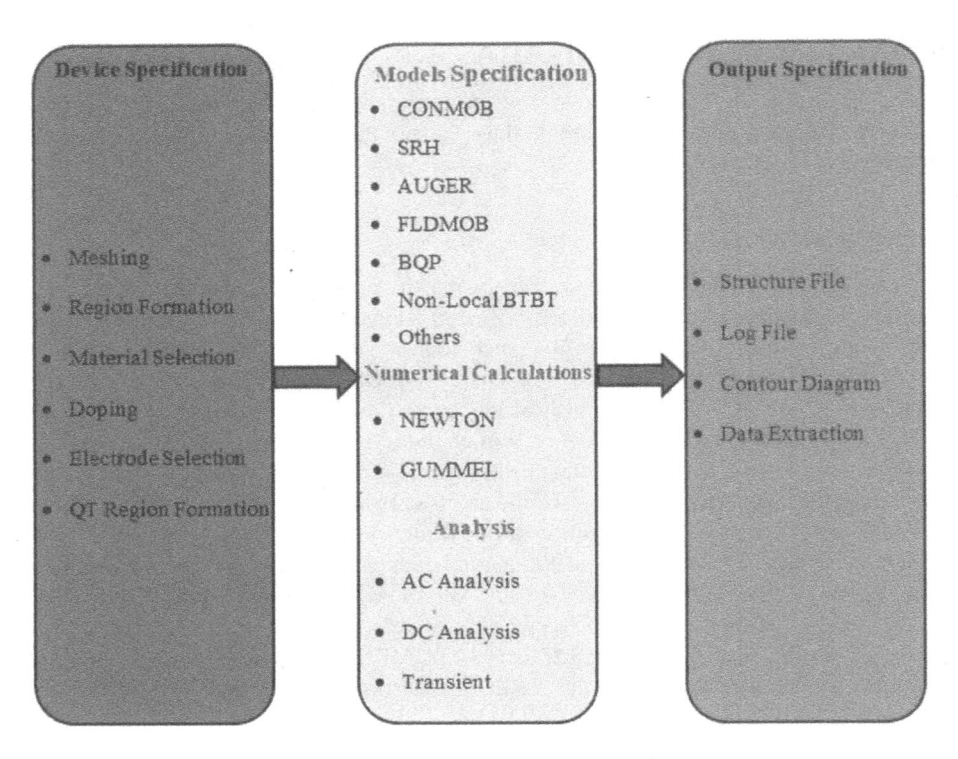

FIGURE 15.7   Block diagram of the design of device and its characterization.

15.3.3 Models Specification

There are different models used to obtain the device characterization: CONMOB, SRH, AUGER, FLDMOB, BQP, and non-local BTBT.

- *Concentration-dependent low-field mobility model (CONMOB):* To activate this model, CONMOB is utilized as a part of the MODELS statement. This model provides information about low-field mobility of electrons and holes at 300K for silicon and gallium arsenide [39].
- *Shockley-Read-Hall recombination model (SRH):* The SRH parameter is utilized as a part of the MODELS articulation for initiating this model. There are a couple of client quantifiable parameters that are utilized as a part of the MATERIAL proclamation, such as TAUN0 and TAUP0, which are electron and hole lifetime parameters. This model indicates the recombination of electron and holes through the SRH recombination strategy happening inside the device [39].
- *Fermi-Dirac model (FERMI):* This model is similar to Fermi-Dirac measurements. It is used in those regions that are degenerately doped yet have decreased convergences of carriers [39]. To initiate this model FERMI is utilized as a part of the MODELS articulation.
- *Bohm quantum potential model (BQP):* This model is utilized for the hydrodynamic models and energy balance, where the semi-classical potential is changed by the quantum potential. The Bohr quantum potential model has two benefits over the density gradient model: it has better convergence and it has better calibration with the Schrödinger-Poisson

model. BQP.NGAMMA and BQP.NALPHA allow setting γ and α parameters for electrons, respectively. Similarly, BQP.PGAMMA and BQP.PALPHA allow setting γ and α values for holes, respectively [39].

- *Non-local BTBT model:* This model is used for the tunneling mechanism of the device. In TFETs current is due to tunneling phenomena, so this model is used to obtain the current. In the tunneling region a special type of meshing is required, and this is known as QT meshing in both lateral and longitudinal directions [39].

15.3.3.1 Numerical Calculation

ATLAS uses the METHOD statement to define the numerical method. To find the solution there are two distinct types of systems [39]: Gummel and Newton.

The Gummel and Newton iteration methods are used for 3D simulations, whereas Gummel, Newton, and Block are all used for 2D simulations. The main technique finds solutions for one obscure variable while others are constant. This procedure will proceed until a steady solution is provided. Not at all like the Gummel technique, the Newton technique solves and finds every one of the problems in the meantime. The block technique is in the middle of the Newton and Gummel techniques as it understands a couple of questions in the meantime [39].

15.3.4 Output Files

Output files have a structure file and log file; the structure file shows the structure of the device and the log file shows the characterization of the device, like current capability. The yield records of ATLAS include three unique types [39]:

- *Run-time output:* The output seen at the base of the DeckBuild Window is the run-time output. Any errors that occur amid this yield will be shown in the run-time window.
- *Log files:* These files contain the characteristics curve of the device after simulation. These include the variation of drain current (I_{DS}) with regard to variation in the gate voltage (V_{GS}). AC analysis is possible by mentioning the frequency.
- *Extraction of parameters in DeckBuild:* To extract the various parameters of the device it is necessary to mention EXTRACT statements in the DeckBuild command line after the generation of the log file, i.e., threshold voltage, SS, and I_{ON}/I_{OFF} ratio can be extracted by using the EXTRACT statement. Finally, the output file containing the various parameters will be generated [39].

15.4 SIMULATION RESULTS OF RING TFET ARCHITECTURE

In this section, discussions of the simulation result of the proposed ring TFET architecture are covered. The ATLAS simulator Silvaco TCAD tool is used to obtain the results of the ring architecture device. The device parameter describes the energy band diagram, electron/hole concentration, BTBT rate, and electric field. Analog parameters cover I_{ON}/I_{OFF} current, SS, and threshold voltage.

15.4.1 Device Parameters

The proposed ring architecture device is an n-type device. The doping in the source is P-type and doping in the drain side is N-type. The doping profile in the source/drain region is created with the work-function difference of the electrode material. Figure 15.8(a) verifies the electron concentration across the device length in the proposed device, with various operating regions, i.e., equilibrium state, OFF state, and ON state. At the source side electron concentration is

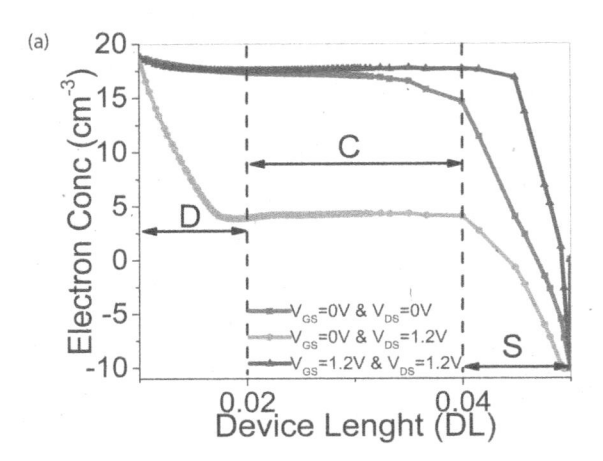

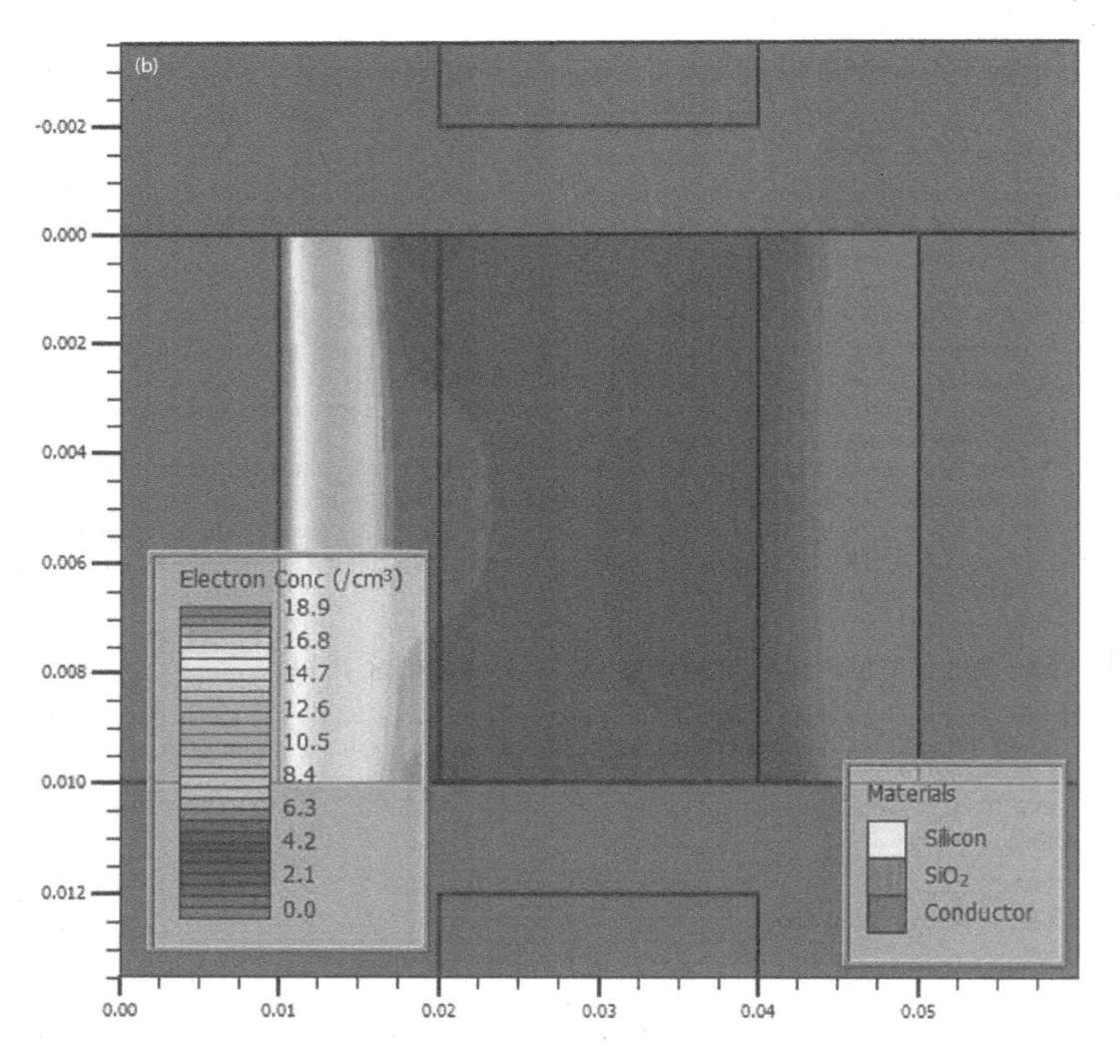

FIGURE 15.8 (a) Electron concentration. (b) Contour graph of electron concentration in the OFF state.

(Continued)

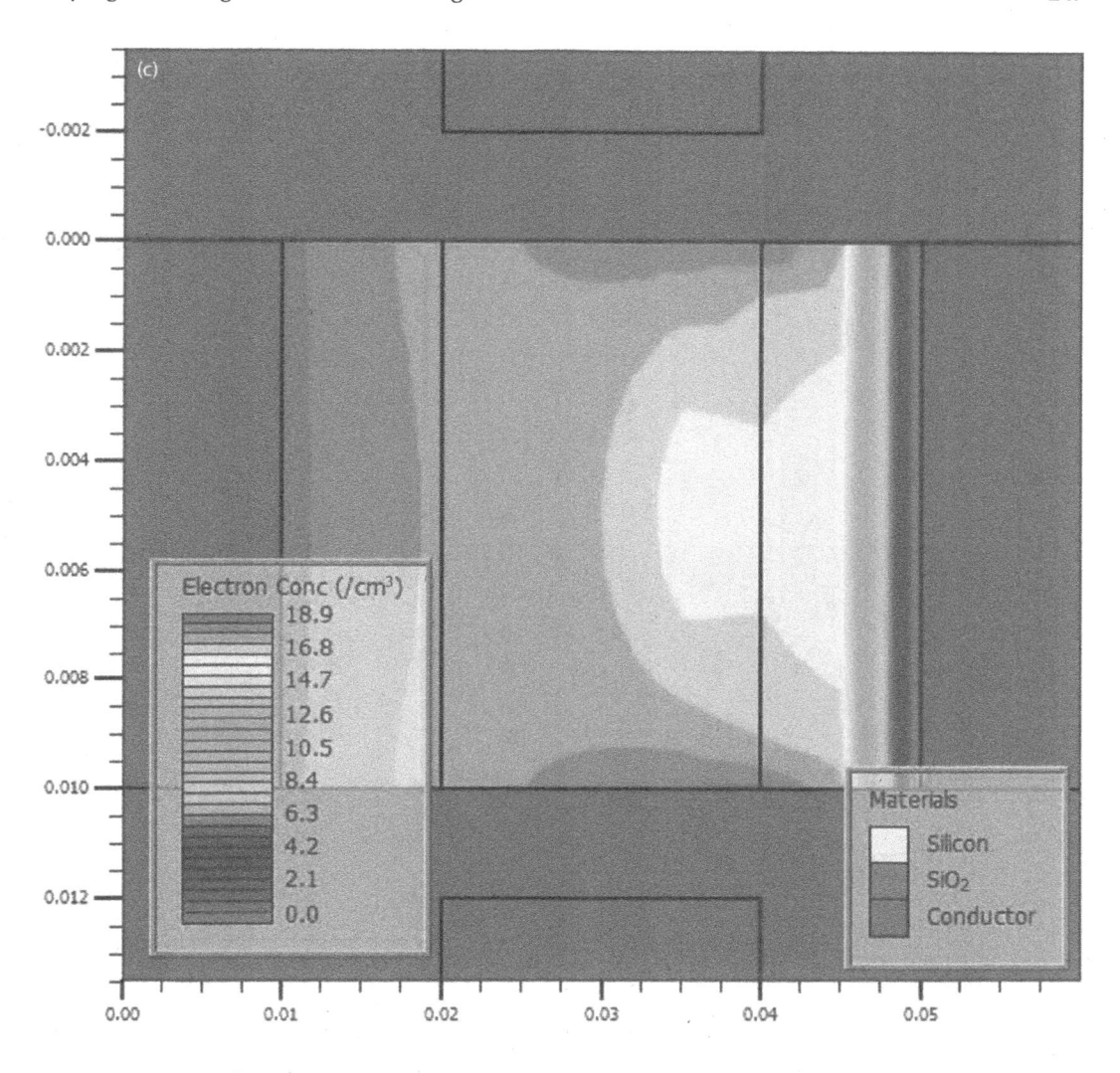

FIGURE 15.8 (*Continued*) (c) Contour graph of electron concentration in the ON state.

maximum and lower at the drain side in the ON state. Figure 15.8(b) illustrates the contour graph of electron concentration in the OFF state. In the OFF state the electron concentration in the source and channel is lower and slightly higher at the drain side. Figure 15.8(c) depicts the contour graph of electron concentration in the OFF state. In the ON state, both source and drain voltages are maximum, and electron concentration across the channel and drain sides are maximum and constants.

Figure 15.9(a) verifies the hole concentration across the device length in the proposed device, with various operating regions, i.e., equilibrium state, OFF state, and ON state. At the source side hole concentration is maximum and lower at the drain side in the ON state. Figure 15.9(b) depicts the contour graph of hole concentration in the OFF state. In the OFF state the hole concentration on the source side is higher and is lower at the channel and drain sides. Figure 15.9(c) illustrates the contour graph of hole concentration in the ON state. In the ON state, both source

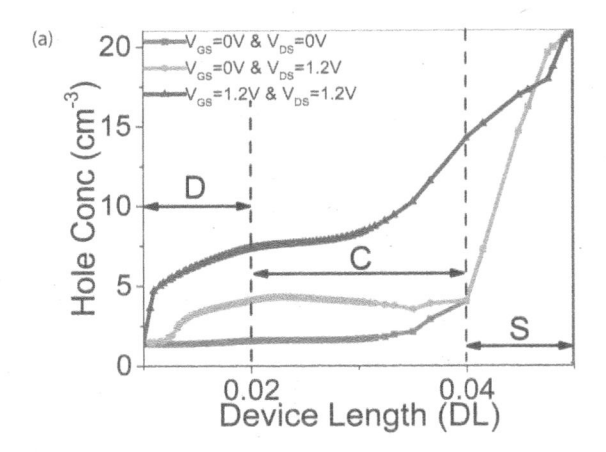

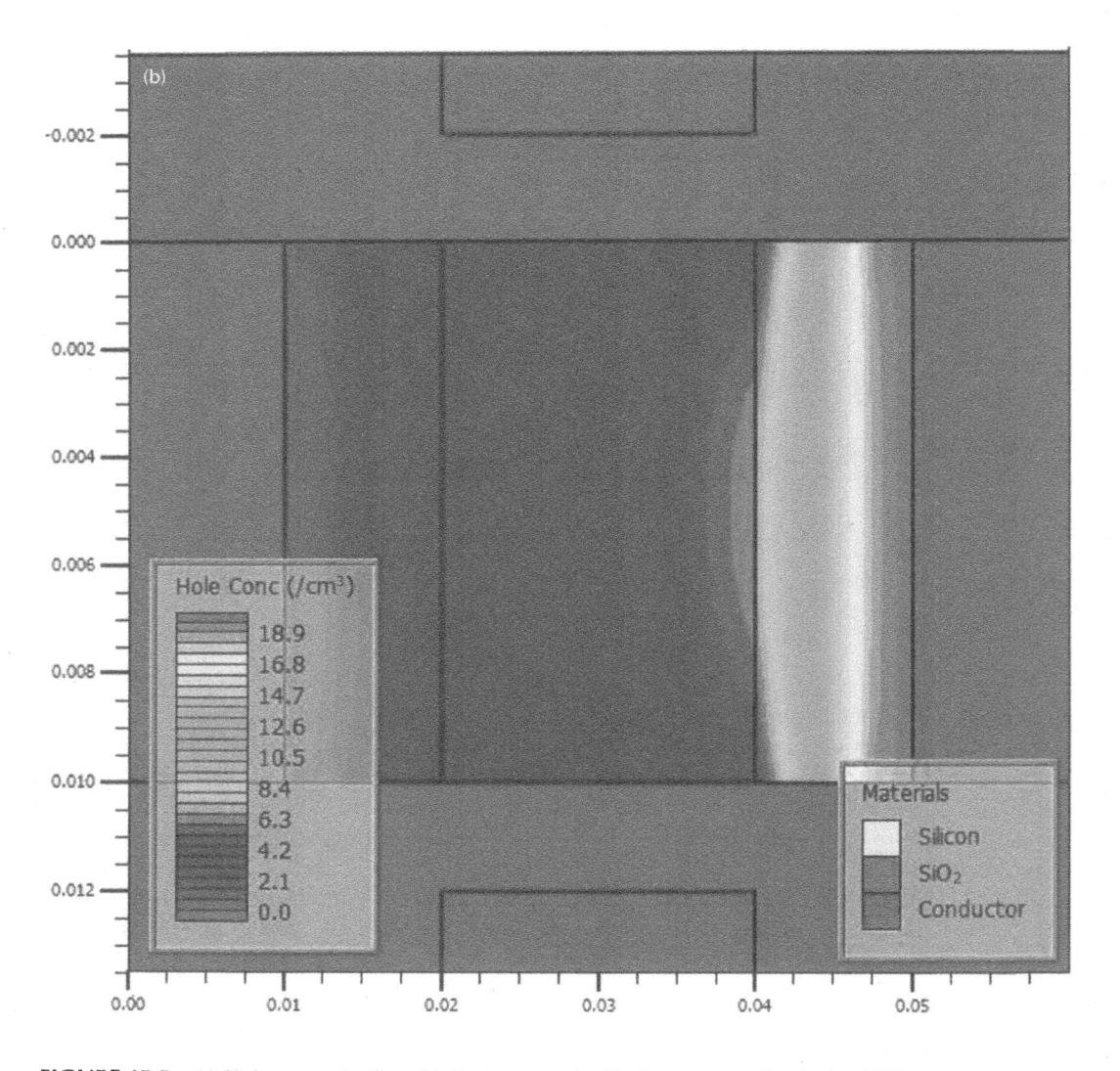

FIGURE 15.9   (a) Hole concentration. (b) Contour graph of hole concentration in the OFF state.

(Continued)

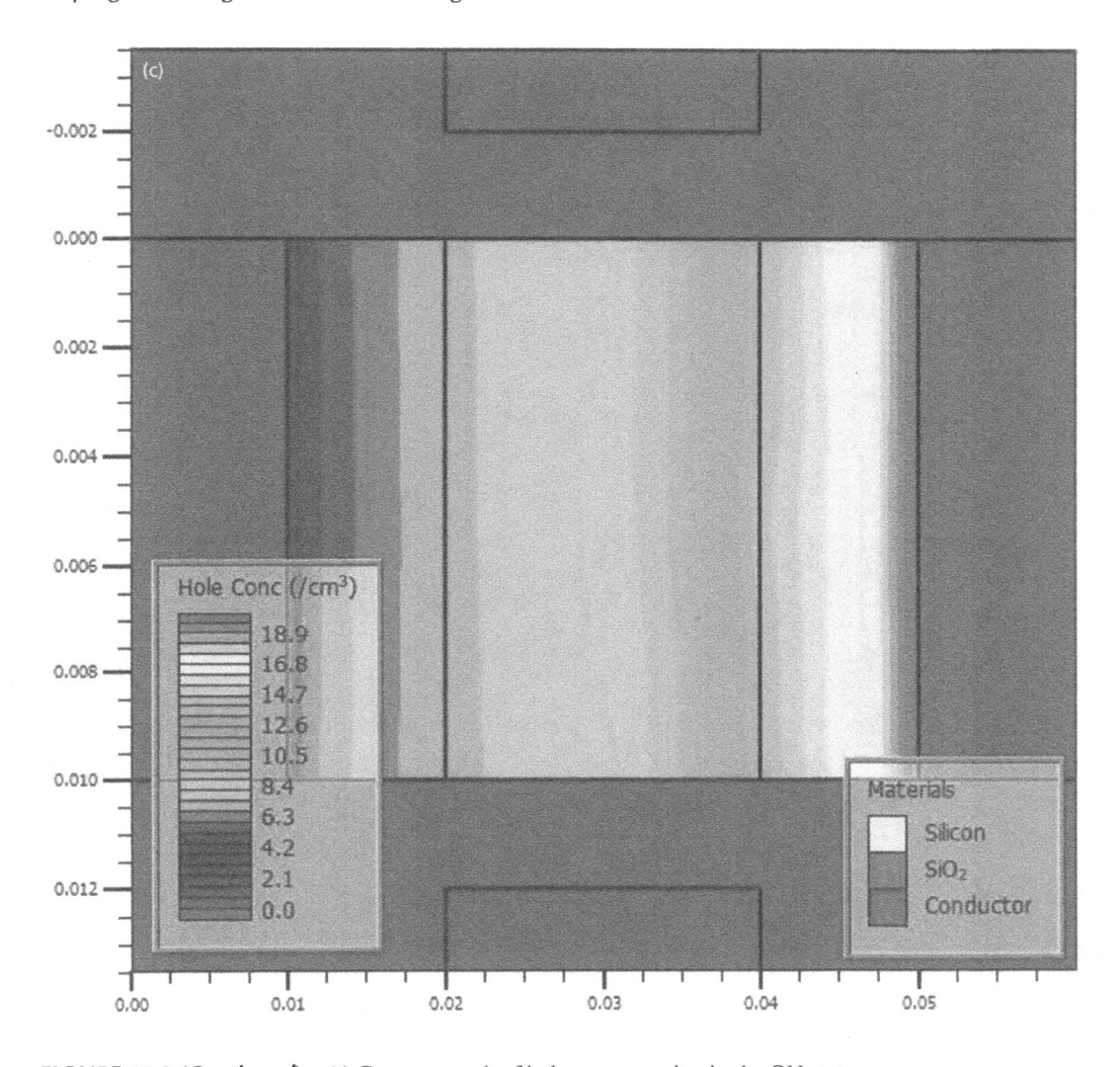

FIGURE 15.9 (*Continued*) (c) Contour graph of hole concentration in the ON state.

and drain voltages are maximum, and hole concentration across the channel and drain sides are lower and constants.

Figure 15.10(a) verifies the potential variation across the device length in the proposed ring architecture device. The supply drain voltage across the device is 1.2 V. Figure 15.10(b and c) illustrates the contour diagram of potential in the OFF and ON states, respectively. In the ON state the potential increases while moving from channel side to drain side. The potential is maximum at the drain side.

The operation of the TFET depends on the BTBT. The BTBT rate operates with the help of an energy band diagram. The energy band diagram is plotted in all operating regions, i.e., equilibrium, OFF, and ON regions. The energy band diagram plotted in Figure 15.11 concerns device length from the drain region to the source region. Figure 15.11(a) depicts the variation of energy in the equilibrium state. In the equilibrium region, both gate voltage (V_{GS}) and drain voltage (V_{DS}) is zero. In the equilibrium region, all the external supply voltage is kept at zero. In the N-type

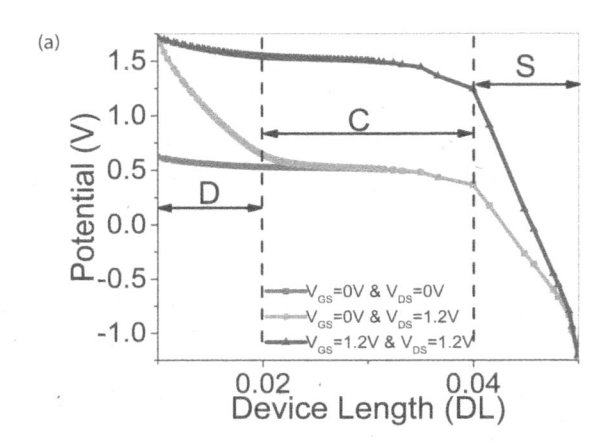

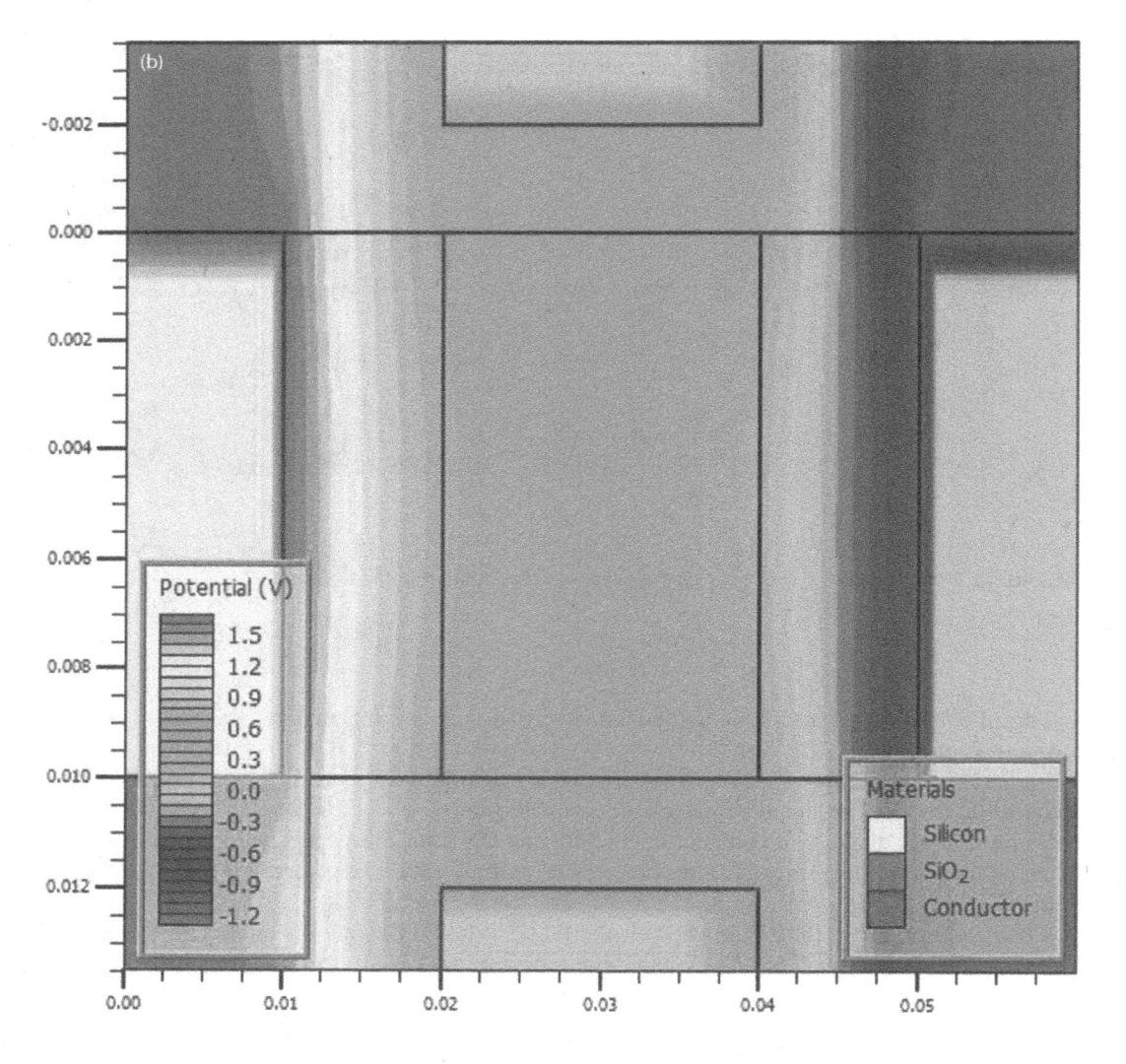

FIGURE 15.10 (a) Potential variation. (b) Contour graph of potential variation in the OFF state.

(Continued)

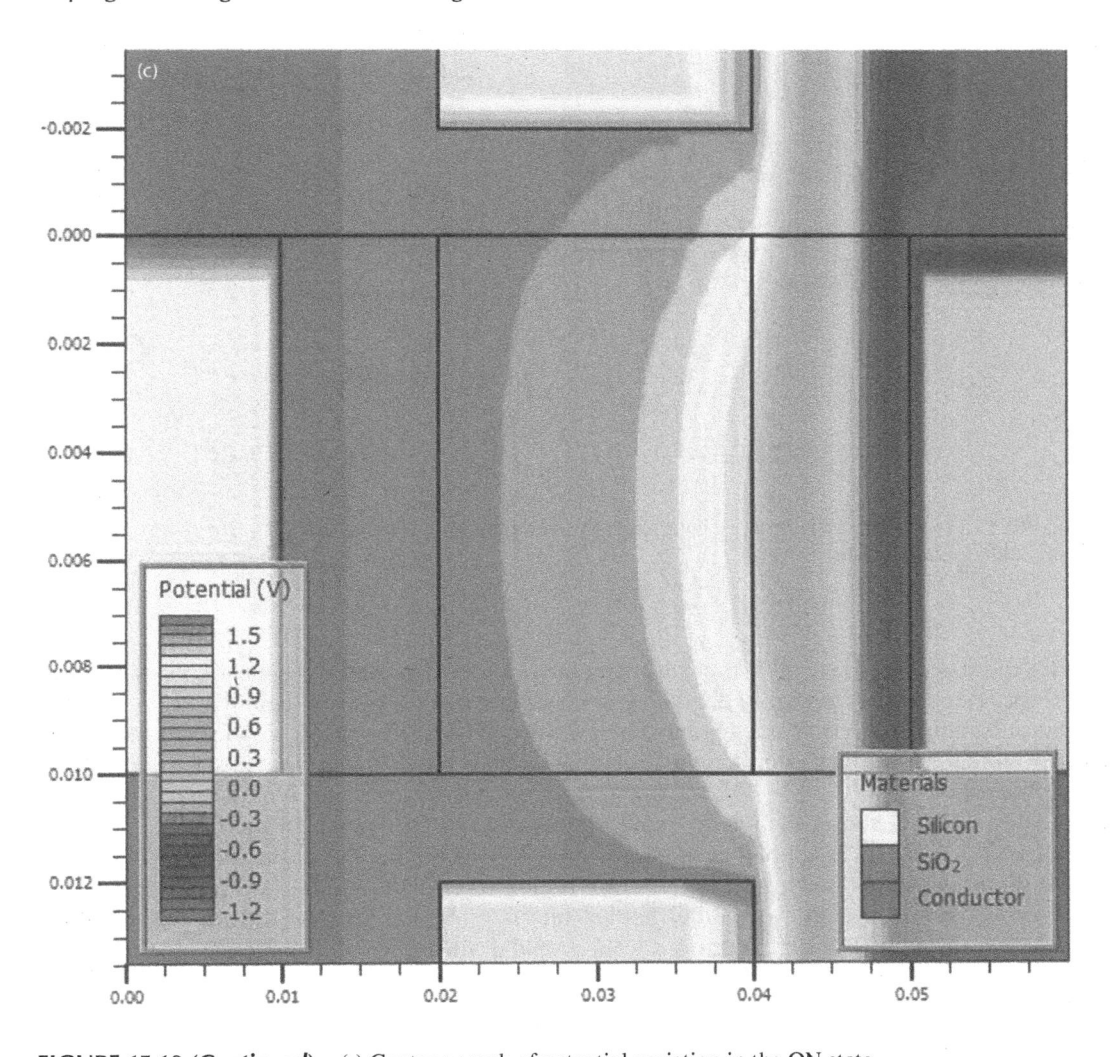

FIGURE 15.10 (*Continued*) (c) Contour graph of potential variation in the ON state.

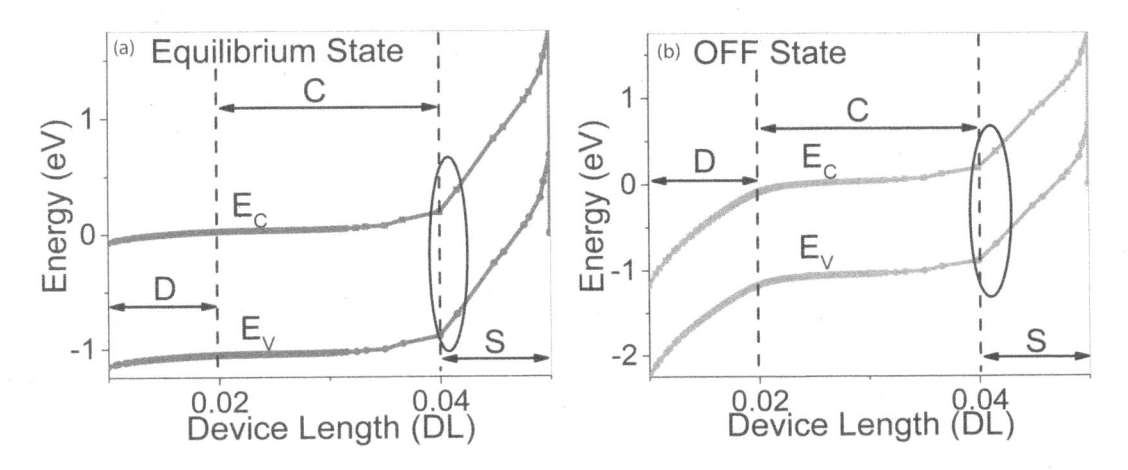

FIGURE 15.11 Energy band diagram variation. (a) Equilibrium state. (b) OFF state.

(*Continued*)

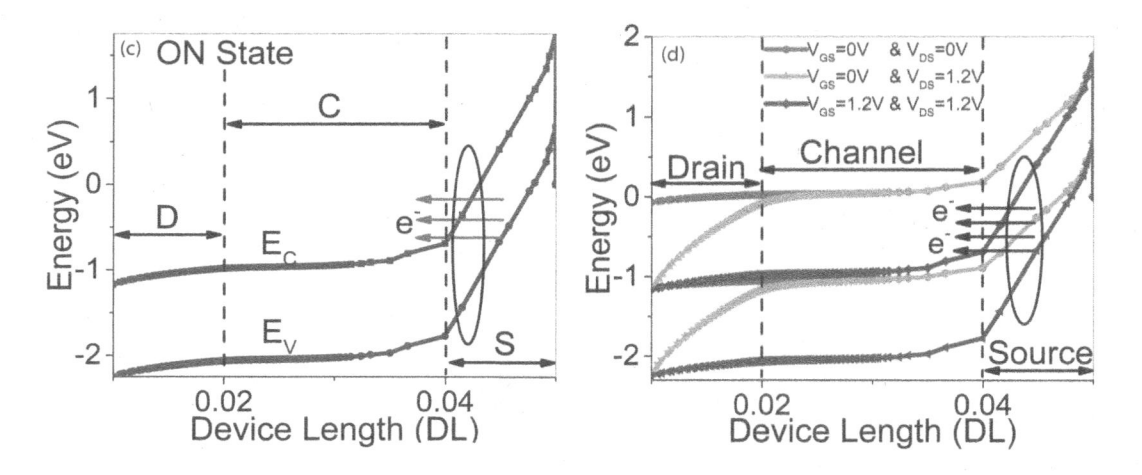

FIGURE 15.11 (*Continued*)   Energy band diagram variation. (c) ON state. (d) Complete energy band diagram in all operating states.

TFET the source region (*S*) is P-type, drain region (*D*) is N-type, and a channel region (*C*) is kept intrinsic. Due to this, the energy band diagram is higher compared with the intrinsic region. The valance band (E_V) of the source region is below the conduction band (E_C) of the channel region, and at the source/channel interface the tunneling width is much higher. In this state tunneling is not possible.

Figure 15.11(b) shows the energy band diagram in the OFF state. In the OFF state, drain voltage (V_{DS} = 1.2 V) is applied at the drain side. After applying the drain voltage, the energy band diagram shifted downward in the drain region (*D*). The tunneling width is large, so the electron cannot tunnel at the source/channel interface. Figure 15.11(c) illustrates the energy band diagram in the ON state; both gate voltage and drain voltage are applied (V_{GS} = 1.2 V and V_{DS} = 1.2 V). Due to lesser tunnel width in ON state, valance band (E_V) electron's tunnel to conduction band (E_C). The tunneling of an electron is shown in Figure 15.11(c) with the tunneling region. Figure 15.11(d) depicts the complete energy band diagram of the proposed device in the various operating states.

Figure 15.12(a) shows the mobility of the proposed ring device. The electron mobility describes the easiness of electron drift in the device. Mobility is a very important parameter for semiconductors and device development. The mobility is defined in two ways i.e. electron mobility and hole mobility. The electron mobility is higher in the equilibrium state at the drain channel interface. As the supply voltage increases, the collision in the device is increasing, thus, electron mobility is reduced. In the ON state the electron mobility is higher in the channel. On the source and drain sides the electron mobility is less. Figure 15.12(b and c) shows the contour diagram of electron mobility in the OFF and ON states.

Figure 15.13(a) illustrates the hole mobility in all possible operating states. The hole mobility is higher in the equilibrium state at the drain channel interface. As the supply voltage increases, the collision in the device increases; thus, the hole mobility reduces. In the ON state the hole mobility is higher in the channel. On the source and drain sides the hole mobility is less. Figure 15.13(b and c) depict the contour diagram of hole mobility in the OFF and ON states. The electron mobility is greater than the hole mobility.

Figure 15.14 illustrates the electron quasi Fermi level (QFL) of the proposed ring device in all possible operating states. The QFL defines the population of the displaced electrons in the conduction

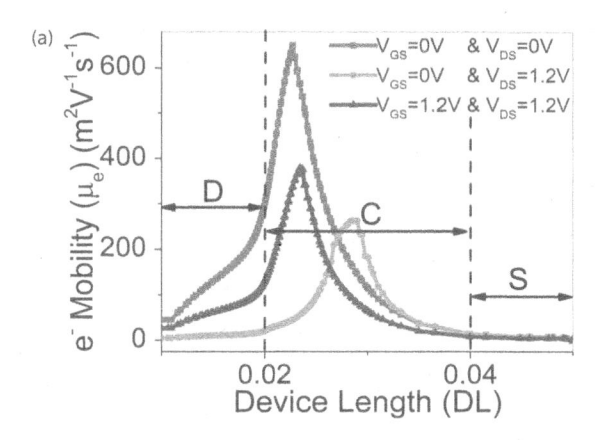

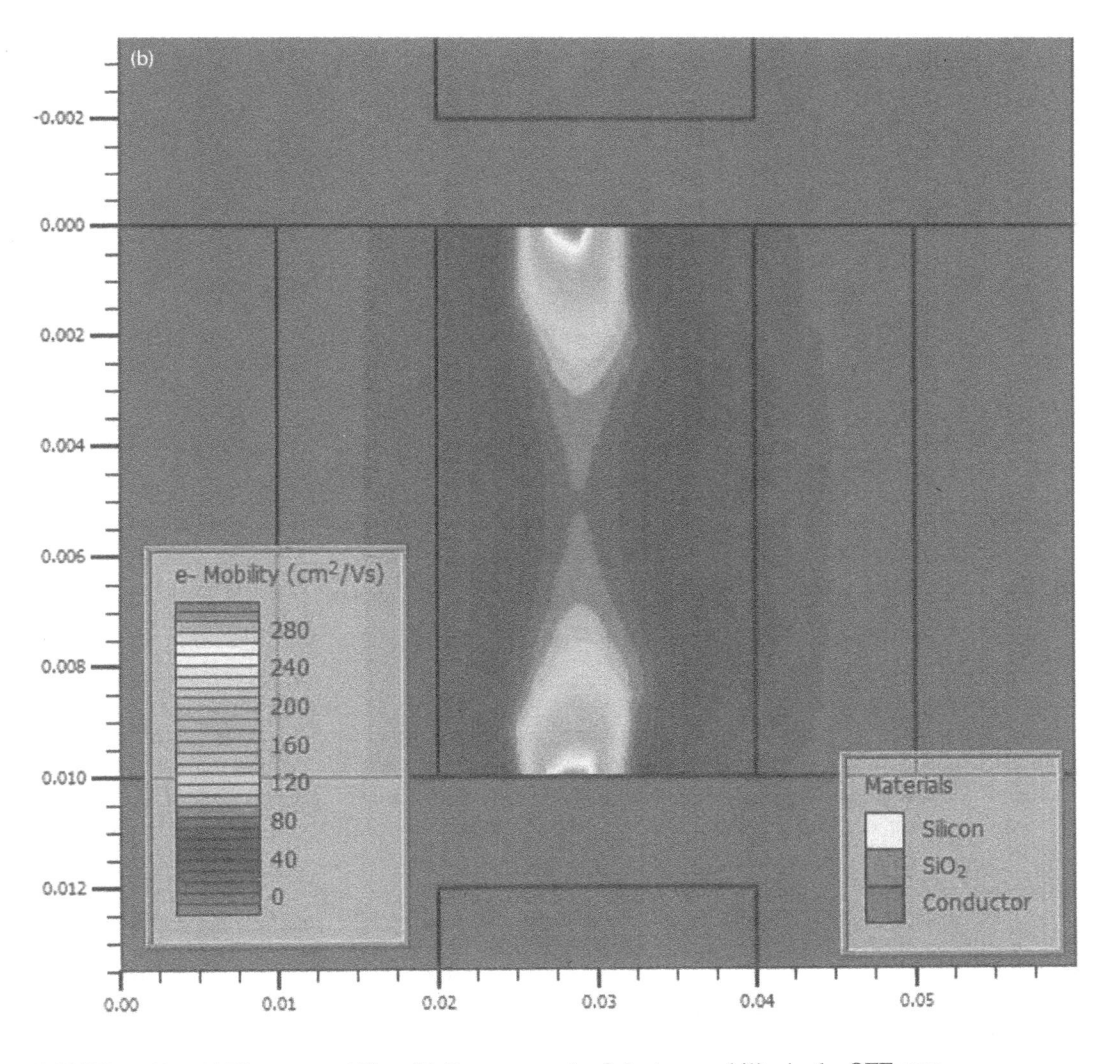

FIGURE 15.12 (a) Electron mobility. (b) Contour graph of electron mobility in the OFF state.

(Continued)

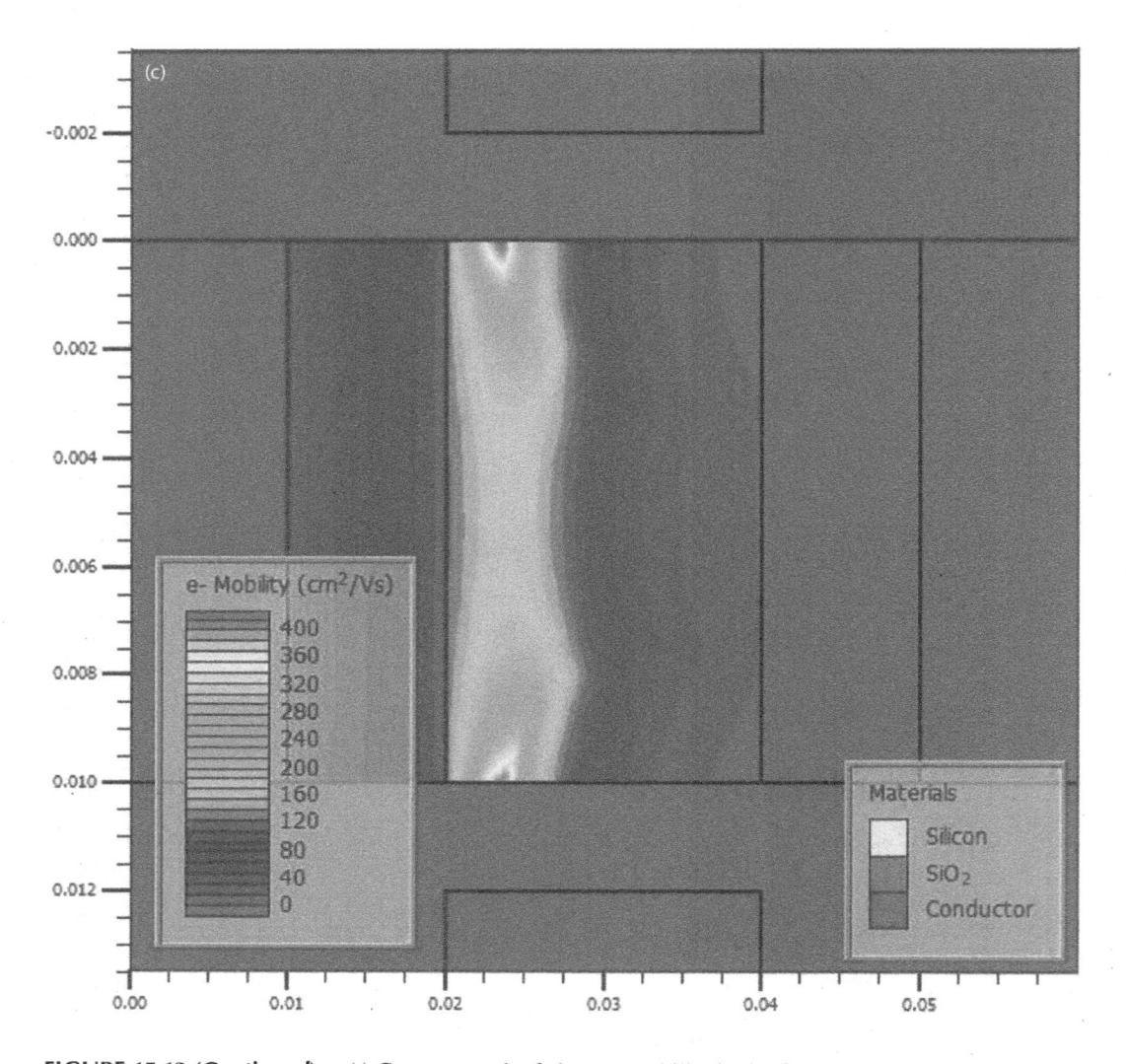

FIGURE 15.12 (*Continued*)   (c) Contour graph of electron mobility in the ON state.

band and valance band with equilibrium conditions. Figure 15.14(a) shows the population of displaced electrons in the OFF and ON states. The population of displaced electrons occurs due to the external supply voltage. In the OFF state the electrons are more displaced from their equilibrium positions. The electrons are more displaced in the channel and drain regions. Figure 15.14(b and c) are depicted in the contour diagram of the electron QFL in the OFF and ON states, respectively. In the ON state a greater population of electrons is displaced at the source side and a smaller electron population is displaced at the drain side.

Figure 15.15(a) shows the population of displaced holes in the OFF and ON states. The population of displaced holes occurs due to the external supply voltage. In the OFF state the holes are more displaced from their equilibrium positions at the drain side. The holes are more displaced in the channel and drain regions. Figure 15.15(b and c) illustrate the contour diagram of the hole QFL in OFF and ON states, respectively. In the OFF state the hole populations displaced are large at the source side and are smaller from the channel side to drain side. In the ON state a larger

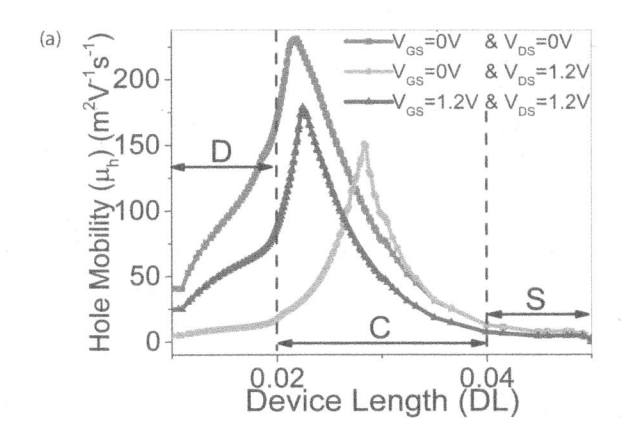

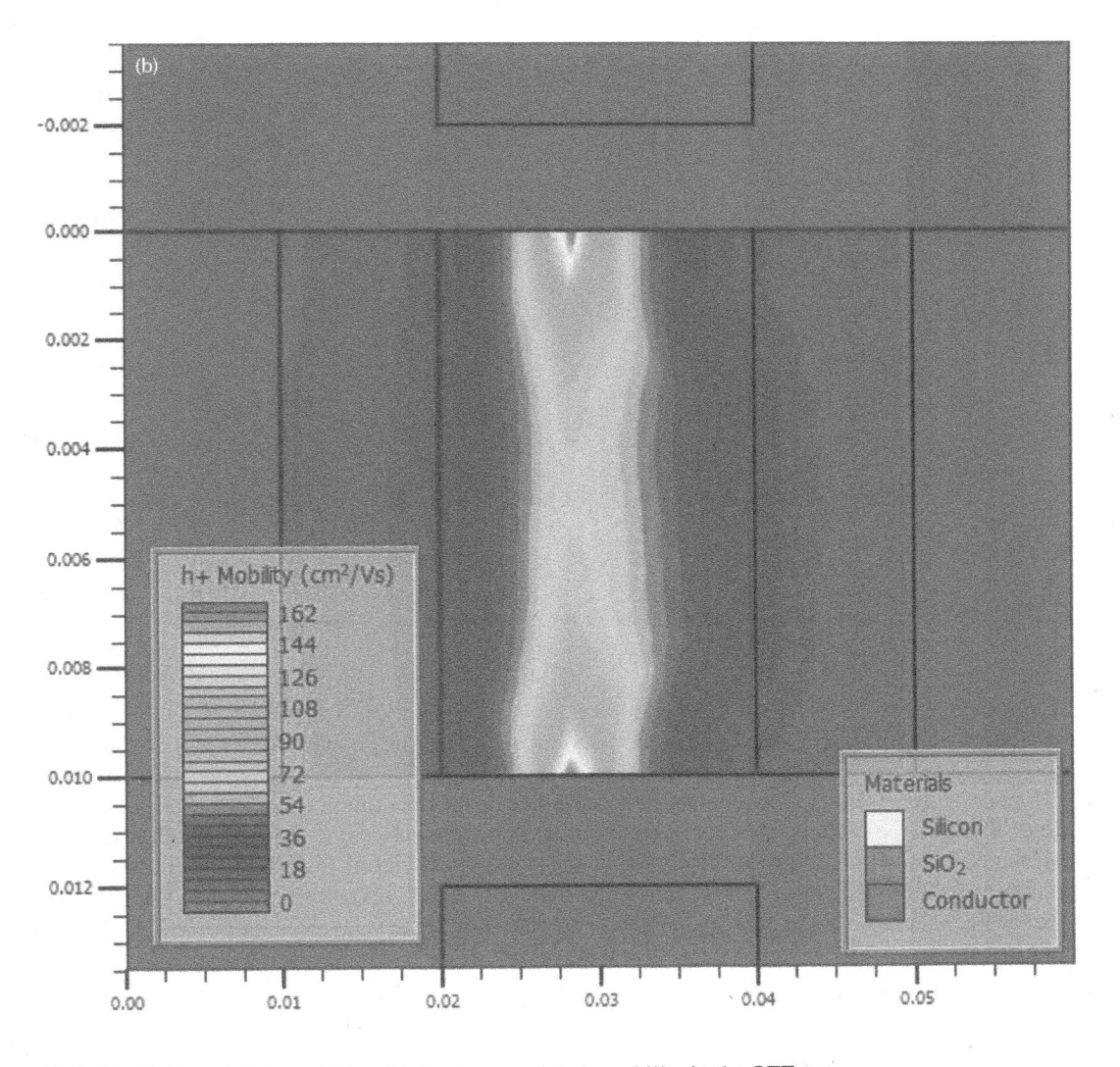

FIGURE 15.13 (a) Hole mobility. (b) Contour graph hole mobility in the OFF state.

(Continued)

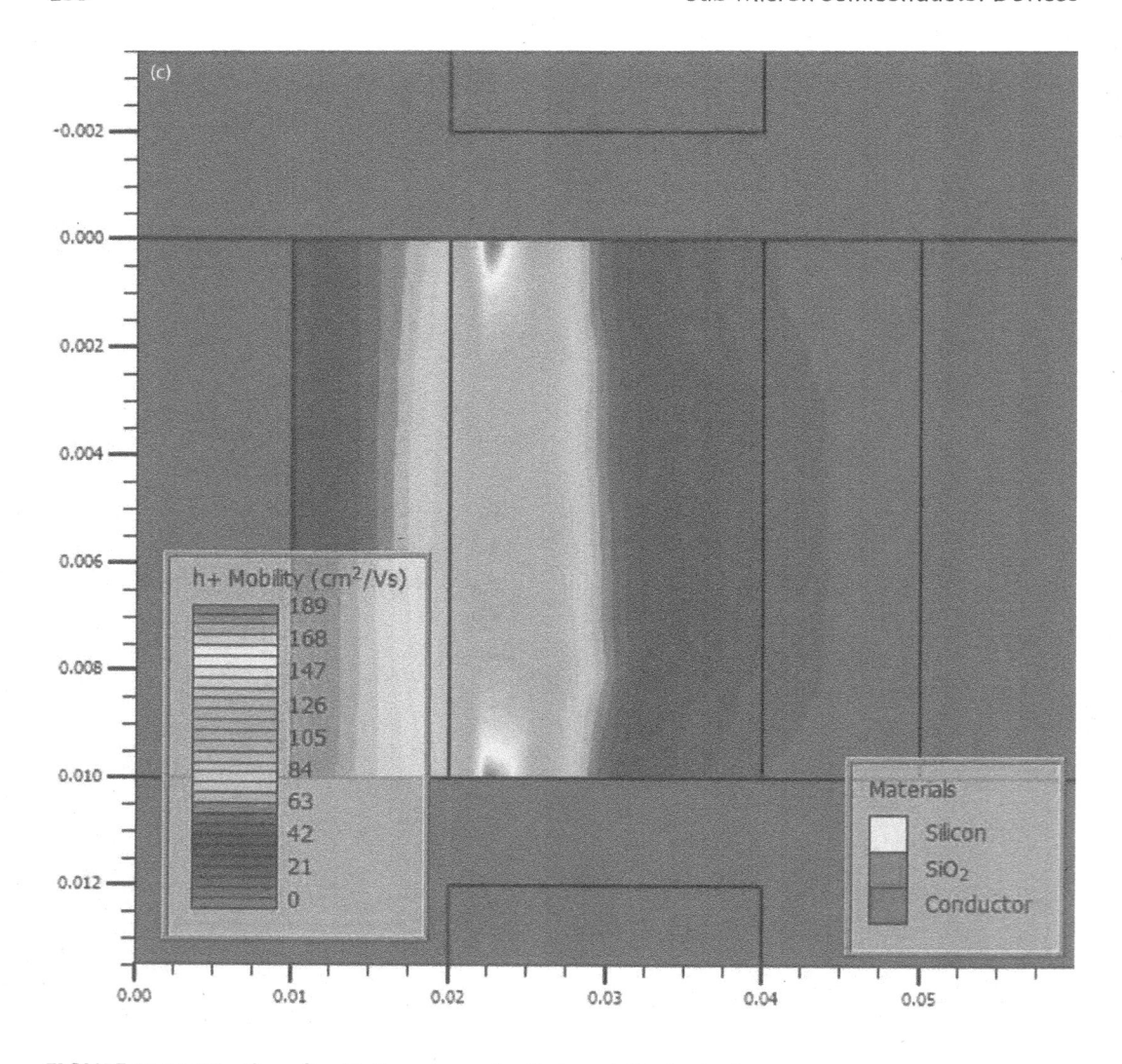

FIGURE 15.13 (*Continued*) (c) Contour graph of hole mobility in the ON state.

population of holes is displaced at the source side and a smaller population of holes is displaced at the drain side.

Figure 15.16(a) depicts the electric field variation in all operating states. The electric field is larger at the source channel interface. At the source channel interface the tunneling width is smaller, and the electric field is larger at the source channel interface. Figure 15.16(b and c) illustrates the contour diagram of the electric field in the OFF and ON states, respectively. In the OFF state the electric field variation at the source channel is very small, so there is less leakage current. In the ON state the electric field is larger at the source channel interface.

Figure 15.17(a) shows the electron BTBT rate in all operating states. The tunneling rate is larger at the source channel interface. The electric field is larger at the source channel interface, so the tunneling rate is also larger at the source channel interface. Figure 15.17(b and c) illustrates the contour diagram of the electron BTBT rate in the OFF and ON states, respectively. In the OFF state the tunneling rate is a little high at the drain channel interface, which is responsible for the leakage current.

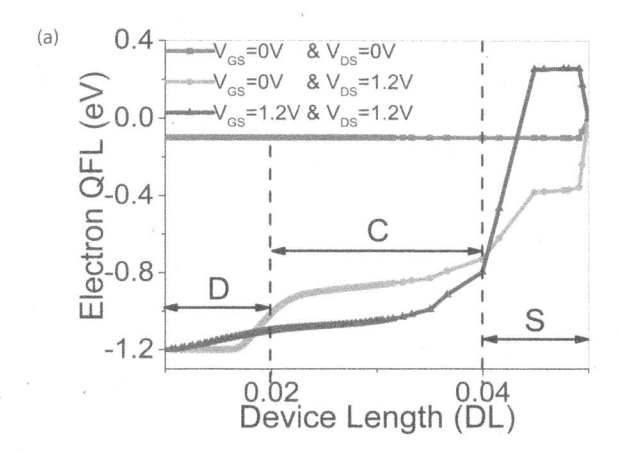

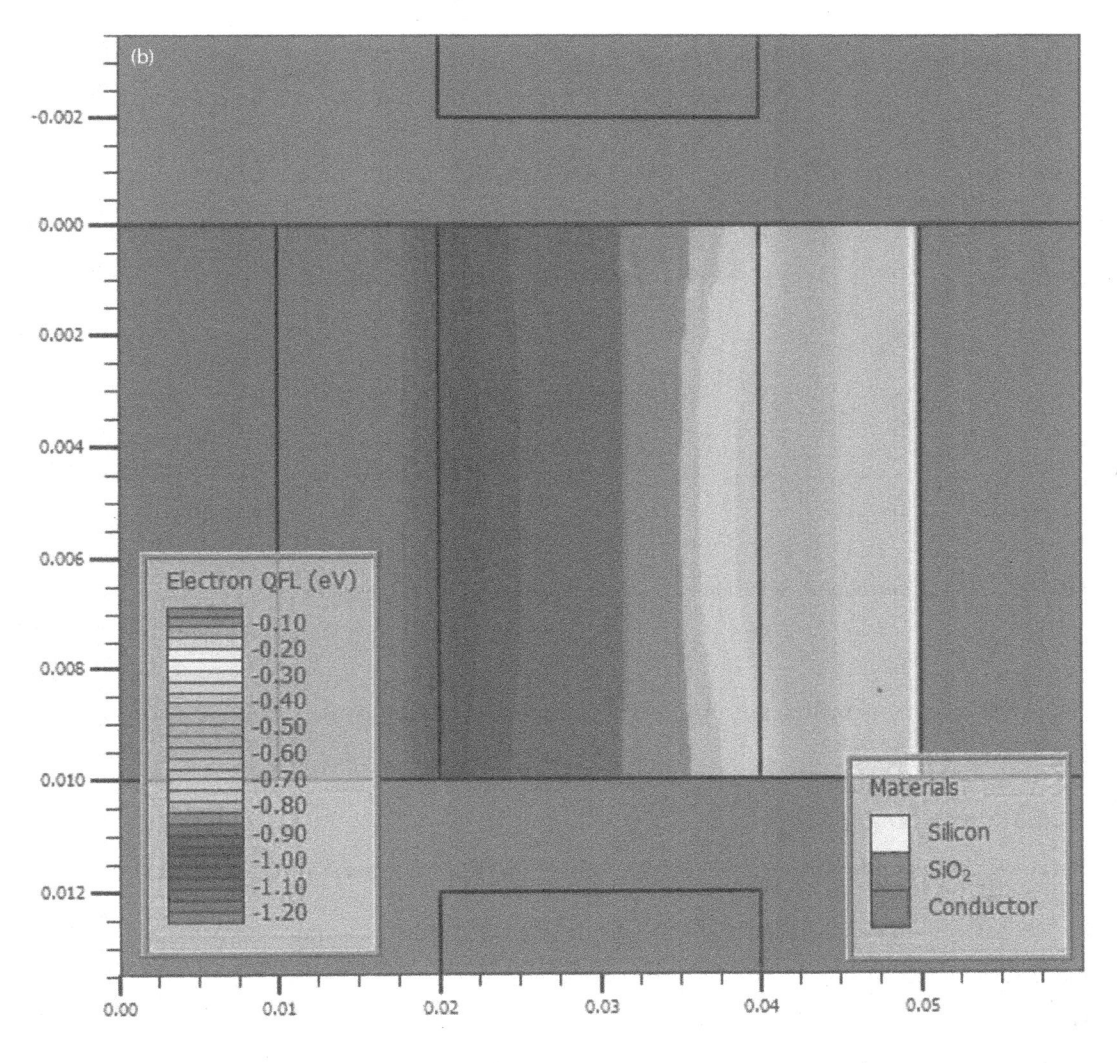

FIGURE 15.14 (a) Electron QFL. (b) Contour graph of electron QFL in the OFF state.

(Continued)

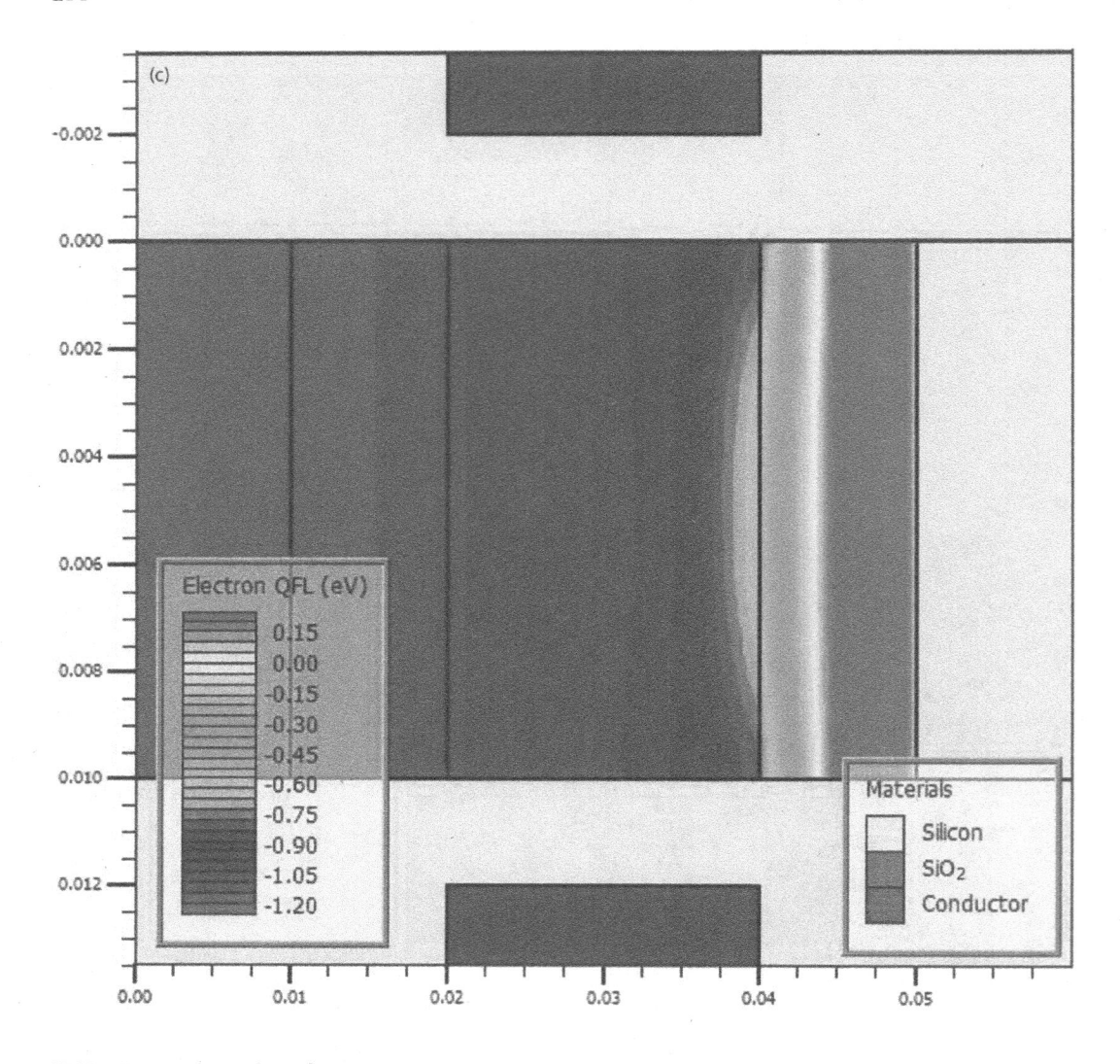

FIGURE 15.14 (*Continued*) (c) Contour graph of electron QFL in the ON state.

In the ON state the tunneling rate is higher at the source channel interface, which is responsible for the higher drain current.

15.4.2 ANALOG PARAMETERS

In analog parameters drain current with gate voltage and drain current with drain voltage have been discussed.

The current-voltage characteristics (V–I characteristics) produce the transistor behavior. The V–I characteristics of the TFET device in different bias conditions provide the proper device behavior. The TFET device operates in the three different transition regions depending on bias conditions. Biasing is applied at the source, gate, and drain electrodes. The operating regions are equilibrium [$V_{GS} = 0$ V and $V_{DS} = 0$ V], OFF transition state [$V_{GS} = 0$ V and $V_{DS} = 1.2$ V], and ON transition state

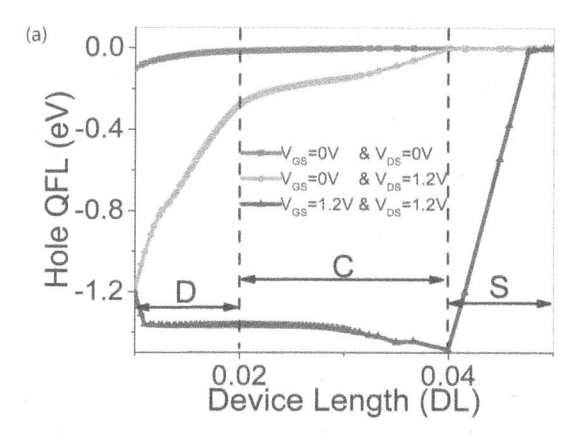

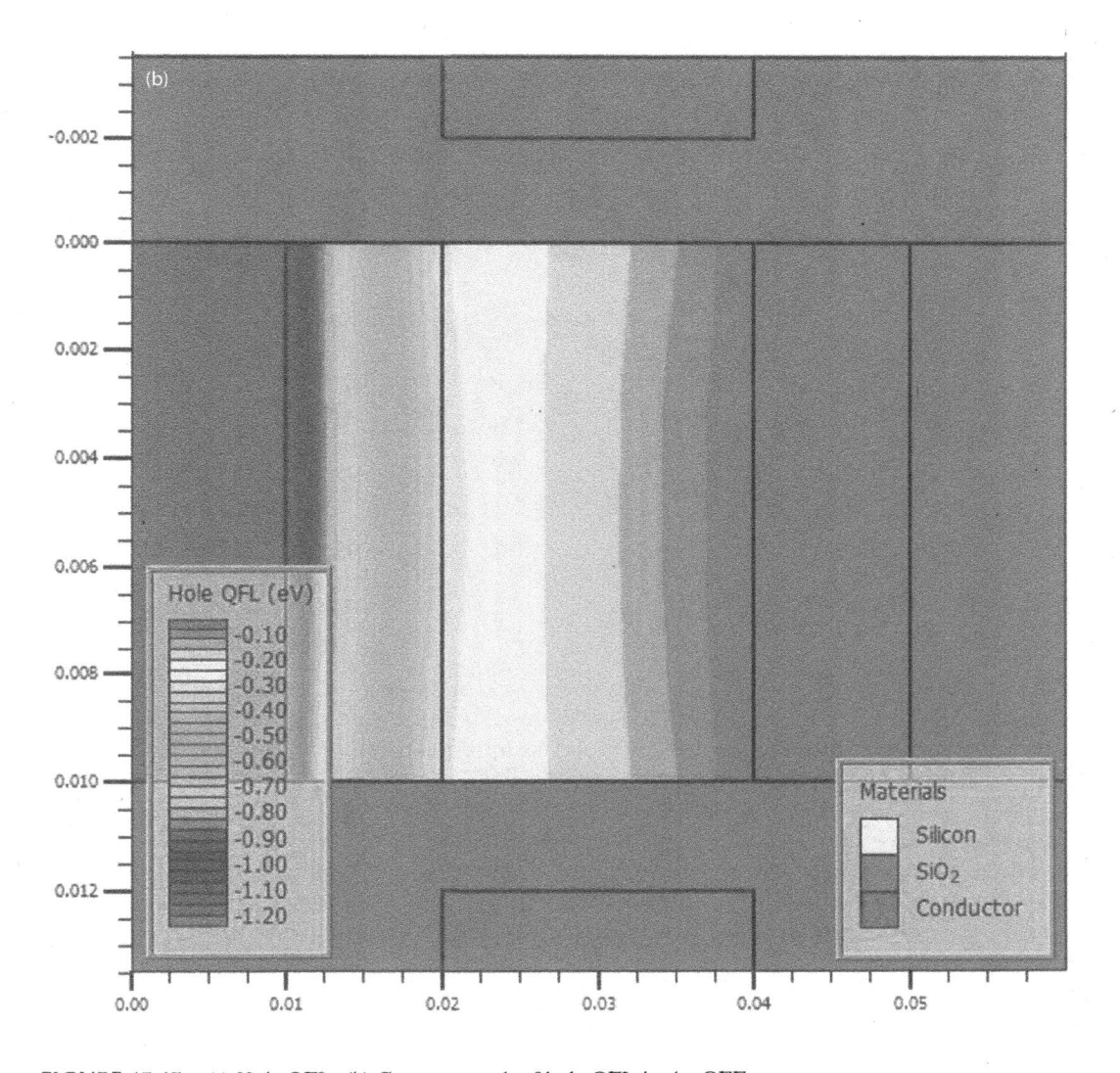

FIGURE 15.15 (a) Hole QFL. (b) Contour graph of hole QFL in the OFF state.

(Continued)

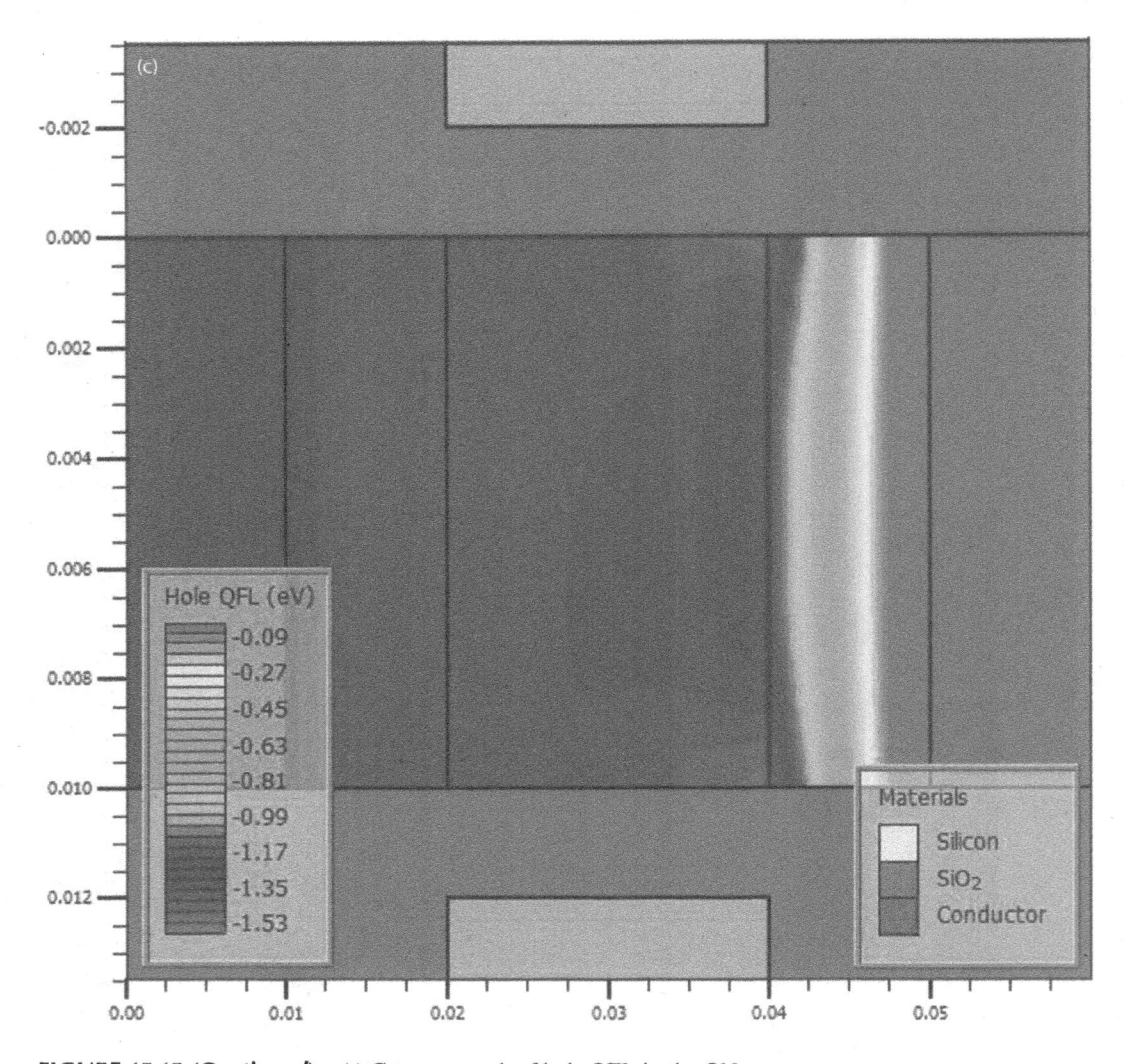

FIGURE 15.15 (Continued)   (c) Contour graph of hole QFL in the ON state.

[V_{GS} = 1.2 V and V_{DS} = 1.2 V]. The two current characteristics are transfer and output. The transfer characteristics describe the graph between the drain current (I_{DS}) and gate-to-source voltage (V_{GS}) with a constant drain-to-source voltage (V_{DS}). As the drain voltage (V_{DS}) of the device increases, the drain current (I_{DS}) of the device increases.

Figure 15.18 illustrates the variation of transfer characteristics (I_{DS} vs. V_{GS}) in both linear and logarithmic scales. The output characteristics describe the plot between drain current (I_{DS}) and drain voltage (V_{DS}) at the constant gate-to-source voltage (V_{GS}). Figure 15.18(a) depicts the drain current with gate voltage in a linear state with constant drain voltage. As the drain voltage increases, the drain current of the device increases. Figure 15.18(b) shows the drain current variation with gate voltage in log scale. Figure 15.19(a and b) represent the output characteristics of the CP-ring-TFET device with different constant gate voltage (V_{GS}) in both linear and log scale, respectively. The drain current (I_{DS}) increases with increasing drain voltage (V_{DS}), but after a certain drain voltage (V_{DS}) the drain current (I_{DS}) does not increase; it reaches the saturation state. Figure 15.19(b) shows the saturated drain current (I_{DS}) for V_{GS} = 0.3 V, after drain voltage (V_{DS}) greater than 0.3 V (V_{DS} > 0.3 V).

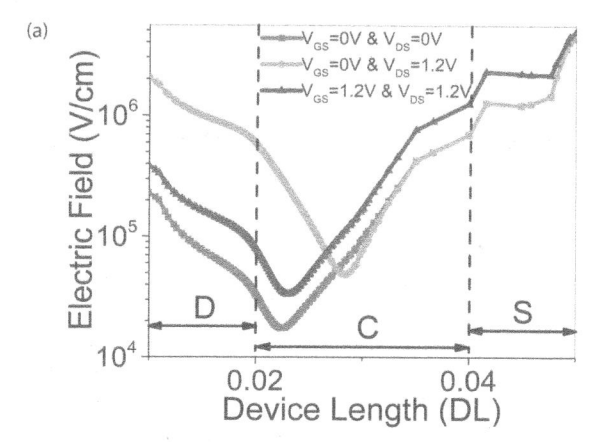

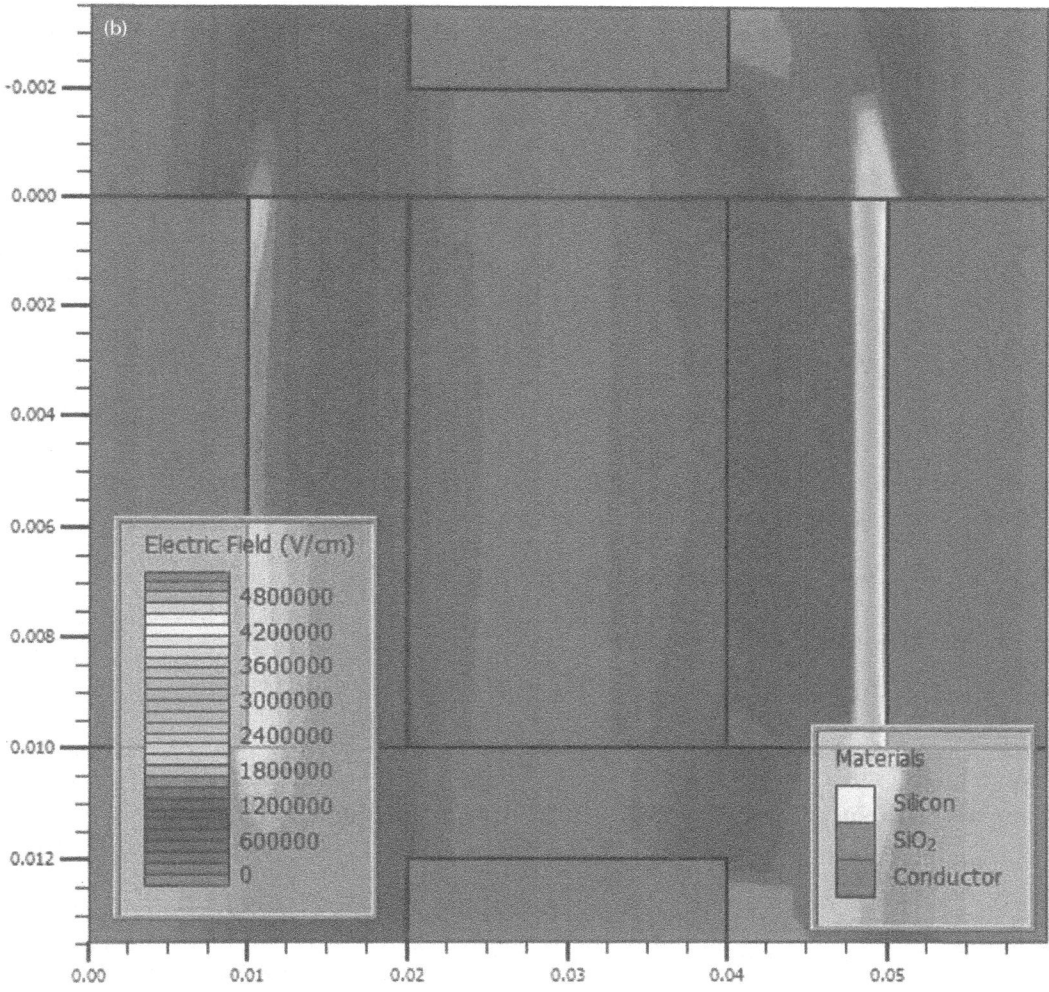

FIGURE 15.16 (a) Electric field variation. (b) Contour graph of electric field in the OFF state.

(Continued)

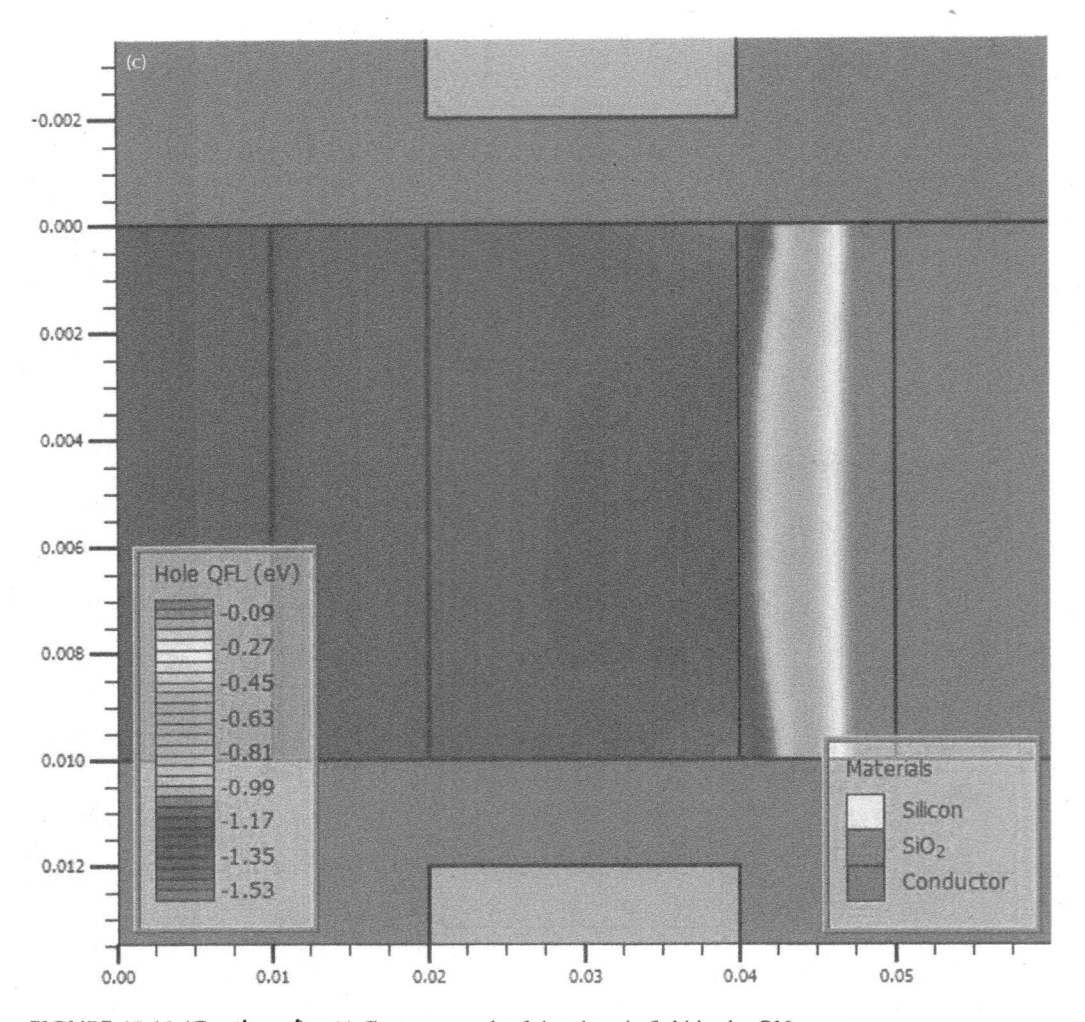

FIGURE 15.16 (*Continued*)   (c) Contour graph of the electric field in the ON state.

15.5   COMPARISON OF RESULTS

Comparison of the proposed device with existing device are shown in Table 15.2.

TABLE 15.2
Comparison of Results from the Proposed Device and Existing Device

	I_{ON} (A/um)	Subthreshold Slope (mV/dec)	I_{OFF} (A/um)	V_{th} (V)	I_{ON}/I_{OFF}
Ring-TFET	2×10^{-5}	14	6×10^{-19}	0.4	3×10^{13}
[40]	2.36×10^{-7}	4.88	5×10^{-19}	1.08	4.72×10^{12}
[41]	2.59×10^{-8}	36.24	7.59×10^{-17}	0.47	3.418×10^{8}
[42]	5×10^{-6}	60	4.54×10^{-19}	0.8	1.1×10^{13}
[43]	2.4×10^{-6}	120	2.4×10^{-12}	0.5	1×10^{6}
[44]	1.4×10^{-5}	36.1	1.6×10^{-15}	0.64	8.75×10^{9}
[45]	5×10^{-7}	12.9	1×10^{-13}	0.52	5×10^{6}

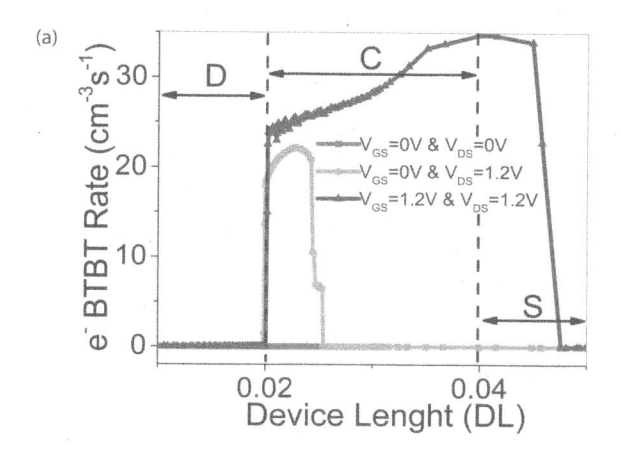

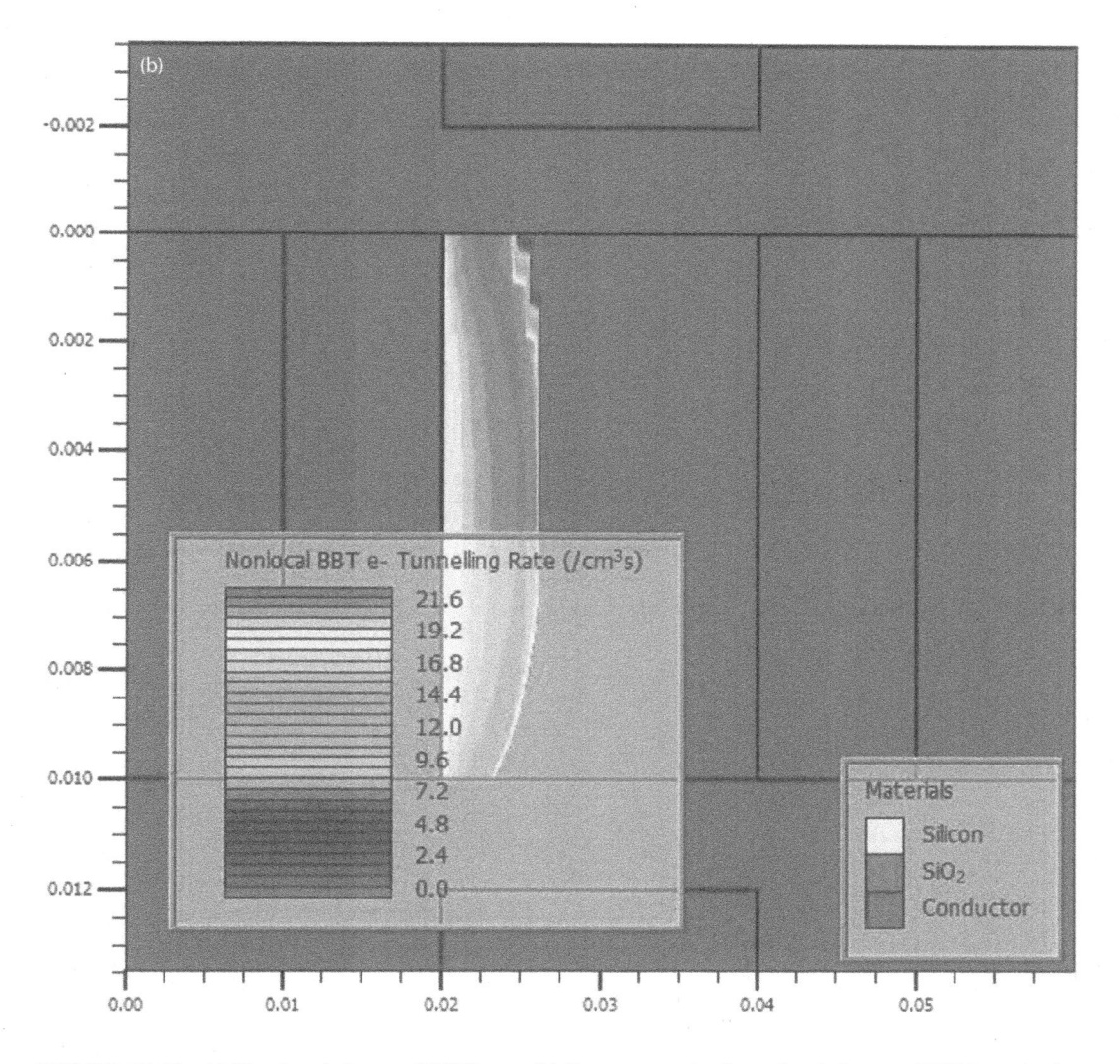

FIGURE 15.17 (a) Non-local electron BTBT rate. (b) Contour graph of non-local electron BTBT rate in the OFF state.

(Continued)

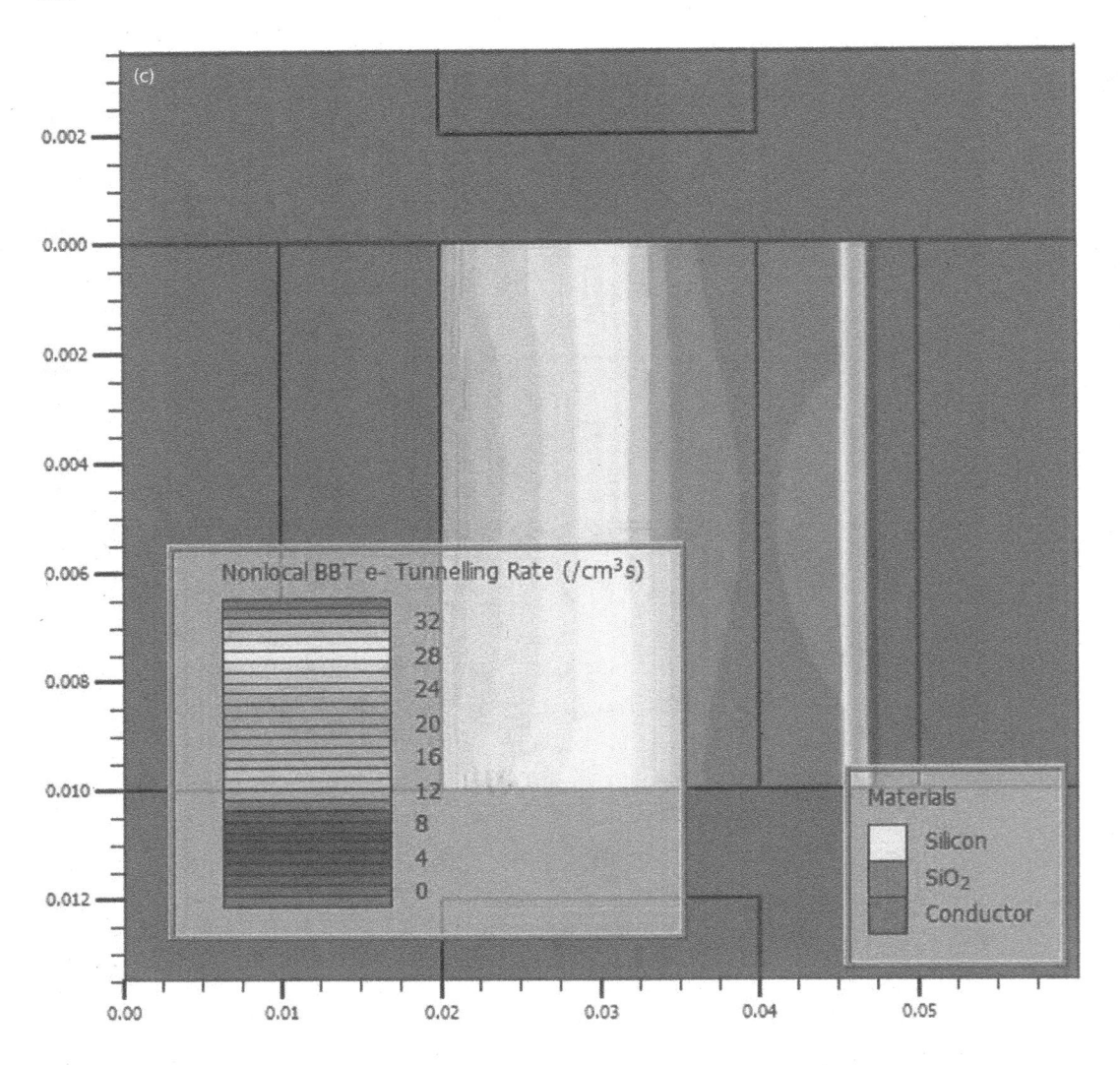

FIGURE 15.17 (*Continued*)   (c) Contour graph of non-local electron BTBT rate in the ON state.

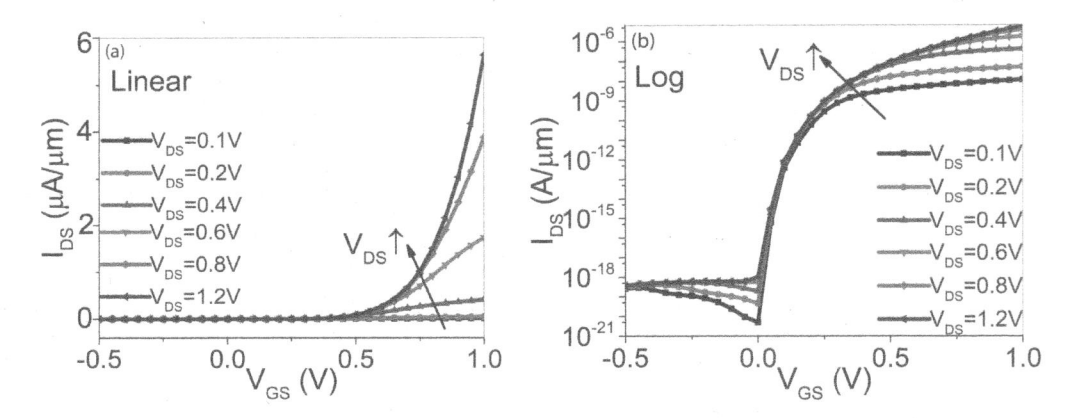

FIGURE 15.18   (a) Drain current variation with different V_{DS} variations (linear scale). (b) I_{DS} with different V_{DS} variations (log scale).

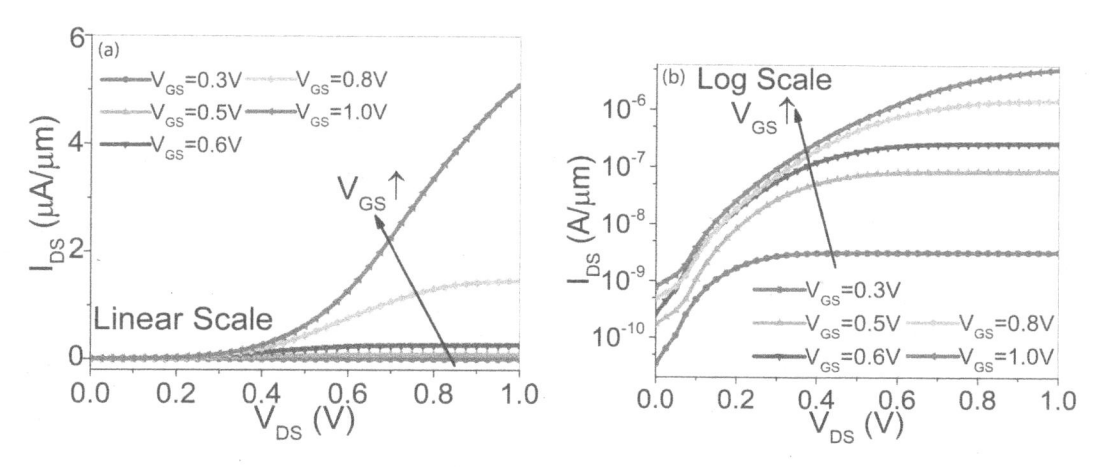

FIGURE 15.19 (a) Drain current variation with different V_{GS} variations (linear scale). (b) I_{DS} with different V_{GS} variations (log scale).

15.6 SUMMARY

The proposed ring architecture TFET shows a much improved device and analog parameters with existing devices. The proposed device shows a drain current of 20 $\mu A/\mu m$. This device also shows the less leakage current. Due to less leakage current, the proposed device shows a higher I_{ON}/I_{OFF} current ratio, which is helpful for biomedical applications. The proposed device also shows better switching speed for the digital circuit applications. The performance of the proposed device can be further enhanced by exploring other fields of the proposed area. The following are ideas that could be explored in the future:

- Development of compact modeling of dopingless CP-based ring architecture TFETs
- Development of different types of sensors such as gas, pressure, etc.
- Designing a 2D material-based FET/TFET architecture
- Development of a circuit with the help of ring architecture TFETs

REFERENCES

1. D. Kahng, and M. M. Atalla, "Silicon-silicon dioxide field induced surface device," Solid State Device Research Conference, June 1960.
2. D. Kahng, "A historical perspective on the development of MOS transistors and related devices," IEEE Transactions on Electron Devices, vol. 23, pp. 655–657, Jul. 1976.
3. W. F. Brinkman, D. E. Haggan, and W. W. Troutman, "A history of the invention of the transistor and where it will lead us," IEEE Journal of Solid-State Circuits, vol. 32, pp. 1858–1865, Dec. 1997.
4. C. A. Mack, "Fifty years of Moore's law," IEEE Transactions on Semiconductor Manufacturing, vol. 24, pp. 202–207, May 2011.
5. J. Bardeen, and W. H. Brattain, "Physical principles involved in transistor action," Physical Review, vol. 75, no. 8, pp. 1208, 1949.
6. A. B. Garrett, "The discovery of the transistor: W. Shockley, J. Bardeen, and W. Brattain," Journal of Chemical Education, vol. 40, no. 6, pp. 302, 1963.
7. R. G. Roozbahani, "BJT-BJT, FET-BJT, and FET-FET," IEEE Circuits and Devices Magazine, vol. 20, no. 6, pp. 17–22, 2004.
8. U. Zillmann, and F. Herzel, "An improved SPICE model for high-frequency noise of BJTs and HBTs," IEEE Journal of Solid-State Circuits, vol. 31, no. 9, pp. 1344–1346, 1996.
9. C. H. Wann, K. Noda, T. Tanaka, M. Yoshida, and C. Hu, "A comparative study of advanced MOSFET concepts," IEEE Transactions on Electron Devices, vol. 43, no. 10, pp. 1742–1753, 1996.

10. R. H. Yan, A. Ourmazd, and K. F. Lee, "Scaling the Si MOSFET: from bulk to SOI to bulk," IEEE Transactions on Electron Devices, vol. 39, no. 7, pp. 1704–1710, 1992.

11. B. Hoefflinger. ITRS: The international technology roadmap for semiconductors. In Chips 2020, pp. 161–174. Springer, Berlin, Heidelberg, 2011.

12. R. G. Dreslinski, M. Wieckowski, D. Blaauw, D. Sylvester, and T. Mudge, "Near-threshold computing: reclaiming moore's law through energy efficient integrated circuits," Proceedings of IEEE, vol. 8, pp. 253–256, Feb. 2010.

13. D. A. Antoniadis, and A. Khakifirooz, "MOSFET performance scaling: limitations and future options," 2008 IEEE International Electron Devices Meeting, pp. 1–4, Dec. 2008.

14. V. Barkhordarian, "Power MOSFET basics," Powerconversion and Intelligent Motion, vol. 22, no. 6, 1996.

15. S. Devadas, K. Keutzer, and J. White, "Estimation of power dissipation in CMOS combinational circuits using boolean function manipulation," IEEE Transactions on Computer-Aided Design of Integrated Circuits and Systems, vol. 11, no. 3, pp. 373–383, 1992.

16. R. X. Gu, and M. I. Elmasry, "Power dissipation analysis and optimization of deep submicron CMOS digital circuits," IEEE Journal of Solid-State Circuits, vol. 31, no. 5, pp. 707–713, 1996.

17. S. Borkar, "Getting gigascale chips: challenges and opportunities in continuing Moore's law," Queue, vol. 1, no. 7, p. 26, 2003.

18. D. J. Wouters, J. P. Colinge, and H. E. Maes, "Subthreshold slope in thin -film SOI MOSFETs," IEEE Transactions on Electron Devices, vol. 37, no. 9, pp. 2022–2033, 1990.

19. J. P. Colinge, "Subthreshold slope of thin-film SOI MOSFETs," IEEE Electron Device Letters, vol. 7, no. 4, pp. 244–246, 1990.

20. C. W. Lee, A. N. Nazarov, I. Ferain, N. D. Akhavan, R. Yan, P. Razavi, R. Yu, R. T. Doria, and J. P. Colinge, "Low subthreshold slope in junctionless multigate transistors," Applied Physics Letters, vol. 96, no. 10, pp. 102–106, 2010.

21. B. Yu, L. Chang, S. Ahmed, H. Wang, S. Bell, C. Y. Yang, C. Tabery, C. Ho, Q. Xiang, T. J. King, and J. Bokor, "FinFET scaling to 10 nm gate length," Digest International Electron Devices Meeting, pp. 251–254, Dec. 2002.

22. T. Dang, L. Anghel, and R. Leveugle, "CNTFET basics and simulation," International Conference on Design and Test of Integrated Systems in Nanoscale Technology, pp. 28–33, Sep. 2006.

23. L. Shen, S. Heikman, B. Moran, R. Coffie, N. Q. Zhang, D. Buttari, I. P. Smorchkova, S. Keller, S. P. DenBaars, and U. K. Mishra, "AlGaN/AlN/GaN high-power microwave HEMT," IEEE Electron Device Letters, vol. 22, no. 10, pp. 457–459, 2001.

24. N. Kumar, and A. Raman, "Novel design approach of extended gate-on-source based charge-plasma vertical-nanowire TFET: proposal and extensive analysis," IEEE Transactions on Nanotechnology, vol. 19, pp. 421–428, 2020.

25. D. Bimberg, M. Grundmann, and N. N. Ledentsov, Quantum dot heterostructures. John Wiley & Sons, Berlin, 1999.

26. S. M. Turkane, and A. K. Kureshi, "Review of tunnel field effect transistor (TFET)," International Journal of Applied Engineering Research, vol. 11, no. 7, pp. 4922–4929, 2016.

27. D. Singh, S. Pandey, K. Nigam, D. Sharma, D. S. Yadav, and P. Kondekar, "A charge-plasma-based dielectric-modulated junctionless TFET for biosensor label-free detection," IEEE Transactions on Electron Devices, vol. 64, no. 1, pp. 271–278, 2016.

28. A.K. Gupta, A. Raman, and N. Kumar, 2019, "Cylindrical nanowire-TFET with core-shell channel architecture: design and investigation," Silicon, pp. 1–8. https://doi.org/10.1007/s12633-019-00331-1

29. A. K. Gupta, and A. Raman, "Electrostatic-doped nanotube TFET: proposal, design, and investigation with linearity analysis," Silicon, vol. 13, pp. 2401–2413, 2021.

30. N. Kumar, and A. Raman, "Performance assessment of the charge-plasma-based cylindrical GAA vertical nanowire TFET with impact of interface trap charges," IEEE Transactions on Electron Devices, vol. 66, no. 10, pp. 4453–4460, 2019.

31. A. K. Gupta, and A. Raman, "Performance analysis of electrostatic plasma-based dopingless nanotube TFET," Applied Physics A, vol. 126, no. 7, pp. 1–10, 2020.

32. N. Kumar, and A. Raman, "Design and investigation of charge-plasma-based work function engineered dual-metal-heterogeneous gate Si-Si 0.55 Ge 0.45 GAA-cylindrical NWTFET for ambipolar analysis," IEEE Transactions on Electron Devices, vol. 66, no. 3, pp. 1468–1474, 2019.

33. K. Boucart, and A. M. Ionescu, "Double-gate tunnel FET with high-κ gate dielectric," IEEE Transactions on Electron Devices, vol. 54, pp. 1725–1733, July 2007.

34. S. Saurabh, and M. J. Kumar, "Novel attributes of a nanoscale dual material gate tunnel field effect transistor," IEEE Trans. on Electron Devices, vol. 58, pp. 404–410, Feb. 2011.
35. S. J. Koester, I. Lauer, A. Majumdar, J. Cai, J. Sleight, S. Bedell, P. Solomon, S. Laux, L. Chang, S. Koswatta, W. Haensch, P. Tomasini, and S. Thomas, "Are Si/SiGe tunneling field-effect transistors a good idea?," ECS Transactions, vol. 33, no. 6, pp. 357–361, 2010.
36. L. Wang, E. Yu, Y. Taur, and P. Asbeck, "Design of tunneling fieldeffect transistors based on staggered heterojunctions for ultralow-power applications," IEEE Electron Device Letters, vol. 31, pp. 431–433, May 2010.
37. Q. Zhang, T. Fang, H. Xing, A. Seabaugh, and D. Jena, "Graphene nanoribbon tunnel transistors," IEEE Electron Device Letters, vol. 29, pp. 1344–1346, Dec. 2008.
38. J. T. Smith, C. Sandow, S. Das, R. A. Minamisawa, S. Mantl, and J. Appenzeller, "Silicon nanowire tunneling field-effect transistor arrays: improving subthreshold performance using excimer laser annealing," IEEE Trans. on Electron Devices, vol. 58, pp. 1822–1829, July 2011.
39. Atlas User's Manuel. Silvaco, 2008, p. 5. https://silvaco.com/search/Atlas+Manual
40. B. R. Raad, S. Tirkey, D. Sharma, and P. Kondekar, "A new design approach of dopingless tunnel FET for enhancement of device characteristics," IEEE Trans. Electron Devices, vol. 64, no. 4, pp. 1830–1836, Apr. 2017.
41. J. Patel, D. Sharma, S. Yada, A. Lemtur, and P. Suman, "Performance improvement of nano wire TFET by hetero-dielectric and heteromaterial: at device and circuit level," Microelectronics Journal, vol. 85, pp. 72–82, Mar. 2019.
42. N. Kumar, U. Mushtaq, S. I. Amin, and S. Anand, "Design and performance analysis of dual-gate all around core-shell nanotube TFET," Superlattices Microstructures, vol. 125, pp. 356–364, Jan. 2019.
43. K. E. Moselund, H. Schmid, C. Bessire, M. T. Bjork, H. Ghoneim, and H. Riel, "InAs–Si nanowire heterojunction tunnel FETs," IEEE Electron Device Letters, vol. 33, no. 10, pp. 1453–1455, Oct. 2012.
44. S. Yadav, R. Madhukar, D. Sharma, M. Aslam, D. Soni, and N. Sharma, "A new structure of electrically doped TFET for improving electronic characteristics," Applied Physics, vol. 124, no. 7, pp. 517, Jul. 2018.
45. H. Asai, T. Mori, T. Matsukawa, J. Hattori, K. Endo, and K. Fukuda, "Steep switching less than 15 mV dec−1 in silicon-on-insulator tunnel FETs by a trimmed-gate structure," Japanese Journal of Applied Physics, vol. 58, 2019.

16 Hybrid Intelligent Technique-Based Doping Profile Optimization in a Double-Gate Hetero-Dielectric TFET

Sagarika Choudhury, Krishna Lal Baishnab,
Brinda Bhowmick, and Koushik Guha

CONTENTS

16.1 INTRODUCTION

The complex process of semiconductor device design requires precise models and efficient optimizers. The process of device design includes tuning of essential parameters to achieve the specified target. The conventional design technique (Goswami and Bhowmick 2014; Goswami et al. 2016; Sahu et al. 2020) involves simulating the device in CAD-based tools (Bennett 2014), validating the results, comparing with the specification, and retuning if required. Thus, the efficiency of the design process mainly depends on the efficiency of design engineers. Now, as the device dimensions are shrinking, appropriate utilization of computer simulation-aided and optimization-aided manufacturing techniques will lower the number of test keys and hit-and-trial errors. Moreover, scaling has resulted in numerous short channel effects in conventional metal oxide semiconductor field-effect transistors (MOSFETs). Further, subthreshold swing (SS) cannot be reduced below 60 mV/dec due to the thermionic mode of operation (Nilsson et al. 2006; Knoch et al. 2007; Koswatta et al. 2009). Thus, tunnel field-effect transistors (TFETs) working on the band-to-band-tunneling (BTBT) (Toh et al. 2007; Seabaugh and Zhang 2010; Verhulst et al. 2010) mechanism is found to be the best alternative to conventional MOSFETS and can achieve lower SS values. This chapter presents a device simulation–based optimization method for the TFET design. This work utilizes the analysis of a hetero-dielectric (HD) double-gate (DG) n-type TFET (Bhowmick et al. 2012). The model for the proposed structure is obtained by analytically solving a two-dimensional (2D) Poisson's equation with HD gate oxide (Bhowmick et al. 2012). The goal is to minimize the computational cost and complexity for obtaining optimized device dimension by using several high-performance evolutionary algorithms. Thus, the proposed technique requires an objective function and the best suited algorithm to find the optimum device dimension. Hence, the derived expression for the electric field is then used as an objective function because it contains most of the necessary physical parameters of the device. The higher the electric field, the higher is the tunneling efficiency at the junction (Toh et al. 2007; Verhulst et al. 2010). Thus, by computing the electric field at the tunnel junction we can compute the current ratio as a function of change in doping concentration. The parameters, which significantly influence the device performance, are sorted out by performing sensitivity analysis and they are selected as design variables to be tuned by optimization techniques. As doping across various regions significantly affects device performance, the influence of the doping profile variation is studied in this chapter. Moreover, the width of the tunnel junction also plays a crucial role in device operation. So, the length of the source, drain, and channel are also considered as design variables. The main goal is to maximize the current ratio by doping profile optimization. This technique aids in overcoming the random fluctuation in semiconductor devices. The doping profile is recomputed in the proposed device to decrease the device parameter fluctuations due to the random doping profile variation.

Various optimization algorithms such as evolutionary algorithms and swarm-based approaches, numerical techniques, and hybrid approaches are then used to find the optimized values for the design variables within the specified range.

An analysis of the results reflects that the hybrid intelligent optimization algorithm provides the best result with higher accuracy and better convergence. The proposed hybrid technique makes use of a combination of evolutionary algorithms and swarm intelligence-based techniques. The performances of the algorithms are then validated by performing complexity, convergence, and accuracy analysis. Further, the accuracy of the model is validated by re-simulating the device in TCAD and comparing it with the hit-and-trial-based device performance. This chapter has been segmented into the following sections. First the structure is elaborated in Section 16.2, and then in the subsequent section 6.3 the mathematical model of the proposed device is discussed. Further, the optimization algorithms that are used are reviewed in Section 16.4. Next, the methodology of optimization is discussed in Section 16.5, and finally, the result and conclusion are presented in Section 16.6 and 16.7.

16.2   THE PROPOSED DG TFET

The HD TFET structure to be optimized is shown in Figure 16.1. The source is designed with germanium material with p-doping, the drain is an n-type, and the channel is intrinsic. The optimized doping concentrations of the regions are found by using optimization algorithms. The bias across source and drain are 1.2 V and 0.7 V. The length L_1 is the length of channel.

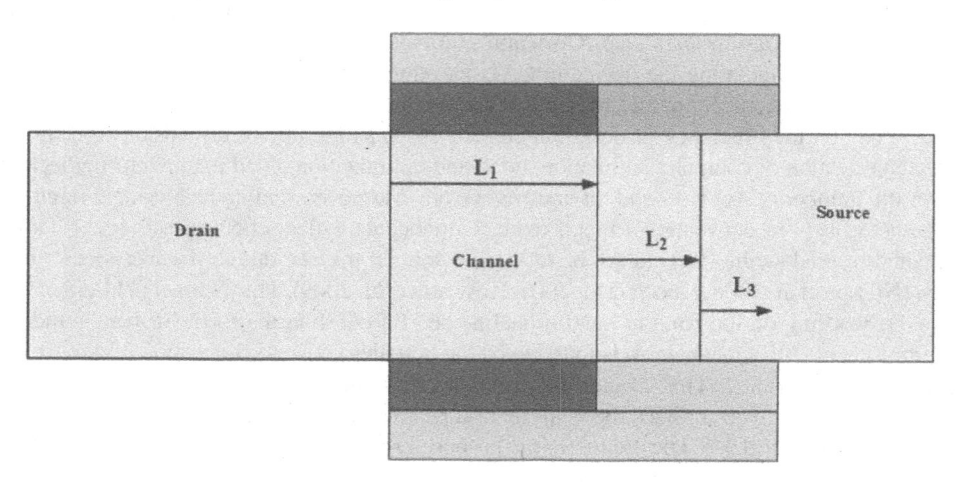

FIGURE 16.1   The hetero-dielectric double-gate TFET.

16.3 THE MATHEMATICAL MODEL OF THE DEVICE

The mathematical model of a standard DG hetero-junction TFET (Bhowmick et al. 2012) is used as a reference to perform the analysis. The 2D Poisson's equation is represented as Equation 16.1

$$\frac{\partial^2 \psi}{\partial^2 x} + \frac{\partial^2 \psi}{\partial^2 y} = \frac{qN_A}{\varepsilon} \tag{16.1}$$

where the potential is represented, q is the charge, N_A is the acceptor doping, and permittivity of medium is ε. The solution of Equation 16.1 is given by

$$\psi(x, y) = a_0(x) + a_1(x)y + a_2(x)y^2 \tag{16.2}$$

Assuming the boundary condition as in Equation 16.1, the coefficients a_0 and a_1 are solved and can be solved for the structure.

The expression of surface potential is found as

$$\psi_s(x) = B\exp(px) + C\exp(-px) + \psi_k \tag{16.3}$$

where ψ_k is the surface potential along one dimension, and B and C are constants given by the following equations

$$B = \frac{1}{K} \left\{ \begin{array}{l} 2\left(-\psi_k - \frac{qN_D L_1^2}{2\varepsilon}\right)\sinh p\,(L_2 - L_3) - 2(\varphi_1 - \psi_k) \\[2mm] -L_1 p(\varphi_1 - \psi_k)e^{-p(L_1 - L_2)} + q\frac{N_D L_1^2}{\varepsilon}e^{p(L_2 - L_3)} \\[2mm] +2(\varphi_2 + V_{DS} - \psi_k)\sinh p(L_3 - L_1) \end{array} \right\} \tag{16.4}$$

$$C = \frac{(\varphi_2 + V_{DS} - \psi_k)e^{pL_3} - e^{pL_2}(\varphi_1 - \psi_k)}{K} \tag{16.5}$$

$$K = 2\ \sinh p(L_2 - L_3) \tag{16.6}$$

$$p^2 = \frac{\varepsilon^*}{\varepsilon \tau t_s} \tag{16.7}$$

By applying an external potential at $x = 0$ and $x = L_3$, potential φ_1 and φ_2 are determined. From the previous analysis the expression of the electric field is then determined

$$E_x = \frac{d\psi_s(x)}{dx} + \frac{\varepsilon^*\left\{\dfrac{d\psi_s(x)}{dx} - v\right\}}{\varepsilon \tau}y - \frac{1}{2}\frac{\varepsilon^*\left\{\dfrac{d\psi_s(x)}{dx} - v\right\}}{\varepsilon \tau t_s}y^2 \tag{16.8}$$

$$E_y = a_1(x) + 2a_2(x)y \tag{16.9}$$

Thus, the objective function is given by

$$E = \sqrt{E_x^2 + E_y^2} \tag{16.10}$$

Thus, the electric field can be found by optimizing Equation 6.10.

16.4 THE NEED FOR THE OPTIMIZATION ALGORITHM

The design process is known to have three major approaches to optimize a structure. A brief discussion on each of these approaches is discussed next and depicted in Figure 16.2.

- *Design experimentally:* Experimental designing of any semiconductor technology involves huge fabrication cost and complexity; thus, it is not practically feasible to experimentally design any device without appropriately simulating and testing the important parameters.
- *Design with simulated models:* The current research on device level focuses on simulating the structures at the software level, testing the performance at the circuit level, and then after checking all the important parameters performing fabrication. The use of simulated models has advantages, such as high speed and low cost. However, it has been observed that device simulation involves optimizing the structure by the hit-and-trial method. Thus, this design process requires human involvement for finding the optimum structure by hit and trial, which may result in inaccuracy.
- *Design with optimization algorithms (Rao 2009):* The primary goal of this work is to use a program to find the best design without affecting the device properties instead of using the hit-and-trial technique. This program is an optimization algorithm that commands the computer to change the parameters to maximize or minimize the objective function.

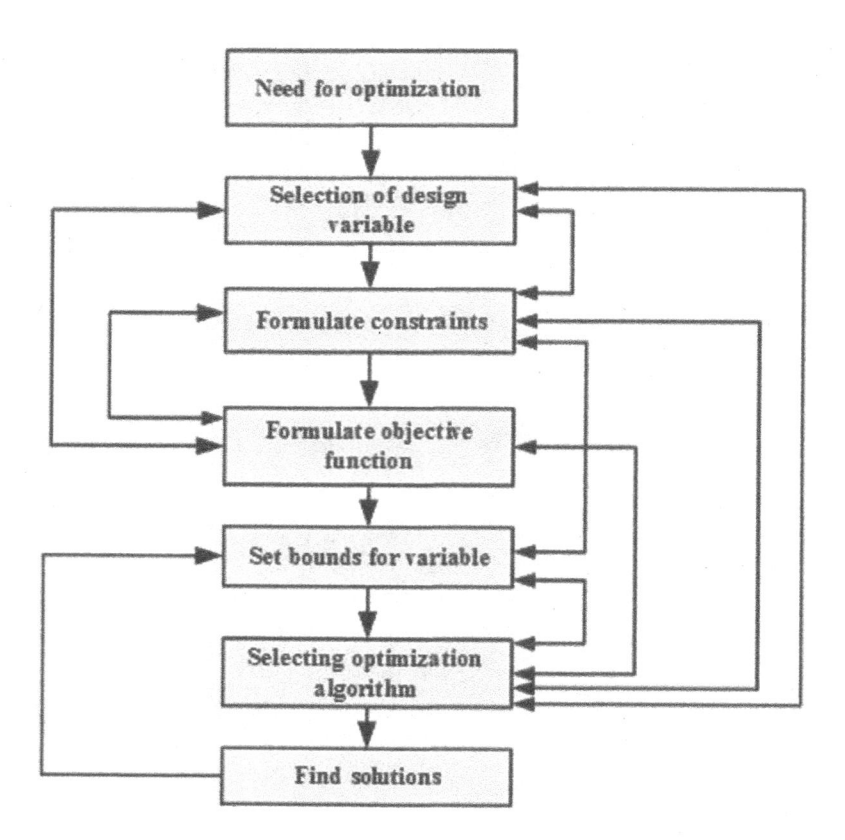

FIGURE 16.2 Flowchart of optimal design.

16.4.1 CLASSIFICATION OF ALGORITHMS

Optimization algorithms can be mostly categorized as conventional (deterministic) or modern (stochastic) algorithms.

- *Conventional (deterministic) algorithms:* This approach finds the same solution in multiple runs if the initial point is not changed. They have no random component and also get stuck at local optima (Rao 2009).
- *Modern (stochastic) algorithms:* These techniques are equipped with random components and find different solutions in each run even when the initial solution is the same. These algorithms are run multiple times until the satisfaction of the end condition and it estimates global optima. The main focus of this chapter will be on modern algorithms. The stochastic algorithms can again be classified into individual-based and population-based algorithms (Li et al. 2001; Rao 2009; Basu et al. 2015).

16.4.2 THE INTELLIGENT ALGORITHMS

Evolutionary algorithms (Yu and Gen 2010) are population-based metaheuristics. The solutions provided by this technique are approximate in nature. These techniques help in solving real-time problems that cannot be solved by conventional mathematical techniques (Li et al. 2001). They can be considered as a section of evolutionary computation. These are stochastic in nature. They are useful in situations in which approximate solutions are desired, but they become extremely exhaustive when used for exact solutions. However, these algorithms require two basic criteria to be fulfilled: they require an objective function and the design variables must be chosen judiciously. Over the years many researchers have proposed many varieties of evolutionary algorithms (Saha et al. 2018) on the nature and food-searching behavior of various biologically inspired organisms.

In this chapter, we have employed different metaheuristic algorithms (Basu et al. 2015) for solving our problem statement and presented a comparative study of the performance of these algorithms in regard to solving practical problems.

16.4.3 HYBRID DEPSO

The DEPSO algorithm (Zhang and Xie 2003) is the hybrid version of the DE and PSO algorithms (Zhang and Xie 2003). The DE performs much better with high population diversity and improved local search ability and has some advantages, such as its ability to maintain the diversity of the population and the ability to explore local searches, but the algorithm fails to memorize the previous process and use the global information about the search space. To gain the advantages of both PSO (Kennedy and Eberhart1995) and DE (Storn and Price 1995), the hybrid DEPSO algorithm has been designed. In Figure 16.3 the PSO algorithm is incorporated into DE.

16.5 THE NEW APPROACH PROBLEM FORMULATION

In this work the designer needs to make an approximate model by solving Poisson's equation and then find the suitable parameter, formulate the problem, and use an optimization technique to determine the best design. The goal is to find the best design out of all available options and the best suitable algorithm. This approach is fast and less error prone, but this process needs a great deal of consideration while designing it. The various steps involved in the process are discussed in the following sections.

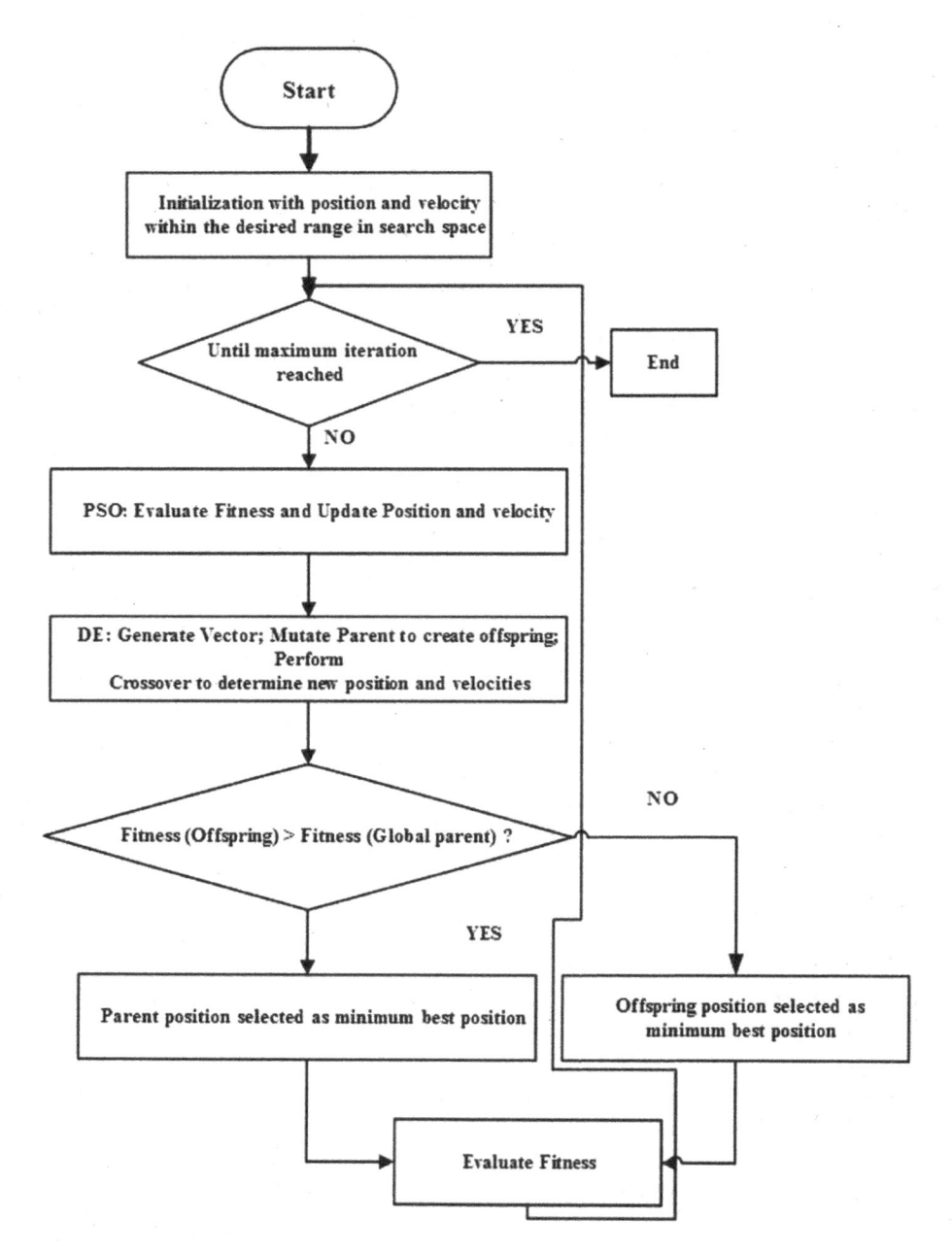

FIGURE 16.3 Flowchart of the DEPSO algorithm.

16.5.1 Selection of Design Variables and Their Ranges

The selection of design variables significantly affects the device performance. The work studies the influence of doping concentration in device performance. To find the influence of the doping concentration, it is varied analytically keeping all other parameters as fixed. Tables 16.1 and 16.2 show the impact of dimension and doping concentration variation. It has been observed that a small change in doping concentration results in a large variation of current ratio. Moreover, the length of the tunnel junction also plays a crucial role in device operation. It has also been found that

TABLE 16.1

Sensitivity Analysis to Assess Impact of Dimension Variation in DG TFET

Parameters	Lengths (nm)	Optimized x (0-40nm) by DEPSO (0-40 nm)	Optimized Electric Field (DEPSO) (V/cm)	Optimized x (0-40nm) by DE (0-35 nm)	Optimized Electric Field (DE) (V/cm)	Optimized x (0-40nm) by PSO (0-35) nm	Optimized Electric Field (PSO) (V/cm)
	$L_1 = 0.9x, L_2 = 0.08x, L_3 = 0.02x$	35	2.8×10^6	37	2.72×10^6	36	2.67×10^6
	$L_1 = 0.09x, L_2 = 0.8x, L_3 = 0.01x$	38	1.9×10^6	39	1.8×10^6	40	1.87×10^6
	$L_1 = 0.08x, L_2 = 0.8x, L_3 = 0.02x$	33	1.75×10^6	32	1.7×10^6	32	1.5×10^6

TABLE 16.2

Sensitivity Analysis to Assess Impact of Doping Profile Variation in DG TFET

Parameters	Range (cm^{-3})	Optimized Electric Field (DEPSO) (V/cm)	Optimized Electric Field (DE) (V/cm)	Optimized Electric Field (PSO) (V/cm)
Doping (cm^{-3})	$N_A = 10^{21}$ $N_D = 7 \times 10^{18}$ $N_C = 5 \times 10^{16}$	1×10^5	1.4×10^5	0.9×10^5
	$N_A = 10^{20}$ $N_D = 5 \times 10^{18}$ $N_C = 2 \times 10^{16}$	2.8×10^6	2×10^6	1.8×10^6
	$N_A = 10^{19}$ $N_D = 1 \times 10^{18}$ $N_C = 1 \times 10^{16}$	1×10^6	0.8×10^6	0.6×10^6

increasing L_1 and L_3 and decreasing L_2 results in an increased electric field across the junction. So, the length of source, drain, and channel are also considered as design variables. Table 16.3 reflects the design parameters and their allowed range as per the International Technology Roadmap for Semiconductors (ITRS) requirement.

Table 16.1 shows the impact of length variation. DEPSO-based optimization performed better than PSO- and DE-based optimization for the problem, whereas PSO had mostly extreme results. DEPSO still performed better taking optimum values for most of the variables. Further, more

TABLE 16.3

Range and Variables to be Optimized

Variables	Range
L_1 (nm)	$(0.01–0.99) \times x$
L_2 (nm)	$(0.01–0.99) \times x$
L_3 (nm)	$(0.01–0.99) \times x$
N_A (cm^{-3})	$10^{18}–10^{21}$
N_D (cm^{-3})	$1 \times 10^{18}–7 \times 10^{18}$
N_C (cm^{-3})	$1 \times 10^{16}–5 \times 10^{18}$

algorithms will be applied to the problem to test the performance and see which algorithms perform the best for surface potential optimization. It is evident from Table 16.1 that decreasing L_2 increases the electric field at the tunnel junction; thus, the lengths are chosen as design variables.

The analysis in Table 16.2 shows that increasing the doping concentration increases the electric field up to a certain limit. Further increase results in a sharp fall of electric field due to a decrease in tunneling across the junction and results in low current ratio. Similarly, the decrease in doping concentration also results in marked decline in the field. Thus, it is necessary to find the optimum doping concentration.

16.5.2 ELECTRIC FIELD OPTIMIZATION OF THE PROPOSED STRUCTURE

In a TFET design, the main aim is to improve the current ratio. It is clearly known (Toh et al. 2007; Verhulst et al. 2010; Raad et al. 2018; Barah et al. 2019) that increasing the electric field across the tunneling junction can significantly improve tunnel efficiency, thus, current will improve. Again the tunnel junction field depends on certain factors such as doping concentration, junction width, gate overlapping, and so forth. Thus, by optimizing some of the parameters we can improve the efficiency (Choudhury et al. 2020). So, the expression of electric field is used from reference papers (Bhowmick et al. 2012), and many metaheuristic algorithms are used to optimize the expression. Intelligent algorithms such as DEPSO (Zhang and Xie 2003), PSO (Kennedy and Eberhart 1995), DE (Storn and Price 1995), WOA (Mirjalili and Lewis 2016), and HBPSO (Hao et al. 2014) are more efficient in solving real-world problems compared with the traditional numerical techniques (Li 2001) or the hit-and-trial-based optimization (Toh et al. 2007; Verhulst et al. 2010) approaches.

16.6 EXPERIMENTAL RESULTS

The optimization of the objective function discussed in this section is performed in MATLAB® software on a high-performance system. The optimized values of design variables obtained through the process are listed in Table 16.4. Further, the convergence and time complexity analyses of all the algorithms are performed. Based on the analyses shown in Figure 16.4 and Table 16.4, it can be concluded that DEPSO performs best with highest convergence among all algorithms. Further, the device is simulated with the previously mentioned dimensions and doping concentrations. The percentage error between the optimized value of the electric field and simulated result is listed in Table 16.5. The lowest value error in electric field computation is achieved by using the DEPSO algorithm. Figures 16.5 and 16.6 show the plot of the electric field and drain current of the

TABLE 16.4

Comparative Analysis for Different Algorithms for Electric Field Optimization Computed after 20 Individual Runs

Variables	Range	DEPSO	DE	PSO	WOA	HBPSO
L_1 (nm)	$(0.01\text{-}0.99) \times x$	**0.735×34**	0.0100×34.86	0.0200×35.6	0.0100×40	0.0100×35.4
L_2 (nm)	$(0.01\text{-}0.99) \times x$	**0.20×34**	0.955×34.86	0.8800×35.6	0.9800×40	0.9800×35.4
L_3 (nm)	$(0.01\text{-}0.99) \times x$	**0.10×34**	0.035×34.6	0.0100×35.6	0.0100×40	0.0200×35.4
x (nm)	$0\text{-}40$	**34**	34.86	3356	40	35.4
N_A (cm^{-3})	$10^{18}\text{-}10^{21}$	**10^{20}**	10^{19}	10^{20}	10^{21}	10^{20}
N_D (cm^{-3})	$1 \times 10^{18}\text{-}7 \times 10^{18}$	**5×10^{18}**	1.9×10^{18}	5.5×10^{18}	7×10^{18}	1×10^{18}
N_C (cm^{-3})	$1 \times 10^{16}\text{-}5 \times 10^{18}$	**1×10^{16}**	1.5×10^{16}	2×10^{16}	5×10^{16}	1×10^{16}
Electric field (V/cm)		**2.83×10^6**	2.75×10^6	2.87×10^6	2.74×10^6	2.88×10^6

Note: Bold values represent the best values obtained.

TABLE 16.5
Error Analysis for Different Algorithms for Electric Field

Algorithms	Algorithm-Based Electric Field	TCAD Simulated Value (V/cm)	Deviation (%)
DEPSO	2.83×10^6	2.8×10^6	1.07
DE	2.75×10^6	2.8×10^6	1.7
PSO	2.87×10^6	2.8×10^6	2.5
WOA	2.74×10^6	2.8×10^6	2.1
HBPSO	2.88×10^6	2.8×10^6	2.8

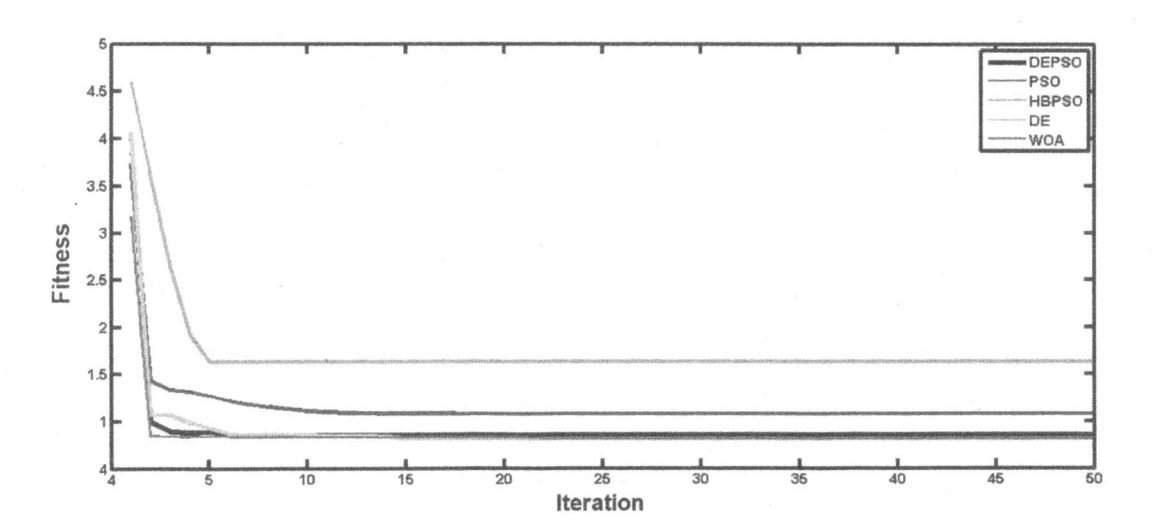

FIGURE 16.4 Convergence plot.

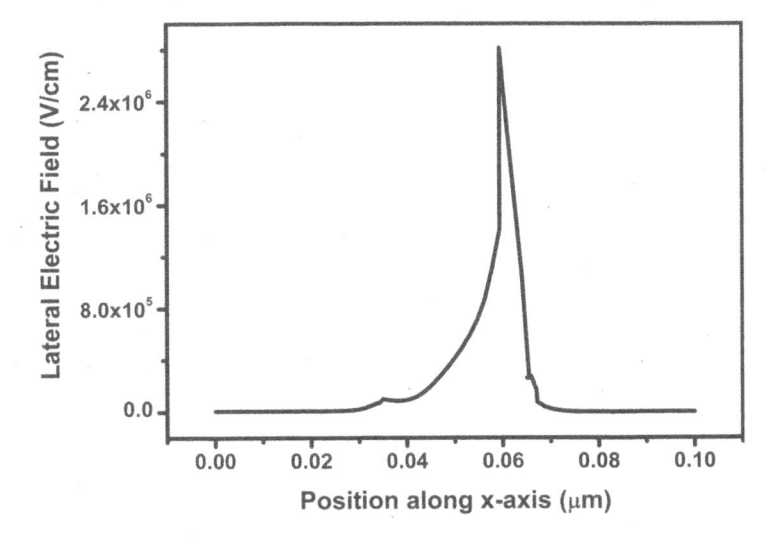

FIGURE 16.5 Electric field.

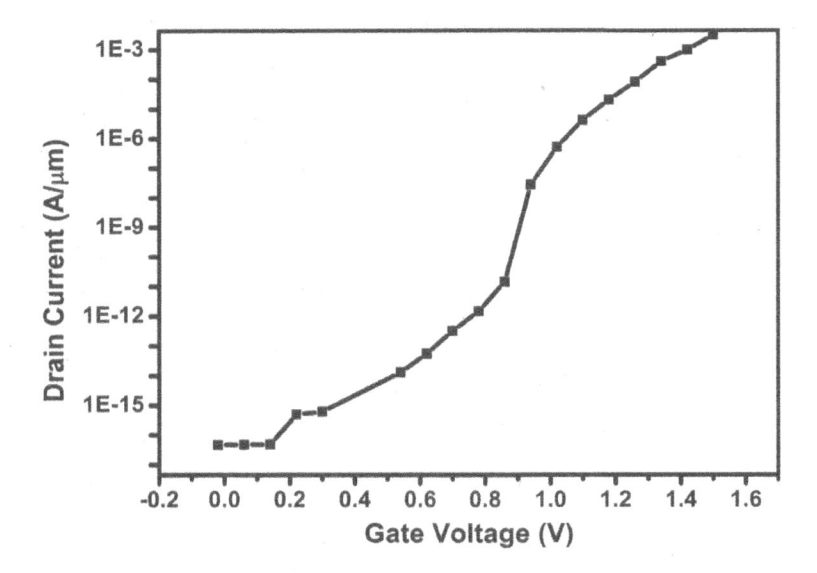

FIGURE 16.6 Drain current plot of optimized structure.

optimized device. Figure 16.7 depicts the comparative plot of the simulated electric field versus modeled field. Moreover, the values obtained through the optimization technique are again validated by designing TFET in TCAD software. Table 16.6 shows the improved electric field computation compared with the TCAD simulation without optimization.

Thus, the previous analysis shows that although the proposed device has appreciable values of current similar to the conventional TFET devices, the use of optimization algorithms shows remarkable improvement in the values of drain current and electric field as depicted in Table 16.6. The optimized values of the electric field, ON current, and OFF current are found to be 2.8×10^6, 4×10^{-3}, and 4×10^{-17}.

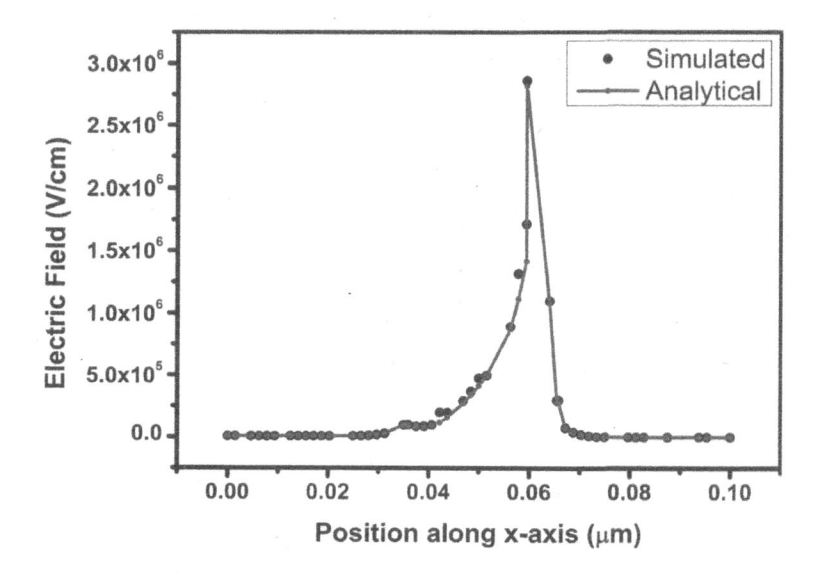

FIGURE 16.7 Comparison plot of simulated and analytical fields.

TABLE 16.6
Validation of DEPSO Algorithms for Electric Field

Parameters	DEPSO-Based Results	TCAD Simulated Results (Without Optimization) (Bhowmick et al. 2012)	TCAD Simulated Results (With Optimization) This Work
Electric field (V/cm)	2.86×10^6	2.5×10^6	2.8×10^6
ON current (A)	-	4.8×10^{-4}	4×10^{-3}
OFF current (A)	-	4.8×10^{-15}	4×10^{-17}

16.7 CONCLUSION

This chapter mainly discussed the impact of doping concentration in device performance and using optimization algorithms to improve efficiency. The technique marks the use of numerous algorithms, and the best accuracy and efficiency is obtained in the case of DEPSO. A comparison with the conventional design process shows that the use of the optimization technique is an excellent approach to tune the parameters and achieves the target design.

As fabricating a semiconductor device is a complex and costly process, having an efficient optimizer can help reduce complexity. This technique outperforms the state-of-the-art design techniques and provides best accuracy along with exceptional computational efficiency. A current ratio of 1×10^{13} and average SS of 45 mV/dec is achieved by optimizing the proposed structure.

REFERENCES

Barah, D., A. K. Singh, and B. Bhowmick. 2019. TFET on selective buried oxide (SELBOX) substrate with improved ION/IOFF ratio and reduced ambipolar current, *Silicon*, 11, 973–981. https://doi.org/10.1007/s12633-018-9894-0

Basu, S., M. Sharma, and P. S. Ghosh. 2015. Metaheuristic applications on optimisation problems: a survey, *OPSEARCH*, 52, 3, 530–561. doi: 10.1007/s12597-014-0190-5

Bennett, H. S. 2014. International Technology Roadmap for Semiconductors Radio Frequency and Analog/Mixed-Signal Technologies. http://www.https://www.nist.gov/publications/2014-international-technology-roadmap-semiconductors-radio-frequency-and-analogmixed

Bhowmick, B., and S. Baishya. 2014. A physics-based model for electrical parameters of double gate hetero-material nano scale tunnel FET, *International Journal of Applied Information Systems (IJAIS)*, 1, 3, 2249–0868. doi: 10.5120/ijais12-450142

Choudhury, S., B. Bhowmick, and K. L. Baishnab. 2020. A double-gate heteromaterial tunnel FET optimized using an evolutionary algorithm, *Journal of Computational Electronics*, 19, 277–282. https://doi.org/10.1007/s10825-019-01426-z

Goswami, R., and B. Bhowmick. 2014. Hetero-gate-dielectric gate-drain underlap nanoscale TFET with a δp+Si1-xGexlayer at source-channel tunnel junction, *2014 International Conference on Green Computing Communication and Electrical Engineering (ICGCCEE)*, Coimbatore, India, 1–5. doi: 10.1109/ICGCCEE.2014.6922302

Goswami, R., B. Bhowmick, and S. Baishya. 2016. Physics-based surface potential, electric field and drain current model of a δp+ Si1–xGex gate–drain underlap nanoscale n-TFET, *International Journal of Electronics*, 103, 9, 1566–1579. doi: 10.1080/00207217.2016.1138514

Hao, L., X. Gang, Y. G. Ding, and Y. B. Sun. 2014. Human behavior-based particle swarm optimization, *The Scientific World Journal*, 2014, 194706. https://doi.org/10.1155/2014/194706

Kennedy, J., and R. C. Eberhart. 1995. Particle swarm optimization. *Proceedings of the ICNN'95–International Conference on Neural Networks*, Perth, Australia, 1942–1948.

Knoch, J., S. Mantl, and J. Appenzeller. 2007. Impact of the dimensionality on the performance of tunneling FETs: Bulk versus one-dimensional devices, *Solid-State Electronics*, 51, 572–578.

Koswatta, S. O., M. S. Lundstrom, and D. E. Nikonov. 2009. Performance comparison between p-i-n tunneling transistors and conventional MOSFETs, *IEEE Transactions on Electronic Devices*, 56, 456–465.

Li, Y., J. Liu, T. Chao, and S. M. Sze. 2001. A new parallel adaptive finite volume method for the numerical simulation of semiconductor devices, Computer Physics Communications, 142, 285–289.

Mirjalili, S., and A. Lewis. 2016. The whale optimization algorithm, *Advances in Engineering Software*, 95, 51–67. doi: 10.1016/j.advengsoft.2016.01.008

Nilsson, P. 2006. Arithmetic reduction of the static power consumption in nanoscale CMOS, *13th IEEE International Conference on Electronics, Circuits and Systems*, Nice, France, 656–659.

Raad, B. R., D. Sharma, and S. Tirkey. 2018. Source engineered tunnel FET for enhanced device electrostatics with trap charges reliability, *Microelectronic Engineering*, 194, 79–84.

Rao, S. S. 2009. *Engineering Optimization Theory and Practice*. Wiley, Hoboken, NJ.

Saha, C., N. Agbu, R. Jinks, and M. Nazmul Huda. 2018. Review article of the solar PV parameters estimation using evolutionary algorithms, Solar and Photoenergy Systems, 2, 2, 63–75.

Sahu, S. A., R. Goswami, and S. K. Mohapatra. 2020. Characteristic enhancement of hetero dielectric DG TFET using SiGe pocket at source/channel interface: proposal and investigation, *Silicon*, 12, 3, 513–520. doi: 10.1007/s12633-019-00159-9

Seabaugh, C., and Q. Zhang. 2010. Low-voltage tunnel transistors for beyond CMOS logic, Proceedings of the IEEE, 98, 2095–2110.

Storn, R., and K. Price. 1995. Differential evolution – a sample and efficient adaptive scheme for global optimization over continuous spaces, *International Computer Science Institute*, 4, 1–11.

Toh, E. H., G. H. Wang, G. Samudra, and Y. C. Yeo. 2007. Device physics and design of double-gate tunneling field-effect transistor by silicon film thick- ness optimization, *Applied Physics Letters*, 90, 263507.

Verhulst, A. S., W. G. Vandenberghe, D. Leonelli, R. Rooyackers, A. Vandooren, G. Pourtois, S. D. Gendt, M. M. Heyns, and G. Groeseneken. 2010. Boosting the on-current of Si-based tunnel field-effect transistors, *ECS Transactions*, 33, 6, 363–372.

Yu, X., and M. Gen. 2010. *Introduction to Evolutionary Algorithms*, Springer-Verlag, London.

Zhang, W. J., and X. F. Xie. 2003. DEPSO: Hybrid particle swarm with differential evolution operators, *Proceedings of the IEEE International Conference on System, Security and Assurance*, Washington, DC, 3816–3821.

17 Graphene Nanoribbon Devices
Advances in Fabrication and Applications

Juan M. Marmolejo-Tejada, Jaime Velasco-Medina, and Andres Jaramillo-Botero

CONTENTS

17.1 INTRODUCTION

The advent of nanostructured materials has opened the possibility of implementing novel electronic devices for a wide range of applications. This includes alternatives for silicon-based integrated circuits and sensors with improved capabilities that can result in smaller and faster devices with added functionalities and improved performance compared with state-of-the-art commercial counterparts.

Graphene, a two-dimensional (2D) atomic-scale honeycomb lattice made of carbon atoms, is one of the most studied materials in the nanoscale because of its electronic, magnetic, optical, and mechanical properties. Although several road maps and review articles describe the main benefits of graphene-based devices, along with graphene and graphene nanoribbon (GNR) fabrication techniques, defects and effects on physicochemical properties, simulation frameworks for modeling nanoelectronic devices, and applications [1–10], several challenges are yet to be solved to materialize their potential applications across a number of fields, which include energy production and storage, high-frequency electronic devices in high-speed information processing and communications, lightweight composites, high-end instrumentation, sensors and metrology, flexible and transparent electronics, photonics, biomedicine, molecular devices, spintronics, quantum electronics, etc. [2, 8, 11, 12]. Persisting challenges include the need for defect-free materials in many applications, and the unavailability of cost-effective and highly scalable fabrication processes.

Carrier mobility in graphene is higher than in Si and other semiconducting materials (up to $10^6 cm^2 V^{-1} s^{-1}$ in suspended graphene [13]), although it presents zero bandgap in its 2D sheet form. This means that replacing a semiconductor channel material in a typical field-effect transistor (FET) for graphene could result in very high OFF-state currents because the device cannot be switched off, and low ON/OFF current ratios (I_{on}/I_{off}). Such high I_{off} means higher leakage current when the device is in the OFF state and higher power consumption. Then, this needs to be minimized to reach the standards of complementary metal oxide semiconductor (CMOS) technology.

Despite its zero bandgap electronic structure, graphene (and other 2D materials) have very high surface-to-volume ratios, enabling large amounts of reactive sites for adsorbing target molecules and allowing real-time detection and measurement of different kinds of biological markers (biomarkers). Furthermore, surface functionalization of graphene has also been explored for improving selectivity in sensing devices by means of differentiable electronic signatures.

Bandgap engineering of graphene has been addressed to enable applications where non-zero bandgap is required [1, 7]. Some strategies include:

1. Constraining large-area graphene along one dimension to form GNRs [14],
2. Applying a gate bias to bilayer graphene [15],
3. Patterning graphene nanomeshes (GNMs) [16–18],
4. Straining graphene [19], and
5. Applying chemical modifications to graphene [20, 21].

GNR electronic transport properties are highly dependet on their size and the shape of their edges, which are usually labeled as zigzag (ZGNR) or armchair (AGNR) [14]. Ideal GNRs are assumed to be periodic across their length and can be identified as N-ZGNR and N-AGNR (N corresponds to the number of carbon atoms along the width), as shown in Figure 17.1(a and b) for a 12-AGNR and 5-ZGNR, respectively. Moreover, other GNR morphologies are possible, such as chevron and cove, as discussed later.

To characterize their electronic transport properties, several computational approaches have been used and results have been applied and validated against experimental implementations. For instance, tight-binding (TB) calculations without e^- spin show that ZGNRs and AGNRs with $N = 3p + 2$ (p is an integer) are metallic, while AGNRs with $N = 3p$ or $N = 3p + 1$ are semiconducting. Furthermore, the addition of e^- spin shows that ZGNRs have a non-zero bandgap, inversely proportional to their width due to energy splitting from spin-polarized edge states [22]. Furthermore, *ab initio* calculations show that all AGNR families are semiconducting, with the $N = 3p + 2$ family having the smallest bandgap among all armchair families [1].

When contrasting these results with experiments, different measurements on very narrow GNRs (1.4–20 nm) show they have semiconducting properties [23–25]. For instance, 7-nm-wide ZGNRs and narrower have antiferromagnetic (semiconducting) behavior, with bandgaps ranging from 200 to 300 meV, whereas wider ZGNRs are ferromagnetic (metallic), with parallel-aligned spin states at both edges. Additionally, GNR structures with predominant zigzag edges exhibit smaller bandgap than structures with predominant armchair edges [26], although there is an inverse relation between the GNR's width and its bandgap, independent of crystallographic direction [23].

Unfortunately, bandgap engineering in GNRs results in reduced carrier mobilities, similar to conventional semiconductors [27]. Very narrow ribbons, with >0.5 eV bandgaps, would be necessary for obtaining high I_{on}/I_{off} ratios, around 10^4-10^7; however, such structures can have carrier mobilities reduced by two orders of magnitude and more [27, 28]. Although this could be a major drawback in integrated circuit applications, various approaches have been explored for improving

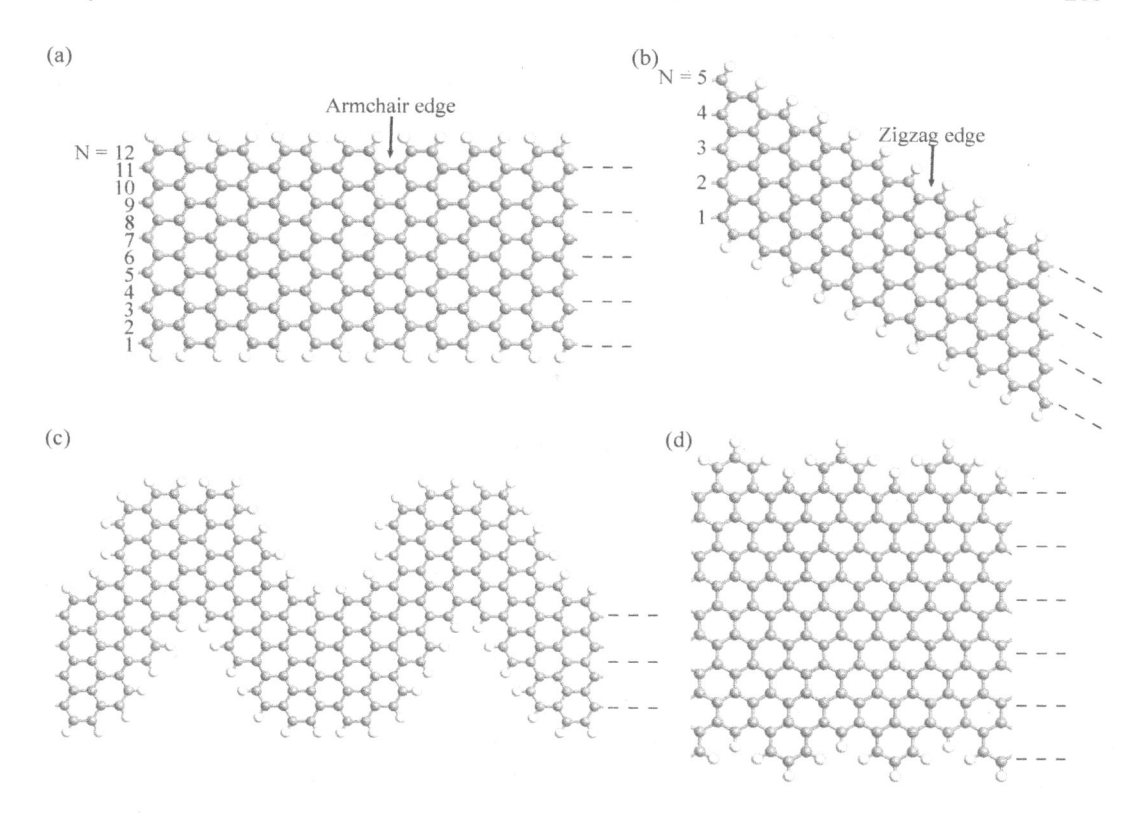

FIGURE 17.1 Atomic structure of GNR types. (a) 12-AGNR, (b) 5-ZGNR, (c) chevron-type GNR, and (d) cove-type GNR.

the mobility and overall performance of GNR-based transistor devices, such as uniform and substitutional doping [29].

This chapter provides an overview and summary of recent experimental and theoretical advances on GNR-based devices for applications in electronic circuits and sensors. Section 17.2 discusses recent fabrication techniques, defects, and structural modifications. Section 17.3 focuses on the performance of devices and that have been explored for achieving improved performance of devices and Section 17.4 discusses advances on electronic circuits and sensing applications. Persisting challenges are presented in Section 17.5 and concluding remarks in Section 17.6.

17.2 GRAPHENE NANORIBBON FABRICATION

Fabrication of materials and devices in the nanoscale are divided in top-down or bottom-up techniques. Conventional microelectronics employs top-down approaches for depositing and etching materials to pattern devices and wires onto desired substrates with very high reproducibility and throughput. However, these techniques usually lack atomic-scale precision, making it very difficult to obtain materials and devices with highly reproducible features. On the other hand, bottom-up approaches use self-assembling capabilities of particular molecules for spontaneous or directed assembly of different nanostructures.

17.2.1 TOP-DOWN VERSUS BOTTOM-UP TECHNIQUES

Several approaches have been used for the fabrication of GNRs. Table 17.1 summarizes representative and most recent top-down and bottom-up techniques for the fabrication of GNRs, with their

TABLE 17.1

Room-Temperature GNR Characteristics of Selected Fabrication Techniques I_{on}/I_{off}

Technique	Advantages	Disadvantages	Min. Width (nm)	Carrier Mobility ($cm^2V^{-1}s^{-1}$)	Bandgap (meV)	I_{on}/I_{off}	Reference
Exfoliation and chemical treatment	- Nearly pristine GNRs - Smooth edges	- Non-uniform shapes	<10	~100–200	>300	Up to 10^7	[24]
Metal-assisted etching	- Aligned GNRs - Large-scale production	- Large and non-uniform width	19	N/A	~100	~7–10 up to 5000	[25]
CVD and block copolymer lithography	- Possible integration to existing technologies	- Low bandgap opening and switching capabilities	9–12	~70–120	58–78 (100K)	~5–70	[31]
Flattened CNTs	- Smooth and straight edges	- Requires improved yield	1.4	2443	494	10^4	[32]
Etching of aligned MWCNTs	- Large aspect ratio ~90 - Ohmic junctions	- GNRs are double- or triple-layer - Low on-state current	~80	350 (holes)	N/A	27	[33]
Focused ion beam etching	- Smooth edge ribbons	- Variations in edge structures	200	371.6	0	up to 10^3	[34]
Solid-liquid-solid-guided growth from silicon nanowires	- Programmable geometries with long length	- Requires substrate transfer	~50	~600	N/A	~270	[35]
Convergent electron beam nanosculpting	- Devices with zigzag and mixtures of zigzag and armchair edges	- Kinked ribbons	1.2–3.7	N/A	100–930	N/A	[36]
Cyclodehydrogenation of precursor monomers	- Well-defined edges - Atomically precise width	- Limited to substrates compatible with precursors	0.74	N/A	1600	N/A	[37]
On-surface synthesis of GNR heterojunctions	- Deterministic growth and formation of heterojunctions	- Large bandgap for switching applications	3	N/A	2100 and 2450	N/A	[38]
2D arrays of self-assembled GNRs	- Narrow GNRs with well-defined armchair or cove-type edges	- Hybrid structures with metallic and semiconducting behavior	<7.34	~10^{-5} –0.42	200–2870	1.6–22.4	[39]

Notes: I_{ON}/I_{OFF} was measured on top-, back- or side-gated FET devices, according to the cited references.

corresponding performance metrics, advantages, disadvantages, and measured I_{on}/I_{off} ratios on top-gated or back-gated FET devices.

Generally, manufacturing techniques need to be as simple and cost-effective as possible; however, there are significant challenges in the large-scale implementation of laboratory procedures (prototypes) and in the elaboration of higher quality-grade samples (high crystallinity and ultralow defect density) with high aspect ratios (length/width) for continued progress in the field [2, 8, 30].

17.2.1.1 Top-Down Approaches

Chemical vapor deposition (CVD) and epitaxy techniques have been employed for wafer-scale production of graphene on different substrates [40] However, the biggest challenge consists in patterning GNRs with atomic precision for obtaining devices with reproducible features under ambient conditions [41]. Nonetheless, several researchers have proved the possibility of patterning room-temperature stable, narrow GNRs (10 nm and below) [23–25, 41, 42]. Alternative techniques have been explored, although their integration to conventional semiconductor processes for large-scale manufacturing has been difficult to realize, such as unzipping multiwalled carbon nanotubes (MWCNTs) [9] and direct laser writing [4], among others.

Fab-compatible processes are preferred for integration with current semiconductor technologies. Then, integration of conventional processes, such as CVD and thermal annealing, are expected to allow commercial implementations [31]. This could be further exploited by combining bottom-up strategies, as described next.

17.2.1.2 Bottom-Up Approaches

Bottom-up approaches have gained recent attention due to the possibility of producing atomically precise GNR morphologies with tuned electronic properties by selecting appropriate precursor molecules. For instance, several approaches have built on the work from Cai et al. for obtaining defect-free AGNRs on Au(111) and Ag(111) surfaces from the 10,10′-dibromo-9,9′-bianthryl precursor monomer molecule [37]. Widths as low as 0.74 nm and bandgaps that range from 0.5 to 2.6 eV can be obtained using this precursor molecule [37, 43], polyphenylene [30], perylene precursors [44], and combinations of phenylene, naphthalene, and anthracene, and cyclodehydrogenation with iron trichloride ($FeCl_3$) precursors [45], among many others.

Surface-assisted and solution-based synthetic approaches have been used for producing chevron-type GNRs [38, 46] (see Figure 17.1(c)), with the possibility of designing atomically precise intrinsic/extrinsic heterojunctions from single precursors and further chemical treatment, such as sulfur dopant atoms for tuning bandgap energies (from ~2 eV down to 0.8 eV) [47]. Moreover, phenols and acenes have been explored for obtaining cove-type GNRs and 2D arrays [39] (see Figure 17.1(d)).

Bottom-up synthesis of ZGNRs has also been reported, such as 6-ZGNRs on Au(111) substrates, although system stability is still difficult to obtain [5]. Inclusion of zigzag extensions to 7-AGNRs has been used for controlling the magnetic properties of GNR-based systems that may be useful for spin-related applications [48].

17.2.2 FABRICATION DEFECTS

Precise control over the structure and minimization of defects are fundamental for avoiding undesirable alterations to the electronic properties of GNRs. Nonetheless, the particularity of each application dictates its own quality needs and associated costs for industrial production [2, 8], which may also include strategies for overcoming manufacturing defects and enhancing the material's physical properties [6]. In general, structural disorders and fabrication defects on graphene are identified as follows [11]:

1. Surface ripples from substrate imperfections or thermal fluctuations,
2. Topological lattice defects, including single and double vacancies, Stone-Wales (SW) defects, and 555-777 defects,
3. Impurity states due to undesired adatoms,
4. Trapped charges in the surface or substrate, and
5. Edge variations.

Up to now, no top-down technique has been able to comply with all the requirements for obtaining defect-free GNRs. Bottom-up techniques have shown the possibility of obtaining atomically precise structures with well-defined edges. However, most manufacturing processes employ metal-surface-assisted chemical reactions that require substrate transferring for final applications and may result in surface defects and difficult large-scale implementations. In this sense, on-surface synthesis techniques from monomer precursors are the most promising alternatives for controllably tuning the electronic properties of GNRs, particularly, when directly grown on insulating substrates. For instance, a recent effort from Kolmer et al. shows the fabrication of well-defined AGNRs on semiconducting metal oxide surfaces (TiO_2), allowing direct applications on electronic devices and beyond [49].

17.2.2.1 Edge Variations

Variations in fabrication processes, particularly top-down approaches, result in structures with unidentical dimensions, edge definition, and symmetry, producing significant differences in the electronic properties of devices due to carrier scattering [14]. One of the most critical issues in producing high-quality GNRs is edge uniformity. Edge roughness, caused by the absence of carbon atoms in the edge, significantly decreases carrier mobility (especially in very narrow GNRs) [50] and conductivity (through variable bandgap across ribbons) [51], and deteriorates the I_{on}/I_{off} ratio and frequency response of GNR-FETs [28].

Therefore, electronic device applications would require atomically precise edge structures to obtain reproducible features. Bottom-up fabrication techniques are capable of guaranteeing precise edge definition by tuning specific monomer precursors to the desired application. Nonetheless, practical implementations still require additional efforts for avoiding the more expensive (and difficult to integrate in the large scale) synthesis on metal substrates.

17.2.2.2 Surface Defects

Growing graphene sheets on a substrate and transferring them to another substrate usually results in undesired surface roughness due to impurities in the interface, affecting the strength of C-C bonds within the structure and significantly decreasing current flow and performance of devices. Although GNR bandgap is unaffected [52], carrier mobility and conductivity of devices are drastically reduced, particularly in longer and narrower GNRs [50]. Therefore, direct synthesis of GNRs on clean metal−oxide surfaces would be preferable for minimizing resulting roughness from substrate transferring [49].

17.2.2.3 Topological Lattice Defects

Although graphene's lattice is usually well ordered and its surface is relatively inert, oxygen and nitrogen molecules could be adsorbed under ambient conditions, making it necessary to employ relatively high annealing temperatures of around 800°C [53]. Different types of vacancies have been observed within graphene structures. Some manufacturing processes can produce single-vacancy (SV) and double-vacancy (DV) defects. Computational results indicate that FETs based on AGNRs with vacancies have reduced current flow capabilities and very low I_{on}/I_{off} ratios [54]. Therefore, most of graphene manufacturing approaches aim to avoid or minimize these defects by increasing the growth temperature, although their intentional use could be employed in sensor applications by providing specific sites for adsorbing target molecules [53].

Other topological lattice defects that have been observed include SW deformations, which occur when a C-C bond rotates 90 degrees in the octagon of divacant graphene and forms two heptagonal and two pentagonal carbon bonds, instead of four hexagonal carbon bonds, which also result in reduced electronic transport capabilities [55]. Similarly, the the 555-777 defect results as the C-C bond in the octagon of divacant graphene rotates 90 degrees, forming three pentagons and three heptagons, but also affecting the ribbon's electronic properties, according to its location and symmetry within the structure [55].

17.2.3 Chemically Modified Graphene Nanoribbons

Alternatively, chemical modifications have been explored for bandgap engineering of graphene. For instance, hydrogenation and fluorination of graphene sheets alter the sp^2 hybridization to sp^3, allowing a bandgap opening of a few electronvoltes without reducing current flow capabilities [20], as long as large carrier mobilities and low scattering defects can be maintained [2].

Alternative approaches include chemical doping with heteroatoms, such as boron and nitrogen, replacing carbon atoms in the backbone of graphene for obtaining extrinsic semiconducting behavior, such as n-type GNRs, that could be useful for complementary logic devices [56].

17.2.4 Graphene Nanomeshes

As manufacturing of atomically precise GNRs becomes necessary for applications in electronics, difficulties in process integration, repeatability, and cost-effectiveness have paved the way to alternatives in bandgap engineering of graphene. Starting from large-area graphene, GNMs have been explored by patterning periodic nanopores with different shapes and sizes that form some type of GNR network. These nanopores can be circular [16, 17, 57, 58], square [18], triangular [59, 60], rhomboid [59], hexagonal [60], or irregular [58], which also determines the structure's electronic properties by controlling the neck width and periodicity (edge-to-edge and center-to-center distance between neighboring nanopores, respectively) [16]. GNM fabrication techniques include:

1. Electron beam lithography and reactive ion etching [57],
2. Block copolymer lithography and O_2 plasma etching [16],
3. Nanoimprint lithography and O_2 plasma etching [17],
4. Nanosphere lithography [61],
5. Bottom-up assembly of chevron-type nanopores [62], and
6. Helium ion beam milling [63].

17.3 GRAPHENE NANORIBBON DEVICES

Most GNR-based device implementations have been realized on top- or back-gated FET devices, as presented in Table 17.1. Figure 17.2 shows schematic representations of these devices in which the channel consists of single GNR structures, GNR arrays, or GNMs. Although back-gated FETs are typically used for proof-of-concept purposes rather than practical implementations of integrated circuits [3], they are particularly useful in sensing applications in which the channel can be functionalized with specific ligands for improved selectivity to target analytes.

In addition to FET devices, different device architectures have been proposed for exploiting tunneling effects, ambipolar behavior, and negative resistance properties. These devices and their calculated performance metrics are summarized in Table 17.2.

17.3.1 Field-Effect Devices

Typical top-, back- or double-gated FET devices use an intrinsic channel with heavily doped or metallic source and drain contacts that could be further connected to large-area graphene for lowering contact resistance. As seen in Figure 17.2, GNRs with different topologies have been explored, including GNR arrays that simplify manufacturing processes and improve current flow capabilities. The typical gate insulator is SiO_2; however, high-k dielectric materials have been explored for reducing undesirable leakage currents in the gate terminal.

Novel device architectures have also been explored to achieve low-power requirements, comparable to silicon-based CMOS technologies. These include additional gate control for bandgap

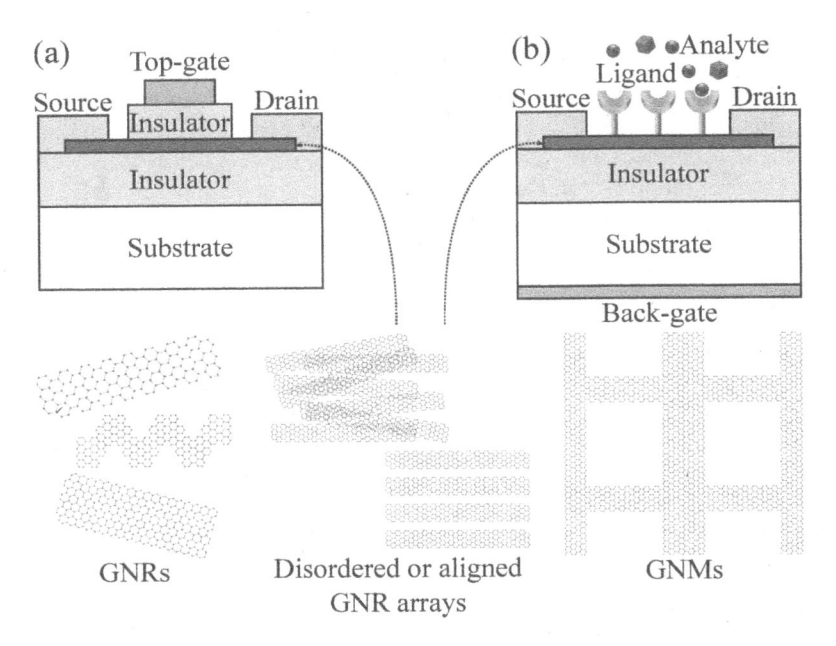

FIGURE 17.2 GNR-based devices. (a) Top-gated FET device for electronic circuit applications. (b) Back-gated FET with ligand-functionalized channel for sensing applications. The red block represents the device channel, which can be formed by single GNRs, GNR arrays, or GNMs.

modulation [42, 64], homojunctions and heterojunctions [21, 65, 68], chemical treatment [69], and multiterminal devices [70], among others that, combined with field-effect operation, allow improved I_{on}/I_{off} ratios, subthreshold swing, and sensing capabilities. For instance, hybrid graphene/hexagonal boron nitride (hBN) nanorribbons have been predicted to allow FET devices with I_{on}/I_{off} ratios up to 1.42×10^4 [21]. Furthermore, graphene structures can be exploited as on-chip interconnect material for ohmic contacts [34, 35], where short, near-metallic wires (5 nm-long, 5-AGNR with ~0.1 eV bandgap) have been demonstrated [71].

17.3.2 TUNNELING-EFFECT DEVICES

Quantum tunneling has also been explored for improving the performance of traditional FET devices and overcoming short-channel difficulties. Tunneling field-effect transistors (TFETs) use

TABLE 17.2
Simulation Performance of GNR-Based Devices

Device	Subthreshold Slope (mV/dec)	I_{on}/I_{off}	Reference
GNR-based negative differential resistance device	N/A	$>10^5$	[64]
Vertical graphene-hexagonal boron nitride FET	170	1.8×10^4	[65]
Bandgap modulation with transverse field	37	$\sim 10^3$	[42]
Homojunction tunnel FET	60	10^5	[66]
Negative differential resistance on chevron-type GNRs	38	2.1×10^6	[67]

band-to-band tunneling to achieve ON/OFF states. Then, a typical TFET is constructed with n-type source/drain, intrinsic channel, and a p-type drain/source, although such a doping profile could be obtained through additional gate control that modifies the concentration of major carriers in different regions along the transport direction [72]. This allows and suppresses electron tunneling, allowing I_{on}/I_{off} ratios above 10^5 (due to very low OFF-state currents) and very steep subthreshold characteristics [66].

Theoretical results show that bilayer GNRs allow higher ON-state current than single-layer GNR devices with $I_{on}/I_{off} = 8.8 \times 10^3$ and fast switching speed [73]. Tunneling barriers are also possible in homojunctions and heterojunctions. For instance, short small gap ($3p+2$) and large gap ($3p$ and $3p+1$) AGNR junctions have been used as quantum wells and potential barriers, respectively, to improve the performance of GNR-based TFET devices [74]. Moreover, graphene/hBN structures have been used for designing high-performance TFET devices, comparable to modern devices, where 1.8×10^4 I_{on}/I_{off} ratio and 170 mV/dec subthreshold swing are estimated [65].

17.3.3 SPINTRONIC DEVICES

By adding e^- spin, it is possible to obtain very low power circuits due to the non-dissipative nature of coherent spin rotation, although high control of spin and long spin lifetimes are simultaneously required [2]. Spin-$\uparrow$ and spin-$\downarrow$ electrons have equal propagation along the same direction in conventional electronics, which results in a net charge current $I = I^\uparrow + I^\downarrow \neq 0$, that is, spin unpolarized $I^S = \frac{\hbar}{2q}\left(I^\uparrow - I^\downarrow\right) = 0$. Nonetheless, opposite spin electrons can be polarized to obtain unequal spin-$\uparrow$ and spin-$\downarrow$ currents, resulting in $I \neq 0$ and $I^S \neq 0$, where the limit is a pure spin current once spin-$\uparrow$ and spin-$\downarrow$ electrons are propagated equally in the opposite direction, so that $I = 0$ and $I^S \neq 0$ [75]. Research in spintronic devices aims to find effective mechanisms for generating, manipulating and detecting spin currents for all-spin devices in traditional electronics and quantum information processing [2].

One of the most popular methods to generate pure spin currents is the spin Hall effect (SHE). This is possible in materials that have very strong coupling between the orbital and the spin degree of freedom of electrons, known as spin-orbit coupling (SOC). As observed in Figure 17.3(a),

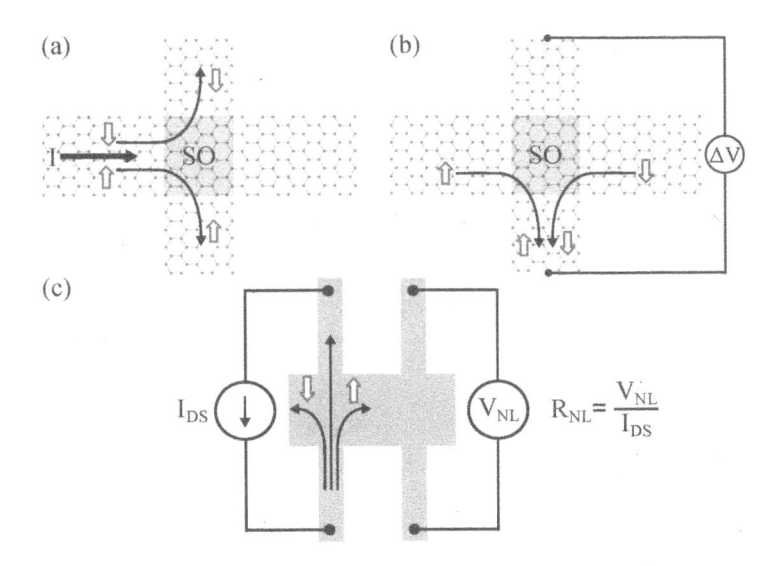

FIGURE 17.3 (a) Spin current generation due to SHE phenomenology. (b) Spin current detection due to iSHE phenomenology. (c) Non-local spin Hall measurement.

spin-orbit (SO) interactions within a sample carrying a longitudinal charge current produce a transverse separation of spin-$\uparrow$ and spin-$\downarrow$ electrons. The inverse SHE (iSHE) occurs on a sample with SO interactions carrying longitudinal spin current that generates transverse charge current (see Figure 17.3(b)); therefore, this is one of the preferred methods to detect spin currents. Both effects are combined in the device shown in Figure 17.3(c), where spin current is generated in the central region by driving a charge current (I_{DS}) between the left-side terminals due to SHE. This spin current is measured by the right side terminals due to iSHE, which is known as the non-local voltage (V_{NL}). The non-local resistance ($R_{NL} = V_{NL}/I_{DS}$) characterizes this type of device, which may use different materials and heterostructures for improved effectiveness in converting charge to spin current and vice versa.

Graphene allows SO interaction that results in SHE at low energy, enabling both charge and spin currents, in gapless edge states [76]. Although spatial confinement is useful for improving SOC effects [77], SO interaction is very low in pristine graphene. Because of this, researchers have looked for methods to enhance the SO coupling in graphene, such as the addition of small concentrations of covalently bonded hydrogen atoms to graphene's surface [78], nickel, and gold, where very large SHE has been observed [79, 80]. Furthermore, heterostructures using graphene/hBN [81, 82] and graphene/transition metal dichalcogenide (TMDC) [83] devices have shown the effectiveness of multiterminal systems for obtaining high R_{NL} differences, due to Zeeman SHE, proximity-induced Rashba-Edelstein effects, or the formation of valley Hall topological currents [84], which could be employed for transistor devices with very high I_{on}/I_{off} ratios and low power consumption.

17.4 GRAPHENE NANORIBBON CIRCUITS AND BIOSENSORS

Several GNR-based devices that were discussed in Section 17.3 have been proposed or manufactured for particular applications in electronic circuits and sensors. In this section, we summarize some of the most significant advances in this area.

17.4.1 ELECTRONIC CIRCUITS

As stated in the International Technology Roadmap for Semiconductors (ITRS), fundamental limits in scalling of CMOS technology have motivated the search for alternative materials, operating principles, and device architectures that may be integrated into current digital systems for added functionality and improved device capabilities. For this purpose, short-gate length GNR-FETs are expected to have better performance than silicon MOSFETs due to less dramatic short-channel effects, such as direct source-drain tunneling [3].

Conventional architectures for logic circuits require non-zero bandgap materials (preferably ~1 eV, similar to Si, for reliable switching), without compromising carrier mobilities. This would be essential for achieving high ON-state currents and large I_{on}/I_{off} ratios (>10^6), for which voltage gain $A_v > 10$ is preferable [2]. Experimental approaches have employed large-area graphene (0.25–20 μm) for implementing inverter and other logic gate circuits [85] that have shown high-power dissipation and different input/output voltage levels.

To obtain similar functionalities without sacrificing performance, atomically precise GNR structures would be mandatory, although other proposed bandgap strategies or device architectures still need experimental validation in low-power, large-scale, cascading logic circuits. Theoretical calculations of all-graphene circuits have shown very promising performance metrics; however, most of the proposed architectures rely on perfectly smooth and clean GNR structures that are still challenging to produce on a large scale. Nonetheless, these efforts are expected to materialize as manufacturing techniques advance. This may include the implementation of heterostructues with nanomesh patterns for controllable bandgap engineering and improved ON-state currents and I_{on}/I_{off} ratios [63].

17.4.2 BIOSENSORS

Biological conditions, health states on organisms, and environmental hazardous substances can be identified through the safe, fast, and reliable detection of biomarkers by means of label-free sensors. Particularly, the absence of chemical labels (fluorescent or radioactive groups) helps to simplify experiments, improve real-time observations and cost-effectiveness, and minimize detection errors [68, 70]. This is particularly useful for monitoring normal biological processes, detecting pathogenic processes, or scanning pharmacological responses, among others. Furthermore, the Internet of Things (IoT) paradigm envisions the interconnection of sensor networks for real-time transferring of information between devices and end users, allowing low-latency diagnostics applications. For this purpose, graphene-based devices have already exhibited promising overall performance and are rapidly becoming attractive alternatives for electrochemical sensors and biosensors [9].

Other applications include pharmaceutical and biomedical research, monitoring of food quality control, plant growth monitoring, and many others, where very expensive and slow-response laboratory instrumentation is usually employed. For instance, low-cost and high-resolution biosensors in agricultural applications represent an essential tool for assessing crop response to different stresses by measuring primary and secondary metabolites, organic acids, and toxins, among others. Drug delivery and gene therapy have also been explored, although caution is needed because the uptake and multiple exposure to graphene-based nanomaterials have proved to cause toxicity to human body cells by means of DNA damage, apoptosis, autophagy, or necrosis [10].

As depicted in Figure 17.2(b), surface functionalization is used for improving their selectivity capabilities, where the graphene's aromatic structure is useful for non-covalent self-assembling of pyrene-based moieties [9, 68, 70], and nanoparticle surface dispersion [86, 87] for improved sensing capabilities. This is further exemplified in Figure 17.4, where the channel of an AGNR-FET device is functionalized with 1-pyrenebutyric acid (PyBA) linkers through $\pi - \pi$ interactions on the pyrene end, and covalently bind UDP-glucose molecules (an intermediate reactant in the synthesis of sucrose in a plant cell's cytoplasm) on the carboxylic end, showing differentiable electronic signature, directly related to the concentration of UDP-glucose molecules on the device's surface [68, 70].

GNRs and reduced graphene oxide nanoribbons (GONRs) have been explored for attaching specific antibodies and detecting folic acid, methyl parathion, morphine, diclofenac, lymphoma cancer cells, and other target molecules, when used in modified screen printed electrodes (SPEs) [88, 89]. GNMs and porous graphene materials have also been used for sensing applications, such as electrochemical immunosensors [86] and DNA sequencing [90], in which conductance differences are used to determine different nucleobases located inside the nanopore.

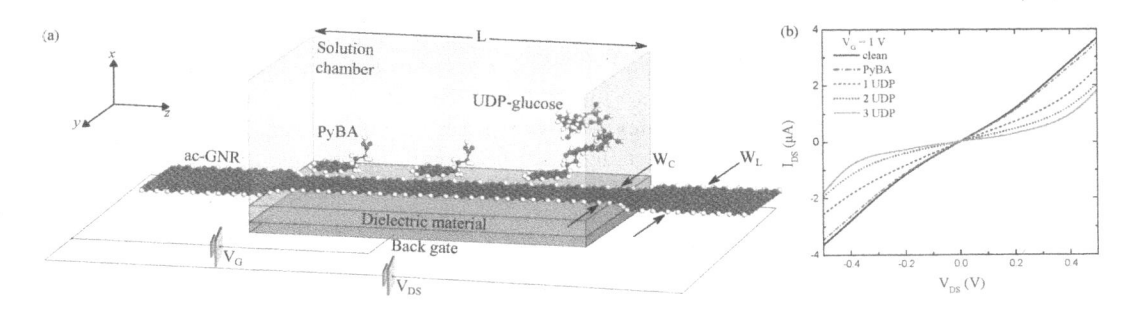

FIGURE 17.4 (a) All-AGNR-FET sensor. The surface is functionalized with PyBA molecules for the selective bounding of UDP-glucose molecules. (b) IV characteristics with zero (red dot-dashed line), one (dotted blue line), two (dotted green line), and three (solid orange line) UDP-glucose molecules compared with the bare device (solid black line). (Reproduced from [68].)

17.5 PERSISTING CHALLENGES

GNR fabrication techniques have been constantly progressing during the last few years, showing some improvements in controllably obtaining defect-free materials. For instance, bottom-up techniques using monomer precursors are useful for precisely producing very narrow structures with tunable bandgaps that may allow applications in switching devices, but at the expense of reduced carrier mobilities. None of the methodologies thus far are still suitable for large-scale production in an industrial environment [3]; however, recent advances on direct synthesis of GNRs on insulating surfaces could significantly boost electronic device applications [49].

Novel circuit designs employing different physical operational principles have been proposed and some proof-of-concept implementations have been realized; however, industrial applications still require some years of development for reaching sufficient maturity. Accordingly, it seems unlikely that GNR-based devices will replace CMOS circuits in the near future; however, there is a big potential for large-scale implementations of highly sensitive sensor devices that could enable applications with real-time response in a wide variety of fields. Nonetheless, technological integration and power supply remain challenging. Furthermore, toxicity concerns need to be addressed for avoiding harmful interactions with living organisms and the environment.

The reliable operation of electronic devices needs careful consideration of thermal dissipation, particularly when packaging millions of devices into single chips for everyday computing applications. GNR-based spintronic devices allow low-power consumption due to the non-dissipative nature of coherent spin rotation. Theoretical and experimental works have shown significant progress in generating, manipulating, and detecting spin currents. Furthermore, the recent advances and commercial availability of non-volatile spin-transfer torque magnetic random-access memory (STT-RAM) devices show the viability of massive device integration for spin-based graphene circuits.

17.6 CONCLUDING REMARKS

This chapter overviews the progress in GNR fabrication and recent advances in device modeling and implementation for electronic applications. Although computational results show the suitability of GNR-based devices for digital circuit design, these applications require very high purity and low-defect density structures. This is expected to be achieved through progressive improvements in manufacturing techniques and the development of novel designs that are more resilient to defects, while providing consistent performance. Top-down and bottom-up manufacturing approaches offer high throughput and precision, respectively. In this regard, hybrid processes that include conventional fab-compatible steps and enable scalable, reliable, and cost-effective production of devices with highly reproducible features are more promising for practical applications.

In addition to the excellent electronic properties of GNRs, their large surface area and inertness enable sensing applications with very low detection limits. Nonetheless, applications are restricted to off-body devices due to toxicity issues of intake and continued exposure to graphene-based materials. Furthermore, GNRs can be further exploited as low-resistance on-chip interconnects and the possible realization of all-graphene devices with superior performance.

The design and verification of novel devices that operate under different physical principles beyond typical FET, such as tunneling effect and spintronics, will significantly boost the realization of GNR-based devices on a large scale. As more experiments on single chip integration become available, slow integration of graphene into industrial implementations, rather than full replacement of silicon-based technologies, seems more realistic in the near future.

ACKNOWLEDGMENTS

J.M.M-T acknowledges support from the World Bank; the Colombian Ministry of Science, Technology and Innovation; the Colombian Ministry of Education; the Colombian Ministry of Industry and Tourism; and ICETEX within the Colombian Scientific Ecosystem program OMICAS:

Optimización Multiescala In-silico de Cultivos Agrícolas Sostenibles (Infraestructura y validación en Arroz y Caña de Azúcar). This is anchored at the Pontificia Universidad Javeriana in Cali and funded under grant FP44842-217-2018 and OMICAS Award ID: 792-61187.

REFERENCES

1. Sudipta Dutta, and Swapan K. Pati. Novel properties of graphene nanoribbons: A review. *Journal of Materials Chemistry*, 20:8207–8223, 2010.
2. Andrea C. Ferrari, Francesco Bonaccorso, Vladimir Fal'ko, Konstantin S. Novoselov, Stephan Roche, Peter Boggild, Stefano Borini, Frank H. L. Koppens, Vincenzo Palermo, Nicola Pugno, Jose A. Garrido, Roman Sordan, Alberto Bianco, Laura Ballerini, Maurizio Prato, Elefterios Lidorikis, Jani Kivioja, Claudio Marinelli, Tapani Ryhanen, Alberto Morpurgo, Jonathan N. Coleman, Valeria Nicolosi, Luigi Colombo, Albert Fert, Mar Garcia-Hernandez, Adrian Bachtold, Gregory F. Schneider, Francisco Guinea, Cees Dekker, Matteo Barbone, Zhipei Sun, Costas Galiotis, Alexander N. Grigorenko, Gerasimos Konstantatos, Andras Kis, Mikhail Katsnelson, Lieven Vandersypen, Annick Loiseau, Vittorio Morandi, Daniel Neumaier, Emanuele Treossi, Vittorio Pellegrini, Marco Polini, Alessandro Tredicucci, Gareth M. Williams, Byung Hee Hong, Jong-Hyun Ahn, Jong Min Kim, Herbert Zirath, Bart J. van Wees, Herre van der Zant, Luigi Occhipinti, Andrea Di Matteo, Ian A. Kinloch, Thomas Seyller, Etienne Quesnel, Xinliang Feng, Ken Teo, Nalin Rupesinghe, Pertti Hakonen, Simon R. T. Neil, Quentin Tannock, Tomas Lofwander, and Jari Kinaret. Science and technology roadmap for graphene, related two-dimensional crystals, and hybrid systems. *Nanoscale*, 7:4598–4810, 2015.
3. Zhansong Geng, Bernd Hähnlein, Ralf Granzner, Manuel Auge, Alexander A. Lebedev, Valery Y. Davydov, Mario Kittler, Jorg Pezoldt, and Frank Schwierz. Graphene nanoribbons for electronic devices. *Annalen der Physik*, 529(11):1700033, 2017.
4. Narendra Kurra, Qiu Jiang, Pranati Nayak, and Husam N. Alshareef. Laser-derived graphene: A three-dimensional printed graphene electrode and its emerging applications. *Nano Today*, 24:81–102, 2019.
5. Junzhi Liu, and Xinliang Feng. Synthetic tailoring of graphene nanostructures with zigzag-edged topologies: Progress and perspectives. *Angewandte Chemie*, 59(52):23386–23401, 2020.
6. Lili Liu, Miaoqing Qing, Yibo Wang, and Shimou Chen. Defects in graphene: Generation, healing, and their effects on the properties of graphene: A review. *Journal of Materials Science & Technology*, 31(6):599–606, 2015.
7. Juan M. Marmolejo-Tejada, and Jaime Velasco-Medina. Review on graphene nanoribbon devices for logic applications. *Microelectronics Journal*, 48:18–38, 2016.
8. K.S. Novoselov, V. I. Fal'ko, L. Colombo, P.R. Gellert, M.G. Schwab, and K. Kim. A roadmap for graphene. *Nature*, 490(7419):192–200, Oct 2012.
9. Umamaheswari Rajaji, Rameshkumar Arumugam, Shen-Ming Chen, Tse-Wei Chen, Tien-Wen Tseng, Sathishkumar Chinnapaiyan, Shih-Yi Lee, and Wen-Han Chang. Graphene nanoribbons in electrochemical sensors and biosensors: a review. *International Journal of Electrochemical Science*, 13:6643–6654, 2018.
10. Pravin Shende, Steffi Augustine, and Bala Prabhakar. A review on graphene nanoribbons for advanced biomedical applications. *Carbon Letters*, 30:465–475. Feb 2020.
11. D.S.L. Abergel, V. Apalkov, J. Berashevich, K. Ziegler, and Tapash Chakraborty. Properties of graphene: A theoretical perspective. *Advances in Physics*, 59(4):261–482, 2010.
12. Haomin Wang, Hui Shan Wang, Chuanxu Ma, Lingxiu Chen, Chengxin Jiang, Chen, Xiaoming Xie, An-Ping Li, and Xinran Wang. Graphene nanoribbons for quantum electronics. *Nature Reviews Physics*, 3:791–802, Sep 2021.
13. A. Geim. Graphene update. *Bulletin of the American Physical Society*, 55(2), 2010.
14. Kyoko Nakada, Mitsutaka Fujita, Gene Dresselhaus, and Mildred S. Dresselhaus. Edge state in graphene ribbons: Nanometer size effect and edge shape dependence. *Physics Review B*, 54:17954–17961, Dec 1996.
15. Jeroen B. Oostinga, Hubert B. Heersche, Xinglan Liu, Alberto F. Morpurgo, and Lieven M.K. Vandersypen. Gate-induced insulating state in bilayer graphene devices. *Nature Materials*, 7(2):151–157, Feb 2008.
16. Jingwei Bai, Xing Zhong, Shan Jiang, Yu Huang, and Xiangfeng Duan. Graphene nanomesh. *Nature Nanotechnology*, 5(3):190–194, Mar 2010.
17. Xiaogan Liang, Yeon-Sik Jung, Shiwei Wu, Ariel Ismach, Deirdre L. Olynick, Stefano Cabrini, and Jeffrey Bokor. Formation of bandgap and subbands in graphene nanomeshes with sub-10 nm ribbon width fabricated via nanoimprint lithography. *Nano Letters*, 10(7):2454–2460, 2010.

18. Ryutaro Sako, Naomi Hasegawa, Hideaki Tsuchiya, and Matsuto Ogawa. Computational study on band structure engineering using graphene nanomeshes. *Journal of Applied Physics*, 113(14):143702, 2013.

19. Zhen Hua Ni, Ting Yu, Yun Hao Lu, Ying Wang, Yuan Ping Feng, and Ze Xiang Shen. Uniaxial strain on graphene: Raman spectroscopy study and band-gap opening. *ACS Nano*, 2(11):2301–2305, 2008.

20. Gyungseon Seol, and Jing Guo. Assessment of graphene nanomesh and nanoroad transistors by chemical modification. In *2011 IEEE International Electron Devices Meeting (IEDM)*, pages 2.3.1–2.3.4, 2011.

21. Van-Truong Tran, J. Saint-Martin, and P. Dollfus. Modulation of bandgap and current in graphene/bn heterostructures by tuning the transverse electric field. In *2014 International Workshop on Computational Electronics (IWCE)*, pages 1–2, June 2014.

22. Young-Woo Son, Marvin L. Cohen, and Steven G. Louie. Half-metallic graphene nanoribbons. *Nature*, 444(7117):347–349, Nov 2006.

23. Melinda Y. Han, Barbaros Özyilmaz, Yuanbo Zhang, and Philip Kim. Energy band-gap engineering of graphene nanoribbons. *Physical Review Letters*, 98:206805, May 2007.

24. Xiaolin Li, Xinran Wang, Li Zhang, Sangwon Lee, and Hongjie Dai. Chemically derived, ultrasmooth graphene nanoribbon semiconductors. *Science*, 319(5867):1229–1232, 2008.

25. Pablo Solis-Fernandez, Kazuma Yoshida, Yui Ogawa, Masaharu Tsuji, and Hiroki Ago. Dense arrays of highly aligned graphene nanoribbons produced by substrate-controlled metal-assisted etching of graphene. *Advanced Materials*, 25(45):6562–6568, 2013.

26. Kyle A. Ritter, and Joseph W. Lyding. The influence of edge structure on the electronic properties of graphene quantum dots and nanoribbons. *Nature Materials*, 8(3):235–242, Mar 2009.

27. Frank Schwierz. Graphene transistors. *Nature Nanotechnology*, 5(7):487–496, Jul 2010.

28. F. Schwierz. Graphene transistors: status, prospects, and problems. *Proceedings of the IEEE*, 101(7):1567–1584, July 2013.

29. N.D. Akhavan, G. Jolley, G.U. Membreno, J. Antoszewski, and L. Faraone. Study of uniformly doped graphene nanoribbon transistor (gnr) fet using quantum simulation. In *2012 Conference on Optoelectronic and Microelectronic Materials Devices (COMMAD)*, pages 67–68, 2012.

30. Lukas Dossel, Lileta Gherghel, Xinliang Feng, and Klaus Mullen. Graphene nanoribbons by chemists: Nanometer-sized, soluble, and defect-free. *Angewandte Chemie International Edition*, 50(11):2540–2543, 2011.

31. Javier Arias-Zapata, Jerome Daniel Garnier, Hasan-al Mehedi, Antoine Legrain, Bassem Salem, Gilles Cunge, and Marc Zelsmann. Engineering self-assembly of a high-x block copolymer for large-area fabrication of transistors based on functional graphene nanoribbon arrays. *Chemistry of Materials*, 31(9):3154–3162, 2019.

32. Changxin Chen, Yu Lin, Wu Zhou, Ming Gong, Zhuoyang He, Fangyuan Shi, Xinyue Li, Justin Zachary Wu, Kai Tak Lam, Jian Nong Wang, Fan Yang, Qiaoshi Zeng, Jing Guo, Wenpei Gao, Jian-Min Zuo, Jie Liu, Guosong Hong, Alexander L. Antaris, Meng-Chang Lin, Wendy L. Mao, and Hongjie Dai. Sub-10-nm graphene nanoribbons with atomically smooth edges from squashed carbon nanotubes. *Nature Electronics*, 4(9):653–663, Sep 2021.

33. Abhay A. Sagade, and Ameya Nyayadhish. A carbon nanotube–graphene nanoribbon seamless junction transistor. *Nanoscale Advances*, 2(2):659–663, 2020.

34. Jianping Wang, and Quan Wang. Investigating electrical properties of controllable graphene nanoribbon field effect transistors. *Physica B: Condensed Matter*, 583:412022, 2020.

35. Chuan Liu, Bing Yao, Taige Dong, Haiguang Ma, Shaobo Zhang, Junzhuan Wang, Jun Xu, Yi Shi, Kunji Chen, Libo Gao, et al. Highly stretchable graphene nanoribbon springs by programmable nanowire lithography. *npj 2D Materials and Applications*, 3(1):1–9, 2019.

36. Chunmeng Liu, Jiaqi Zhang, Manoharan Muruganathan, Hiroshi Mizuta, Yoshifumi Oshima, and Xiaobin Zhang. Origin of nonlinear current-voltage curves for suspended zigzag edge graphene nanoribbons. *Carbon*, 165:476–483, 2020.

37. Jinming Cai, Pascal Ruffieux, Rached Jaafar, Marco Bieri, Thomas Braun, Stephan Blankenburg, Matthias Muoth, Ari P. Seitsonen, Moussa Saleh, Xinliang Feng, Klaus Mullen, and Roman Fasel. Atomically precise bottom-up fabrication of graphene nanoribbons. *Nature*, 466(7305):470–473, Jul 2010.

38. Christopher Bronner, Rebecca A. Durr, Daniel J. Rizzo, Yea-Lee Lee, Tomas Marangoni, Alin Miksi Kalayjian, Henry Rodriguez, William Zhao, Steven G. Louie, Felix R. Fischer, et al. Hierarchical on-surface synthesis of graphene nanoribbon heterojunctions. *ACS Nano*, 12(3):2193–2200, 2018.

39. Takahiro Kojima, Takahiro Nakae, Zhen Xu, Chinnusamy Saravanan, Kentaro Watanabe, Yoshiaki Nakamura, and Hiroshi Sakaguchi. Bottom-up on-surface synthesis of two-dimensional graphene nanoribbon networks and their thermoelectric properties. *Chemistry: An Asian Journal*, 14(23):4400–4407, 2019.

40. C.Y. Sung. Post Si CMOS graphene nanoelectronics. In *2011 International Symposium on VLSI Technology, Systems and Applications (VLSI-TSA)*, pages 1–2, 2011.
41. A.K. Geim, and K.S. Novoselov. The rise of graphene. *Nature Materials*, 6(3):183–191, Mar 2007.
42. Lieh-Ting Tung, M.V. Mateus, and E.C. Kan. Tri-gate graphene nanoribbon transistors with transverse-field bandgap modulation. *IEEE Transactions on Electron Devices*, 61(9):3329–3334, Sep 2014.
43. C. Bronner, F. Leyssner, S. Stremlau, M. Utecht, P. Saalfrank, T. Klamroth, and P. Tegeder. Electronic structure of a subnanometer wide bottom-up fabricated graphene nanoribbon: end states, band gap, and dispersion. *Physical Review B*, 86:085444, Aug 2012.
44. Takashi Kitao, Michael W.A. MacLean, Kazuki Nakata, Masayoshi Takayanagi, Masataka Nagaoka, and Takashi Uemura. Scalable and precise synthesis of armchair-edge graphene nanoribbon in metal–organic framework. *Journal of the American Chemical Society*, 142(12):5509–5514, 2020.
45. Kyung Tae Kim, Jae Woong Jung, and Won Ho Jo. Synthesis of graphene nanoribbons with various widths and its application to thin-film transistor. *Carbon*, 63:202–209, 2013.
46. Jinming Cai, Carlo A. Pignedoli, Leopold Talirz, Pascal Ruffieux, Hajo Sode, Liangbo Liang, Vincent Meunier, Reinhard Berger, Rongjin Li, Xinliang Feng, Klaus Mullen, and Roman Fasel. Graphene nanoribbon heterojunctions. *Nature Nanotechnology*, 9(11):896–900, Nov 2014. Letter.
47. Yun Cao, Jing Qi, Yan-Fang Zhang, Li Huang, Qi Zheng, Xiao Lin, Zhihai Cheng, Yu-Yang Zhang, Xinliang Feng, Shixuan Du, Sokrates T. Pantelides, and Hong-Jun Gao. Tuning the morphology of chevron-type graphene nanoribbons by choice of annealing temperature. *Nano Research*, 11(12):6190–6196, Dec 2018.
48. Qiang Sun, Xuelin Yao, Oliver Groning, Kristjan Eimre, Carlo A Pignedoli, Klaus Müllen, Akimitsu Narita, Roman Fasel, and Pascal Ruffieux. Magnetically coupled spin states in armchair graphene nanoribbons with asymmetric zigzag edge extensions, *Nano Letters*, 20(9):6429–6436, 2020.
49. Marek Kolmer, Ann-Kristin Steiner, Irena Izydorczyk, Wonhee Ko, Mads Engelund, Marek Szymonski, An-Ping Li, and Konstantin Amsharov. Rational synthesis of atomically precise graphene nanoribbons directly on metal oxide surfaces. *Science*, 369(6503):571–575, 2020.
50. Tian Fang, Aniruddha Konar, Huili Xing, and Debdeep Jena. Mobility in semiconducting graphene nanoribbons: phonon, impurity, and edge roughness scattering. *Physics Review B*, 78:205403, Nov 2008.
51. H. Zeng, J. Zhao, and J.W. Wei. Electronic transport properties of graphene nanoribbons with anomalous edges. *The European Physical Journal Applied Physics*, 53:2, 2011.
52. M. Sanaeepur, A.Y. Goharrizi, and M.J. Sharifi. Performance analysis of graphene nanoribbon field effect transistors in the presence of surface roughness. *IEEE Transactions on Electron Devices*, 61(4):1193–1198, April 2014.
53. Paulo T. Araujo, Mauricio Terrones, and Mildred S. Dresselhaus. Defects and impurities in graphene-like materials. *Materials Today*, 15(3):98–109, 2012.
54. Sheng Chang, Yajun Zhang, Qijun Huang, Hao Wang, and Gaofeng Wang. Effects of vacancy defects on graphene nanoribbon field effect transistor. *Micro & Nano Letters*, 8(11):816–821, November 2013.
55. Gun-Do Lee, C.Z. Wang, Euijoon Yoon, Nong-Moon Hwang, Doh-Yeon Kim, and K.M. Ho. Diffusion, coalescence, and reconstruction of vacancy defects in graphene layers. *Physical Review Letters*, 95:205501, Nov 2005.
56. Kyung Tae Kim, Jong Won Lee, and Won Ho Jo. Charge-transport tuning of solution-processable graphene nanoribbons by substitutional nitrogen doping. *Macromolecular Chemistry and Physics*, 214(23):2768–2773, 2013.
57. J. Eroms, and D. Weiss. Weak localization and transport gap in graphene antidot lattices. *New Journal of Physics*, 11(9):095021, 2009.
58. Hideyuki Jippo, Mari Ohfuchi, and Chioko Kaneta. Theoretical study on electron transport properties of graphene sheets with two- and one-dimensional periodic nanoholes. *Physics Review B*, 84:075467, Aug 2011.
59. Wei Liu, Z.F. Wang, Q.W. Shi, Jinlong Yang, and Feng Liu. Band-gap scaling of graphene nanohole superlattices. *Physics Review B*, 80:233405, Dec 2009.
60. Mihajlo Vanevic, Vladimir M. Stojanovic, and Markus Kindermann. Character of electronic states in graphene antidot lattices: flat bands and spatial localization. *Physics Review B*, 80:045410, Jul 2009.
61. Alexander Sinitskii, and James M. Tour. Patterning graphene through the self-assembled templates: toward periodic two-dimensional graphene nanostructures with semiconductor properties. *Journal of the American Chemical Society*, 132(42):14730–14732, 2010.
62. Peter H. Jacobse, Ryan D. McCurdy, Jingwei Jiang, Daniel J. Rizzo, Gregory Veber, Paul Butler, Rafał Zuzak, Steven G. Louie, Felix R. Fischer, and Michael F. Crommie. Bottom-up assembly of nanoporous graphene with emergent electronic states. *Journal of the American Chemical Society*, 142(31):13507–13514, 2020.

63. F. Liu, M. Muruganathan, S. Ogawa, Y. Morita, M. Schmidt, and H. Mizuta. Half-meshed and fully-meshed suspended graphene for transport gap engineering. In *2020 IEEE Silicon Nanoelectronics Workshop (SNW)*, pages 69–70, 2020.

64. Yasin Khatami, Jiahao Kang, and Kaustav Banerjee. Graphene nanoribbon based negative resistance device for ultra-low voltage digital logic applications. *Applied Physics Letters*, 102(4):043114, 2013.

65. N. Ghobadi, and M. Pourfath. A comparative study of tunneling fets based on graphene and GNR heterostructures. *IEEE Transactions on Electron Devices*, 61(1):186–192, Jan 2014.

66. Qin Zhang, Yeqing Lu, C.A. Richter, D. Jena, and A. Seabaugh. Optimum bandgap and supply voltage in tunnel fets. *IEEE Transactions on Electron Devices*, 61(8):2719–2724, Aug 2014.

67. Samuel Smith, Juan-Pablo Llinás, Jeffrey Bokor, and Sayeef Salahuddin. Negative differential resistance and steep switching in chevron graphene nanoribbon field-effect transistors. *IEEE Electron Device Letters*, 39(1):143–146, 2017.

68. A. Jaramillo-Botero, and J. M. Marmolejo-Tejada. All-armchair graphene nanoribbon field-effect uridine diphosphate glucose sensor: First-principles in-silico design and characterization. *IEEE Sensors Journal*, 19(11):3975–3983, 2019.

69. A. Kumar, V. Kumar, S. Agarwal, A. Basak, N. Jain, A. Bulusu, and S.K. Manhas. Nitrogen-terminated semiconducting zigzag GNR FET with negative differential resistance. *IEEE Transactions on Nanotechnology*, 13(1):16–22, Jan 2014.

70. Juan M. Marmolejo-Tejada, and Andres Jaramillo-Botero. Four-terminal graphene nanoribbon sensor devices: in-silico design and characterization. *Computational Materials Science*, 196:110506, 2021.

71. Amina Kimouche, Mikko M. Ervasti, Robert Drost, Simo Halonen, Ari Harju, Pekka M. Joensuu, Jani Sainio, and Peter Liljeroth. Ultra-narrow metallic armchair graphene nanoribbons. *Nature Communications*, 6(1):10177, Dec 2015.

72. M.R. Muller, A. Gumprich, F. Schutte, K. Kallis, U. Kunzelmann, S. Engels, C. Stampfer, N. Wilck, and J. Knoch. Buried triple-gate structures for advanced field-effect transistor devices. *Microelectronic Engineering*, 119:95–99, 2014.

73. A.S. Azman, Z. Johari, and R. Ismail. Performance evaluation of dual-channel armchair graphene nanoribbon field-effect transistor. In *2014 IEEE International Conference on Semiconductor Electronics (ICSE)*, pages 138–141, Aug 2014.

74. N. Hasegawa, R. Sako, H. Tsuchiya, and M. Ogawa. Band structure and electron transport in multi-junction graphene nanoribbons. In *2012 IEEE Silicon Nanoelectronics Workshop (SNW)*, pages 1–2, June 2012.

75. B.K. Nikolic, L.P. Zarbo, and S. Souma. Spin currents in semiconductor nanostructures: a nonequilibrium green-function approach. *The Oxford Handbook on Nanoscience and Technology: Frontiers and Advances*, Volume I, Chapter 24, pages 814–866, 2010.

76. C.L. Kane, and E.J. Mele. Quantum spin hall effect in graphene. *Physical Review Letters*, 95:226801, Nov 2005.

77. Anjan Soumyanarayanan, Nicolas Reyren, Albert Fert, and Christos Panagopoulos. Emergent phenomena induced by spin-orbit coupling at surfaces and interfaces. *Nature*, 539(7630):509–517, 2016.

78. Jayakumar Balakrishnan, Gavin Kok Wai Koon, Manu Jaiswal, A. H. Castro Neto, and Barbaros Ozyilmaz. Colossal enhancement of spin-orbit coupling in weakly hydrogenated graphene. *Nature Physics*, 9(5):284–287, 2013.

79. Dinh Van Tuan, Frank Ortmann, David Soriano, Sergio O. Valenzuela, and Stephan Roche. Pseudospin-driven spin relaxation mechanism in graphene. *Nature Physics*, 10(11):857–863, 2014.

80. D. Van Tuan, J.M. Marmolejo-Tejada, X. Waintal, B.K. Nikolić, S.O. Valenzuela, and S. Roche. Spin hall effect and origins of nonlocal resistance in adatom-decorated graphene. *Physical Review. Letters*, 117:176602, Oct 2016.

81. R.V. Gorbachev, J.C.W. Song, G.L. Yu, A.V. Kretinin, F. Withers, Y. Cao, A. Mishchenko, I.V. Grigorieva, K.S. Novoselov, L.S. Levitov, and A.K. Geim. Detecting topological currents in graphene superlattices. *Science*, 346(6208):448–451, 2014.

82. J.M. Marmolejo-Tejada, J.H. Garca, M.D. Petrović, P.-H. Chang, X.-L. Sheng, A. Cresti, P. Plecháč, S. Roche, and B.K. Nikolić. Deciphering the origin of nonlocal resistance in multiterminal graphene on hexagonal-boron-nitride with ab initio quantum transport: fermi surface edge currents rather than fermi sea topological valley currents. *Journal of Physics: Materials*, 1(1):015006, Sep 2018.

83. Jose H. Garcia, Aron W. Cummings, and Stephan Roche. Spin hall effect and weak antilocalization in graphene/transition metal dichalcogenide heterostructures. *Nano Letters*, 17(8):5078–5083, 2017. PMID: 28715194.

84. A. Cresti, B. K. Nikolić, J. H. García, and S. Roche. Charge, spin and valley hall effects in disordered graphene. *La Rivista del Nuovo Cimento*, 39(12):587–667, Dec 2016.

85. Floriano Traversi, Valeria Russo, and Roman Sordan. Integrated complementary graphene inverter. *Applied Physics Letters*, 94(22):223312, 2009.

86. Lavanya Jothi, Saravana Kumar Jaganathan, and Gomathi Nageswaran. An electrodeposited au nanoparticle/porous graphene nanoribbon composite for electrochemical detection of alpha-fetoprotein. *Materials Chemistry and Physics*, 122514, 2019.

87. Ruizhong Zhang, Chia-Liang Sun, Yu-Jen Lu, and Wei Chen. Graphene nanoribbon-supported PTPD concave nanocubes for electrochemical detection of TNT with high sensitivity and selectivity. *Analytical Chemistry*, 87(24):12262–12269, 2015. PMID: 26568380.

88. Li-Na Feng, Zhi-Ping Bian, Juan Peng, Fang Jiang, Guo-Hai Yang, Ying-Di Zhu, Di Yang, Li-Ping Jiang, and Jun-Jie Zhu. Ultrasensitive multianalyte electrochemical immunoassay based on metal ion functionalized titanium phosphate nanospheres. *Analytical Chemistry*, 84(18):7810–7815, 2012. PMID: 22913388.

89. Pegah Hashemi, Abbas Afkhami, Behzad Baradaran, Raheleh Halabian, Tayyebeh Madrakian, Fabiana Arduini, Tien-Anh Nguyen, and Hasan Bagheri. Well-orientation strategy for direct immobilization of antibodies: development of the immunosensor using the boronic acid-modified magnetic graphene nanoribbons for ultrasensitive detection of lymphoma cancer cells. *Analytical Chemistry*, 92(16):11405–11412, 2020.

90. Hatef Sadeghi, L. Algaragholy, T. Pope, S. Bailey, D. Visontai, D. Manrique, J. Ferrer, V. Garcia-Suarez, Sara Sangtarash, and Colin J. Lambert. Graphene sculpturene nanopores for DNA nucleobase sensing. *The Journal of Physical Chemistry B*, 118(24):6908–6914, 2014. PMID: 24849015.

18 Design and Analysis of Various Neural Preamplifier Circuits

Swagata Devi, Koushik Guha, and Krishna Lal Baishnab

CONTENTS

18.1 INTRODUCTION

In 2016, approximately 276 million people were affected by neurological disorders, and of those, 9 million have died (Feigin et al. 2019). An article published in 2014 mentioned that based on the statistical data collected, 30 million of the current population of 1.27 billion have been diagnosed with neurological disorders in India. It also reported that the neurological disorders had been more widespread in rural regions compared with the urban population (a ratio of 1.9:1) (Gourie-Devi 2014). Epilepsy, Alzheimer's disease, Parkinson's disease, multiple sclerosis, and migraines are the most common neurological disorders worldwide. Another article cited that globally, 50 million people have been diagnosed with epilepsy alone, with a majority of them dwelling in developing countries and about 10 million residing in India (Santhosh et al. 2014). In 1993, the International League Against Epilepsy (ILAE) defined epilepsy as a condition characterized by recurrent seizures that take place with no abrupt recognized reasons (Hauser and Kurland 1975). The treatment of epileptic seizures involves recording intracranial electroencephalography (iEEG) signals, which have a frequency range of 14–250 Hz, comprising γ peaks or β-γ peaks (Edakawa et al. 2016). Also, currently high-frequency oscillations (HFOs) are considered as a new potential biomarker for the detection of the same and have a frequency range less than 1 kHz (Zijlmans et al.). These signals are characterized by very low amplitude and are easily corrupted by the electrode skin interfaces or electromyography (EMG) signals. Various treatment procedures have been developed over time for curing epileptic seizures, such as anti-epileptic drug intake and surgical procedures to remove the affected zones of the brain (Guenot et al. 2004). However, it has been observed from statistical reports that most of the patients have become resistant to the anti-epileptic drugs and stopped responding to them over time. The surgical treatment procedure involves abstraction of the epileptic zones in the brain from which the seizures originate, but there is high risk involved with these surgeries. The removal of epileptic zones may simultaneously result in removal of the nearby regions, which may disrupt the normal functions of the body (Téllez-Zenteno et al. 2005). Over the last couple of decades, with advancements in technology, there has been a rise in the demand for neurostimulation devices. These devices can be categorized as vagus nerve stimulation (VNS) and deep brain stimulation (DBS), and they control the seizures before their clinical onset and prevent them from occurring.

They are implanted into the brain or the vagus nerve, in the front end of the neural amplifiers, to record the neural activities.

The design of front-end amplifier circuits requires minimum heating as it may cause damage to the implanted regions or the vital regions nearby. Moreover, while recording the signals these circuits should be immune to the surrounding noises and interference to avoid any false detection in the original data. The circuits involved in recording the data from the electrodes need to operate at low power and consume low noise and area. The primary purpose of the low noise amplifier (LNA) in a neural amplifier is to record these signals and produce a magnified signal with minimum input referred noise (IRN). There have been numerous reported works by various researchers in which different circuits have been implemented as an LNA, and they have shown remarkable results. The performance parameters have been enhanced by novel ideas and through modification in existing designs, as observed. One of the most important methods is choosing the appropriate circuit structure; Harrison and Charles (2003) used a capacitive feedback technique to restrict the gain through bandwidth reduction, which settled around 39.5 dB, with a power of 80 μW and noise efficiency factor (NEF) of 4. Wattanapanitch et al. (2007) used a current scaling technique in the standard folded cascode design to limit the output swing and measured an NEF of 2.67. Qian et al. (2011) utilized both current-splitting and output current scaling techniques to attain low power, and they reported a gain of 39.4 dB with a power consumption of 2.4 μW.

In this chapter, three different unique circuit designs have been analyzed and their design parameters are compared to measure their performance efficiency. The circuits have been discussed elaborately in the upcoming sections. The first circuit is a modified recycling folded cascode (RFC) amplifier that operates at $2 \times V_{DD}$ and it implements voltage doubling; the second circuit is a single-stage fully differential amplifier circuit, which has a flip voltage follower (FVF) structure in its biasing circuitry; and the third circuit is a modified RFC amplifier. The circuit simulations for all three designs have been executed in a 180-nm CMOS SCL Cadence environment. This technology is used because this work is carried out keeping future fabrication in mind. The cost of fabrication available at our resources is compatible with this particular technology; thus, the simulation results have been performed in a 180-nm CMOS SCL Cadence environment. The results have been discussed and compared with the recent works in the relevant area for comprehension and analysis.

18.2 CIRCUIT TOPOLOGIES

18.2.1 CIRCUIT 1: SINGLE STAGE AMPLIFIER WITH VOLTAGE DOUBLING

This design utilizes the concept of voltage doubling in the standard structure of an RFC amplifier, which results in recycling of currents in the devices to effectively increase the effective transconductance without any extra consumption of power. The mounting of the output transistors will enhance the output resistance and doubling the supply voltages will result in elevating the output swing (Qu 1995; Rosa and Lisanby 2012). The structure in Figure 18.1 illustrates a voltage doubling circuit, operating at $2 \times V_{DD}$ supply. M0´ is added to the biasing circuitry to increase output resistance; hence, the common-mode rejection ratio is used. The output transistors Ma3, Mb3, Ma4, and Mb4 have been stacked with Ma3´, Mb3´, Ma4´, and Mb4´, whereas M7 and M8 have been stacked with M7´ and M8´. Hence, the effective DC gain has been increased and the output swing is elevated (Khatwani and Tiwari 2013).

18.2.2 CIRCUIT 2: MODIFIED SINGLE-STAGE FULLY DIFFERENTIAL AMPLIFIER

The design in Figure 18.2 is a single-stage fully differential amplifier that implements a FVF in its biasing circuitry, and the common mode feedback structure is shown in Figure 18.3. The output swing of the circuit is restricted as all the devices operate in the subthreshold region. Also, the gain for the circuit is limited as there is no change in the threshold voltage (Cromwell et al. 2014). An

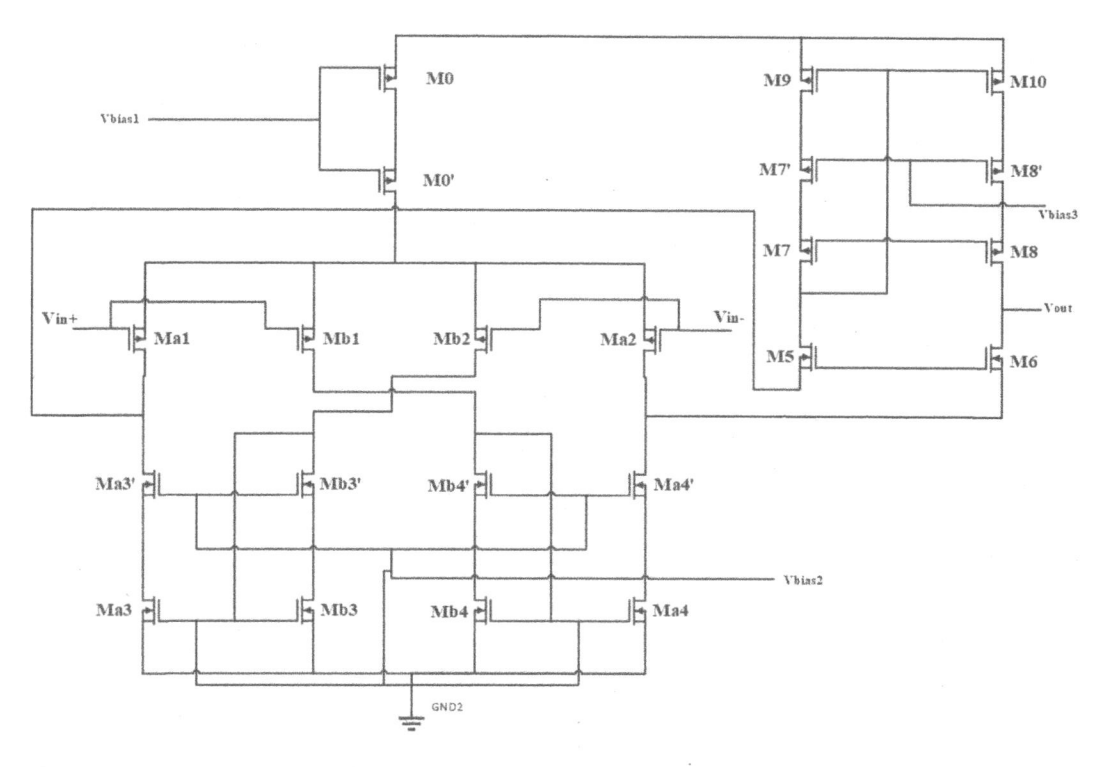

FIGURE 18.1 Modified RFC with voltage doubling.

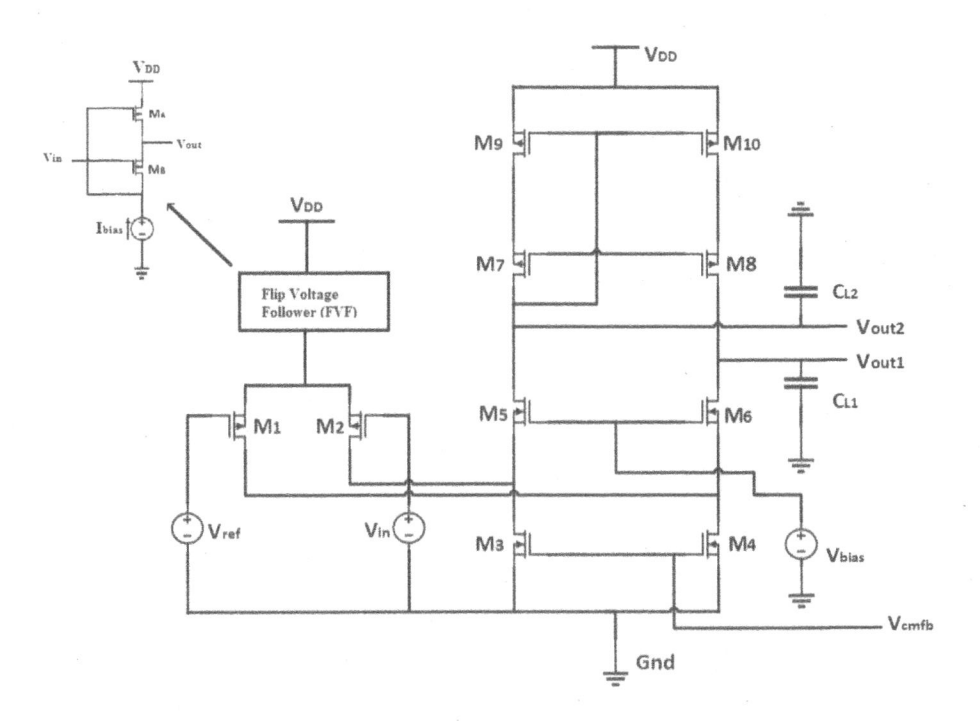

FIGURE 18.2 Modified single-stage fully differential amplifier.

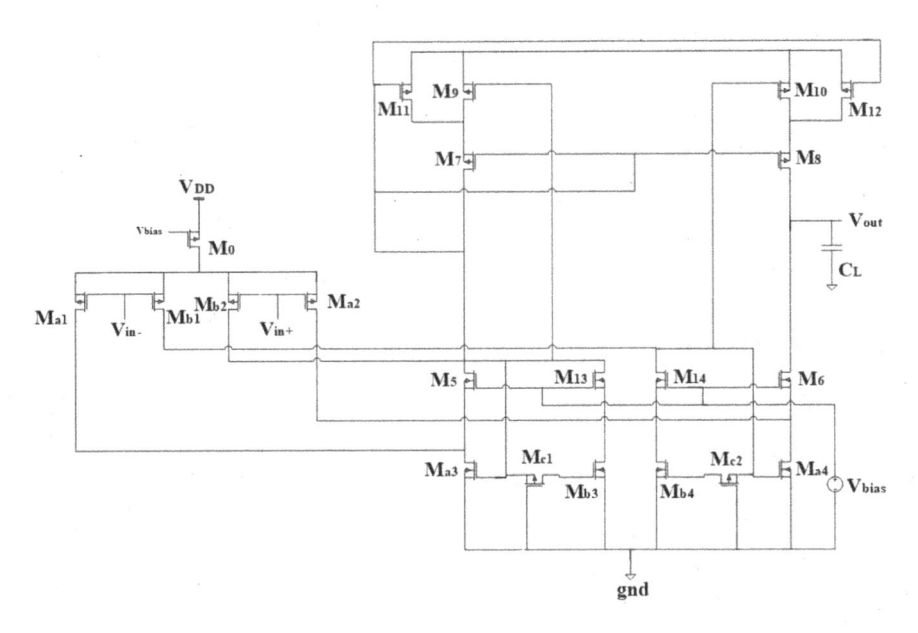

FIGURE 18.3 Modified recycling folded cascode amplifier.

FVF comprises two cascoded transistors that are placed in the biasing circuit instead of a standard source follower. Due to the low impedance at the output node (g_{mA} or g_{mB}), there is larger current sinking capability, which is unchanged by the noise. On the other hand, as these circuits work at low voltages, they are applicable for power efficient circuits. The inclusion of FVF in the design provides an advantage over transconductance, gain bandwidth, and voltage swing.

18.2.3 Circuit 3: Modified Recycling Folded Cascode Amplifier

There is degradation of the phase margin when the folded cascode circuit is modified into an RFC, which will result in stability degradation of the amplifier in a closed loop. To compensate for this, in this design, two transistors (Mc1 and Mc2) have been introduced. These transistors work in the deep triode region and are placed between the gate terminals of Ma3 and Mb3 and Ma4 and Mb4. A transverse connection is made between Mb3–M9 and Mb4–M10 through which M9 and M10 can work as input driving transistors (Zijlmans et al., 2017). Also, to maintain the circuit as a single-ended architecture two transistors, M11 and M12, are cascoded with M9 and M10, such that a current ratio of $(k-1){:}1$ runs all the way through M9:M11 and M10:M12 (Kassiri et al. 2017). The circuit is shown in Figure 18.3.

18.3 SIMULATION RESULTS AND DISCUSSION

The circuit design simulations have been performed in a 180-nm CMOS SCL Cadence environment. The representation in Table 18.1 is a comparative index for the performance parameters of the three simulated circuits. The DC gain for transconductance amplifiers is expressed in terms of transconductance (G_m) and the output impedance (R_o). In circuit 1, the expression for G_m (overall transconductance) is given in Equation (18.1).

$$G_m\big|_{cir1} = g_{ma1}(1+K) \tag{18.1}$$

$$R_o\big|_{cir1} = \left[\left(g_{m5}+g_{m5b}\right)r_{o5}\left[\left(r_{o1}\big\|\left(g_{ma3}+g_{mba3}\right)r_{oa3}\cdot r_{oa3}\right)\right]\right]\Big\|\left[g_{m7}r_{o7}g_{m7'}r_{o7'}r_{o9}\right] \tag{18.2}$$

TABLE 18.1

Comparison of the Performance Parameters of Simulated Circuits

Parameters	Circuit 1	Circuit 2	Circuit 3
DC gain (dB)	75	47.97	47.26
−3-dB bandwidth (Hz)	519.51	500	500
Phase (degrees)	48.48	89.52	82
Power consumed (μW)	3.92	1.07	2.26
Slew rate (V/ms)	1.15	26.85	8.54
Input-referred noise (μV_{rms})	3.2	14.7	8.69
GBW (Hz)	1.45 M	485.28 k	120.62 k

where g_{ma1} represents transconductance of the transistor Ma1, and due to the large output resistance of the design as expressed in Equation (18.2), it shows the highest DC gain among the three designs.

The reason behind the high value in the output impedance is its decreased bias current and the stacked devices. The value for K plays a crucial part in shaping the phase margin of the circuits and it is preferred such that $\omega_{p3} > \omega_u$ for high-speed applications. In case of low-speed applications, the value of K does not restrict the phase margin, but for high-speed applications it is generally chosen between 2 and 4 to lessen the phase margin degradation. The gain versus frequency graph is illustrated in Figure 18.4, where a comparison is drawn among the three simulated designs (Yao et al. 2003; Mohseni and Najafi 2004). The second circuit has a FVF in the design that increases the driving capability of the circuit. The sinking current is high as the output impedance is low. The circuit works in the weak inversion region and operating voltage is low; hence, the threshold voltage does not change but the gain is limited to 47.97 dB. Similarly, the third circuit design also operates in the same region of operation, thereby restricting the gain at 47.26 dB.

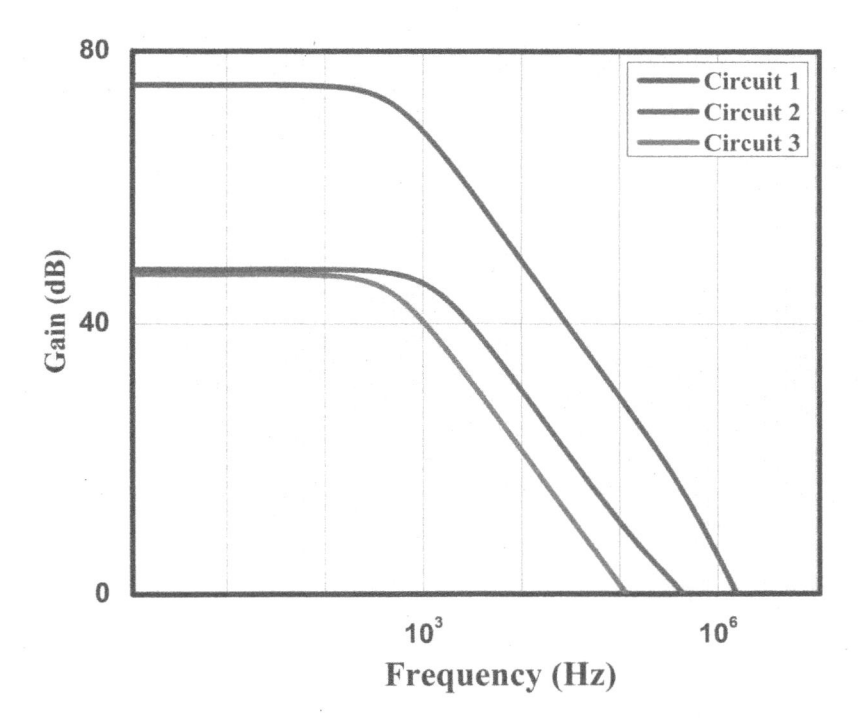

FIGURE 18.4 Gain versus frequency graph.

All the operational transconductance amplifier designs have one dominant pole, ω_{p1}. This pole is related to the capacitive load C_L at the output node (Yen et al. 2004; Gosselin et al. 2007). Also, ω_{p2} is a nondominant pole associated with the source of the transistor M5 and is given by g_{m5}/C_{gs5}. The voltage doubling circuit has a mirror pole zero combination: the pole ω_{p3} is given by $g_{mb3}/(1 + K)C_{gsb3}$ and the zero ω_{z1} is given by $(1 + K)\,\omega_{p3}$. The additional high-frequency pole ω_{p4} is associated with the source of M3a′ and is expressed as $g_{ma3}/C_{gsa3'}$. The reduced device sizes induce reduced parasitic capacitances; therefore, this circuit maintains a phase margin of 48.48 degrees. In the second circuit design, apart from the previously mentioned dominant and nondominant poles, there is an additional pole due to the presence of FVF. However, it is pushed to a very high frequency and the design exhibits an excellent phase of 89.92 degrees. There is degradation in the phase for the conventional design of the RFC and its stability. To compensate for this degradation, two transistors, Mc1 and Mc2, working in the deep triode region are added between the gates of the input transistors and a crossover connection was introduced between the output transistors to implement M9 and M10 as input driving transistors. An extra pole ω_{p2} and a zero ω_{z1} have been added, which are expressed as $\frac{g_{m,3b}}{kC_{GB3,3b}}$ and $\frac{1}{R_{on}\,kC_{GB3,3b}}$. The value of R_{on} should be selected in such a manner that the pole and zero cancel each other and the phase margin is enhanced and exhibits a phase of 82 degrees.

The noise in an operational transconductance amplifier is a combination of the thermal and flicker noise; it is shown in the Equation (18.3):

$$\overline{i_o^2} = \left[4K_B T\gamma g_m + \frac{K_F g_m{}^2}{C_{ox} LWf} \right]\Delta f \tag{18.3}$$

The former term represents thermal noise and the latter represents the flicker noise, respectively, where K_B and T are the Boltzmann constant and temperature in Kelvin, respectively. C_{ox} denotes gate oxide capacitance per unit area, K_F signifies process-dependent parameter constant, W and L are the device width and channel length, γ represents bias dependent parameter, and f denotes frequency (Mohseni et al. 2004; Wu and Xu 2005). The overall phase comparison for three circuits is shown in figure 18.5.

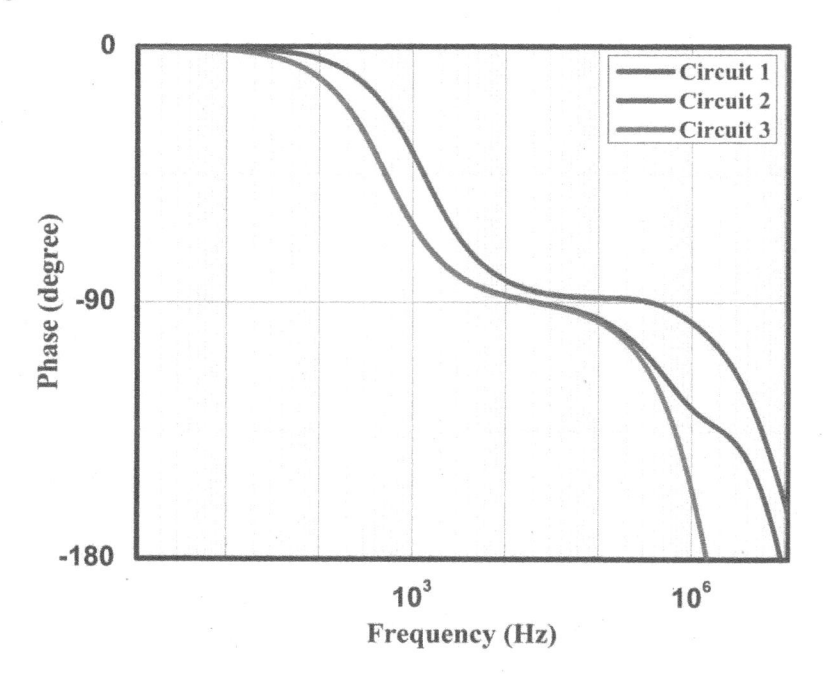

FIGURE 18.5 Phase versus frequency graph.

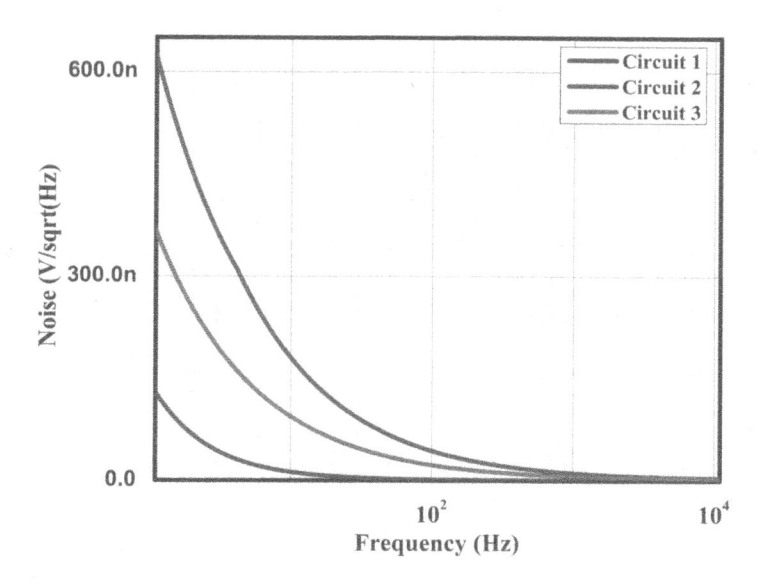

FIGURE 18.6 Representation of spectral noise density versus frequency.

The comparative representation of spectral noise density for the three circuits is illustrated in Figure 18.6. The IRN for the voltage doubling circuit is the combination of the equivalent noise contribution by the transistors Ma1, Mb1, Ma3, Mb3, and M9, and then doubling it. The overall noise contribution by circuit 1 is 3.2 μV_{rms}. As the second and third circuit, both are operating in the weak inversion region, and the flicker noise is dominant over thermal noise. The extra Si and SiO$_2$ energy levels add to the increasing flicker noise. The noise contributions for circuits 2 and 3 are a result of the noise contribution by M1a, M3 (or M3a for circuit 3), and M9. The measured IRNs are measured as 14.7 and 8.69 μV_{rms}. The design in circuit 2 is functioning at a very low supply voltage and the use of the FVF results in increased current, which is proportional to the input voltage; therefore, it has the minimum consumed power among the three designs (Amaya et al. 2015). The simulated results for circuit 3 reveal a good power noise trade-off with a consumed power of 2.26 μW and a noise of 8.69 μV_{rms}, and a DC gain of 47.26 dB with a phase margin of 82 degrees. Table 18.2 illustrates a comparison of the results of the proposed circuit with some recent works put forward by various researchers.

TABLE 18.2
Comparison of Simulated Results with Recent Works

Paper	This Work	Harrison and Charles (2003)	Wattanapanitch et al. (2007)	Mohseni and Najafi (2004)	Gosselin et al. (2007)
Application	Epileptic Seizures	Neural Recording	Neural Recording	Neural Recording	Bioamplifier
Input amplitude (V)	10–500 μm	50–500 μm	50–500 μm	50–500 μm	50–500 μm
Voltage supply (V)	1.2	2.5	0.5	1.5	1.8
Gain (dB)	47.26	39.3	40.85	39.3	49.52
Phase margin (°)	82	52	–	75.5	90
Noise (μV_{rms})	8.69	2.2	3.06	7.8	5.6
Bandwidth (Hz)	500	~7.2k	45–5.32 k	0.1–9.1 k	98.4–9.1 k
Power (μW)	2.26	80	7.6	115	8.4

18.4 CONCLUSION

As discussed previously, it has been observed that for the modified RFC amplifier operating at $2 \times V_{DD}$, there is an increase in the overall transconductance of the circuit. This enhancement is achieved through the current recycling technique without any additional power expense, the output impedance of the circuit is increased, and the output swing of the circuit is enhanced. In the single-stage fully differential amplifier circuit with an FVF structure in its biasing circuitry, all the MOSFETS involved in the structure are working in the weak inversion region. This leads to the enhancement of transconductance, slew rate, and gain-bandwidth product of the design (Majidzadeh et al. 2011). The modified RFC amplifier introduces two MOS transistors operating in the deep triode regions and acting as resistors between the gates of Ma1 and Mb1 and Ma2 and Mb2 devices to compensate for the degradation in the phase margin of the circuit. The circuit simulations for all three designs have been performed in a 180-nm CMOS SCL Cadence environment for the ease of compatibility and cost of fabrication with the available resources. The simulation results reveal that the fully differential amplifiers are more resistant to external interference due to its reduced susceptibility to common mode noise; additionally they provide high-output swing. The measured values for IRN are comparatively less; however, there is increased power due to the additional common mode feedback circuitry. The improvised RFC circuits have increased swing and transconductance without any additional increase in consumed power. Furthermore, this chapter also focused on the comparison of the simulated results with the recent works in the relevant area for comprehension and analysis (Saidulu et al. 2016).

REFERENCES

Amaya, R. J., A. Rodriguez-Perez, and M. Delgado-Restituto. 2015. A low noise amplifier for neural spike recording interfaces. *Sensors* 15, no. 10: 25313–25335.

Cromwell, L., F. J. Weibell, and E. A. Pfeiffer. 2014. *Biomedical Instrumentation and Measurements.* 2nd ed. Prentice-Hall of India, New Delhi.

Edakawa, K., T. Yanagisawa, H. Kishima, R. Fukuma, S. Oshino, H. M. Khoo, M. Kobayashi, M. Tanaka, and T. Yoshimine. 2016. Detection of epileptic seizures using phase–amplitude coupling in intracranial electroencephalography. *Scientific Reports* 6, 25422.

Feigin, V. L., E. Nichols, T. Alam, M. S. Bannick, E. Beghi, N. Blake, and W. J. Culpepper et al. 2019. Global, regional, and national burden of neurological disorders, 1990–2016: A systematic analysis for the global burden of disease study 2016. *The Lancet Neurology* 18, no. 5: 459–480.

Gosselin, B., M. Sawan, and C. A. Chapman. 2007. A low-power integrated bioamplifier with active low-frequency suppression. *IEEE Transactions on Biomedical Circuits and Systems* 1, no. 3: 184–192.

Gourie-Devi, M. 2014. Epidemiology of neurological disorders in India: Review of background, prevalence and incidence of epilepsy, stroke, Parkinson's disease and tremors. *Neurology India* 62, no. 6: 588.

Guenot, M. 2004. Surgical treatment of epilepsy: Outcome of various surgical procedures in adults and children. *Revue Neurologique* 160: 5S241–5S250.

Harrison, R. R., and C. Charles. 2003. A low-power low-noise CMOS amplifier for neural recording applications. *IEEE Journal of Solid-State Circuits* 38, no. 6: 958–965.

Hauser, W. A., and L. T. Kurland. 1975. The epidemiology of epilepsy in Rochester, Minnesota, 1935 through 1967. *Epilepsia* 16, no. 1: 1–66.

Kassiri, H., S. Tonekaboni, M. T. Salam, N. Soltani, K. Abdelhalim, J. L. P. Velazquez, and R. Genov. 2017. Closed-loop neurostimulators: A survey and a seizure-predicting design example for intractable epilepsy treatment. *IEEE Transactions on Biomedical Circuits and Systems* 11, no. 5: 1026–1040.

Khatwani, P., and A. Tiwari. 2013. A survey on different noise removal techniques of EEG signals. *International Journal of Advanced Research in Computer and Communication Engineering* 2, no. 2: 1091–1095.

Majidzadeh, V., A. Schmid, and Y. Leblebici. 2011. Energy efficient low-noise neural recording amplifier with enhanced noise efficiency factor. *IEEE Transactions on Biomedical Circuits and Systems* 5, no. 3: 262–271.

Mohseni, P., and K. Najafi. 2004. A fully integrated neural recording amplifier with DC input stabilization. *IEEE Transactions on Biomedical Engineering* 51, no. 5: 832–837.

Qian, C., J. Parramon, and E. Sanchez-Sinencio. 2011. A micropower low-noise neural recording front-end circuit for epileptic seizure detection. *IEEE Journal of Solid-State Circuits* 46, no. 6: 1392–1405.

Qu, H. 1995. *Self-adapting algorithms for seizure detection during EEG monitoring.* Ph.D. thesis, McGill University, Montreal, Canada.

Rosa, M. A., and S. H. Lisanby. 2012. Somatic treatments for mood disorders. *Neuropsychopharmacology* 37, no. 1: 102–116.

Saidulu, B., A. Manoharan, and K. Sundaram. 2016. Low noise low power CMOS telescopic-OTA for bio-medical applications. *Computers* 5, no. 4: 25.

Santhosh, N. S., S. Sinha, and P. Satishchandra. 2014. Epilepsy: Indian perspective. *Annals of Indian Academy of Neurology* 17, no. Suppl 1: S3.

Téllez-Zenteno, J. F., R. Dhar, and S. Wiebe. 2005. Long-term seizure outcomes following epilepsy surgery: A systematic review and meta-analysis. *Brain* 128, no. 5: 1188–1198.

Wattanapanitch, W., M. Fee, and R. Sarpeshkar. 2007. An energy-efficient micropower neural recording amplifier. *IEEE Transactions on Biomedical Circuits and Systems* 1, no. 2: 136–147.

Wu, H., and Y. P. Xu. 2005. A low-voltage low-noise CMOS instrumentation amplifier for portable medical monitoring systems. *In The 3rd International IEEE-NEWCAS Conference*, Quebec, Canada, 295–298.

Yao, L., M. Steyaert, and W. Sansen. 2003. A 0.8-V, 8-/spl mu/W, CMOS OTA with 50-dB gain and 1.2-MHz GBW in 18-pF load. *In ESSCIRC 2004-29th European Solid-State Circuits Conference (IEEE Cat. No. 03EX705)*, Estoril, Portugal, 297–300.

Yen, C. J., W. Y. Chung, and M. C. Chi. 2004. Micro-power low-offset instrumentation amplifier IC design for biomedical system applications. *IEEE Transactions on Circuits and Systems I: Regular Papers* 51, no. 4: 691–699.

Zijlmans, M, G. A. Worrell, M. Dümpelmann, T. Stieglitz, A. Barborica, M. Heers, A. Ikeda, N. Usui, and M. L. V. Quyen. 2017. How to record high-frequency oscillations in epilepsy: A practical guideline. *Epilepsia* 58, no. 8: 1305–1315.

Zijlmans, M, J. Jacobs, R. Zelmann, F. Dubeau, and J. Gotman. 2009. High frequency oscillations and seizure frequency in patients with focal epilepsy. *Epilepsy Research* 85, no. 2–3: 287–292.

19 Design and Analysis of Transition Metal Dichalcogenide-Based Feedback Transistor

Prateek Kumar, Maneesha Gupta, Kunwar Singh, Ashok Kumar Gupta, and Naveen Kumar

CONTENTS

19.1 INTRODUCTION

Moore was proved correct in 1975 in saying that the "transistor density in integrated circuit doubles after every 24 months" [1]. The technique, which was utilized to decrease the size of the transistor, is known as scaling. Scaling is done by a factor of "S," where S can be between 1.2 and 1.4. Scaling was verified positively in reduced power dissipation, lesser area, and improved speed [2, 3]. Things turned drastically when transistors entered the nanometer regime. At nanometer scale, short channel effects (SCEs) like velocity saturation, hot electron effect, punch through, avalanche breakdown, and drain lowering hampered the performance of the devices noticeably, resulting in lower subthreshold swing (SS), poor I_{ON}/I_{OFF} ratio, and higher power dissipation [4]. Failure of the metal oxide semiconductor field-effect transistor (MOSFET) has caused researchers to search for new semiconductor devices with better characteristics in the nanometer region.

With the passing quarter century, researchers have found devices with different working principles, like tunnel field-effect transistors (TFETs) [5, 6], junctionless transistors [7], nanotubes [8], and feedback transistors [9–15]. The basic working principle of TFET is band-to-band tunneling (BTBT). In 2007, Choi demonstrated the first experimental TFET, which had an SS less than 52.8 mV/dec and proved that practical devices with an SS less than 60 mV/dec are possible [16]. Although the proposed device suffered from less ON-state current, it was suggested that ON-state current can be improved with an increased source-channel doping gradient, using low bandgap material and reduced gate oxide thickness. Dewey et al. [17] introduced the III–V heterojunction-based device. The proposed device offered a better ON state compared with the homojunction transistor. Comparing power dissipation at a 13-nm channel length, the authors found that power dissipation is 10-fold less in a TFET compared with MOSFET. After the first discovery, different forms of TFETs have been proposed, for example, vertical TFETs and multigate TFETs, a gate all around TFETs, and charge plasma-based TFETs. TFETs now have found applications in practical forms such as electronic switches and sensors, etc. Although the TFET is a promising device for the future, there are factors that are holding TFETs back, including complex fabrication, large miller capacitances, low ON-state current, and the effect of random dopant fluctuations (RDFs) [18].

With the discovery of graphene, nanotube-based devices have nourished significantly [8]. Nanotube FETs consist of two types: (1) Schottky barrier–controlled FET and (2) MOSFET-like nanotubes. In the Schottky barrier-controlled type, the majority of carriers define conductivity by tunneling through the Schottky barrier, whereas in later transmissions of charge carriers conductivity is unipolar and is controlled by gate-source biasing. A carbon nanotube has the ability to replace a silicon FET on a molecular level. Work done by Collins et al. in the article "Nanotubes for Electronics" indicated that nanotube-based processors can be 1000 times as fast as today's processors [19]. Also, nanotubes have found diverse applications in the fields of light-emitting diodes, nanoscopic lasers, and diodes. Tans et al. have also proposed future devices using nanotubes in the form of the single-electron transistor. The effect of a magnetic field on the density of states of nanotubes is called the Zeeman effect and is also studied in Ref. [20]. Nanotubes depend on the Schottky tunneling effect of contact material (metallic [M]/semiconducting [S]) on the M/S channel of the nanotube creating the M-M, M-S, S-M, and S-S interfaces. This is studied in the work "Crossed Nanotube Junctions" [21]. On reviewing nanotubes, Che stated that while nanotubes possess several advantages, they have poor transport characteristics and fabrication. The junctionless transistor is another device that can be used to replace MOSFETs in the future [22]. As there is no doping involved, junctionless transistors have significantly less fabrication issues [7]. To develop, they need either a complete source to drain the uniform doping or no doping at all. For earlier cases fabricated devices suffer from higher OFF-state current, and later devices suffer from poor ON-state current. Feedback transistors are the latest addition in the list of the devices that can be used in the future to replace MOSFETs. There are various feedback diodes proposed so far. Feedback FETs have working principles similar to the gated p-i-n diode. Proposed feedback FETs have complex operations or multigate electrode structures, but in this chapter the single-gated feedback FET is proposed and is thoroughly examined.

A new material was required for nanoscale devices as silicon feedback created deplorable results. Researchers provided the solution in the form of two-dimensional (2D) materials or single-layer materials. Two-dimensional materials are those in which charge carriers are free to transport in a plane and are confined in the third direction. So far around 700 2D materials have been found to be stable; out of these the most common materials are graphene [23], transition metal dichalcogenides (TMDCs) [24, 25], and black phosphorus (BP) [26]. Graphene was discovered and has shown tremendous application in the fields of electronics, sensors, batteries, etc., but transistors designed from graphene exhibit very poor transfer characteristics as graphene has a zero bandgap structure similar to metals of transistors designed to attain high OFF-state current. The bandgap of graphene can be tailored, but it decreases the mobility of the charge carriers, which makes them unsuitable for semiconducting applications.

TMDC materials are found in the form MX_2, where M represents transition metals like molybdenum, tungsten, and so forth, and X denotes chalcogens like sulfur, selenium, etc. Most common TMDC materials are molybdenum disulfide (MoS_2), tungsten disulfide (WS_2), molybdenum ditelluride ($MoTe_2$), molybdenum diselenide ($MoSe_2$), and tungsten diselenide (WSe_2). TMDCs are extremely promising for nanoscale devices as a single layer of TMDC materials is only 6.5 Å thick. In TMDC materials, each atom is bound to each other by the van der Waals force. TMDC materials can be fabricated using different techniques, the most common of which are exfoliation, chemical vapor deposition, and molecular beam epitaxy. These materials have a direct bandgap in a few-layered configuration, which make them suitable even for optoelectronic applications. In the work "2D Transition Metal Dichalcogenides," Manzeli et al. thoroughly reviewed TMDC materials and put the limelight on spin orbiting and the behavior of material under high-frequency applications [25]. Although there are transistors designed using TMDC materials, most of them suffer from a poor I_{ON}/I_{OFF} ratio because of poor mobility of charge carriers of TMDC materials.

BP is an allotrope of phosphorus. It has a configuration from a monolayer to five layers. In BP, mobility of charge carriers is very high and it increases from 350 cm²/V-sec for the monolayer configuration to 2755 cm²/V-sec for the five-layer configuration. BP has excellent electronic properties that are very difficult to fabricate, which leads to a very expensive high-end product [26].

In this work, the feedback transistor is designed using $MoTe_2$ as the channel material. $MoTe_2$ is preferred as it has the lowest bandgap among all the materials, which is desired for low power applications.

19.2 DEVICE STRUCTURE AND BASIC WORKING

The proposed device is shown in Figure 19.1. It is 122 nm in length, out of which the channel length is 120 nm and both drain and source spacers are 1 nm each. Complete width of the device is 6 nm, which is divided as 5 nm and 1 nm between $MoTe_2$ and SiO_2, which is used as dielectric gate oxide. Of the 120-nm channel length, the gate and drain region have p-type doping of 10^{18} cm⁻³ and the remaining channel is an n type with a concentration of 10^{19}cm⁻³.

The device properties, such as the energy curve, channel potential, electric field, and recombination rate, are studied across the cutline AA′. Basic working of the device is explained using Figure 19.2. In the channel region, the gate terminal and the drain terminal are p-type doped because the energy curve forms a hump, as shown in the figure. At the OFF state, that is, gate voltage (V_{GS}) and drain voltage (V_{DS}) = 1 V, charge carriers do not have sufficient energy to move from the source terminal to the drain terminal. When applying $V_{GS} = V_{DS} = 1$ V, electrons start to accumulate below the gate and drain terminals because the energy curve shifts downward. Also, due to the electric field from the terminals, electrons are high on energy; hence, transport to the source side to drain side contributes to the conduction. Due to the sudden flow of carriers, there is a sudden rise in drain current (I_{DS}) because the designed device provides very low subthreshold slope.

In Table 19.1, parameters used to stimulate the device are listed. Different models used to stimulate the device are conmob, consrh, trap.auger, fldmob, and BGN print.

19.3 RESULTS

For the proposed device, different device properties (energy curve, channel potential, electric field, and recombination rate), analog properties (drain current-gate voltage characteristics [I_{DS}-V_{GS}], total capacitance, and cutoff frequency), and linearity properties are examined.

In Figure 19.3, the energy curve across the channel is plotted. It shows that the OFF state below the gate and drain region energy curve has shifted upward, which prohibits flow of the charge carriers leading to very low OFF-state current. In the ON state, the energy curve below the gate and drain is shifting downward, making transmission of charge carriers feasibly possible.

Curve for potential across the channel (black line) and electric field (red line) across the channel is plotted in Figure 19.4. At $V_{GS} = V_{DS} = 1$ V, the entire diode can be considered as two p-n junction

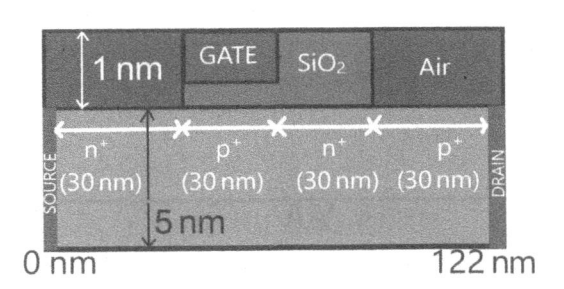

FIGURE 19.1 $MoTe_2$-based feedback FET.

TABLE 19.1

Parameters of the MoTe$_2$-Based Feedback FET

Parameters	Feedback Transistor (nm)
Device length	122
Source spacer length	1
Drain spacer length	1
Channel radius	5
Gate dielectric	3
Gate depth	1

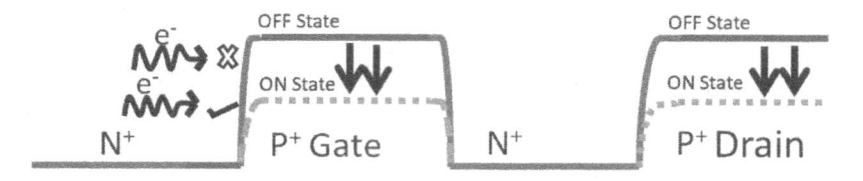

FIGURE 19.2   Basic working of a feedback transistor is explained. Red line is the OFF state and the blue line is the ON state.

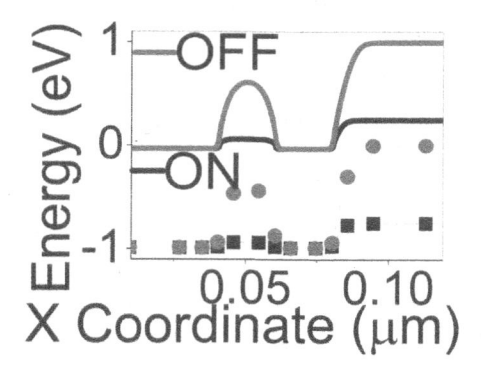

FIGURE 19.3   Energy curve for the proposed device along cutline AA′ (shown in Figure 19.1). Red line indicates OFF state, whereas ON state is indicated by the black line.

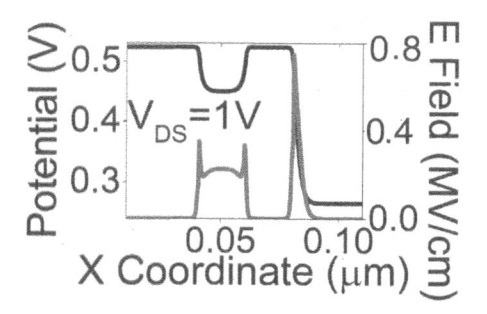

FIGURE 19.4   Curve for potential and electric field across the channel is plotted. Black line indicates the left y-axis/channel potential, whereas the red line indicates the right y-axis/electric field.

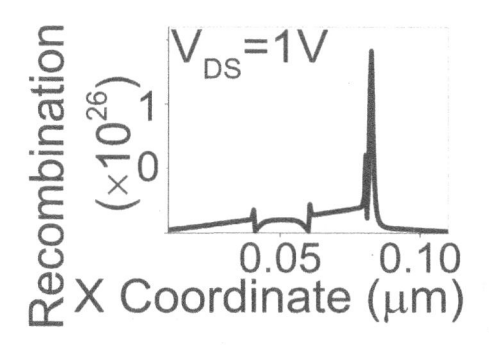

FIGURE 19.5 Recombination rate across the channel.

diodes in the forward bias condition. There are two peaks in the channel potential curve; both of the peaks are at the p-n interface indicating the barrier potential. Similarly, the electric field is driven from potential across the channel. Spikes in the electric field curve are at potential variation points. Electric field formation in the channel is responsible for sweeping the charge carrier from the source to drain. Interestingly, the electric field peak is at the p-n interface only, which is similar to the p-n junction diode.

In a p-n junction diode, when an electron enters the r-region diode, maximum probability of the recombination is at the junction. Similarly for the feedback transistor, the recombination rate is highest at the junctions as shown in Figure 19.5. Recombination is higher at the drain-side terminal, because a higher number of charge carriers are entering from the source side.

The curve for I_{DS}-V_{GS} on a log and linear scale is shown in Figure 19.6. Flow of I_{DS} is negligible for the region and V_{GS} is negative. As V_{GS} enters the positive region, the conduction band shifts downward and a sudden flow takes place. The same is supported by the I_{DS}-V_{GS} curve. Subthreshold slope obtained for the device is 1.7 mV/dec, which is very low compared with MOSFET slope at 60 mV/dec. This indicates that the proposed device can be used more efficiently for the applications in which high switching speed is required. A comparison has been made with previously designed feedback transistors in Table 19.2.

The curve for transconductance is plotted in Figure 19.7. Transcondcutance shows the rate of change in I_{DS}-V_{GS} chacracteristics. Figure 19.6 shows that a sudden spike in curve at voltage transition is $V_{GS} \approx 0$ V, hence, the curve for g_m shows a spike at $V_{GS} \approx 0$ V. As g_m is highest just after $V_{GS} \approx 0$ V, the transistor has maximum amplification ability at that point only.

Total capacitance is vital in defining AC characteristics of the device. Mostly it depends on the source-gate capacitance as the source terminal is n doped and has a higher number of charge carriers.

TABLE 19.2
Feedback FET Designed in this Work Is Compared with Previously Designed Feedback FETs

Reference	SS (mV/dec)	I_{OFF} (A/μm)
[9]	42	10^{-12}
[10]	3.7	10^{-9}
[11]	18	10^{-11}
[12]	10	10^{-11}
This work	1.7	10^{-15}

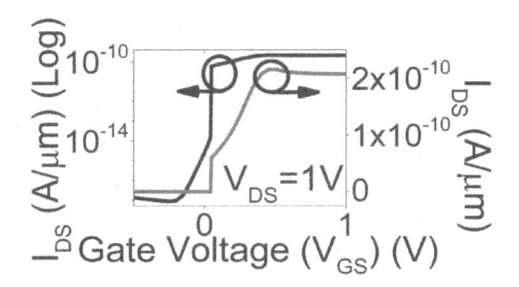

FIGURE 19.6 I_{DS}-V_{GS} characteristic for the proposed device on log scale (black line) and linear scale (red line).

Total capacitance is plotted in Figure 19.8. Cutoff frequency (f_T) is given as $\frac{g_m}{2\pi c_{gg}}$. Figure 19.7 shows that the peak of g_m is at $V_{GS} \approx 0$ V; hence, f_T is also defined at that point only. The curve for f_T is plotted in Figure 19.8. Maximum value achieved for f_T is 2.75 GHz.

The curve for g_{m2} and g_{m3} is plotted in Figure 19.9. Both are higher-order differentials of g_m; hence, the peak curve is achieved at similar points. As g_{m3} defines the zero crossing point (ZCP), it is important to note that ZCP is near to zero, which indicates that the proposed device should be an ideal choice for low power applications.

Linearity parameters are calculated on the basis of the following equations [27]:

$$VIP3 = \sqrt{24 \times g_{m1}/g_{m3}} \tag{19.1}$$

where $g_{m1} = \left(\partial I_d / \partial V_{gs}\right)$, $g_{m2} = \left(\partial^2 I_d / \partial V_{gs}^2\right)$, and $g_{m3} = \left(\partial^3 I_d / \partial V_{gs}^3\right)$.

$$IIP3 = \frac{2}{3}\frac{g_{m1}}{g_{m3} \times R_s} \tag{19.2}$$

$$HD2 = 0.5V_a\frac{\left(\dfrac{dg_{m1}}{dV_{GT}}\right)}{2g_{m1}} \tag{19.3}$$

$$HD3 = 0.25V_a\frac{\left(\dfrac{d^2g_{m1}}{dV^2_{GT}}\right)}{6g_{m1}} \tag{19.4}$$

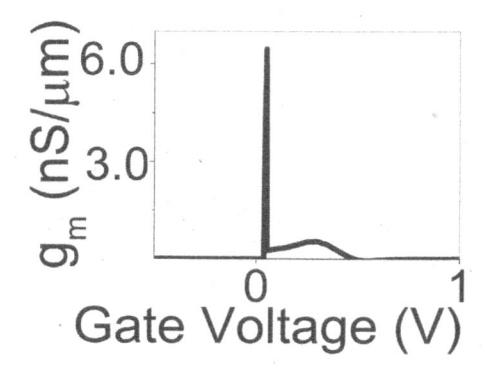

FIGURE 19.7 Variation in g_m with respect to gate voltage (V_{GS}).

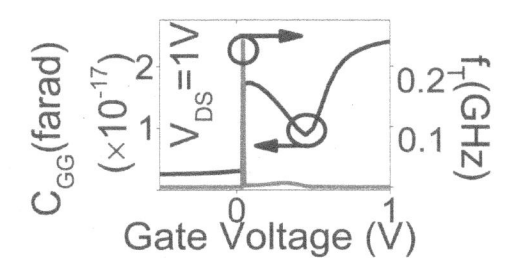

FIGURE 19.8   Curve for total capacitance and cutoff frequency is plotted. Black line represents total capacitance and the red line axis represents cutoff frequency.

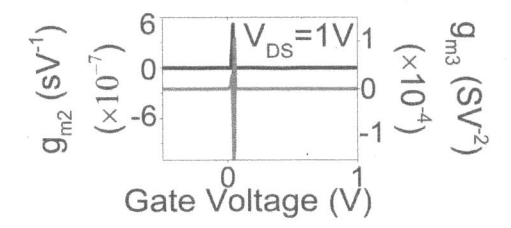

FIGURE 19.9   Curve for second-order transconductance (black line) and third-order transconductance (red line).

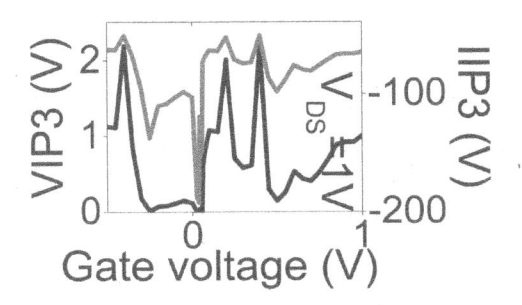

FIGURE 19.10   Curve for *VIP*3 and *IIP*3 against gate voltage is plotted: black line, VIP3 and red line, IIP3.

The curve for *VIP*3 and *IIP*3 is plotted in Figure 19.10. As both are inversely proportional to g_{m3} (Equations 19.1 and 19.2), the higher value of *VIP*3 and *IIP*3 indicates better linearity. At operating voltage $V_{GS} = V_{DS} = 1$ V and *VIP*3 and *IIP*3 are 1.03 V and −65.04 V, respectively; however, the highest value attained for *VIP*3 is at $V_{GS} = 0.2$ V and is 2.03 V, and the highest value attained for *IIP*3 is also at the same potential and the value is −51.28 V.

The curve for *HD*2 and *HD*3 is depicted in Figure 19.11. For better properties, harmonic distortion should be low as it is directly proportional to g_{m2}. At operating voltage $V_{GS} = V_{DS} = 1$ V, and the value obtained for *HD*2 and *HD*3 is −33.16 and −52.06, respectively. The lowest value of −65.2 dBm achieved for *HD*2 is at $V_{GS} = 0.2$ V and for *HD*3 (−340.69 dBm) is achieved at $V_{GS} = 0.5$ V.

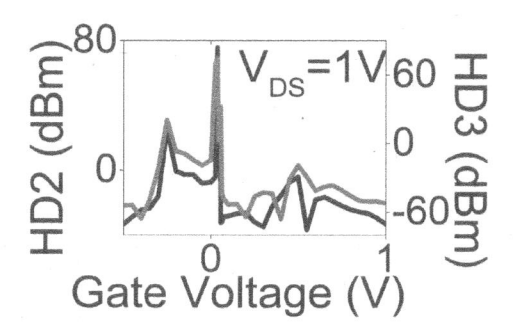

FIGURE 19.11   Curve for *HD*2 (black line) and *HD*3 (red line) against V_{GS} is plotted.

19.4   CONCLUSION

In this work, the feedback transistor using MoTe$_2$ as channel material has been designed for fast switching and low power applications. The feedback transistor designed provides a subthreshold slope of 1.7 mV/dec and OFF-state current of 10^{-15} A/μm. For the proposed device, different device, analog, and linearity properties are examined. It was found that feedback transistors can be used to replace MOSFETs in the future. There is also a scope of improved ON-state current in the future.

REFRENCES

1. Moore, G. E. "Cramming more components onto integrated circuits," Intel Electronics Magazine, 1965, Retrieved April 1, 2020: https://www.cs.utexas.edu/~fussell/courses/cs352h/papers/moore.pdf
2. Muralidhar, R., I. Lauer, J. Cai, D. J. Frank, and P. Oldiges. "Toward ultimate scaling of MOSFET," IEEE Transactions on Electron Devices, vol. 63, no. 1, pp. 524–526, 2015.
3. Liu, H., A. T. Neal, and P. D. Ye. "Channel length scaling of MoS2 MOSFETs," ACS Nano, vol. 6, no. 10, pp. 8563–8569, 2012.
4. Bulucea, C., F.-C. Wang, and P. Chaparala. "Field-effect transistor for alleviating short-channel effects," U.S. Patent, 6,548,842, issued April 15, 2003.
5. Bhuwalka, K. K., S. Sedlmaier, A. K. Ludsteck, C. Tolksdorf, J. Schulze, and I. Eisele. "Vertical tunnel field-effect transistor," IEEE Transactions on Electron Devices, vol. 51, no. 22004, pp. 279–282, 2004.
6. Ionescu, A. M., and H. Riel. "Tunnel field-effect transistors as energy-efficient electronic switches," Nature, vol. 479, no. 7373, pp. 329–337, 2011.
7. Goto, K.-I., and Z. Wu. "Non-uniform channel junction-less transistor," U.S. Patent, 8,487,378, issued July 16, 2013.
8. Barraud, S., R. Coquand, M. Casse, M. Koyama, J. M. Hartmann, V. Maffini-Alvaro, and C. Comboroure. "Performance of omega-shaped-gate silicon nanowire MOSFET with diameter down to 8 nm," IEEE Electron Device Letters, vol. 33, no. 11, pp. 1526–1528, 2012.
9. Lee, M., Y. Jeon, M. Kim, and S. Kim. "Flexible semi-around gate silicon nanowire tunnel transistors with a sub-kT/q switch," Journal of Applied Physics, vol. 117, no. 224502, 2015.
10. Choi, W. Y., J. Y. Song, B. Y. Choi, J. D. Lee, Y. J. Park, and B. G. Park. "70 nm impact-ionization metal-oxide semiconductor (I-MOS) devices integrated with tunneling field-effect transistors (TFETs)," IEDM Technical Digest, pp. 975–978, 2005.
11. Jo, J., W. Y. Choi, J. D. Park, J. W. Shim, H. Y. Yu, and C. Shin. "Negative capacitance in organic/ferroelectric capacitor to implement steep switching MOS devices," Nano Letters, vol. 15, pp. 4553–4559, 2015.
12. Dirani, H. E., Y. Solaro, P. Fonteneau, C. A. Legrand, D. Marin-Cudraz, D. Golanski, P. Ferrari, and S. Cristoloveanu. "A band-modulation device in advanced FDSOI technology: sharp switching characteristics," Solid-State Electron, vol. 125, pp.103–113, 2016.
13. Lee, C., J. Sung, and C. Shin. "Understanding of feedback field-effect transistor and its applications," Applied Sciences, vol. 10, no. 9, 2020.
14. Singh, D., and G. C. Patil. "Performance analysis of feedback field-effect transistor-based biosensor," IEEE Sensors Journal, vol. 20, no. 22, pp. 13269–13276, 2020.

15. Güneş, F., and O. Yurttakal. "Full flexible performance characterization of a feedback applied transistor with LNA applications," International Journal of Circuit Theory and Applications, vol. 48, pp. 56–71, 2020.

16. Choi, W. Y., B.-G. Park, J. D. Lee, and T.-J. K. Liu. "Tunneling field-effect transistors (TFETs) with sub-threshold swing (SS) less than 60 mV/dec," IEEE Electron Device Letters, vol. 28, no. 8, pp. 743–745, 2007.

17. Dewey, G., B. Chu-Kung, J. Boardman, J. M. Fastenau, J. Kavalieros, R. Kotlyar, and W. K. Liu et al. "Fabrication, characterization, and physics of III–V heterojunction tunneling field effect transistors (H-TFET) for steep sub-threshold swing." In 2011 International Electron Devices Meeting, pp. 33–6. IEEE, 2011.

18. Choi, W. Y., and H. K. Lee. "Demonstration of hetro-gate-dielectric tunneling field-effect transistors," Nano Convergence, vol. 3, pp. 1–15, 2016.

19. Collins, P. G., and P. Avouris. "Nanotubes for electronics," Scientific American, vol. 283, no. 6, pp. 62–69, 2000.

20. Tans, S. J., A. R. M. Verschueren, and C. Dekker. "Room-temperature transistor based on a single carbon nanotube," Nature, vol. 393, no. 6680, pp. 49–52, 1998.

21. Fuhrer, M. S., J. Nygård, L. Shih, M. Forero, Y.-G. Yoon, H. J. Choi, Jisoon Ihm, S. G. Louie, A. Zettl, and P. L. McEuen. "Crossed nanotube junctions," Science, vol. 288, no. 5465, pp. 494–497, 2000.

22. Che, Y., H. Chen, H. Gui, J. Liu, B. Liu, and C. Zhou. "Review of carbon nanotube nanoelectronics and macroelectronics," Semiconductor Science and Technology, vol. 29, no. 7, pp. 073001, 2014.

23. Dan, L., and R. B. Kaner. "Graphene-based materials," Nature Nanotechnology, vol. 3, no. 101, pp. 101–105, 2008.

24. McDonnell, S. J., and R. M. Wallace. "Atomically-thin layered films for device applications based upon 2D TMDC materials," Thin Solid Films, vol. 616, pp. 482–501, 2016.

25. Manzeli, S., D. Ovchinnikov, D. Pasquier, O. V. Yazyev, and A. Kis. "2D transition metal dichalcogenides," Nature Reviews Materials, vol. 2, no. 8, pp. 17033, 2017.

26. Morita, A. "Semiconducting black phosphorus," Applied Physics A, vol. 39, no. 4, pp. 227–242, 1986.

27. Chaujar, R., R. Kaur, M. Saxena, M. Gupta, and R. S. Gupta. "TCAD assessment of Gate Electrode Workfunction Engineered Recessed Channel (GEWE-RC) MOSFET and its multi-layered gate architecture, Part II: Analog and large signal performance evaluation," Superlattices Microstructure, vol. 46, no. 4, pp. 645–655, 2009.

20 Reduced Graphene-Metal Phthalocyanine-Based Nanohybrids for Gas-Sensing Applications

Aman Mahajan and Manreet Kaur Sohal

CONTENTS

20.1 INTRODUCTION

Day by day, the environment around us is becoming polluted with various combustible, flammable, toxic, and hazardous gases emitted from factories, automobiles, and shipping and healthcare industries. The need to detect these gases for the safety of earth inhabitants has ushered in the advancement of gas-sensing technology, which can now sense parts per billion (ppb) levels of these dangerous gases. The device that detects these noxious gases is called a gas sensor. Based on the operation mechanism, these sensors are classified into catalytic, electrochemical, thermal conductivity, optical, semiconductor, acoustic wave, and chemiresistive types.

Chemiresistor gas sensors are predominantly used because they are inexpensive, portable, and consume minimum amounts of power. They operate on the reversible gas adsorption process principle, which changes the active layer's electrical resistance whenever the analyte gas adsorbs on the sensing layer. The resistance of n- and p-type active layers may decrease or increase depending on the reducing (CO, C_2H_6O, H_2, etc.) or oxidizing (NO_2, O_3, SO_2, etc.) nature of the target gas. The magnitude of change in resistance is proportional to the concentration of target gas present in the sensor's vicinity. A key indicator of a high-performance sensor is if its sensing material has a large specific adsorption area.

Nanomaterials have a high surface area to volume ratio, which facilitates the gas molecules' adsorption and improves gas sensitivity. Nanostructured metal oxide (MO) semiconductors, because they are inexpensive together with a lightweight and compact design, are widely used as active layer materials in chemiresistors. MOs, such as SnO_2-, ZnO-, CeO_2-, and TiO_2-based gas sensors, have been used for various gases and volatile organic compound (VOC) detection. However, these

DOI: 10.1201/9781003126393-20

TABLE 20.1

List of Graphene-Based Chemiresistive Gas Sensors

Material Used	Selectivity	Reference
ZnO/graphene	NO_2	Alfano et al. (2019)
rGO	Dimethyl methylphosphonate	Alizadeh and Soltani (2016)
SnO_2/GO	Ethanol	Pargoletti et al. (2019)
Pd/graphene	H_2	Lee et al. (2019)
Ag/polymer/aminoanthraquinone/GO	NO_2	Li et al. (2019)
Boron and nitrogen/GO hydrogel	NO_2	Wu et al. (2019)
Graphene	NO_2 and NH_3	Schiattarella et al. (2017); Travan and Bergmann (2019); Rigoni et al. (2017)
WO_3/Pt-rGO nanosheets	Acetone	Chen et al. (2018)
Polyurethane nanofibers-rGO	NO_2	Duy et al. (2017)
Graphene	NO_2	Ricciardella et al. (2017)
Sulfonated rGO hydrogel	NO_2 and NH_3	Wu et al. (2017)
MoO_3/rGO	H_2S	MalekAlaie et al. (2015)
Tannic acid-rGO	NH_3	Yoo et al. (2014)
CuPc/rGO	Cl_2	Kumar et al. (2017)

sensors suffer from issues of cross-selectivity and high operating temperatures and high power consumption. To overcome these limitations, room temperature (RT) sensors have been developed that are safer, more economical, and consume less power.

In the past few years, various organic materials (like porphyrins and phthalocyanines [Pcs]), conducting polymers (such as polythiophene, polyamine [PANI], polypyrrole), and carbon-based materials have been explored for their RT sensing applications. Among them, graphene for gas sensors has recently attracted serious attention mainly due to its extraordinary electrical conductivity, high thermal as well as chemical stability, and excellent mechanical strength.

Also, graphene's derivatives, i.e., graphene oxide (GO) and reduced graphene oxide (rGO), show a high specific adsorption surface area (Tarcan et al. 2020). Compared with pristine graphene, the GO can be produced in high yield from inexpensive graphite. Nevertheless, the oxygen-related functional groups decrease its conductivity immensely at RT and restrict its application in the gas-sensing field. To stabilize the conductivity, the reduction of GO is performed. By adjusting the reduction content, the sensing characteristics of rGO can be optimized. To further improve its sensing performance, various graphene-based nanocomposites with metals, MOs, polymers, and small organic molecules have been studied. A list of previously reported graphene-based chemiresistive sensors is tabulated in Table 20.1.

In this chapter, a comprehensive review of synthesis techniques of rGO, nanohybrids, and rGO-based gas sensors has been given. At last, a facile way to develop ultrasensitive graphene-based sensors through functionalization has been illustrated. It has been shown that by carefully modulating the properties of graphene sensors, excellent performance of sensors can be realized.

20.2 GRAPHENE

A monoatomic layer of sp^2-hybridized C atoms that are densely packed in a hexagonal structure is called graphene (Figure 20.1). Since its discovery (Novoselov et al. 2005), graphene is the most widely explored nanomaterial and it has attracted extensive attention from researchers for its extremely diversified applications because of its exceptionally high electrical conductivity with low noise and high mechanical strength together with zero bandgap. Additionally, it has a surprisingly

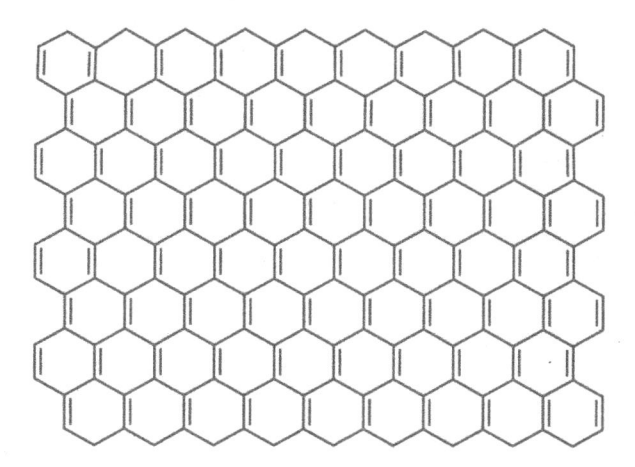

FIGURE 20.1 Monoatomic layer of graphene.

high surface area, high electron mobility of 2×10^4 cm^{2}V^{-1}s^{-1}, and high charge carrier density of 10^{13} cm^{-2}. Because of these qualities, graphene has been used in numerous applications including energy storage, Li-ion batteries, solar cells, supercapacitors, micro-electro-mechanical systems, gas sensors, and biosensors.

Graphene is considered an excellent sensing material, mainly due to its high specific adsorption surface area equivalent to 26^{30} m^{2}g^{-1}, excellent electrical properties, and thermal and chemical stability. Graphene delivers the largest adsorption area per unit volume, about twofold higher than that of carbon nanotubes (CNTs), as every atom on the graphene monolayer sheet is capable of interacting with the analyte gas. Even molecule-level adsorption of a gas on a graphene layer can bring about a quantifiable change in graphene's resistance (Schedin et al. 2007). Further, its low electrical resistivity of 10^{-6} Ω with high RT carrier mobility and high carrier density aids in smooth electronic movement between adsorbed gas molecules and electrodes.

20.2.1 GRAPHENE OXIDE

Despite all of these results, graphene production is expensive, which limits its commercial use. So, as a low-cost alternative, GO, which is almost like graphene, has been explored for gas-sensing applications. Unlike graphene, its surface is linked to different functional groups like epoxide, carbonyl, hydroxyl, and phenol (Figure 20.2(a)). It is usually prepared from graphite oxide, which

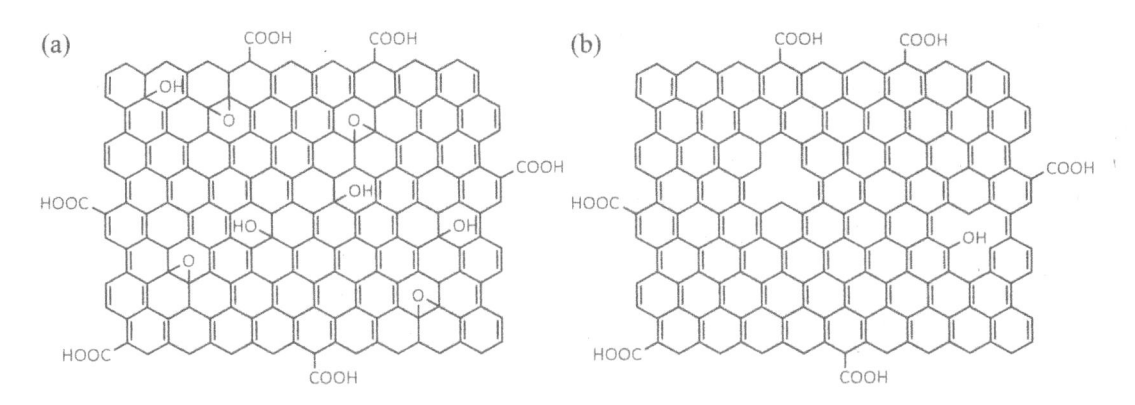

FIGURE 20.2 (a) Graphene oxide and (b) reduced graphene oxide.

is structurally similar to graphite, but with extra-large irregular spacing between sheets. Various techniques, such as self-assembly, chemical vapor deposition (CVD), epitaxial growth, and the arc discharge method in addition to micromechanical and electrochemical exfoliation, the chemical synthesis method, and the Hummers method have been used to synthesize GO.

However, due to an increase in oxygen-containing functional groups, a huge decrease in the conductivity of GO has been observed, which is unfavorable for gas sensing (Wang et al. 2016). Thus, to stabilize the conductivity, which is a prerequisite for efficient gas sensors, the reduction of GO has been carried out.

20.3 REDUCED GRAPHENE OXIDE

20.3.1 SYNTHESIS METHODS OF REDUCED GRAPHENE OXIDE

The different reduction methods of GO to produce rGO (Figure 20.2(b)) are briefly discussed next:

- *Chemical reduction*: This is an inexpensive and facile way to synthesize large quantities of rGO at RT. In this method, a solution of GO and water is ultrasonicated to form a homogeneous dispersion. An appropriate strong chemical base is added to the solution to obtain the rGO in the form of precipitates. Reduction of GO has been carried out by means of different reducing agents, like amino acids, hydrazine hydrate, phenylhydrazine, strongly alkaline solutions, glucose, hydroxylamine, hydroquinone, pyrrole, sodium borohydride, urea, and ascorbic acid (Chua and Pumera 2014). The oxygen-containing functional groups are mostly eradicated during the reduction procedure, and π-electron conjugation of graphite is partially restored. The resultant rGO largely resembles graphene sheets. However, it still contains some residual oxygen atoms and structural defects.
- *Thermal reduction*: Thermal reduction is the process of reducing GO in a vacuum or inert gas or reducing gas. It involves multistep eradication of oxygen-containing functional groups from GO. During rapid heating, the oxygen substituents attached to the graphene decompose into CO or CO_2 gases, which then exert enormous pressure on stacked graphene sheets. When these gases are evolved, the sheets are exfoliated to form monomolecular graphene sheets. In comparison, the rGO obtained has fewer oxygen substitutions. Nevertheless, the pressure exerted by the gases produced on stacked carbon layers and the removal of carbon atoms damage the resulting graphene sheets and induce structural defects that consequently affect the electronic properties of rGO produced.
- *Microwave and photoreduction*: Graphitic precursor materials can be quickly reduced and exfoliated using the energy-efficient microwave and photoirradiation techniques. In this method, the reduction of either dry GO or GO in suspension solution can be performed to obtain rGO. Microwaves transfer the energy directly to the reactants and raise its temperature instantaneously (Chen et al. 2010). This sudden increase in internal temperature causes the ultrafast reduction of GO to form rGO. As a result, the reaction time is shortened to just seconds and reaction efficiency is improved.
- Likewise, photoreduction is performed by irradiating GO with ultraviolet (UV) light, sunlight, or a KrF excimer laser. However, in this technique, high-quality rGO can be obtained without any addition of photocatalyst.
- *Photocatalyst reduction*: Photocatalysts like TiO_2 can also help to carry out reduction of GO by removing the various unwanted oxygen-related functionalization attached to the GO surface. Primarily, on UV irradiation, holes and electrons are produced on the TiO_2 surface due to charge separation. The ethanol (in GO-ethanol solution) readily reacts with the holes to produce ethoxy radicals, while electrons interact with the carbon sheets and, thus, reduce the attached functional groups.

- *Solvothermal/hydrothermal reduction*: Solvothermal or hydrothermal reduction is performed in sealed containers under high pressure and moderate temperature. In the solvothermal method, by increasing the pressure, the solvent's temperature is brought above its boiling point, which plays the role of a reducing agent. The hydrothermal technique acts as a green chemistry alternative to organic chemicals, as it uses water as a solvent. Under these supercritical conditions, the reactivity of the solvent escalates and eventually shrinks the reaction time.

20.3.2 rGO-Based Gas Sensors

Compared with graphene, rGO can be easily processed into ultrathin sensing sheets and via chemical grafting; it can be functionalized with different chemical groups to tune the sensing characteristics. The dependence of sensing properties on annealing temperature and degree of reduction has been investigated on GO reduced with thermal annealing in an inert gas (Ar) atmosphere (Lu et al. 2009). The prepared sensors have been reported to detect NO_2 and NH_3 at RT. For further improvement of a gas sensor's response, sensitivity, and selectivity with lower Detection limits (DLs), the nanocomposites of rGO with various organic and inorganic molecules, metals, MOs, and polymers have been explored worldwide. For instance, when combining rGO with carbon nanodots, detection of NO_2 gas at a small concentration of 5 ppm (74.5% response) with a DL of 10 ppb has been achieved (Hu et al. 2017). The rGO nanocomposites with Pt, Ag, and Au metal nanoparticles prepared via the chemical reduction method have been shown to sense NH_3 gas at a 1-ppm concentration (Shin et al. 2013). WO_3/rGO nanocomposites decorated with palladium (Pd) metal have been shown to detect H_2 gas in a different concentration range 20–10,000 ppm (Esfandiar et al. 2014). The SnO_2/rGO nanospheres and CuO/rGO nanoflowers have been used to detect NH_3 and formaldehyde gas molecules at RT (Zhang et al. 2017). H_2S sensors fabricated using SnO_2 quantum wire-rGO-based nanocomposites showed a DL of 43 ppb at RT (Song et al. 2016). Qin et al. (2020) have developed rGO a nanosheet/polypyrrole composite for RT portable detection of NH_3.

Functionalizing rGO with organic materials like porphyrins (Sakthinathan et al. 2017) and Pcs (Guo et al. 2019) is a great strategy to improve sensing characteristics. However, limited research has been conducted and published on the effect of organic materials linking on rGO.

20.4 PHTHALOCYANINES (Pcs)

Pcs have emerged as vital semiconductors in the advanced technology fields of optoelectronics, nonlinear optics, photovoltaics, organic semiconductors, liquid crystal displays, and gas sensors (Gounden et al. 2020). The wide popularity of these materials can be partially ascribed to their high level of processability, non-toxicity, and low cost. Pc is an $18\text{-}\pi$ conjugated macrocycle composed of four nitrogen-linked isoindole units in a usual square planar structure (Figure 20.3). The high degree of π-electron delocalization and closeness of the energy levels of frontier orbitals allow the Pc molecule to easily donate and absorb an electron, making it suitable for various semiconductor applications. Pc electronic and surface properties can be easily tuned by varying this central metal atom alone. In contrast to metal-free Pcs, metallophthalocyanines (MPcs; M = Zn, Cu, Co, Ag, etc.) are highly chemically, thermally, and environmentally stable.

Also, there are an additional 16 substitutional sites available at peripheral (2, 3, 9, 10, 16, 17, 23, 24) or/and non-peripheral sites (1, 4, 8, 11, 15, 18, 22, 25) on the central metal atom (see Figure 20.3) (Mukherjee et al. 2016). The functionalization changes the electron density at peripheral nitrogen sites and renders Pcs with highly tunable properties. In principle, these functionalities can be separated into two groups: (1) electron withdrawing and (2) electron releasing. Chemical units, including sulfonyl, carboxyl, or fluor groups, comprise the former group, whereas functions such as amino, alkoxy, or alkyl groups, represent the latter group, respectively. Through manipulation of central metal atom and substitutional groups, Pcs can be tailored to selectively detect various toxic gases

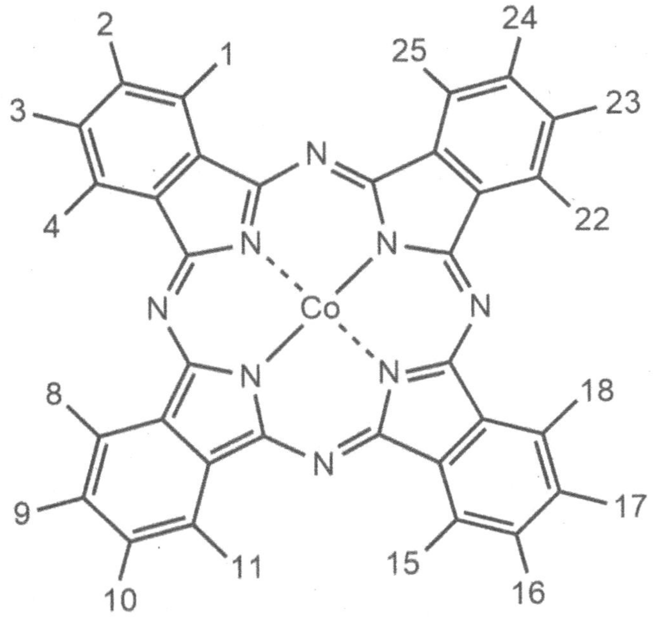

FIGURE 20.3   Representative phthalocyanine ring.

with high response. The π-electrons present at the axial position of the macrocycle aids in interaction with analyte gases, which are eventually manifested as variations in conductivity, mass, and optical properties. Thus, Pcs can be used as chemiresistive, solid-state ionic, or field-effect transistor (FET) sensors for oxidizing and reducing gases, capacitive, quartz crystal microbalance, Surface plasmon resonance (SPR), Surface acoustic wave (SAW), or ellipsometry gas sensors.

20.4.1   Phthalocyanines as Gas Sensors

The consequence of NO_2 adsorption on electrical characteristics of metal-free Pc (H_2Pc) has been studied by Shinbo et al. (2002). They observed that $ITO/H_2Pc/Al$ heterostructure sensors' resistance increased significantly as the adsorbed NO_2 captured electrons from the Pc layer. The gas-sensing performance of the metal-free Pc (H_2Pc) and MPc (M = Mn, Cu, Ni, Zn, Co and Pb) single crystals has been studied, and on exposure to NO_2 and dinitrogen tetroxide (N_2O_4), an increase in conductivity by a factor of 10^8 has been observed (van Ewyk et al. 1980). Sub-ppm levels of NO_2 have also been detected using CuPc chemiresistors (Chia et al. 2016). Moreover, it has been observed that the type of morphology of NiPc film depends on the temperature of the substrate during deposition and post-deposition annealing, and that affects the sensing characteristics due to variation of adsorption sites and conducting path (Liu et al. 2004).

Altindal et al. (2001) prepared thin films using two amino and two alkyl sulfonyl groups using peripherally substituted soluble Pcs. They observed that the sensing response increases linearly with the rising concentration (0.05–0.15 ppm) of Cl_2 and Br_2. The effect of fluorine substitution on the sensing response of $MPcF_x$ (M = Co, Zn, Cu; x = 0, 4, 16) films toward NH_3 has also been thoroughly investigated (Klyamer et al. 2018). The $MPcF_4$ films have been found to be more sensitive among others due to lower binding energies (BE) of the $M-NH_3$ bond and the response has been observed to decrease in order of Co > Zn > Cu. Additionally, hexadecafluorinated MPc (M = Zn, Cu, Co) hybrids with Single-walled carbon nanotubes (SWCNTs)/multi-walled carbon nanotubes (MWCNTs) have been employed for ppb-level sensing of Cl_2 (Sharma et al. 2017, 2018a,b, 2019). Although many groups have performed extensive investigations on CNT/MPc hybrids, sporadic studies have been reported on rGO/MPc hybrids.

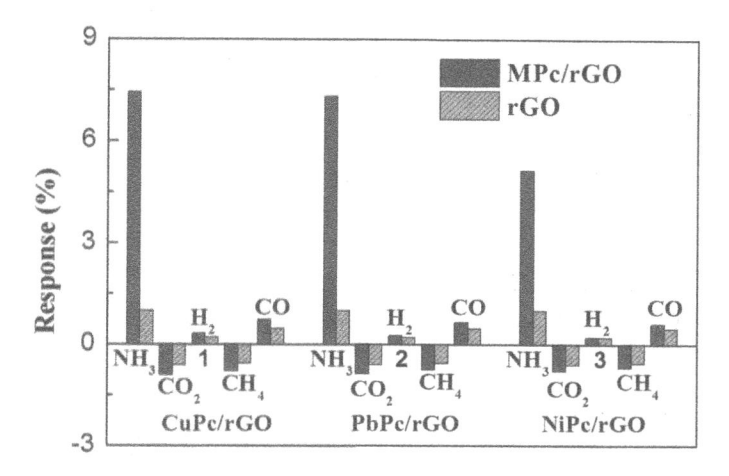

FIGURE 20.4 Selectivity behavior of rGO and MPc/rGO (M = Cu, Ni, Pb) toward various gases. (Adapted from Li et al. 2015. Published by Springer Nature.)

20.4.2 rGO/Pc Hybrid-Based Gas Sensors

It has been proposed that synergetic effects from linking MPc with rGO via covalent or noncovalent functionalization enhance the electrical conductivity of the latter and improve the processibility of the former. The first such mutual interaction of rGO and MPc affecting the sensing characteristics toward NH_3 was observed by Zhou et al. (2014). The hybrid sensors showed relatively higher sensitivity with smaller response and recovery time in contrast to rGO. Further studies have been performed using rGO functionalized with tetra-α-iso-pentyloxyphthalocyanine copper/nickel/lead (Cu/Ni/Pb Pc) (Li et al. 2015). Hybrid sensors have been observed to show enhancement of more than sixfold compared with that of a pure rGO sensing device toward 800 ppb of NH_3 (Figure 20.4).

To demonstrate the effect of a substituted Pc molecule on NH_3 sensing properties of rGO, Guo et al. (2019) prepared four different rGO/RPcCo nanohybrids where substituent group R = tetra-β-carboxylphenoxyl (cpoPcCo), tetra-β-phenoxyl (poPcCo), tetra-β-(4-carboxy-3-methoxyphenoxy) (cmpoPcCo), and tetra-β-(3-methoxyphenoxy) (mpoPcCo). They coupled rGO with RPcCo using the Hummers method as depicted in Figure 20.5(a). They investigated the sensing properties of rGO, RPcCo, and rGO/RPcCo nanohybrid sensors by exposing them to 17 different test gases and

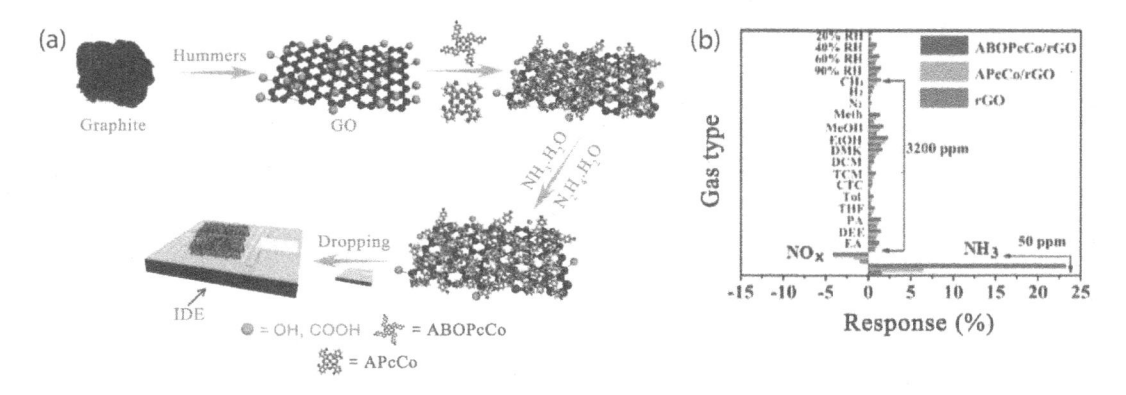

FIGURE 20.5 (a) Schematic of the Hummers method for synthesis RPcCo/rGO nanohybrids and (b) selectivity behavior of rGO, RPcCo, and RPcCo/rGO nanohybrid-based sensors toward different gases. (Adapted from Guo et al. 2019. Published by The Royal Society of Chemistry.)

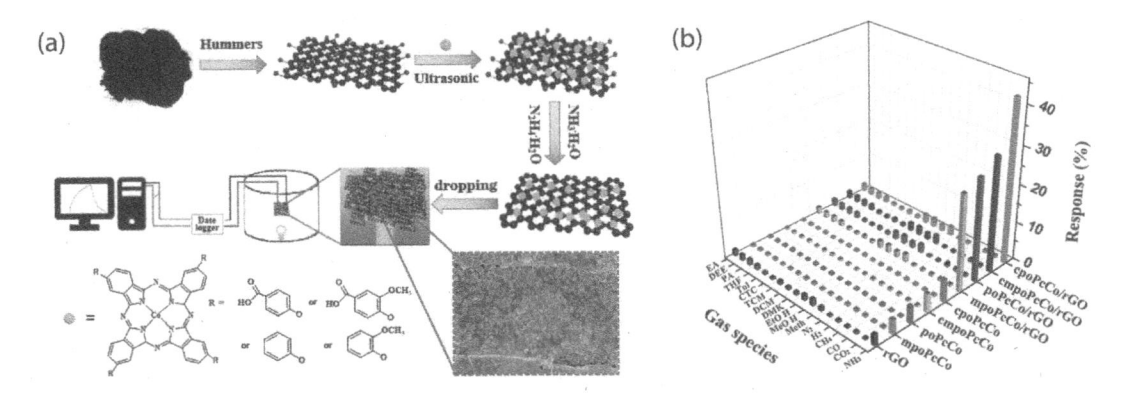

FIGURE 20.6 (a) Representation of the synthesis process of rGO, ABOPcCo/rGO, APcCo/rGO, and FPcCo/rGO nanohybrids and (b) selective response behavior of hybrids toward various gases. (Adapted from Wang et al. 2018. Published by The Royal Society of Chemistry.)

found that hybrid sensors showed a superior response to NH_3 (Figure 20.5(b)). It was observed that the rGO/cpoPcCo sensor exhibited a higher response toward NH_3 compared with other sensors with a DL of 3.7 ppb.

The same group has also studied the rGO sensing characteristics by anchoring rGO with different amino substituents such as tetra-α-aminophthalocyanine cobalt (APcCo) and tetra-α-(p-aminobenzyloxy) phthalocyanine cobalt (ABOPcCo) using a self-assembly reaction (Figure 20.6(a)) and compared it with substituent-free phthalocyanine cobalt (FPcCo/rGO) (Wang et al. 2018). The ABOPcCo/rGO sensor showed excellent sensing characteristics toward 50 ppm NH_3 approximately 7 and 15 times better than FPcCo/rGO- and APcCo/rGO-based sensors. They proposed that substituent groups on PcCo affected the charge transfer between hybrids and NH_3, as APcCo's amino group suppressed it, while the ABOPcCo's aminophenoxy group facilitated it (Figure 20.6(b)). Using Density functional theory (DFT) studies, they found a smaller bond length between the Co of ABOPcCo and NH_3 (Figure 20.7), suggesting stronger adsorption and better charge transport in compared with other rGO hybrids.

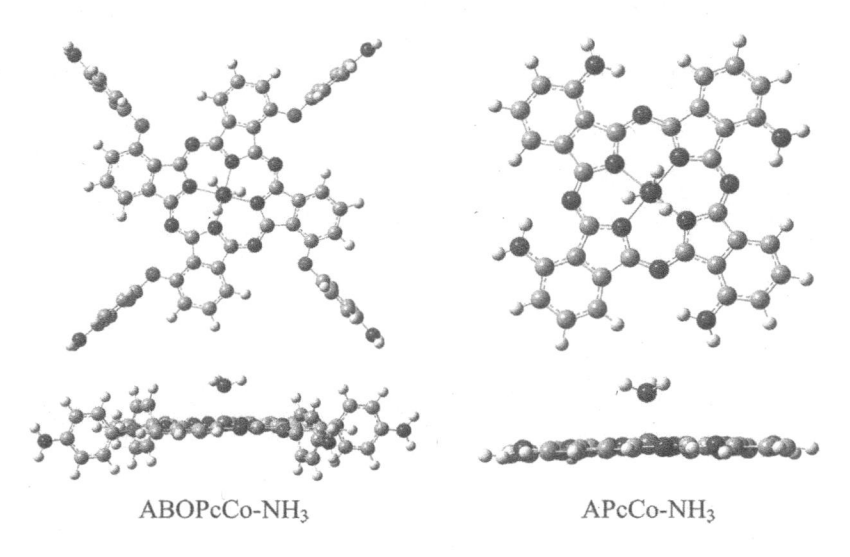

ABOPcCo-NH$_3$ APcCo-NH$_3$

FIGURE 20.7 Structures of NH_3-PcCos from the top and side view as estimated from DFT studies. (Adapted from Wang et al. 2018. Published by The Royal Society of Chemistry.)

MPcs/rGO with different metal ions (Cu^{2+}, Zn^{2+}, Co^{2+}, Ni^{2+}, and Pb^{2+}) have been investigated. On the basis of increasing ionic radii, these metal ions can be arranged as Pb^{2+} (119 pm) > Zn^{2+} (74 pm) > Cu^{2+} (73 pm) > Ni^{2+} (69 pm) > Co^{2+} (65 pm), whereas on the basis of d-electrons they can be arranged as Pb^{2+} (10) = Zn^{2+} (10) > Cu^{2+} (9) > Ni^{2+} (8) > Co^{2+} (7).

Thus, metals with smaller ionic radii have lesser d-electrons, which can contribute to π-electrons in the conjugated ring, leading to more accepter power of the Pc macrocycle and more sensitivity (Li et al. 2015). Further, the structure of the Pc macrocycle is also found to significantly affect the sensing response. Most of the Pc molecules have planar structures except PbPc, which forms four pyramid structures because of the limitation of central space for the larger Pb^{2+} ion in the Pc ring. This type of structure is favorable for gas adsorption due to its relatively weak conjugating power, which allows the central metal atom to easily adsorb target gases. As in the case of 1,8,15,22-tetra-(4-tert-butylphenoxyl)-metallophthalocyanine (TBPOMPc, M = Cu, Ni, and Pb)/rGO hybrids, the maximum response observed for TBPOPbPc/rGO has been linked to the pyramidal structure of the Pc macrocycle (Yu et al. 2017). Also, the sensing response and selectivity of a sensor have also been found to depend on the type of interaction between the target gas and central metal atom of MPc. The $r_{M\text{-}G}$ (bond length of metal-adsorbed gas) parameter determines the degree of gas adsorption on MPc. A stronger adsorption interaction has been found to occur when the bond length is smaller than estimated from BE of the MPc-gas system. The more negative the BE, the stronger the gas molecule is bonded to MPc. The DFT calculations in case of the MPc-NH_3 (M = Co, Zn, Cu, Ni, Pd, and Pb) system revealed that the BE follows the order of CoPc (−52.70 kJ/mol) < ZnPc (−51.27 kJ/mol) < CuPc (−7.09 kJ/mol) < PbPc (2.87 kJ/mol) < PdPc (9.79 kJ/mol) < NiPc (9.43 kJ/mol). Because negative BE indicates an exothermic process, Co, Zn, or Cu Pcs formed stable structures with NH_3 and showed a higher response compared with the rest for which NH_3 adsorption is endothermic. A literature review of different rGO/Pc-based sensors is provided in Table 20.2. From Table 20.2, it can be concluded that Pc molecules remarkably enhance the sensitivity of pristine rGO down to the ppb level.

20.5 GAS-SENSING PROPERTIES OF rGO/MPc HYBRIDS

In this section, synthesis and gas-sensing applications of hybrids, namely rGO/CuPc, rGO/F_{16}CuPc, and rGO/F_{16}ZnPc, are demonstrated to understand the possible reasons behind the improved gas-sensing performance.

TABLE 20.2
List of Previously Reported Pc/rGO-Based Gas Sensors

Sensor	Selectivity	Detection Limit (ppb)	Reference
CuPc/rGO	NH_3	400	Zhou et al. (2014)
1,8,15,22-tetra-(4-*tert*-butylphenoxyl)-metallophthalocyanine (TBPOMPc, M = Cu, Ni, and Pb)/rGO	NH_3	300	Yu et al. (2017)
Tetra-α-iso-pentyloxyphthalocyanine (CuPc, NiPc, PbPc)/rGO	NH_3	800	Li et al. (2015)
Tetra-α-(p-aminobenzyloxy)phthalocyanine cobalt (ABOPcCo)/rGO	NH_3	78	Wang et al. (2018)
cpoPcCo/rGO	NH_3	3.7	Guo et al. (2019)
CuPc/rGO	Cl_2	1.97	Kumar et al. (2017)
F_{16}CuPc/rGO	Cl_2	1.41	Kumar (2018)
F16ZnPc/rGO	Cl_2	–	Sharma et al. (2019)

The GO has been prepared through the Hummer method mentioned previously (Kumar et al. 2018). The chemical route presented in Figure 20.8 has been followed to prepare hybrids of rGO/MPc. Exfoliation of GO has been performed via sonication of a dried GO and Dimethylformamide (DMF) mixture. Simultaneously, the solution of MPc powder in DMF has been prepared and dropwise added into the GO mixture with continuous stirring. The obtained mixture has then been continuously stirred in the dark for 24 hours. Before refluxing at 100°C for a further 24 hours, hydrazine hydrate ($N_2H_4 \cdot xH_2O$) was added to this solution. At last, the resultant is consequently filtered, and the obtained hybrids are dried for 3 hours at 150°C in a vacuum oven. For gas sensor fabrication, rGO/MPc solution is drop-casted on thermally deposited gold electrodes on a glass substrate.

Through π–π stacking, CuPc was observed to form nanoflowers in-plane to the rGO surface in case of rGO/CuPc hybrids using High-resolution transmission electron microscope (HRTEM) (Kumar et al. 2017) (inset, Figure 20.8). The gas-sensing properties of rGO and rGO/CuPc hybrid sensors have been studied by exposing them to 500 ppb of different oxidizing and reducing gases at RT. The rGO/CuPc hybrid sensor (Figure 20.9(a)) (Kumar et al. 2017) has been found to be highly selective toward Cl_2 with a response of 99%. Further, compared with the pure rGO sensor, the rGO/CuPc hybrid sensor showed a higher change in conductance values on incident of 50–3000 ppb of Cl_2 (Figure 20.9(b)) (Kumar et al. 2017). The DL of rGO/CuPc is calculated to be 1.97 toward Cl_2 (Kumar et al. 2017).

To understand the underlying mechanism of Cl_2 sensing, hybrid sensors have been investigated using Raman spectroscopy and electrochemical impedance spectroscopy (EIS) techniques. On exposing the rGO/CuPc hybrid sensor to 1000 ppb of Cl_2, Raman peaks corresponding to the macrocyclic vibrations of CuPc molecules (Saini et al. 2009; Gladkov et al. 2007) showed a downshift of 2 cm^{-1} compared with unexposed Raman spectra (Figure 20.10(a)).

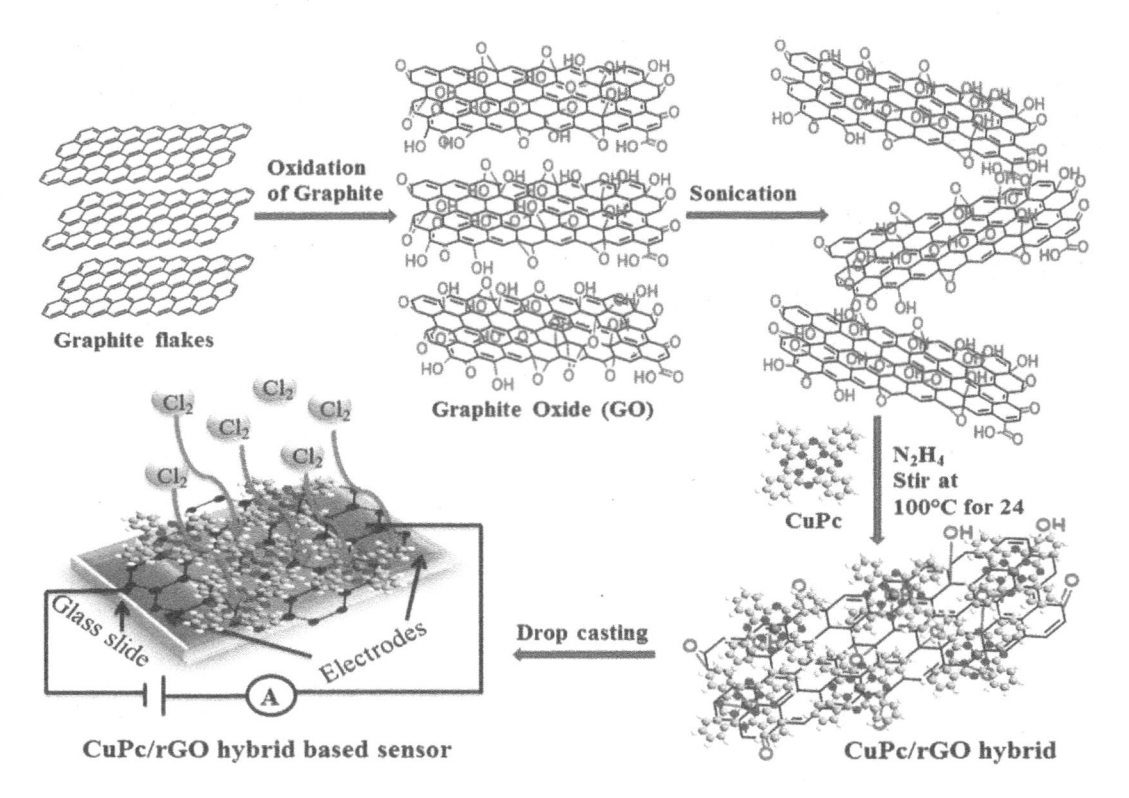

FIGURE 20.8 Schematic of chemical route to synthesize MPc/rGO hybrids. (Adapted from Kumar et al. 2017. Published by The Royal Society of Chemistry.)

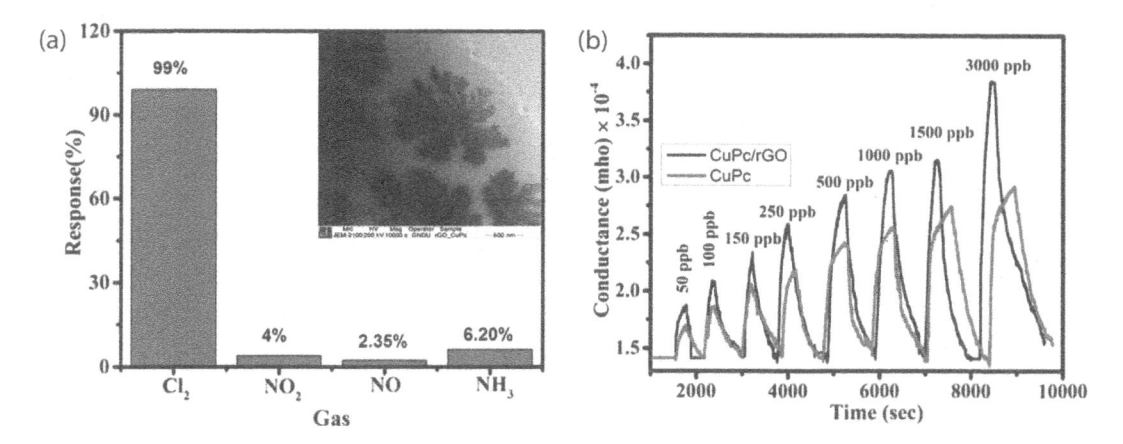

FIGURE 20.9 Inset: HRTEM image of the CuPc/rGO hybrid. (a) Selective response of CuPc/rGO hybrid sensor toward Cl_2 and (b) response curve of rGO and CuPc/rGO hybrid sensor toward various concentrations of Cl_2. (Adapted from Kumar et al. 2017. Published by The Royal Society of Chemistry.)

Further, peaks of 1451 and 1527 cm^{-1} have been observed to shift by 5 and 6 cm^{-1}, respectively, which confirms that Cu metal ions of the rGO/CuPc hybrid mostly interact with Cl_2. EIS studies have been performed to investigate the effect of bulk (R_0) and grain boundary resistance (R_1) of rGO sheets, and capacitance (C_1) across Pc and rGO grains toward Cl_2 sensing. Also, the effect of contact resistance (R_2) and capacitance (C_2) between hybrid and metal electrode has been examined. For CuPc/rGO a single semicircle observed in the Nyquist plot both before and after introduction of a fixed concentration of Cl_2 has been fitted using an equivalent circuit (Figure 20.10(b)). The diameter of the semicircle has been observed to reduce on Cl_2 exposure resulting in a drop in R_0, R_1, and R_2, and an upsurge in C_1 and C_2 values in relation to virgin hybrid samples (Table 20.3). Thus, this variation implies that the grain boundaries between the CuPc ring and rGO sheets act as dominating active sites for Cl_2 physisorption and provide a conducting path for electron movement from a CuPc macrocycle to rGO sheets.

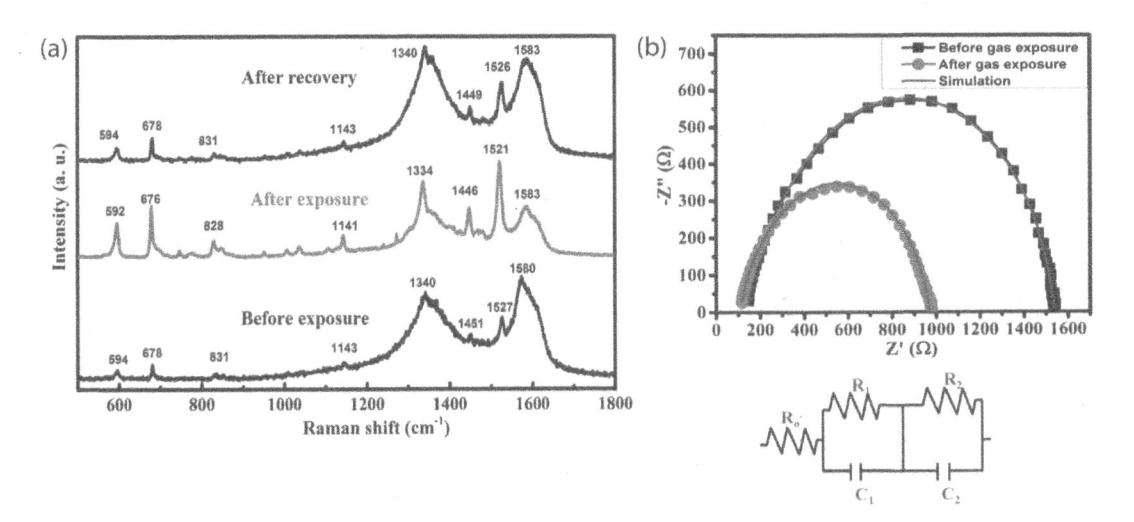

FIGURE 20.10 (a) Raman spectra of CuPc/rGO hybrid in air, on Cl_2 exposure, and after recovery and (b) EIS spectra of CuPc/rGO hybrid in air and after Cl_2 exposure. (Adapted from Kumar et al. 2017. Published by The Royal Society of Chemistry.)

TABLE 20.3

EIS Parameters Calculated for CuPc/rGO-Based and F_{16}CuPc/rGO-Based Gas Sensors Before and After Cl_2 Exposure

	R_0 (Ω)	R_1 (Ω)	R_2 (Ω)	C_1 (nF)	C_2 (nF)
In air (CuPc/rGO)	149	1150	216	9.78	7.71
On Cl_2 (CuPc/rGO)	119	653	180	13.5	9.16
In air (F_{16}CuPc/rGO)	1175	5811	1260	2.1	1.71
On Cl_2 (F_{16}CuPc/rGO)	1123	3774	1190	4.9	3.16

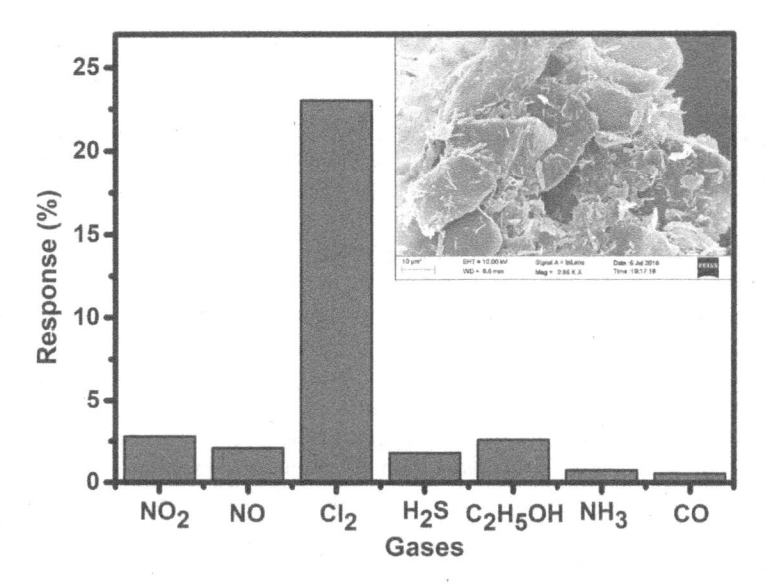

FIGURE 20.11   FESEM image (inset) and selective response characteristics of F16ZnPc/rGO hybrids. (Adapted from Sharma et al. 2019.)

When using substituted CuPc (F_{16}CuPc), similar behavior can be observed; however, due to the presence of a fluorine group better sensing characteristics like high response values, smaller response and recovery periods, and lower DL can be obtained. As fluorination of CuPc alters the electronic structure of MPc, the electron charge transfer between the metal ion and adsorbing gas would be enhanced resulting in better sensitivity.

On changing the Cu ion to Zn (F_{16}ZnPc), the morphology of the Pc molecule has been observed to vary from flowers to rods like the structure on rGO as seen using Field-emission scanning electron microscope (FESEM) (Sharma et al. 2019). The rGO/F_{16}ZnPc-based (inset, Figure 20.11) (Sharma et al. 2019) sensors have been found to be highly selective toward Cl_2 with a response of 23% (Figure 20.11), which is lower than that observed in case of rGO/CuPc sensors. This further confirms that ionic radii of the central metal ion predominantly govern the sensitivity of the MPc molecule and rGO/MPc hybrids.

20.6 SUMMARY

Graphene has demonstrated a high potential to be an excellent chemiresistive-based gas sensor due to its high sensitivity, responsivity, selectivity, reproducibility, low DL, and cost. In this chapter, we have briefly discussed the various methods available to synthesize rGO and its gas-sensing performance. It has been revealed that MPc/rGO-based hybrid sensors exhibit outstanding selectivity, faster recovery at RT with good reversibility, and low DL down to the ppb level. Further, it has been observed that the type of Pc molecule (whether substituted or unsubstituted), and its central metal ion, directly influence the sensing characteristics of hybrid sensors.

REFERENCES

Alfano, B., M. L. Miglietta, T. Polichetti, E. Massera, A. Bruno, G. Di Francia, and P. Delli Veneri. 2019. "Improvement of NO_2 Detection: Graphene Decorated with ZnO Nanoparticles." *IEEE Sensors Journal* 19 (19): 8751–57. doi:10.1109/JSEN.2019.2922412.

Alizadeh, Taher, and Leyla Hamed Soltani. 2016. "Reduced Graphene Oxide-Based Gas Sensor Array for Pattern Recognition of DMMP Vapor." *Sensors and Actuators B: Chemical* 234 (October): 361–70. doi:10.1016/J.SNB.2016.04.165.

Altindal, A., Öztürk, Z.Z., Dabak, S. and Bekaroĝlu, Ö., 2001. Halogen sensing using thin films of crosswise-substituted phthalocyanines. *Sensors and Actuators B: Chemical, 77*(1–2), pp.389–394. https://doi.org/10.1016/S0925-4005(01)00759-6

Chen, Lu, Lei Huang, Youjie Lin, Liman Sai, Quanhong Chang, Wangzhou Shi, and Qi Chen. 2018. "Fully Gravure-Printed WO_3/Pt-decorated rGO Nanosheets Composite Film for Detection of Acetone." *Sensors and Actuators B: Chemical* 255 (February). Elsevier: 1482–90. doi:10.1016/J.SNB.2017.08.158

Chen, Wufeng, Lifeng Yan, and Prakriti R. Bangal. 2010. "Preparation of Graphene by the Rapid and Mild Thermal Reduction of Graphene Oxide Induced by Microwaves." *Carbon* 48 (4): 1146–52. doi: 10.1016/j.carbon.2009.11.037.

Chia, L. S., S. Palale, and P. S. Lee. 2016. "Thickness-Dependent Sensitivity of Copper Phthalocyanine Chemiresistive Nitrogen Dioxide Sensors." In *2016 IEEE Sensors*, 1–3. doi:10.1109/ICSENS.2016.7808405.

Chua, Chun Kiang, and Martin Pumera. 2014. "Chemical Reduction of Graphene Oxide: A Synthetic Chemistry Viewpoint." *Chemical Society Reviews* 43 (1): 291–312. doi:10.1039/C3CS60303B.

Duy, Le Thai, Tran Quang Trung, Adeela Hanif, Saqib Siddiqui, Eun Roh, Wonil Lee, and Nae-Eung Lee. 2017. "A Stretchable and Highly Sensitive Chemical Sensor Using Multilayered Network of Polyurethane Nanofibres with Self-Assembled Reduced Graphene Oxide." *2D Materials* 4 (2). IOP Publishing: 25062. doi:10.1088/2053-1583/aa6783.

Esfandiar, Ali, Azam Irajizad, Omid Akhavan, Shahnaz Ghasemi, and Mohammad Reza Gholami. 2014. "Pd–WO_3/Reduced Graphene Oxide Hierarchical Nanostructures as Efficient Hydrogen Gas Sensors." *International Journal of Hydrogen Energy* 39 (15): 8169–79. doi: 10.1016/j.ijhydene.2014.03.117.

Gladkov, L.L., Gromak, V.V. and Konstantinova, V.K., 2007. Interpretation of resonance Raman spectra of Zn-phthalocyanine and Zn-phthalocyanine-d16 based on the density functional method. *Journal of Applied Spectroscopy*, 74(3), pp.328–332. https://doi.org/10.1007/s10812-007-0053-4

Gounden, Denisha, Nolwazi Nombona, and Werner E. van Zyl. 2020. "Recent Advances in Phthalocyanines for Chemical Sensor, Non-Linear Optics (NLO) and Energy Storage Applications." *Coordination Chemistry Reviews* 420: 213359. doi: 10.1016/j.ccr.2020.213359.

Guo, ZhiJiang, Bin Wang, Xiaolin Wang, Yong Li, Shijie Gai, Yiqun Wu, and XiaoLi Cheng. 2019. "A High-Sensitive Room Temperature Gas Sensor Based on Cobalt Phthalocyanines and Reduced Graphene Oxide Nanohybrids for the ppb-Levels of Ammonia Detection." *RSC Advances* 9 (64): 37518–25. doi:10.1039/C9RA08065A.

Hu, Jing, Cheng Zou, Yanjie Su, Ming Li, Nantao Hu, Hui Ni, Zhi Yang, and Yafei Zhang. 2017. "Enhanced NO_2 Sensing Performance of Reduced Graphene Oxide by in Situ Anchoring Carbon Dots." *Journal of Materials Chemistry C* 5 (27): 6862–71. doi:10.1039/C7TC01208J.

Klyamer, Darya, Aleksandr Sukhikh, Sergey Gromilov, Pavel Krasnov, and Tamara Basova. 2018. "Fluorinated Metal Phthalocyanines: Interplay between Fluorination Degree, Films Orientation, and Ammonia Sensing Properties." *Sensors (Basel, Switzerland)* 18 (7): 2141. doi:10.3390/s18072141.

Kumar, Sanjeev. 2018. *Graphene Based Materials for Device Applications*. Guru Nanak Dev University, Amritsar.

Kumar, Sanjeev, Navdeep Kaur, Anshul Kumar Sharma, Aman Mahajan, and R. K. Bedi. 2017. "Improved Cl_2 Sensing Characteristics of Reduced Graphene Oxide when Decorated with Copper Phthalocyanine Nanoflowers." *RSC Advances* 7 (41): 25229–36. doi:10.1039/C7RA02212C.

Kumar, Sanjeev, Rajinder Singh, Aman Mahajan, R. K. Bedi, Vibha Saxena, and D. K. Aswal. 2018. "Optimized Reduction of Graphite Oxide for Highly Exfoliated Silver Nanoparticles Anchored Graphene Sheets for Dye Sensitized Solar Cell Applications." *Electrochimica Acta* 265: 131–39. doi:https://doi.org/10.1016/j.electacta.2018.01.154.

Lee, Nam H., Un-Bong Baek, and Seung-Hoon Nahm. 2019. "Hydrogen Sensing Using Paper Sensors with Pencil Marks Decorated with Palladium." *Sensors* 19(14): 3050. https://doi.org/10.3390/s19143050

Li, F., Peng, H., Xia, D., Yang, J., Yang, K., Yin, F. and Yuan, W., 2019. Highly sensitive, selective, and flexible NO2 chemiresistors based on multilevel structured three-dimensional reduced graphene oxide fiber scaffold modified with aminoanthroquinone moieties and Ag nanoparticles. *ACS applied materials & interfaces*, 11(9), pp. 9309–9316. https://doi.org/10.1021/acsami.8b20462

Li, Xiaocheng, Bin Wang, Xiaolin Wang, Xiaoqing Zhou, Zhimin Chen, Chunying He, Zheying Yu, and Yiqun Wu. 2015. "Enhanced NH_3-Sensitivity of Reduced Graphene Oxide Modified by Tetra-α-Iso-Pentyloxymetallophthalocyanine Derivatives." *Nanoscale Research Letters* 10 (1): 373. doi:10.1186/s11671-015-1072-3.

Liu, C. J., J. J. Shih, and Y. H. Ju. 2004. "Surface Morphology and Gas Sensing Characteristics of Nickel Phthalocyanine Thin Films." *Sensors and Actuators B: Chemical* 99 (2): 344–49. doi:https://doi.org/10.1016/j.snb.2003.11.034.

Lu, Ganhua, Leonidas E. Ocola, and Junhong Chen. 2009. "Reduced Graphene Oxide for Room-Temperature Gas Sensors." *Nanotechnology* 20 (44): 445502. doi:10.1088/0957-4484/20/44/445502.

MalekAlaie, M., M. Jahangiri, A.M. Rashidi, A. HaghighiAsl, and N. Izadi. 2015. "Selective Hydrogen Sulfide (H_{2S}) Sensors Based on Molybdenum Trioxide (MoO_3) Nanoparticle Decorated Reduced Graphene Oxide." *Materials Science in Semiconductor Processing* 38 (October). Pergamon: 93–100. doi:10.1016/J.MSSP.2015.03.034.

Mukherjee D., R. Manjunatha, A. K. Ray, and S. Sampath. 2016. "Phthalocyanines as Sensitive Materials for Chemical Sensors." In *Materials for Chemical Sensing*, edited by Cesar, Paixão T. and Reddy, S. Springer, Cham. doi:10.1007/978-3-319-47835-7_8.

Novoselov, K. S., D. Jiang, F. Schedin, T. J. Booth, V. V. Khotkevich, S. V. Morozov, and A. K. Geim. 2005. "Two-Dimensional Atomic Crystals." *Proceedings of the National Academy of Sciences of the United States of America* 102 (30): 10451 LP–10453. doi:10.1073/pnas.0502848102.

Pargoletti, E., A. Tricoli, V. Pifferi, S. Orsini, M. Longhi, V. Guglielmi, G. Cerrato, L. Falciola, M. Derudi, and G. Cappelletti. 2019. "An Electrochemical Outlook upon the Gaseous Ethanol Sensing by Graphene Oxide-SnO_2 Hybrid Materials." *Applied Surface Science* 483 (July): 1081–89. doi:10.1016/J.APSUSC.2019.04.046.

Pargoletti, Eleonora, and Giuseppe Cappelletti. 2020. "Breakthroughs in the Design of Novel Carbon-Based Metal Oxides Nanocomposites for VOCs Gas Sensing." *Nanomaterials* 10 (8): 1485. doi:10.3390/nano10081485.

Qin, Jieqiong, Jianmei Gao, Xiaoyu Shi, Junyu Chang, Yanfeng Dong, Shuanghao Zheng, Xiao Wang, Liang Feng, and Zhong-Shuai Wu. 2020. "Hierarchical Ordered Dual-Mesoporous Polypyrrole/Graphene Nanosheets as Bi-Functional Active Materials for High-Performance Planar Integrated System of Micro-Supercapacitor and Gas Sensor." *Advanced Functional Materials* 30 (16): 1909756. doi:10.1002/adfm.201909756.

Rigoni, F., R. Maiti, C. Baratto, M. Donarelli, J. MacLeod, B. Gupta, M. Lyu, et al. 2017. "Transfer of CVD-Grown Graphene for Room Temperature Gas Sensors." *Nanotechnology* 28 (41). IOP Publishing: 414001. doi:10.1088/1361-6528/aa8611.

Saini, G.S.S., Singh, S., Kaur, S., Kumar, R., Sathe, V. and Tripathi, S.K., 2009. Zinc phthalocyanine thin film and chemical analyte interaction studies by density functional theory and vibrational techniques. *Journal of Physics: Condensed Matter*, 21(22), p.225006. https://doi.org/10.1088/0953-8984/21/22/225006

Sakthinathan, Subramanian, Subbiramaniyan Kubendhiran, Shen-Ming Chen, Mani Govindasamy, Fahad M. A. Al-Hemaid, M. Ajmal Ali, P. Tamizhdurai, and S. Sivasanker. 2017. "Metallated Porphyrin Noncovalent Interaction with Reduced Graphene Oxide-Modified Electrode for Amperometric Detection of Environmental Pollutant Hydrazine." *Applied Organometallic Chemistry* 31 (9): e3703. doi:10.1002/aoc.3703.

Schedin, F., A. K. Geim, S. V. Morozov, E. W. Hill, P. Blake, M. I. Katsnelson, and K. S. Novoselov. 2007. "Detection of Individual Gas Molecules Adsorbed on Graphene." *Nature Materials* 6 (9): 652–55. doi:10.1038/nmat1967.

Schiattarella, Chiara, Sten Vollebregt, Tiziana Polichetti, Brigida Alfano, Ettore Massera, Maria Lucia Miglietta, Girolamo Di Francia, and Pasqualina Maria Sarro. 2017. "CVD Transfer-Free Graphene for Sensing Applications." *Beilstein Journal of Nanotechnology* 8: 1015–22. doi:10.3762/bjnano.8.102.

Sharma, Anshul Kumar, Aman Mahajan, R. K. Bedi, Subodh Kumar, A. K. Debnath, and D. K. Aswal. 2017. "CNTs Based Improved Chlorine Sensor from Non-Covalently Anchored Multi-Walled Carbon Nanotubes with Hexa-Decafluorinated Cobalt Phthalocyanines." *RSC Advances* 7 (78): 49675–83. doi:10.1039/C7RA08987B.

Sharma, Anshul Kumar, Aman Mahajan, Subodh Kumar, A. K. Debnath, and D. K. Aswal. 2018a. "Tailoring of the Chlorine Sensing Properties of Substituted Metal Phthalocyanines Non-Covalently Anchored on Single-Walled Carbon Nanotubes." *RSC Advances* 8 (57): 32719–30. doi:10.1039/C8RA05529G.

Sharma, Anshul Kumar, Aman Mahajan, Rajan Saini, R. K. Bedi, Subodh Kumar, A. K. Debnath, and D. K. Aswal. 2018b. "Reversible and Fast Responding Ppb Level Cl2 Sensor Based on Noncovalent Modified Carbon Nanotubes with Hexadecafluorinated Copper Phthalocyanine." *Sensors and Actuators B: Chemical* 255: 87–99. doi:10.1016/j.snb.2017.08.013.

Sharma, Anshul Kumar, Manreet Kaur Sohal, R. C. Singh, Aman Mahajan, Arup Biswas, Veerendra K. Sharma, and S. M. Yusuf. 2019. "Sensing Performance of rGO/Phthalocyanine Based Hybrid at Room Temperature." AIP Conference Proceedings 2115:030077. doi:10.1063/1.5112916.

Sharma, Vinay, and Shaikh M. Mobin. 2017. "Cytocompatible Peroxidase Mimic CuO:Graphene Nanosphere Composite as Colorimetric Dual Sensor for Hydrogen Peroxide and Cholesterol with Its Logic Gate Implementation." *Sensors and Actuators B: Chemical* 240 (March): 338–48. doi:10.1016/J.SNB.2016.08.169.

Shin, Jungwoo, Seon-Jin Choi, Inkun Lee, Doo-Young Youn, Chong Ook Park, Jong-Heun Lee, Harry L. Tuller, and Il-Doo Kim. 2013. "Thin-Wall Assembled SnO2 Fibers Functionalized by Catalytic Pt Nanoparticles and Their Superior Exhaled-Breath-Sensing Properties for the Diagnosis of Diabetes." *Advanced Functional Materials* 23 (19): 2357–67. doi:10.1002/adfm.201202729.

Shinbo, Kazunari, Masahiro Minagawa, Hideaki Takasaka, Keizo Kato, Futao Kaneko, and Takahiro Kawakami. 2002. "Electrical and Luminescent Properties Due to Gas Adsorption in Electroluminescent Device of Metal-Free Phthalocyanine." *Colloids and Surfaces A: Physicochemical and Engineering Aspects* 198–200: 905–09. doi:10.1016/S0927-7757(01)01018-4.

Song, Zhilong, Zeru Wei, Baocun Wang, Zhen Luo, Songman Xu, Wenkai Zhang, and Haoxiong Yu, et al. 2016. "Sensitive Room-Temperature H2S Gas Sensors Employing SnO2 Quantum Wire/Reduced Graphene Oxide Nanocomposites." *Chemistry of Materials* 28 (4): 1205–12. doi:10.1021/acs.chemmater.5b04850.

Tarcan, Raluca, Otto Todor-Boer, Ioan Petrovai, Cosmin Leordean, Simion Astilean, and Ioan Botiz. 2020. "Reduced Graphene Oxide Today." *Journal of Materials Chemistry C* 8 (4): 1198–1224. doi:10.1039/C9TC04916A.

Travan, Caterina, and Alexander Bergmann. 2019. "NO2 and NH3 Sensing Characteristics of Inkjet Printing Graphene Gas Sensors." *Sensors* 19 (15): 3379. doi:10.3390/s19153379.

van Ewyk, Robert L., Alan V. Chadwick, and John D Wright. 1980. "Electron Donor–Acceptor Interactions and Surface Semiconductivity in Molecular Crystals as a Function of Ambient Gas." *Journal of the Chemical Society, Faraday Transactions 1: Physical Chemistry in Condensed Phases* 76: 2194–205. doi:10.1039/F19807602194.

Wang, Bin, Xiaolin Wang, Xiaocheng Li, Zhijiang Guo, Xin Zhou, and Yiqun Wu. 2018. "The Effects of Amino Substituents on the Enhanced Ammonia Sensing Performance of PcCo/rGO Hybrids." *RSC Advances* 8 (72): 41280–87. doi:10.1039/C8RA07509C.

Wang, X., X. Li, Y. Zhao, Y. Chen, J. Yu, and J. Wang. 2016. "The Influence of Oxygen Functional Groups on Gas-Sensing Properties of Reduced Graphene Oxide (rGO) at Room Temperature." *RSC Advances* 6 (57): 52339–46. doi:10.1039/C6RA05659H.

Wu, J., Wu, Z., Ding, H., Yang, X., Wei, Y., Xiao, M., Yang, Z., Yang, B.R., Liu, C., Lu, X. and Qiu, L., 2019. Three-dimensional-structured boron-and nitrogen-doped graphene hydrogel enabling high-sensitivity NO2 detection at room temperature. *ACS sensors*, 4(7), pp.1889–1898. https://doi.org/10.1021/acssensors.9b00769

Wu, Jin, Kai Tao, Yuanyuan Guo, Zhong Li, Xiaotian Wang, Zhongzhen Luo, Shuanglong Feng, et al. 2017. "A 3D Chemically Modified Graphene Hydrogel for Fast, Highly Sensitive, and Selective Gas Sensor." *Advanced Science* 4 (3). John Wiley & Sons, Ltd: 1600319. doi:10.1002/advs.201600319.

Yoo, Sweejiang, Xin Li, Yuan Wu, Weihua Liu, Xiaoli Wang, and Wenhui Yi. 2014. "Ammonia Gas Detection by Tannic Acid Functionalized and Reduced Graphene Oxide at Room Temperature." *Journal of Nanomaterials* 2014. doi:10.1155/2014/497384.

Yu, Zheying, Bin Wang, Yong Li, Di Kang, Zhimin Chen, and Yiqun Wu. 2017. "The Effect of Rigid Phenoxyl Substituent on the NH3-Sensing Properties of Tetra-α-(4-Tert-Butylphenoxyl)-Metallophthalocyanine/Reduced Graphene Oxide Hybrids." *RSC Advances* 7 (36): 22599–609. doi:10.1039/C7RA02740K.

Zhang, Dongzhi, Jingjing Liu, Chuanxing Jiang, Aiming Liu, and Bokai Xia. 2017. "Quantitative Detection of Formaldehyde and Ammonia Gas via Metal Oxide-Modified Graphene-Based Sensor Array Combining with Neural Network Model." *Sensors and Actuators B: Chemical* 240: 55–65. doi:10.1016/j. snb.2016.08.085.

Zhou, Xiaoqing, Xiaolin Wang, Bin Wang, Zhimin Chen, Chunying He, and Yiqun Wu. 2014. "Preparation, Characterization and NH₃-Sensing Properties of Reduced Graphene Oxide/Copper Phthalocyanine Hybrid Material." *Sensors and Actuators B: Chemical* 193 (March): 340–48. doi:10.1016/ J.SNB.2013.11.090.

21 Phosphorene Multigate Field-Effect Transistors for High-Frequency Applications

Ramesh Rathinam, Adhithan Pon, and Arkaprava Bhattacharyya

CONTENTS

21.1 INTRODUCTION

According to the International Roadmap for Devices and Systems (IRDS 2020), shrinking the size has driven device designs toward shorter channels for high performance and low-power applications. However, scaling transistors to nanometer sizes will face enormous challenges due to second-order effects. To allow continued scaling of device dimensions, new device structures and alternative channel materials such as germanium, silicon-germanium, and two-dimensional (2D) materials may be used to sustain Moore's law. It suggests that the development of new devices using 2D materials such as graphene, molybdenum disulfide (MoS_2), and phosphorene are promising candidates for improving the performance of submicron devices. Graphene is a zero bandgap material that results in high leakage current and limits it to analog applications only. The transition metal dichalcogenides (TMDs) offer sizable band gaps with a large I_{ON}/I_{OFF} ratio (~10^7) making it suitable for low-power applications. Few-layered black phosphorus (BP) offers the optimum bandgap (0.3 eV) and high carrier mobility that provide the fabrication of transistors with satisfactory I_{ON}/I_{OFF} ratios and better performance. They also provide other advantages in terms of voltage and power gain due to a finite bandgap and are a suitable candidate for future high-frequency applications.

DOI: 10.1201/9781003126393-21

Phosphorene or BP has rapidly gained attention for high-speed and low-power nanoelectronics due to its tunable bandgap (0.3–2 eV) and high carrier mobility (up to ~1000 cm^2/V·s) at room temperature (Li et al. 2014). It is found that flexible BP field-effect transistors (FETs) have better electron and hole mobilities than other more flexible layered semiconducting transistors such as MoS_2 and WSe_2 (Zhu et al. 2016). It is also demonstrated that a phosphorene FET provides a record-high electron mobility (1000 cm^2/V·s) with ambipolar behavior (Das et al. 2014). A complementary metal oxide semiconductor (CMOS) inverter circuit designed using a phosphorene P-type metal oxide semiconductor (PMOS) transistor and a MoS_2 N-type metal oxide semiconductor (NMOS) transistor for future electronic applications has been reported (Das et al. 2014). A reduction in noise level is observed in few-layer BP-FETs due to Al_2O_3 passivation. It is found that a phosphorene MOSFET passivated with Al_2O_3 appears to be more stable for temperature variations, and the presence of traps impact the device performance (Zheng et al. 2018). An improved ambipolar behavior is observed in BP Schottky barrier field-effect transistors (SBFETs) (Cao and Guo 2015).

Ballistic non-equilibrium Green's function (NEGF) method simulations (Liu et al. 2014a) are used to study the quantum effects in phosphorene MOSFETs. The anisotropic band structure leads to considerable performance when compared to other 2D semiconductors. The simulation of 10-nm-channel bilayer phosphorene FETs shows excellent device characteristics such as $I_{ON} > 3$ mA/μm and subthreshold swing (SS) of ~66 mV/dec. FETs using the BP channel have already exhibited a high ON/OFF ratio, low SS, high charge mobility, anisotropic transport behavior, high operating frequencies, relatively high current I_{ON}/I_{OFF} ratios, and excellent current saturation making BP attractive for semiconductor device applications. Transistors made up of BP offers f_{max} ~ 14.5 GHz and f_t ~ 17.5 GHz for a channel length of 0.5 μm (Wang et al. 2014). BP radiofrequency (RF) devices offer the highest saturation velocity, which is an important metric for high-speed and RF flexible nanosystems compared with other 2D semiconductors. A single-gate BPFET at gigahertz frequency offers a peak cutoff frequency (f_t) of 12 GHz and a maximum oscillation frequency (f_{max}) of 20 GHz (Wang et al. 2014) is also reported.

Although BP looks to be a promising material for future CMOS circuits, it has not been analyzed systematically for high-frequency applications. Based on these basic observations, this chapter aims to explore 2D phosphorene-based multigate MOSFETs for future analog/RF applications.

21.2 PHOSPHORENE LAYER

21.2.1 Crystal Structure and Properties

BP is different from other group 15 crystals; it is more stable at room temperature and pressure among other allotropes of phosphorus. BP has an orthorhombic crystal structure with eight atoms per unit cell and each atom is covalently connected to three adjacent phosphorus atoms. The lattice parameters and bond angels are shown in Figure 21.1.

Properties	Type and Values
Structure	Orthorhombic C
Unit cell	a = 3.31 Å; b = 4.43 Å; c = 10.50 Å
Bond angle	103.69; 98.15
Layer spacing	~5 Å

The crystalline symmetric structure of layered BP plays a major role in its electronic properties. The puckered honeycomb nature of BP has a symmetry nature (D_{2h}) in twofold rotation and twofold perpendicular axes in the mirror plane, but has anisotropy in electronic, thermal, and optical properties. Also, the puckered honeycomb nature yields flexibility against tensile and compressive strains.

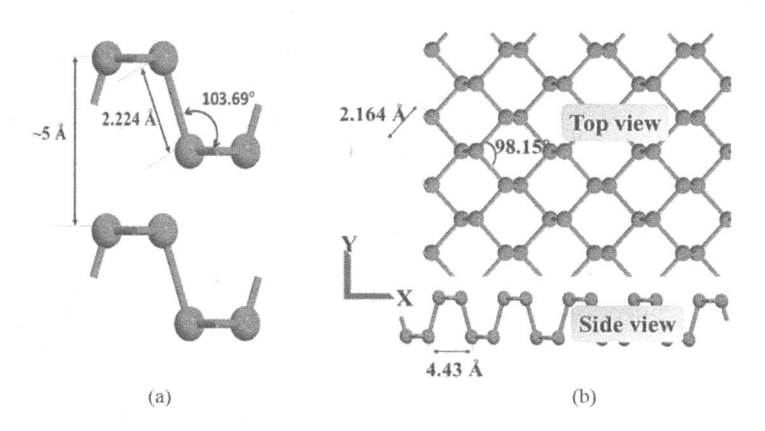

(a) (b)

FIGURE 21.1 (a) Bilayer phosphorene and its top and side view. (b) Lattice parameter.

21.2.2 Fabrication Methods and Characterization

The weaker interlayer affinity of phosphorene enables its synthesis methods, such as mechanical exfoliation (Novoselov et al. 2004; Wang et al. 2015), wet transfer method (liquid-phase exfoliation) (Nicolosi et al. 2013), ultrasonication-based exfoliation, different plasma-assisted method (laser irradiation), and pulsed laser deposition. Figure 21.2 shows the representation for various syntheses, characterization methods, and their application.

- *Mechanical exfoliation:* This technique is broadly used in the synthesis of phosphorene due to its simplicity and quality. A single sheet can be exfoliated using scotch tape (blue Nitto tape). The next important step is to transfer BP flake into a targeted substrate (like SiO_2). However, this process is time-consuming and the exfoliated intrinsic layer is unstable (Novoselov et al. 2004).
- *Wet transfer method (liquid-phase exfoliation):* This is broadly classified into (1) oxidation and a suitable solvent, (2) shear exfoliation (Xu et al. 2015), (3) ion-exchange, (4) ultrasonication-assisted exfoliation, and (v) ion intercalation. All liquid-phase exfoliation methods are more appropriate for industrial applications (Nicolosi et al. 2013). In this method, a thin phosphorene layer is exfoliated from bulk phosphorene directly via a liquid medium.

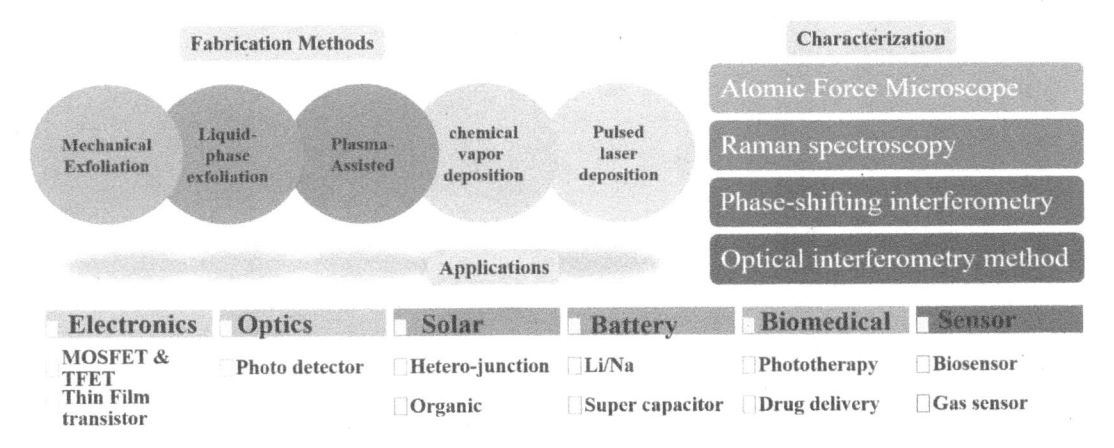

FIGURE 21.2 Schematic representation for various syntheses, characterization methods, and their applications.

- *Pulsed laser deposition:* Amorphous BP (a-BP) is changed into BP with a thickness of 2–10 nm (Yang et al. 2015). A consistent BP flake is obtained by depositing it on the rotating substrate.
- *Characterization:* The quickest way to decide the thickness of the BP flake and the most accurate characterization method is by using an atomic force microscope (AFM) (Liu et al. 2014b). The photoluminescence (PL) method is another way of determining the thickness of the flake (Zhang et al. 2014). Raman spectroscopy is used to determine thickness from diffraction, the intensity ratio between silicon and phosphorene peak at Ag_1, Ag_2, and Bg_2 (Castellanos-Gomez et al. 2014; Lu et al. 2014). The phase-shifting interferometry (PSI) method has been recently developed for the characterization of BP using light-emitting diode (LED) light (Yang et al. 2015). In the optical interferometry method, the optical path length (OPL) is measured by PSI and based on OPL values determining the BP flake number of layers (Yang et al. 2015).

21.2.3 FABRICATION OF BLACK PHOSPHOROUS MOSFET

Figure 21.3 shows the key steps for the fabrication of a back-gate phosphorene MOSFET. Phosphorene flakes are obtained from bulk phosphorous using any one of the previously mentioned methods. In particular, the mechanical exfoliation method is widely used and the BP flake is incorporated on a SiO_2 layer deposited on heavily doped Si substrates. The same silicon substrate is used as the back gate (bottom gate) for controlling the device's operation. The transferred flake is highly reactive to environments (unstable in the air), so it is immediately passivated using Al_2O_3. For metallization, Al_2O_3 was detached from the contact area (dry/wet etching), which does not react with phosphorene. Next, electron-beam lithography is used to shape the source/drain contacts. Usually, a low work-function metal (titanium) is used as a source electrode. The electrical measurements of the fabricated device are performed after the initial calibration. The direct current (DC) characterization is obtained using a parameter analyzer (Agilent B1500 semiconductor) and a probe station (Lakeshore Cryogenic) with micromanipulation probes. The high-frequency characterization is executed using a vector network analyzer (Agilent N5230A).

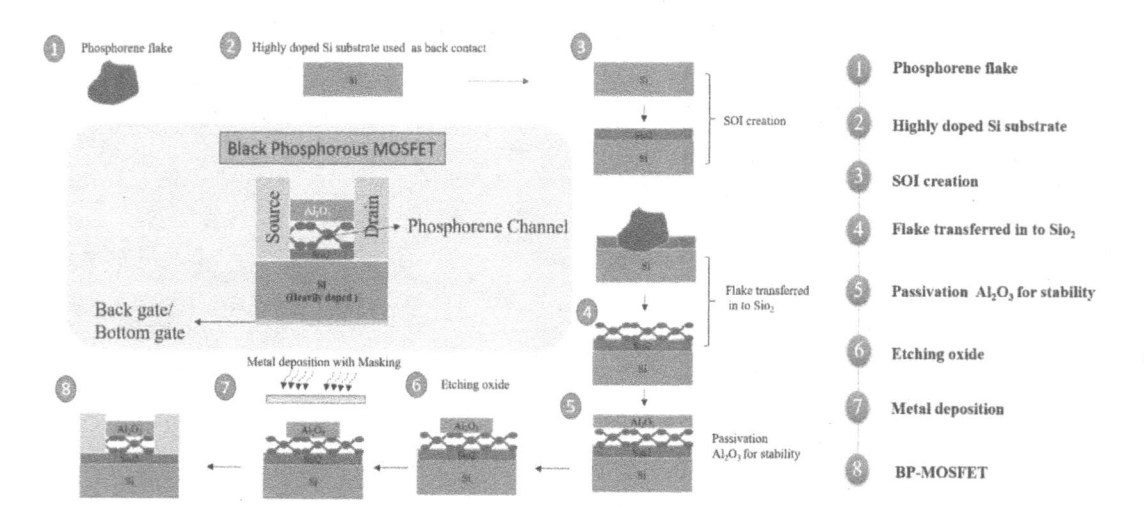

FIGURE 21.3 Schematic representation of black phosphorous MOSFET fabrication flow.

21.2.4 APPLICATIONS

- *Electronics:* The unique tunable bandgap and higher mobility are enabling a BP transistor suitable for both digital and analog applications. BP MOSFET was fabricated successfully for the first time by Li et al. (2014) using a few-layer phosphorene. Also, it was reported that the device has an ambipolar conducting nature with a drain current ratio (I_{ON}/I_{OFF}) ~ 10^5, and mobility is ~1000 cm^2 V^{-1} S^{-1}. The potential of analog/RF performance is evaluated and reported as f_t = 12 GHz and f_{max} = 20 GHz values along with 2×10^3 ON/OFF ratio (Wang et al. 2014). After this remarkable performance, many phosphorene-based MOSFET works were carried out and reported systematically (Buscema et al. 2014; Das et al. 2014; Du et al. 2014; Na et al. 2014; Liu et al. 2016; Si et al. 2017; Robbins and Koester 2017; Yin et al. 2017).

  To improve the performance of the phosphorene MOSFET device, transistors made up of phosphorene nanoribbon (Feng et al. 2018), heterojunctions using both MoS_2 and phosphorene (Xu et al. 2017), thickness engineering (Chen et al. 2017a; Kim et al. 2020), and BP tunneling FETs were also reported (Li et al. 2018; Wu et al. 2019).

- *Optoelectronics:* Traditional semiconductor photodetectors can detect ultraviolet (UV) and visible light due to their high bandgap. Fortunately, BP has a direct narrow (tunable) bandgap; high light absorption efficiency makes it an ideal photodetector (Chen et al. 2017c). The tunable bandgap properties of BP make it suitable for the optical (photodetection) application because it covers a broad spectrum (Castellanos-Gomez 2015). Moreover, the response time (1 ms) is significantly less compared with other 2D materials-based photodetectors (Buscema et al. 2014). Guo et al. (2016) demonstrated for the first time a mid-infrared detector at 3.39 μm with maximum gain, decent noise measurement, and response time. The anisotropic nature and polarization sensitivity of layered BP offers bandwidth from ~400–3750 nm due to intrinsic linear dichroism (Yuan et al. 2015). To further enhance the photodetection performance, photodetectors made up of heterojunction 2D materials are used. In this regard, a 2D heterojunction-based photodetector is created using BP (p-type) and MoS_2 (n-type) (Deng et al. 2014).

- *Biomedical:* BP-based nano FET devices were used as biosensors due to their high sensitivity and selectivity nature. Chen et al. (2017d) have reported that human immunoglobulin G was outstandingly sensed by BP-FET. The sensor response is obtained by changing the electrical resistance in the channel (BP is active material) after introducing the biomolecule. Li et al. (2017) designed a BP-FET sensor with less response time to detect the mercury (Hg^{2+}) content in drinking water. If the level of Hg^{2+} in water increases to 1 part per billion (ppb), it will cause severe health issues like brain damage and kidney failure. The villin headpiece (HP35) protein is sensed easily due to BP puckered morphology compared with the graphene-based sensor (Zhang et al. 2015).

- *Gas sensor:* As the molecule absorption energy, the surface to volume ratio and noise level of BP sensors are at a desirable level. They are suitable for gas sensing compared with other 2D material sensors. So BP devices are inevitable in sensing applications (Donarelli and Ottaviano 2018). The monolayer phosphorene acts as a good sensor for CO, CO_2, NH_3, NO, and NO_2 gas molecules (Kou et al. 2014; Abbas et al. 2015). Lee et al. (2017) reported a suspended BP-FET structure that enhances the sensing due to its two-sided contact (more surface to volume ratio) compared with a single-sided contact sensor. In addition, it has a fast response and recovery time. Also, volatile organic compounds (VOCs) such as ethanol, propionaldehyde, acetone, toluene, and hexane sensed using a single-layer BP, which exhibits strong interaction with VOCs (Ou et al. 2019).

  Other than the previuosly mentioned applications of BP-FET, BP is used for energy storage devices (Yang et al. 2019), batteries (Nagao et al. 2011; Li et al. 2015; Qiu et al. 2017), and organic and perovskite solar cells (Liu et al. 2017; Gong et al. 2020). One of the allotropes of BP (0D), called black phosphorous quantum dot (BPQD) established a significant notice because of its unique properties that are suitable for the previously mentioned applications (Chen et al. 2017b; Zong et al. 2019; Gong et al. 2020). BP also exhibits superconductivity property under high pressure at 10 GPa (Zhang et al. 2017; Alidoust et al. 2019; Sun et al. 2019).

21.3 METHODOLOGY

NEGF is a straightforward method for 2D material FET simulation (Blom et al. 2015; Szabo et al. 2015; Szabó et al. 2018). However, the computational cost of NEGF is very high for the simulation of three-terminal devices. Also, it is not suitable for simulating the RF performance of 2D devices. On the other hand, compact models are trying to simulate the 2D FET characteristics. But it needs an experimental data set such as different bias and temperature conditions for simulating the I_d-V_g and I_d-V_d characteristics. However, scarcity of data and underlying physics limits its accuracy and the consistency is debatable.

Until now, device simulation tools have not available to study the characteristics of 2D FET devices. Many efforts have been made to construct a simulation tool to study the transport mechanism of 2D material devices with reduced computational cost (Nanmeni Bondja et al. 2016). However, the reported work does not include non-idealities studies such as contact resistance (Du et al. 2014; Luo et al. 2017; Mirabelli et al. 2017), Fermi level pinning (Shine and Saraswat 2017), and interface traps (Mirabelli et al. 2017; Pon et al. 2019). To overcome these issues, we introduced a new method that contains both atomistic and technology computer-aided design (TCAD) (TCAD 2012) simulations jointly to perform the 2D FET simulation (Pon et al. 2019). Table 21.1 shows the performance comparison of TCAD + density functional theory (DFT), Compact, and NEGF with its parameter. Green indicates the desirable performance of the methods, and it is found that the TCAD + DFT (hybrid) method is more suitable for simulation of 2D material FET devices.

21.3.1 ATOMISTIC SIMULATION OF PHOSPHORENE LAYERS (DFT)

The characteristics of the single-layer (1L) to few-layer phosphorene (5L) were studied by using the quantum-wise atomistic toolkit (ATK). This method employs meta-generalized gradient approximation (MGGA) and TB09LDA functionals. MGGA functionals provide accurate results over GGA due to the inclusion of exact exchange in DFT to exchange correlation (Tao 2002). This is a hybrid functional developed by Becke to improve the accuracy in DFT formalism (Tao 2002). We have

TABLE 21.1
2D FET Simulation Method Comparison

Method	TCAD + DFT	Compact Model	NEGF
Parameters			
Speed	Moderate	*High	Slow
Accuracy	Better	Low	*Best
Computation cost	*Workable	*Easy	Very high
Complexity	*Workable	Moderate	Very high
Non-ideal simulation	*Yes	*Yes	No
Insight view	*Yes	No	*Yes
Transient and RF simulation	*Yes	*Yes	No
Reliable and consistency	*Very high	High	Medium

*Desirable performance

TABLE 21.2

Electrical Parameters of Monolayer to Few-Layer Phosphorene for Armchair (AC) and Zigzag (ZZ)

Number of Layers	Bandgap (eV)	Armchair		Zigzag	
		m^*_e	m^*_h	m^*_e	m^*_h
1	1.45	0.19	0.2	1.2	7.6
2	1.05	0.2	0.26	1.23	1.42
3	0.79	0.19	0.29	1.23	3.57
4	0.65	0.17	0.31	1.22	1.22
5	0.57	0.16	0.32	1.21	3.36

Note: Numerical values are given in units of electron rest mass m_0.

used $14 \times 14 \times 1$ k-point sampling with a total of 98 irreducible k-points, and the density mesh cutoff was set to 150 Rydberg. Using this simulation setup, we obtained the bandgap of monolayer phosphorene (1.45 eV), which exactly matches with experimental data (Liu et al. 2014b).

The evaluated bandgap values and the respective effective mass (m*) of electrons (m^*_e) and holes (m^*_h) are presented in Table 21.2. It is found that the m* of AC is less compared to the ZZ direction. The bandgap value decreases as the number of layers increases (Figure 21.4) due to quantum confinement in low-dimensional systems, such as nanowires and nanotubes.

21.3.2 SIMULATION OF PHOSPHORENE FET USING TCAD

We modified the effective carrier concentration in the channel using effective carrier concentration 3D DOS (NC_{3D}). The numerical results obtained represent the DOS of the 2D channel (NC_{2D}). To account for the quantum and ballistic effect, we used KVM and quantum confinement models. First, ATK is used to obtain the electrical parameter values. Then the obtained values are exported to the Sentaurus TCAD material file and the multigate device structure is constructed. To account for the high-density carrier concentration in the source/drain regions, the Fermi-Dirac statistics is used. In our simulation, we do not account for the traps and other second-order effects for simplicity (Sentaurus 2018). The detailed method flow, model details, and calibration are given (Rathinam et al. 2020).

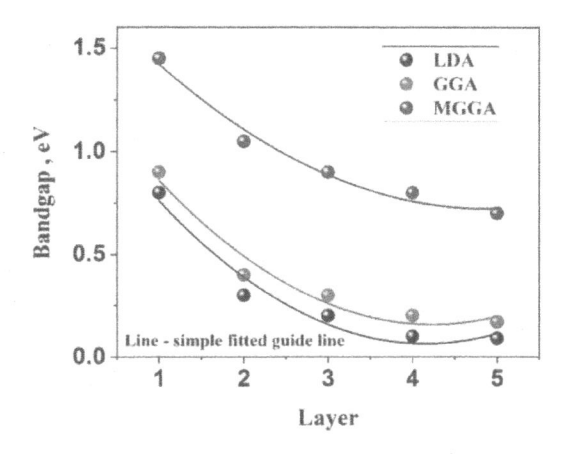

FIGURE 21.4   Variation of the bandgap with a number of layers of phosphorene with various functionals.

21.3.3 COMPACT MODELING

Compact models are required to accurately reproduce the device characteristics for circuit design. There is a demand for accurate compact models for 2D material FET devices. The compact models developed for graphene transistor (Fregonese et al. 2013; Frégonèse et al. 2015) is presented. However, sophisticated compact models are needed for high reliability and better accuracy. The Stanford 2D Semiconductor (S2DS) model was developed for MoS_2 FET and it gives guidelines for other 2D material-based FET. Yarmoghaddam et al. (2020a,b) reported for the first time a compact model for 2D phosphorene FET. It contains the impact of trap, contact resistance, temperature, and ambipolar effects.

21.4 SIMULATION OF PHOSPHORENE FET FOR HIGH FREQUENCY

21.4.1 SINGLE-GATE PHOSPHORENE MOSFET

The schematic diagram of BP single-gate MOSFET (BP-SGMOSFET) is shown in Table 21.3. The phosphorene layer is introduced as a channel material in the device with a gate length (L_G) = 5 nm, source/drain extension ($L_S = L_D$) = 2.5 nm, gate work function (ϕ_m) = 4.25 eV. The device was built along the x/y-direction of the 1-nm/5 nm-thick flake for mono/few-layer phosphene,

TABLE 21.3
Phosphorene Multigate Field-Effect Transistors

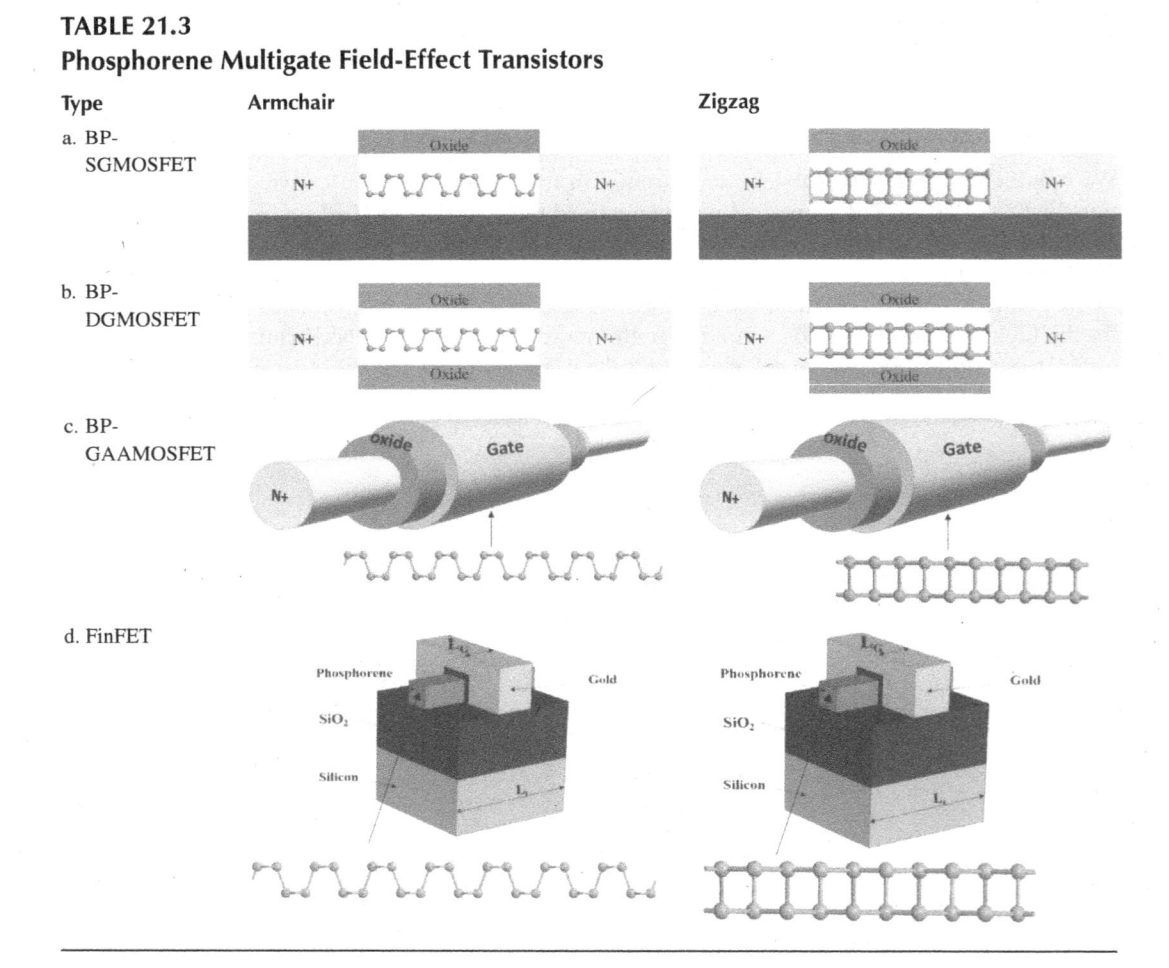

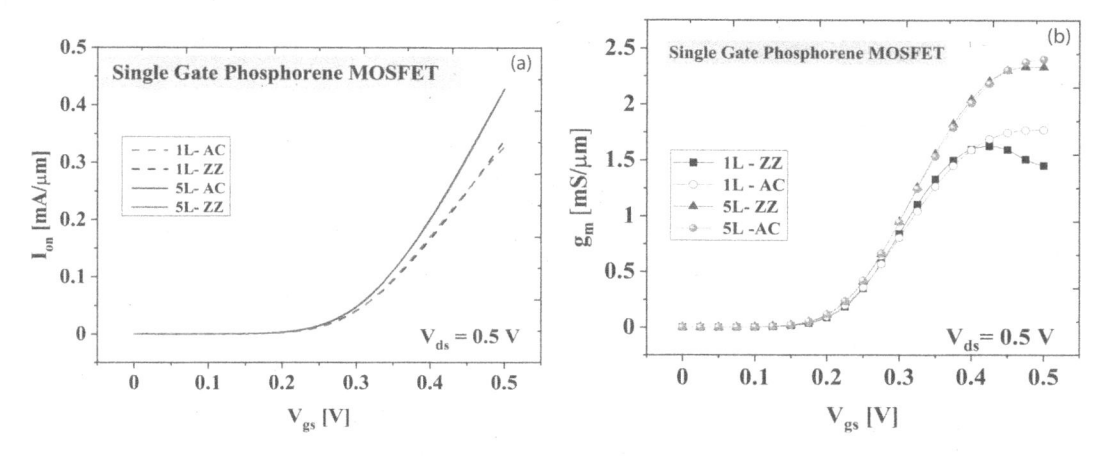

FIGURE 21.5 (a) Transfer characteristics and (b) transconductance of single-gate BP-MOSFET for monolayer and few-layer phosphorene.

respectively, and analyzed for both AC and ZZ directions. The different multigate structures are shown in Table 21.3.

We have extracted the ON current from transfer characteristics (Figure 21.5(a)) at $V_{gs} = V_{ds} = 0.5$ V. An increase in ON current is obtained for few-layer AC/ZZ phosphorene BP-SGMOSFET when compared with its monolayer counterpart due to reduced bandgap (0.5 eV), less effective mass, and increased charge carries in the channel. A comparison of the transconductance (Figure 21.5(b)) shows a peak g_m value of 2.4 mS/μm (2.25 mS/μm) for the AC (ZZ) device at $V_{gs} = 0.5$ V for few-layer BP-SGMOSFET. On the other hand, the monolayer offers less g_m values due to less charge carries in the channel. The BP-SGMOSFET can exhibit reasonably good characteristics (i.e., maximum achievable current ratio > 10^6, $I_{ON} > 0.3$ mA/μm) suitable for digital applications.

The output conductance (g_m) determines the f_t and f_{max} values in short channel devices (<5 nm). The extracted f_{max} and f_t values for monolayer and few-layer AC phosphorene BP-SGMOSFET are 101.3 GHz, 87.8 and 65.7 GHz, and 63.1 GHz, respectively (Figure 21.6(a and b)). It is also found that the AC few-layer SGMOSFET is more desirable for analog/RF applications.

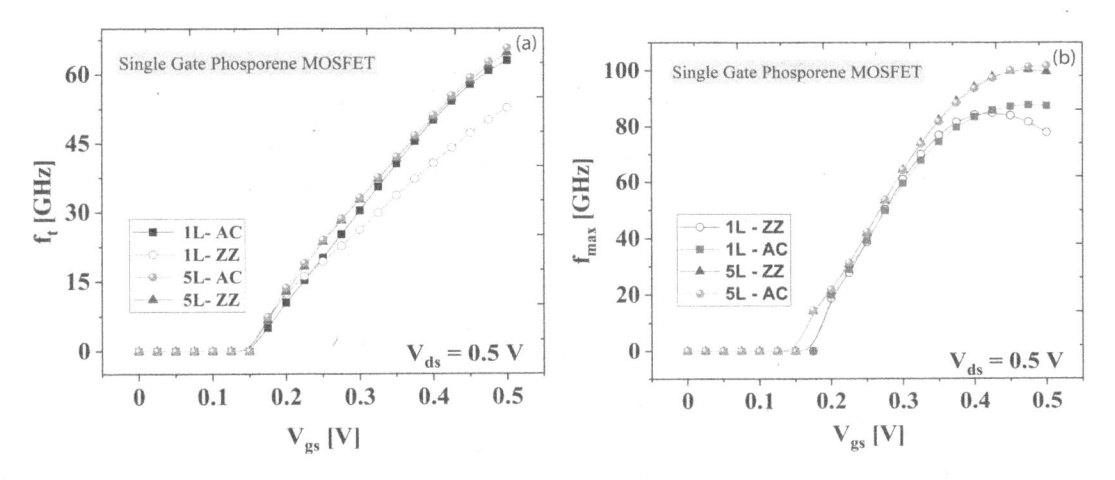

FIGURE 21.6 (a) Cutoff frequency (f_t) and (b) maximum oscillation frequency (f_{max}) of single-gate BP-MOSFET for monolayer and few-layer phosphorene.

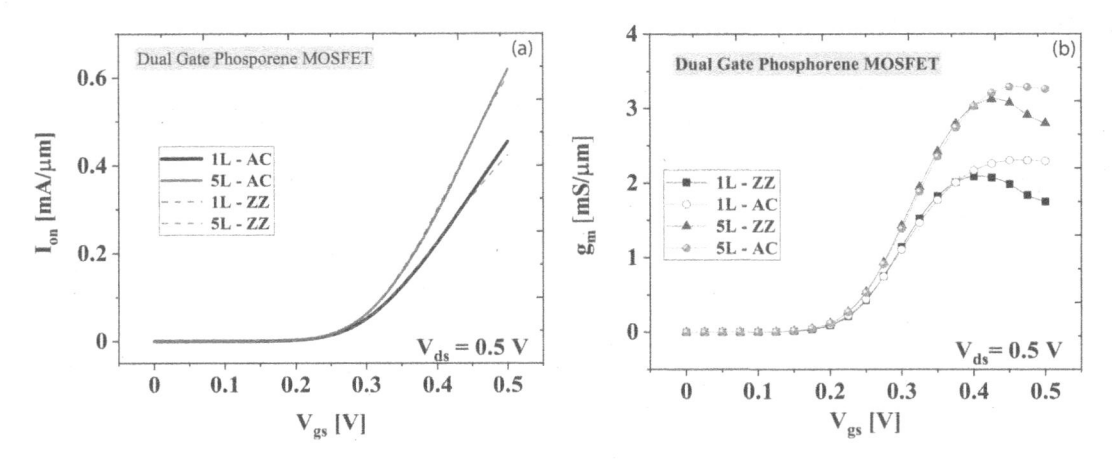

FIGURE 21.7   (a) Transfer characteristics and (b) transconductance of the dual gate for monolayer and few-layer armchair and zigzag phosphorene.

21.4.2   DUAL-GATE PHOSPHORENE MOSFET

The schematic diagram of BP dual-gate MOSFET (BP-DGMOSFET) is shown in Table 21.3. The bottom gate also uses the same gate work function (ϕ_m) = 4.25 eV to fix the leakage current to 100 nA/μm. It is found that the few-layer AC/ZZ phosphorene BP-DGMOSFET gives better ON-current values than the monolayer counterpart performance (Figure 21.7(a)).

A comparison of the transconductance values for monolayer/few-layer BP-DGMOSFET is shown in Figure 21.7(b). A peak g_m value of 3.2 mS/μm/3.1 mS/μm is obtained for a few-layer AC/ZZ device at V_{gs} = 0.5 V. On the other hand, monolayer BP-DGMOSFET offers less g_m values due to less charge carries in the channel. The leakage current values are enhanced compared with BP-SGMOSFET due to its dual-gate structure, but the inclusion of a bottom gate introduces the extra gate capacitance so it degrades the f_t (Figure 21.8(a)) and f_{max} (Figure 21.8(b)) values to some degree. Although BP-DGMOSFET exhibits better characteristics (i.e., maximum achievable current ratio > 10^6, I_{ON} > 0.6 mA/μm), it is found that frequency values are marginally less than the single-gate structure due to the impact of additional gate capacitance.

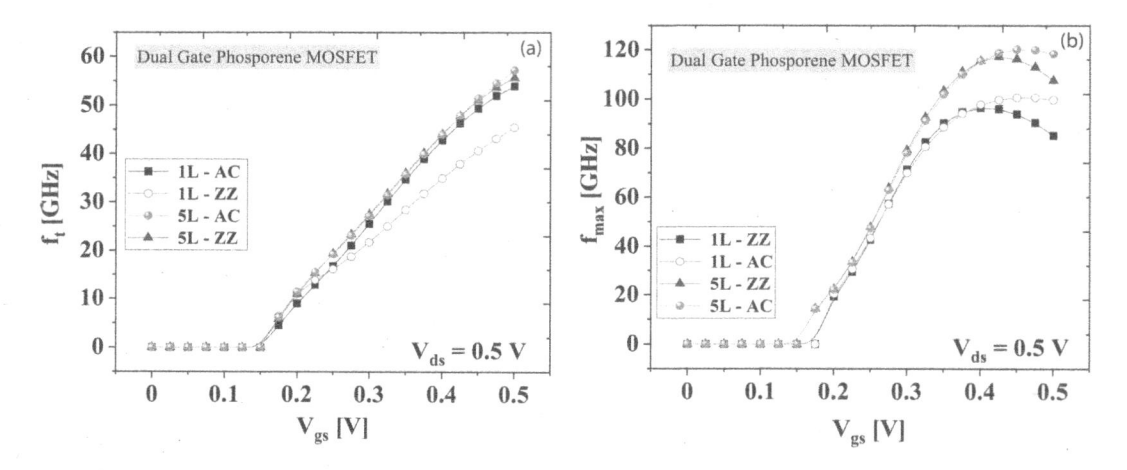

FIGURE 21.8   (a) Cutoff frequency (f_t) and (b) maximum oscillation frequency (f_{max}) performance of dual-gate phosphorene MOSFET for monolayer and few-layer phosphorene.

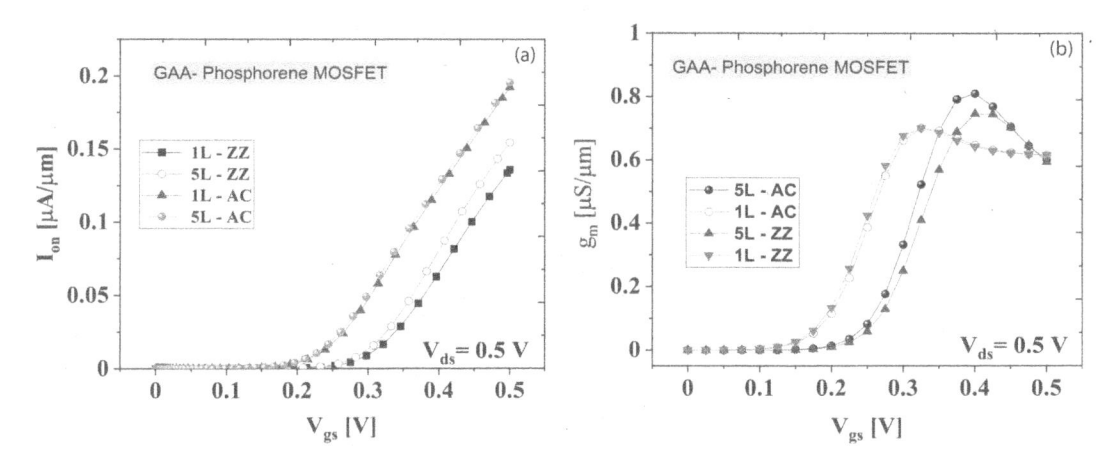

FIGURE 21.9 (a) Transfer characteristics and (b) transconductance of BP-GAAMOSFET for monolayer and few-layer phosphorene.

21.4.3 GATE-ALL-AROUND PHOSPHORENE MOSFET

The schematic diagram of a gate-all-around (GAA) phosphorene MOSFET structure is shown in Table 21.3 and its DC characteristics were studied (Figure 21.9(a)). An increase in ON-current values is obtained due to the previously mentioned reasons (see Section 21.4.2). As expected, we obtained an ultralow leakage current due to more control over the channel offered by the surrounding gate structure. However, due to the increased quantum confinement effect, the device produces less ON current and transconductance values compared with other devices.

A comparison of the transconductance (g_m) values (Figure 21.9(b)) shows peak values of 0.8 $\mu S/\mu m$ (0.7 $\mu S/\mu m$) for the AC (ZZ) device at $V_{gs} = 0.5$ V for few-layer BP-GAAMOSFET. On the other hand, the monolayer device gives less g_m values 0.61 $\mu S/\mu m$ (0.6 $\mu S/\mu m$) for AC (ZZ) due to more quantum confinement effects. As BP-GAAMOSFET exhibits a maximum achievable current ratio ($>10^6$), it is more desirable for low-power applications. It is found that f_t (Figure 21.10(a)) values are slightly less than single-gate/dual-gate structures due to the all-around gate capacitance effect. Figure 21.10(b) shows f_{max} values (almost two times) higher than the single-gate/dual-gate structure and found that it is more suitable for analog/RF applications.

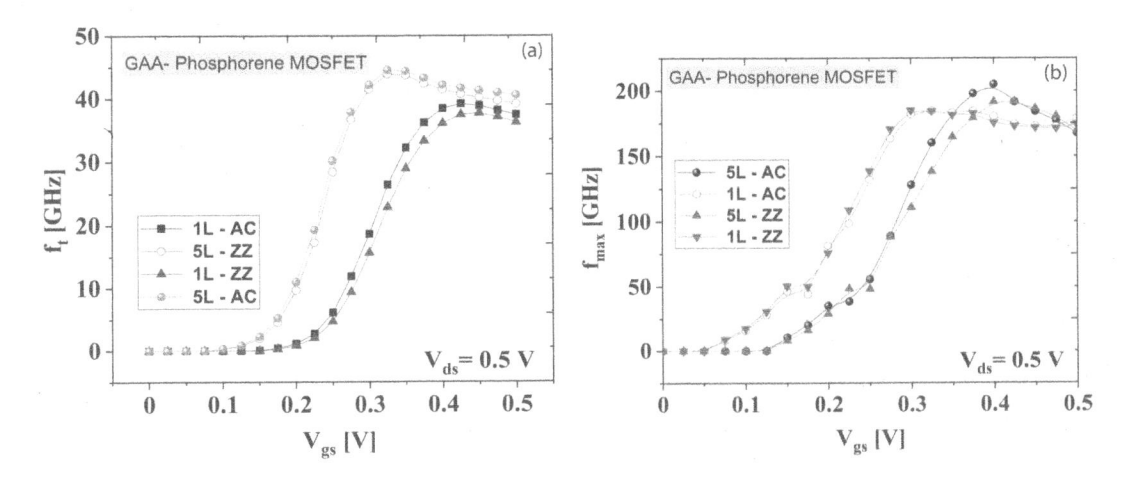

FIGURE 21.10 Comparison of (a) f_t and (b) f_{max} values of GAA for monolayer and few-layer phosphorene.

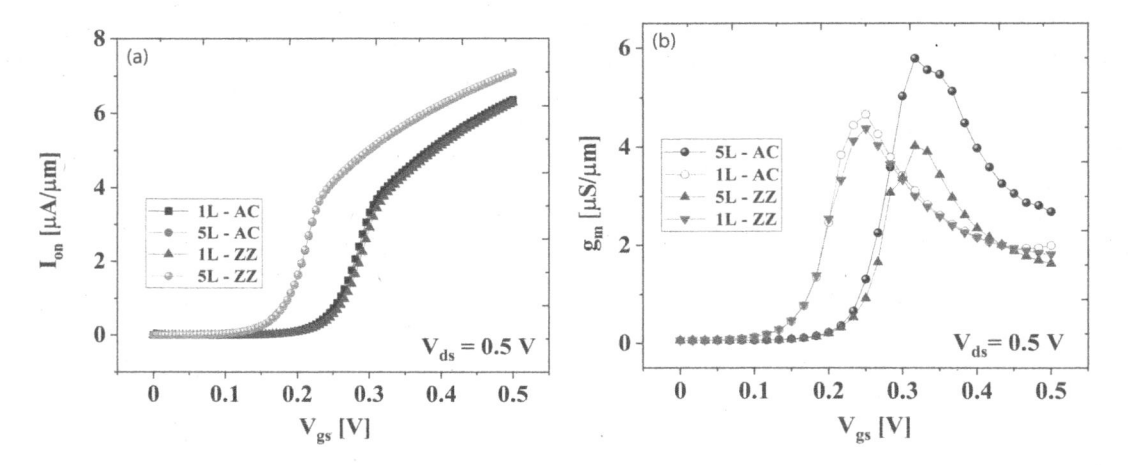

FIGURE 21.11   (a) Transfer characteristics and (b) transconductance of BP-FinFET for monolayer and few-layer phosphorene.

21.4.4   PHOSPHORENE FINFET

The schematic diagram of the BP-FinFET structure is shown in Table 21.3, and Figure 21.11(a) displays its transfer characteristics. The BP-FinFET has a 5-nm channel length (L_g) and the device is built on a silicon-on-insulator (SOI) structure.

The channel direction is along the x-direction and 1-nm-thick/5-nm-thick flakes are used for monolayer/few-layer phosphorene, respectively, for both AC and ZZ directions. A higher ON-current value is obtained in a few-layer AC/ZZ phosphorene BP-FinFET. As predicted, the device provides less leakage current due to more control over the channel by the tri-gate structure of FinFET. Due to more quantum confinement, less ON current and transconductance values are obtained compared with the single- and dual-gate nature. A comparison of the transconductance (Figure 21.11(b)) values shows a peak g_m value of 5.8 $\mu S/\mu m$ (3.9 $\mu S/\mu m$) for the AC (ZZ) device at $V_{gs} = 0.5$ V for few-layer BP-FinFET. On the other hand, the monolayer device offers less g_m values due to its more quantum confinement effect.

Although BP-FinFET exhibits maximum achievable f_{max} values (Figure 21.12(b)) it gives f_t values slightly less than the single-gate/dual-gate structure (Figure 21.12(a)) due to the tri-gate capacitance

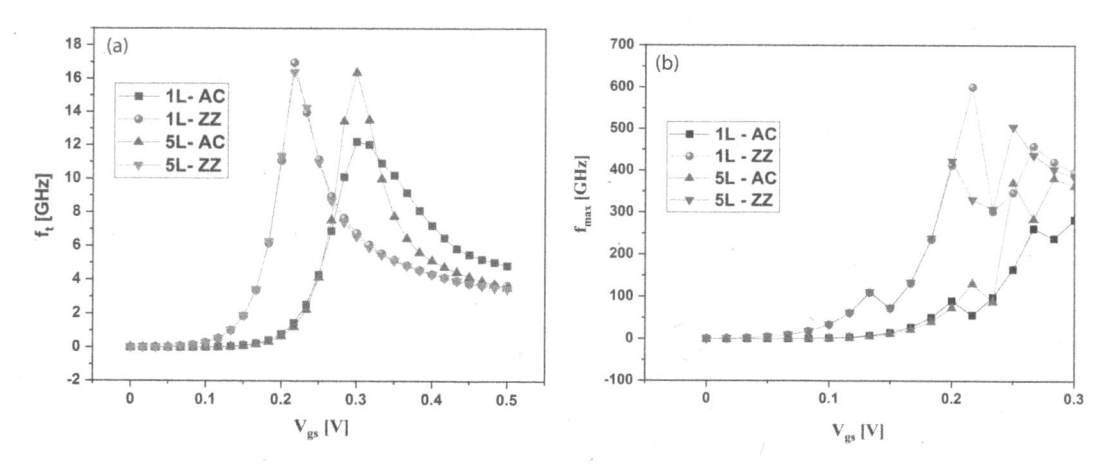

FIGURE 21.12   Comparison of (a) f_t and (b) f_{max} values of BP-FinFET for monolayer and few-layer phosphorene.

TABLE 21.4
Phosphorene Monolayer Multigate MOSFET Figure of Merit

Device	I_{ON} (mA)		I_{OFF} (A/μm)		f_t (GHz)		f_{max} (GHz)	
	AC	ZZ	AC	ZZ	AC	ZZ	AC	ZZ
BP-SGMOSFET	0.33×10^{-3}	0.327×10^{-3}	1.16×10^{-9}	1.4×10^{-9}	63.01	52.78	87.87	84.69
BP-DGMOSFET	0.45×10^{-3}	0.42×10^{-3}	1.0×10^{-9}	1.3×10^{-9}	53.90	45.48	100.58	96.49
BP-GAAMOSFET	1.92×10^{-37}	1.3×10^{-7}	2.7×10^{-12}	2.1×10^{-13}	43.89	37.20	204.7	184.55
BP-FinFET	6.4×10^{-6}	6.31×10^{-6}	0.55×10^{-12}	8.4×10^{-12}	12.21	16.95	640.36	600.68

TABLE 21.5
Phosphorene Few-Layer Performance Multigate MOSFET Figure of Merit

Device	I_{ON} (A/μm)		I_{off} (A/μm)		f_t (GHz)		f_{max} (GHz)	
	AC	ZZ	AC	ZZ	AC	ZZ	AC	ZZ
BP-SGMOSFET	0.42×10^{-3}	0.43×10^{-3}	1.93×10^{-9}	1.92×10^{-9}	65.73	64.76	101.3	100.79
BP-DGMOSFET	0.61×10^{-3}	0.60×10^{-3}	1.64×10^{-9}	1.60×10^{-9}	57.63	55.14	120.17	117.22
BP-GAAMOSFET	1.95×10^{-7}	1.5×10^{-7}	3.11×10^{-12}	1.53×10^{-13}	44.54	39.69	191.6	184.8
BP-FinFET	7.31×10^{-6}	7.3×10^{-6}	0.44×10^{-12}	18.4×10^{-12}	16.35	16.39	458.5	504.5

effect. Figure 21.12(b) shows that f_{max} values are almost five to six times higher than single-gate/dual-gate structures. A comparison of the important figure of merits and their corresponding values for various phosphorene multigate MOSFETs is shown in Tables 21.4 and 21.5.

21.5 BENCHMARK TO OTHER 2D MATERIAL FETS

In our work, we compared the performance of phosphorene dual-gate few-layer AC MOSFET with graphene and MoS_2-based transistors. The graphene and MoS_2 material values are taken from the literature (Schwierz et al. 2015) and their values are tabulated in Table 21.6.

TABLE 21.6
Material Parameters Used for Simulation And Calibration

Parameter	Graphene FET	MoS_2 FET	Monolayer-Phosphorene FET		Few-layer-Phosphorene FET	
			AC	ZZ	AC	ZZ
Affinity (χ)	4.05 eV	4.05	4.1	4.1	3.9-4.1	3.9-4.2
Bandgap	0.1 eV	1.8 eV	1.45	1.45	0.57	0.57
Work function	4.13 eV	4.18	4.2	4.2	4.1-4.2	4.1-4.2
Channel thickness	2 nm	2 nm	2 nm	2 nm	8 nm	8 nm
m_e^*	$0.45 m_0$	$0.45\ m_0$	$0.19\ m_0$	$1.2\ m_0$	$0.16\ m_0$	$0.32\ m_0$
K	6	20	6	15	6	15
α_T	2	1	2	1.5	2	1.5
α_B	0.3	0.7	0.3	0.3	0.3	0.3
Dielectric constant	10-15	8-10	14	14	14	14
Oxide material	HfO_2	HfO_2	HfO_2	HfO_2	HfO_2	HfO_2

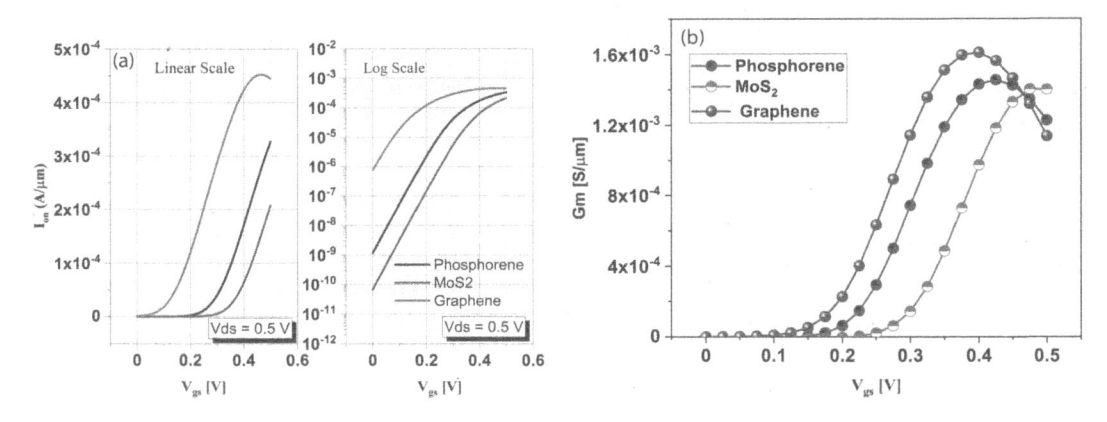

FIGURE 21.13 Comparison of DC characteristics of 2D-material MOSFET. (a) I_{ds}-V_{gs} characteristics. (b) Transconductance values.

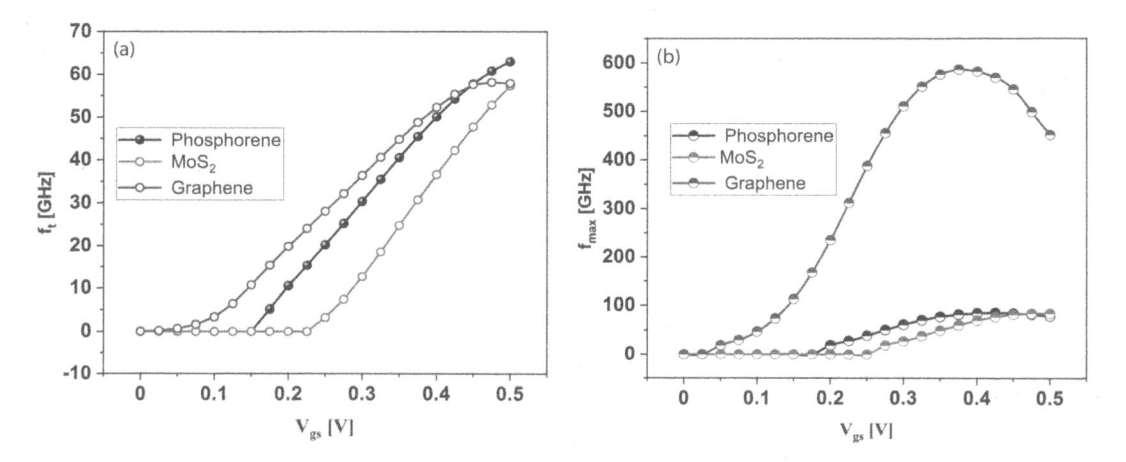

FIGURE 21.14 Comparison of analog/RF performance of 2D-material MOSFETs. (a) f_t and (b) f_{max}.

It is found that the graphene transistor offers more ON current than phosphorene and MoS_2 transistors due to its high mobility and very less effective mass. On the other hand, the leakage current for graphene is very high due to zero bandgap. The MoS_2 transistor has a desirable leakage current due to its high bandgap (Figure 21.13(a)), but its mobility is very low, which is reflected in transconductance values (Figure 21.13(b)), hence, it is not suitable for analog/RF applications.

It is found that phosphorene MOSFET offers f_{max} values = 87.87 GHz compared with the graphene transistor that offers f_{max} = 588 GHz (Figure 21.14(b)). At the same time, the f_t value is obtained around 60 GHz for all transistors (Figure 21.14(a)). It is found that phosphorene-based MOSFETs offer optimum performance for next-generation digital and analog/RF applications.

21.6 SUMMARY

In this chapter, we presented the characteristics of the phosphorene layer and phosphorene FET fabrication procedures. Briefly, we discussed the various methodologies used for simulating phosphorene FET. Also, we discussed the influence of phosphorene on the characteristics of multigate MOSFETs by studying the various figures of merit. It is found that the inclusion of the phosphorene layer in FET devices improves the overall performance of the devices at the submicron level. It is concluded that the incorporation of phosphorene in FET devices shows better performance at high frequencies compared with other 2D material FET devices.

REFERENCES

Abbas, Ahmad N, Bilu Liu, Liang Chen, Yuqiang Ma, Sen Cong, Noppadol Aroonyadet, Marianne Köpf, Tom Nilges, and Chongwu Zhou. 2015. "Black Phosphorus Gas Sensors." *ACS Nano* 9 (5): 5618–24. https://doi.org/10.1021/acsnano.5b01961.

Alidoust, Mohammad, Morten Willatzen, and Antti-Pekka Jauho. 2019. "Control of Superconducting Pairing Symmetries in Monolayer Black Phosphorus." *Physical Review B* 99 (12): 125417. https://doi.org/10.1103/PhysRevB.99.125417.

Blom, Anders, Umberto Martinez Pozzoni, Troels Markussen, and Kurt Stokbro. 2015. "First-Principles Simulations of Nanoscale Transistors." *International Conference on Simulation of Semiconductor Processes and Devices*, 52–55. https://doi.org/10.1109/SISPAD.2015.7292256.

Buscema, Michele, Dirk J Groenendijk, Sofya I Blanter, Gary A Steele, Herre S J Van Der Zant, and Andres Castellanos-Gomez. 2014. "Fast and Broadband Photoresponse of Few-Layer Black Phosphorus Field-Effect Transistors." *Nano Letters* 14 (6): 3347–52.

Cao, Xi, and Jing Guo. 2015. "Simulation of Phosphorene Field-Effect Transistor at the Scaling Limit." *IEEE Transactions on Electron Devices* 62 (2): 659–65. https://doi.org/10.1109/TED.2014.2377632.

Castellanos-Gomez, Andres. 2015. "Black Phosphorus: Narrow Gap, Wide Applications." *Journal of Physical Chemistry Letters* 6 (21): 4280–91. https://doi.org/10.1021/acs.jpclett.5b01686.

Castellanos-Gomez, Andres, Leonardo Vicarelli, Elsa Prada, Joshua O Island, K L Narasimha-Acharya, Sofya I Blanter, and Dirk J Groenendijk, et al. 2014. "Isolation and Characterization of Few-Layer Black Phosphorus." *2D Materials* 1 (2). https://doi.org/10.1088/2053-1583/1/2/025001.

Chen, Fan W, Hesameddin Ilatikhameneh, Tarek A Ameen, Gerhard Klimeck, and Rajib Rahman. 2017a. "Thickness Engineered Tunnel Field-Effect Transistors Based on Phosphorene." *IEEE Electron Device Letters* 38 (1): 130–33. https://doi.org/10.1109/LED.2016.2627538.

Chen, Wei, Kaiwen Li, Yao Wang, Xiyuan Feng, Zhenwu Liao, Qicong Su, Xinnan Lin, and Zhubing He. 2017b. "Black Phosphorus Quantum Dots for Hole Extraction of Typical Planar Hybrid Perovskite Solar Cells." *The Journal of Physical Chemistry Letters* 8 (3): 591–98. https://doi.org/10.1021/acs.jpclett.6b02843.

Chen, Xiaolong, Xiaobo Lu, Bingchen Deng, Ofer Sinai, Yuchuan Shao, Cheng Li, and Shaofan Yuan, et al. 2017c. "Widely Tunable Black Phosphorus Mid-Infrared Photodetector." *Nature Communications* 8 (1): 1672. https://doi.org/10.1038/s41467-017-01978-3.

Chen, Yantao, Ren, Haihui Pu, Jingbo Chang, Shun Mao, and Junhong Chen. 2017d. "Field-Effect Transistor Biosensors with Two-Dimensional Black Phosphorus Nanosheets." *Biosensors and Bioelectronics* 89: 505–10. https://doi.org/10.1016/j.bios.2016.03.059.

Das, Saptarshi, Marcel Demarteau, Andreas Roelofs. 2014. "Ambipolar Phosphorene Field Effect Transistor." *ACS Nano*, 8 (11): 11730–38. https://doi.org/10.1021/nn505868h.

Deng, Yexin, Zhe Luo, Nathan J Conrad, Han Liu, Yongji Gong, Sina Najmaei, Pulickel M Ajayan, Jun Lou, Xianfan Xu, and Peide D Ye. 2014. "Black Phosphorus-Monolayer MoS2 van Der Waals Heterojunction p-n Diode." *ACS Nano* 8 (8): 8292–99. https://doi.org/10.1021/nn5027388.

Donarelli, Maurizio, and Luca Ottaviano. 2018. "2D Materials for Gas Sensing Applications: A Review on Graphene Oxide, MoS_2, WS_2 and Phosphorene." *Sensors (Basel, Switzerland)* 18 (11). https://doi.org/10.3390/s18113638.

Du, Yuchen, Han Liu, Yexin Deng, and Peide D Ye. 2014. "Device Perspective for Black Phosphorus Field-Effect Transistors: Contact Resistance, Ambipolar Behavior, and Scaling." *ACS Nano* 8 (10): 10035–42. https://doi.org/10.1021/nn502553m.

Feng, Xuewei, Lin Wang, Xin Huang, Li Chen, and Kah-Wee Ang. 2018. "Complementary Black Phosphorus Nanoribbons Field-Effect Transistors and Circuits (Invited)." *IEEE Transactions on Electron Devices* 65 (10): 1–7. https://doi.org/10.1109/TED.2018.2848235.

Fregonese, Sebastien, Maura Magallo, Cristell Maneux, Henri Happy, and Thomas Zimmer. 2013. "Scalable Electrical Compact Modeling for Graphene FET Transistors." *IEEE Transactions on Nanotechnology* 12 (4): 539–46. https://doi.org/10.1109/TNANO.2013.2257832.

Frégonèse, Sébastien, Stefano Venica, Francesco Driussi, and Thomas Zimmer. 2015. "Electrical Compact Modeling of Graphene Base Transistors." *Electronics* 4 (4): 969–78. https://doi.org/10.3390/electronics4040969.

Gong, Xiu, Li Guan, Qingwei Li, Yan Li, Tao Zhang, Han Pan, and Qiang Sun, et al. 2020. "Black Phosphorus Quantum Dots in Inorganic Perovskite Thin Films for Efficient Photovoltaic Application." *Science Advances* 6 (15). https://doi.org/10.1126/sciadv.aay5661.

Guo, Qiushi, Andreas Pospischil, Maruf Bhuiyan, Hao Jiang, He Tian, Damon Farmer, Bingchen Deng, et al. 2016. "Black Phosphorus Mid-Infrared Photodetectors with High Gain." *Nano Letters* 16 (7): 4648–55. https://doi.org/10.1021/acs.nanolett.6b01977.

Kim, Seungho, Gyuho Myeong, Wongil Shin, Hongsik Lim, Boram Kim, Taehyeok Jin, Sungjin Chang, Kenji Watanabe, Takashi Taniguchi, and Sungjae Cho. 2020. "Thickness-Controlled Black Phosphorus Tunnel Field-Effect Transistor for Low-Power Switches." *Nature Nanotechnology* 15 (3): 203–6. https://doi.org/10.1038/s41565-019-0623-7.

Kou, Liangzhi, Thomas Frauenheim, and Changfeng Chen. 2014. "Phosphorene as a Superior Gas Sensor: Selective Adsorption and Distinct I–V Response." *The Journal of Physical Chemistry Letters* 5 (15): 2675–81. https://doi.org/10.1021/jz501188k.

Lee, Geonyeop, Suhyun Kim, Sunwoo Jung, Soohwan Jang, and Jihyun Kim. 2017. "Suspended Black Phosphorus Nanosheet Gas Sensors." *Sensors and Actuators, B: Chemical* 250: 569–73. https://doi.org/10.1016/j.snb.2017.04.176.

Li, Hong, Jun Tie, Jingzhen Li, Meng Ye, Han Zhang, Xiuying Zhang, and Yuanyuan Pan, et al. 2018. "High-Performance Sub-10-Nm Monolayer Black Phosphorene Tunneling Transistors." *Nano Research* 11 (5): 2658–68. https://doi.org/10.1007/s12274-017-1895-6.

Li, Likai, Yijun Yu, Guo Jun Ye, Qingqin Ge, Xuedong Ou, Hua Wu, Donglai Feng, Xian Hui Chen, and Yuanbo Zhang. 2014. "Black Phosphorus Field-Effect Transistors." *Nature Nanotechnology* 9 (5): 372–77. https://doi.org/10.1038/nnano.2014.35.

Li, Peng, Dongzhi Zhang, Chuanxing Jiang, Xiaoqi Zong, and Yuhua Cao. 2017. "Ultra-Sensitive Suspended Atomically Thin-Layered Black Phosphorus Mercury Sensors." *Biosensors and Bioelectronics* 98 (May): 68–75. https://doi.org/10.1016/j.bios.2017.06.027.

Li, Weifeng, Yanmei Yang, Gang Zhang, and Yong-Wei Zhang. 2015. "Ultrafast and Directional Diffusion of Lithium in Phosphorene for High-Performance Lithium-Ion Battery." *Nano Letters* 15 (3): 1691–97. https://doi.org/10.1021/nl504336h.

Liu, Fei, Yijiao Wang, Xiaoyan Liu, Jian Wang, and Hong Guo. 2014a. "Ballistic Transport in Monolayer Black Phosphorus Transistors." *IEEE Transactions on Electron Devices* 61 (11): 3871–76. https://doi.org/10.1109/TED.2014.2353213.

Liu, Han, Adam T. Neal, Zhen Zhu, Zhe Luo, Xianfan Xu, David Tománek, and Peide D. Ye. 2014b. "Phosphorene: An Unexplored 2D Semiconductor with a High Hole Mobility." *ACS Nano* 8 (4): 4033–41. https://doi.org/10.1021/nn501226z.

Liu, Shenghua, Shenghuang Lin, Peng You, Charles Surya, Shu Ping Lau, and Feng Yan. 2017. "Black Phosphorus Quantum Dots Used for Boosting Light Harvesting in Organic Photovoltaics." *Angewandte Chemie (International Ed. in English)* 56 (44): 13717–21. https://doi.org/10.1002/anie.201707510.

Liu, Xinke, Kah Wee Ang, Wenjie Yu, Jiazhu He, Xuewei Feng, Qiang Liu, and He Jiang, et al. 2016. "Black Phosphorus Based Field Effect Transistors with Simultaneously Achieved Near Ideal Subthreshold Swing and High Hole Mobility at Room Temperature." *Scientific Reports* 6 (April): 1–8. https://doi.org/10.1038/srep24920.

Lu, Wanglin, Haiyan Nan, Jinhua Hong, Yuming Chen, Chen Zhu, Zheng Liang, Xiangyang Ma, Zhenhua Ni, Chuanhong Jin, and Ze Zhang. 2014. "Plasma-Assisted Fabrication of Monolayer Phosphorene and Its Raman Characterization." *Nano Research* 7 (6): 853–59. https://doi.org/10.1007/s12274-014-0446-7.

Luo, Sheng, Kai Tak Lam, Baokai Wang, Chuang Han Hsu, Wen Huang, Liang Zi Yao, Arun Bansil, Hsin Lin, and Gengchiau Liang. 2017. "Effects of Contact Placement and Intra/Interlayer Interaction in Current Distribution of Black Phosphorus Sub-10-Nm FET." *IEEE Transactions on Electron Devices* 64 (2): 579–86. https://doi.org/10.1109/TED.2016.2635690.

Mirabelli, Gioele, Farzan Gity, Scott Monaghan, Paul K. Hurley, and Ray Duffy. 2017. "Impact of Impurities, Interface Traps and Contacts on MoS2 MOSFETs: Modelling and Experiments." *European Solid-State Device Research Conference*, 288–91. https://doi.org/10.1109/ESSDERC.2017.8066648.

Na, Junhong, Young Tack Lee, Jung Ah Lim, Do Kyung Hwang, Gyu Tae Kim, Won Kook Choi, and Yong Won Song. 2014. "Few-Layer Black Phosphorus Field-Effect Transistors with Reduced Current Fluctuation." *ACS Nano* 8 (11): 11753–62. https://doi.org/10.1021/nn5052376.

Nagao, Motohiro, Akitoshi Hayashi, and Masahiro Tatsumisago. 2011. "All-Solid-State Lithium Secondary Batteries with High Capacity Using Black Phosphorus Negative Electrode." *Journal of Power Sources* 196 (16): 6902–05. https://doi.org/10.1016/j.jpowsour.2010.12.055.

Nanmeni Bondja, Cedric, Zhansong Geng, Ralf Granzner, Jörg Pezoldt, and Frank Schwierz. 2016. "Simulation of 50-Nm Gate Graphene Nanoribbon Transistors." *Electronics* 5 (1): 3. https://doi.org/10.3390/electronics5010003.

Nicolosi, Valeria, Manish Chhowalla, Mercouri G. Kanatzidis, Michael S. Strano, and Jonathan N. Coleman. 2013. "Liquid Exfoliation of Layered Materials." *Science* 340 (6139). https://doi.org/10.1126/science.1226419.

Novoselov, K S, A K Geim, S V Morozov, D Jiang, Y Zhang, S V Dubonos, I V Grigorieva, and A Firsov. 2004. "Electric Field in Atomically Thin Carbon Films." *Science* 306 (5696): 666–69. https://doi.org/10.1126/science.1102896.

Ou, Pengfei, Pengfei Song, Xinyu Liu, and Jun Song. 2019. "Superior Sensing Properties of Black Phosphorus as Gas Sensors: A Case Study on the Volatile Organic Compounds." *Advanced Theory and Simulations* 2 (1): 1800103. https://doi.org/10.1002/adts.201800103.

Pon, Adhithan, Santhia Carmel, Arkaprava Bhattacharyya, and Ramesh Rathinam. 2019a. "Simulation of 2D Layered Material Ballistic FETs Using a Hybrid Methodology." In *2019 IEEE International Conference on Electron Devices and Solid-State Circuits, EDSSC 2019*, 1–3. https://doi.org/10.1109/EDSSC.2019.8754400.

Pon, Adhithan, Kuralla Sivanaga Venkata Poorna Tulasi, and R Ramesh. 2019b. "Effect of Interface Trap Charges on the Performance of Asymmetric Dielectric Modulated Dual Short Gate Tunnel FET." *AEU - International Journal of Electronics and Communications* 102: 1–8. https://doi.org/10.1016/j.aeue.2019.02.007.

Qiu, M, Z T Sun, D K Sang, X G Han, H Zhang, and C M Niu. 2017. "Current Progress in Black Phosphorus Materials and Their Applications in Electrochemical Energy Storage." *Nanoscale* 9 (36): 13384–403. https://doi.org/10.1039/c7nr03318d.

Rathinam, Ramesh, Adhithan Pon, Santhia Carmel, and Arkaprava Bhattacharyya. 2020. "Analysis of Black Phosphorus Double Gate MOSFET Using Hybrid Method for Analogue/RF Application." *IET Circuits, Devices & Systems* 14 (8): 1167–72. https://doi.org/10.1049/iet-cds.2020.0092.

Robbins, Matthew C, and Steven J Koester. 2017. "Black Phosphorus P- and n-MOSFETs with Electrostatically Doped Contacts." *IEEE Electron Device Letters* 38 (2): 285–88. https://doi.org/10.1109/LED.2016.2638818.

Schwierz, F, J Pezoldt, and R Granzner. 2015. "Two-Dimensional Materials and Their Prospects in Transistor Electronics." *Nanoscale* 7 (18): 8261–83. https://doi.org/10.1039/c5nr01052g.

Sentaurus. 2018. *"Manuals."* Mountain View, CA: Synopsys Inc.

Shine, Gautam, and Krishna C Saraswat. 2017. "Analysis of Atomistic Dopant Variation and Fermi Level Depinning in Nanoscale Contacts." *IEEE Transactions on Electron Devices* 64 (9): 3768–74. https://doi.org/10.1109/TED.2017.2720183.

Si, Mengwei, Lingming Yang, Yuchen Du, and Peide D Ye. 2017. "Black Phosphorus Field-Effect Transistor with Record Drain Current Exceeding 1 A/Mm." *Device Research Conference - Conference Digest, DRC* 9 (2016): 2016–17. https://doi.org/10.1109/DRC.2017.7999395.

Sun, Jianping, Prashant Shahi, Miao Gao, Allan H Macdonald, Yoshiya Uwatoko, Tao Xiang, and John B Goodenough. 2019. "Erratum: Pressure-Induced Phase Transitions and Superconductivity in a Black Phosphorus Single Crystal (Proceedings of the National Academy of Sciences of the United States of America (2018) 115 (9935–9940) DOI: 10.1073/Pnas.1810726115)." *Proceedings of the National Academy of Sciences of the United States of America* 116 (3): 1065. https://doi.org/10.1073/pnas.1821331116.

Szabó, Áron, Cedric Klinkert, Davide Campi, Christian Stieger, Nicola Marzari, and Mathieu Luisier. 2018. "Ab Initio Simulation of Band-to-Band Tunneling FETs with Single- and Few-Layer 2-D Materials as Channels." *IEEE Transactions on Electron Devices* 65 (10): 4180–87. https://doi.org/10.1109/TED.2018.2840436.

Szabo, Aron, Steven J Koester, and Mathieu Luisier. 2015. "Ab-Initio Simulation of van Der Waals MoTe2-SnS2 Heterotunneling FETs for Low-Power Electronics." *IEEE Electron Device Letters* 36 (5): 514–16. https://doi.org/10.1109/LED.2015.2409212.

Tao, Jianmin. 2002. "An Accurate MGGA-Based Hybrid Exchange-Correlation Functional." *The Journal of Chemical Physics* 116 (6): 2335–37.

TCAD. 2012. *"Sentaurus Device Manual."* Mountain View, CA: Synopsys Inc.

Wang, Han, Xiaomu Wang, Fengnian Xia, Luhao Wang, Hao Jiang, Qiangfei Xia, Matthew L. Chin, Madan Dubey, and Shu Jen Han. 2014. "Black Phosphorus Radio-Frequency Transistors." *Nano Letters* 14 (11): 6424–29. https://doi.org/10.1021/nl5029717.

Wang, Xiaomu, Aaron M Jones, Kyle L Seyler, Vy Tran, Yichen Jia, Huan Zhao, Han Wang, Li Yang, Xiaodong Xu, and Fengnian Xia. 2015. "Highly Anisotropic and Robust Excitons in Monolayer Black Phosphorus." *Nature Nanotechnology* 10 (6): 517–21. https://doi.org/10.1038/nnano.2015.71.

Wu, Peng, Tarek Ameen, Huairuo Zhang, Leonid A. Bendersky, Hesameddin Ilatikhameneh, Gerhard Klimeck, Rajib Rahman, Albert V. Davydov, and Joerg Appenzeller. 2019. "Complementary Black Phosphorus Tunneling Field-Effect Transistors." *ACS Nano* 13 (1): 377–85. https://doi.org/10.1021/acsnano.8b06441.

Xu, Feng, Binghui Ge, Jing Chen, Lin Huo, Hongyu Ma, Chongyang Zhu, and Weiwei Xia, et al. 2015. "Shear-Exfoliated Phosphorene for Rechargeable Nanoscale Battery." *ArXiv Preprint ArXiv*. http://arxiv.org/abs/1508.07481.

Xu, Jiao, Jingyuan Jia, Shen Lai, Jaehyuk Ju, and Sungjoo Lee. 2017. "Tunneling Field Effect Transistor Integrated with Black Phosphorus-MoS2 Junction and Ion Gel Dielectric." *Applied Physics Letters* 110 (3). https://doi.org/10.1063/1.4974303.

Yang, Jiong, Renjing Xu, Jiajie Pei, Ye Win Myint, Fan Wang, Zhu Wang, Shuang Zhang, Zongfu Yu, and Yuerui Lu. 2015. "Optical Tuning of Exciton and Trion Emissions in Monolayer Phosphorene." *Light: Science and Applications* 4 (7): 1–7. https://doi.org/10.1038/lsa.2015.85.

Yang, Weitao, Long Ye, Fenfa Yao, Chuanhong Jin, Harald Ade, and Hongzheng Chen. 2019. "Black Phosphorus Nanoflakes as Morphology Modifier for Efficient Fullerene-Free Organic Solar Cells with High Fill-Factor and Better Morphological Stability." *Nano Research* 12 (4): 777–83. https://doi.org/10.1007/s12274-019-2288-9.

Yang, Zhibin, Jianhua Hao, Shuoguo Yuan, Shenghuang Lin, Hei Man Yau, Jiyan Dai, and Shu Ping Lau. 2015. "Field-Effect Transistors Based on Amorphous Black Phosphorus Ultrathin Films by Pulsed Laser Deposition." *Advanced Materials* 27 (25): 3748–54. https://doi.org/10.1002/adma.201500990.

Yarmoghaddam, Elahe, Nazila Haratipour, Steven J Koester, and Shaloo Rakheja. 2020a. "A Physics-Based Compact Model for Ultrathin Black Phosphorus FETs - Part I: Effect of Contacts, Temperature, Ambipolarity, and Traps." *IEEE Transactions on Electron Devices* 67 (1): 389–96. https://doi.org/10.1109/TED.2019.2951662.

Yarmoghaddam, Elahe, Nazila Haratipour, Steven J Koester, and Shaloo Rakheja. 2020b. "A Physics-Based Compact Model for Ultrathin Black Phosphorus FETs — Part II : Model Validation Against Numerical." *IEEE Transactions on Electron Devices* 67 (1): 397–405. https://doi.org/10.1109/TED.2019.2955651.

Yin, Demin, Abdulaziz Almutairi, and Youngki Yoon. 2017. "Assessment of High-Frequency Performance Limit of Black Phosphorus Field-Effect Transistors." *IEEE Transactions on Electron Devices* 64 (7): 2984–91. https://doi.org/10.1109/TED.2017.2699969.

Yuan, Hongtao, Xiaoge Liu, Farzaneh Afshinmanesh, Wei Li, Gang Xu, Jie Sun, and Biao Lian, et al. 2015. "Polarization-Sensitive Broadband Photodetector Using a Black Phosphorus Vertical p–n Junction." *Nature Nanotechnology* 10 (8): 707–13. https://doi.org/10.1038/nnano.2015.112.

Zhang, R, J Waters, A K Geim, and I V Grigorieva. 2017. "Intercalant-Independent Transition Temperature in Superconducting Black Phosphorus." *Nature Communications* 8: 15036. https://doi.org/10.1038/ncomms15036.

Zhang, Shuang, Jiong Yang, Renjing Xu, Fan Wang, Weifeng Li, Muhammad Ghufran, and Yong Wei Zhang, et al. 2014. "Extraordinary Photoluminescence and Strong Temperature/Angle-Dependent Raman Responses in Few-Layer Phosphorene." *ACS Nano* 8 (9): 9590–96. https://doi.org/10.1021/nn503893j.

Zhang, Wei, Tien Huynh, Peng Xiu, Bo Zhou, Chao Ye, Binquan Luan, and Ruhong Zhou. 2015. "Revealing the Importance of Surface Morphology of Nanomaterials to Biological Responses: Adsorption of the Villin Headpiece onto Graphene and Phosphorene." *Carbon* 94: 895–902. https://doi.org/https://doi.org/10.1016/j.carbon.2015.07.075.

Zheng, H M, J Gao, S M Sun, Q Ma, Y P Wang, B Zhu, W J Liu, H L Lu, S J Ding, and David W Zhang. 2018. "Effects of Al2O3 Capping and Post-Annealing on the Conduction Behavior in Few-Layer Black Phosphorus Field-Effect Transistors." *IEEE Journal of the Electron Devices Society* 6 (1): 320–24. https://doi.org/10.1109/JEDS.2018.2804481.

Zhu, Weinan, Saungeun Park, Maruthi N. Yogeesh, Kyle M. McNicholas, Seth R. Bank, and Deji Akinwande. 2016. "Black Phosphorus Flexible Thin Film Transistors at Gighertz Frequencies." *Nano Letters* 16 (4): 2301–06. https://doi.org/10.1021/acs.nanolett.5b04768.

Zong, Shenfei, Lingling Wang, Zhaoyan Yang, Hong Wang, Zhuyuan Wang, and Yiping Cui. 2019. "Black Phosphorus-Based Drug Nanocarrier for Targeted and Synergetic Chemophotothermal Therapy of Acute Lymphoblastic Leukemia." *ACS Applied Materials and Interfaces* 11 (6): 5896–902. https://doi. org/10.1021/acsami.8b22563.

22 Analytical Modeling of Reconfigurable Transistors

Ranjith Rajan, Suja Krishnan Jagada, and Rama S. Komaragiri

CONTENTS

22.1 INTRODUCTION

A functionally enhanced or a multifunctional transistor is a promising candidate for further advancing electronics and Moore's law beyond the classical scaling. Reconfigure transistors are such a multifunctional device that can work either as an n-type transistor or a p-type transistor, depending on the polarity of applied bias. The skillful arrangement of the reconfigurable transistors reduces the number of transistors to implement a logic function. These reconfigurable transistors have significant advantages in the design of programmable logic arrays (PLAs). These devices are useful to design the system on a chip based on controllable polarity and run-time reconfiguration of logic circuits and blocks [1]. The reconfigurable transistor provides the new functionality and can deliver high ON current and provide improved scalability, which is an add-on advantage for further developing submicron electronics [2].

A reconfigurable field-effect transistor (RFET) and reconfigurable tunnel field-effect transistor (RTFET) are discussed in this chapter. RFET is a Schottky barrier metal oxide semiconductor FET (SBMOSFET), and its working principle is Schottky barrier tunneling [3, 4]. On the other side, RTFET is a particular type of TFET, and its working principle is band-to-band tunneling (BTBT) [5]. A current characteristics comparison of an RFET and RTFET is discussed in this chapter using a calibrated technology computer-aided design (TCAD) simulation tool.

This chapter explores the analytical modeling of an RFET and an RTFET. The single-gated RFET and double-gated RTFET are selected in this chapter because they are recently designed structures that have different operational mechanisms and number of gates but similarity in functionality. These structures have a short or split gate that is quite different from conventional SBMOSFET and TFET. Thus, the derivation of potential and current in RFET and RTFET also requires some strategies. A new approach, i.e., a combination of parabolic and linear approximation of the potential to model split-gated transistors' potential, is discussed in this chapter. Further, this chapter proposes a novel technique to solve the two-dimensional (2D) Poisson's equation considering exponentially varying charge density in the channel of the device.

22.2 WORKING AND COMPARISON OF RFET AND RTFET

This section presents the workings of and comparisons between an RFET [3] and RTFET [5].

22.2.1 WORKING OF RFET

Structure and band diagram of an RFET when working as an n-type RFET (nRFET) and as a p-type RFET (pRFET) is shown in Figure 22.1(a). RFET has one gated region and an ungated region in the channel (Region-I and Region-II, as shown in Figure 22.1(a)). Source and drain regions of the RFET are made up of metals. A positive drain voltage (V_D) and gate voltage (V_G) must be applied to terminals of the RFET to get nRFET behavior. When V_G is positive, bands in Region-I bend downward and form a thin tunneling barrier at the source-Region-I junction. Due to this thin barrier and sufficiently developed electric field at the junction, there are electrons from the conduction band of the source metal tunnel to the conduction band of the semiconductor, as shown in Figure 22.1(a)-(b). Tunneling the carrier from the conduction (or valence) band of one material to the conduction (or valence) band of another material is called Schottky barrier tunneling. Here, a positive V_G accumulates electrons, and current conduction is due to electron accumulation; thus, the RFET acts as an nRFET. A negative drain voltage (V_D) and a negative gate voltage (V_G) must be applied to terminals of an RFET to get pRFET behavior. When V_G is negative, bands in Region-I bend upward and form a thin tunneling barrier at the source-Region-I junction, as shown in Figure 22.1(a)-(b). A metal with a work function equal to mid-bandgap energy, nickel-silicide (NiS_2), is used as the S/D region in RFET. This material is a suitable supplier of electrons and holes. The thin tunneling barrier and high electric field created at the junction causes the Schottky barrier tunneling of holes from the metal to the semiconductor, as shown in Figure 22.1(a)-(c). Here, the current conduction is due to hole accumulation in the channel; hence, the RFET works as a pRFET. In the case of both nRFET and pRFET, the OFF current of the device (when $V_G = 0$ V) is defined by the thermionic current, conduction of current due to carrier injection across the barrier.

22.2.2 WORKING OF RTFET

Structure and band diagram of RTFET when working as an n-type RTFET (nRTFET) and a p-type RFET (pRTFET) are shown in Figure 22.1(b). In the case of RTFET shown in Figure 22.1(b)-(a), there are two gated regions (I and III), and they are separated by an ungated region (II). The source and drain regions are made up of heavily doped p and n semiconductors, respectively. When the Gate-1 voltage (V_{G1}) is zero at a positive V_D and Gate-2 voltage (V_{G2}), there exists no tunneling barrier in the device, as shown in Figure 22.1(b)-(b). However, an ultralow p-i-n leakage current can flow through the device at this biasing, and it is considered an OFF current of the RTFET. When V_{G1} is positive, bands in Region-I bend downward due to electron accumulation and form a thin tunneling barrier at the p+-Region-I junction. Due to the high electric field and thin tunneling barrier formed at the junction, electrons from the p+ valence band tunnel to the conduction band of Region-I, as shown in Figure 22.1(b)-(b). Tunneling of the carrier from the valance band of one region to the conduction band of other regions is called BTBT. Due to BTBT, current flows through the device. Here, electron accumulation in the channel is the cause of current flow; hence, this mode of operation in RTFET is called the nRTFET mode of operation. When the V_{G2} is negative, and source voltage (V_S) and V_{G1} are negative, the bands in the Region-III bend upward due to hole accumulation. A thin tunneling barrier is created at Region-III-n+ junction, as shown in Figure 22.2(b)-(c). A BTBT of electrons

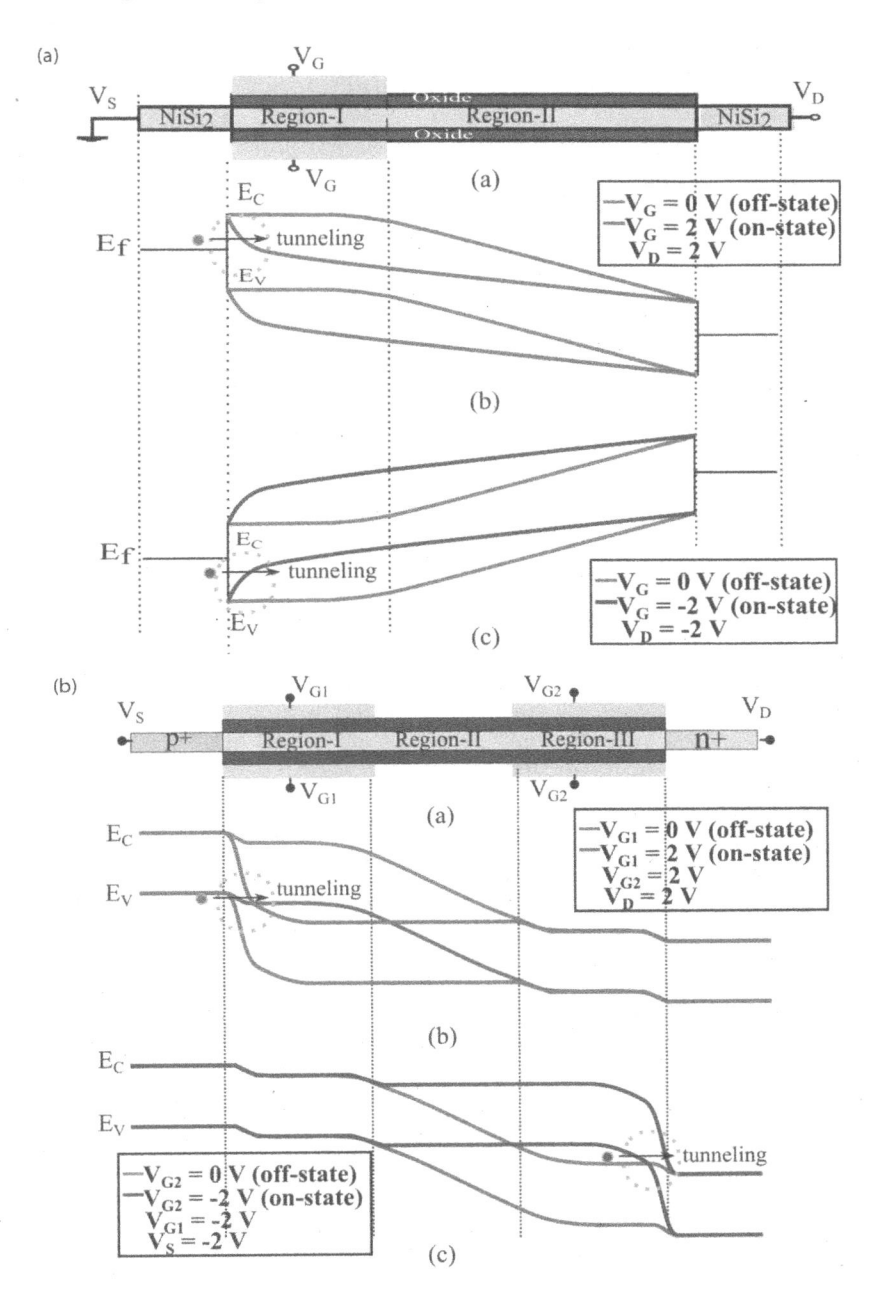

FIGURE 22.1   Structure and band diagrams of an (a) RFET and (b) RTFET.

from the valence band of Region-III to the conduction band of the n+ region occurs, and current conduction is established. Here, the current conduction is due to hole accumulation, and this mode of operation of RTFET is called nRTFET operation.

The same structure given in Ref. [3] of RFET is simulated using Sentaurus® TCAD simulation [6], and the simulation tool has been calibrated against the reported data in Ref. [3]. The comparison of data in Ref. [3] and our simulation is shown in Figure 22.2(a). Current-voltage characteristics of an RFET and an RTFET having oxide thickness (t_{ox}), body thickness (t_{si}), and channel length (L_C)

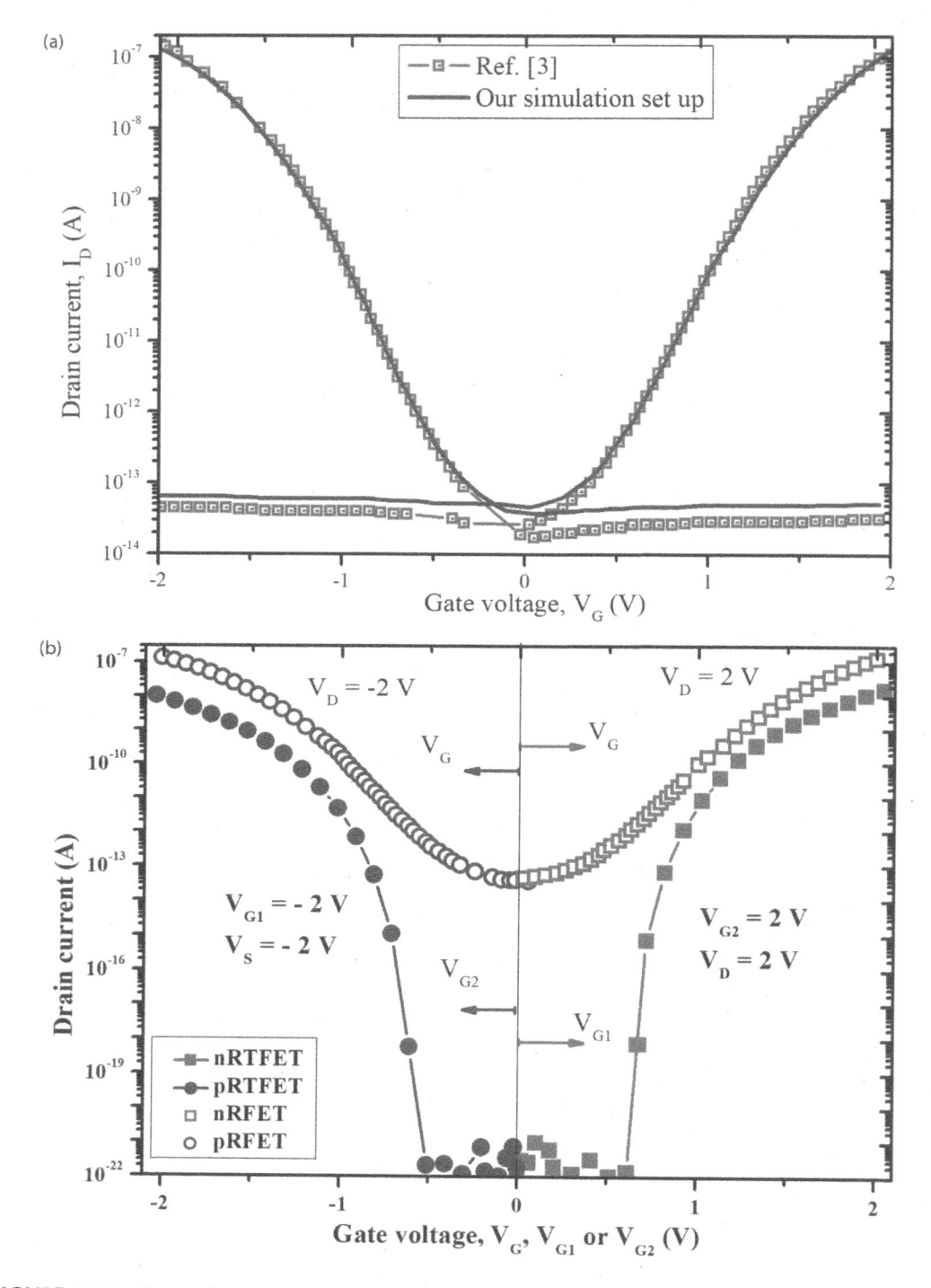

FIGURE 22.2 Current-voltage (a) calibration using the simulated data in Ref. [3], and (b) comparison of RFET and RTFET.

equal to 8, 12, and 220 nm, respectively, are simulated in TCAD, and the comparison is shown in Figure 22.2(b). The dimensions of RFET and RTFET are unchanged throughout this chapter. From Figure 22.2(b), it can be seen that RFET provides more ON current and OFF current than RTFET. Because the barrier height of the metal S/D in RFET is less than the semiconductor bandgap in RTFET, the RFET provides an increased tunneling probability and more ON current than RTFET. On the other hand, the OFF current of the RFET is higher compared with RTFET. The OFF current of RFET depends on the thermally generated carrier emission [4], which is always greater than the OFF current due to p-i-n leakage in RTFET. Also, the RTFET provides a high ON-to-OFF current ratio and low sub-threshold swing (SS) compared with RFET.

22.3 ANALYTICAL MODELING OF RFET

This section discusses the detailed derivation procedure of potential and current in an RFET. This section has derived two potential models, a subthreshold model, a super-threshold model, and a current model. A novel approach to extend a subthreshold model to a super-threshold potential model is explained in this section. A diagram of RFET is shown in Figure 22.3(a). Figure 22.3(a) shows that the Region-I of RFET spans from 0 to L_G and Region-II is from L_G to L_C.

22.3.1 POTENTIAL MODEL

After a rigorous simulation of RFET, it is found that the potential in an RFET varies nonlinearly in the gated region (Region-I) and it varies linearly in the ungated region (Region-II), as shown in Figure 22.3(b). From the observed characteristics in Figure 22.3(b), a method is proposed to solve the potential in the channel of RFET.

To find the potential at Region-I, the 2D Poisson's equation has to be solved. The potential in Region-II is estimated by approximating potential as a linear function of boundary potential along the length of the device. The following assumptions are used to find the potential in the channel region of RFET: (1) the channel is intrinsic or near intrinsic, (2) the thickness of front and back oxide is equal and possess the same dielectric constant, and (3) the charge density in the oxide is zero.

22.3.1.1 Subthreshold Potential Model

A subthreshold potential model is a model that can only be used to find the potential behavior of a device whose range of operation is below the threshold voltage. In a subthreshold potential model, charge density $(\rho(x,y))$ in the channel is taken to be zero if the channel is undoped, and it is equal to a constant if the channel is doped with n-type or p-type impurities. The charge density due to the

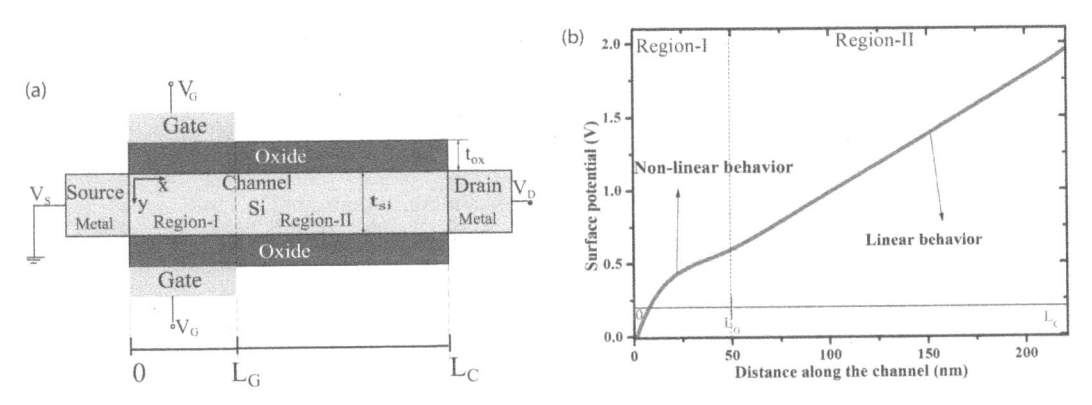

FIGURE 22.3 (a) Structure and dimensions of RFET and (b) simulated potential characteristics of nRFET.

accumulation of electron/holes in the channel is not accounted for in the subthreshold model. As mentioned earlier, the 2D Poisson's equation in Equation (22.1) is solved in Region-I of the RFET.

$$\frac{\partial^2 \psi_I(x,y)}{\partial x^2} + \frac{\partial^2 \psi_I(x,y)}{\partial y^2} = \pm \frac{qN_C}{\epsilon_s} \tag{22.1}$$

where $\psi_I(x,y)$ is the potential distribution in the channel and ϵ_s (Fcm^{-1}) is the permittivity of the semiconductor and N_C (cm^{-3}) is the doping concentration in the channel. In the case of RFET, its channel is normally intrinsic or undoped. However, to derive a general expression for the subthreshold potential model for the both doped and undoped channel, $\rho(x,y)$ in the channel is taken to be 0 or $\pm N_C$. The right side of Equation (22.1) becomes positive if the channel is doped with p-type impurities and negative with n-type impurities. If the channel of RFET is intrinsic, the right side becomes equal to zero. The 2D Poisson's equation is solved using the following top and bottom boundary conditions commonly used for double-gate architecture [7].

$$\psi_I(x,0) = \psi_{I0}(x) = \psi_I(x,t_{si}) = \psi_{I_{si}}(x) = \psi_{Is}(x) \tag{22.2}$$

$$\left.\frac{\partial \psi_I(x,y)}{\partial y}\right|_{y=0} = \frac{\epsilon_{ox}}{\epsilon_s} \frac{\psi_{Is}(x) - V_G'}{t_{ox}} \text{ and } \left.\frac{\partial \psi_I(x,y)}{\partial y}\right|_{y=t_{si}} = -\left.\frac{\partial \psi_I(x,y)}{\partial y}\right|_{y=0} \tag{22.3}$$

where $\psi_{Is}(x)$ is the potential at the surface of the channel; ϵ_{ox} (Fcm^{-1}) and t_{ox} (nm) are permittivity and thickness of the oxide; $V_G' = V_G - V_{FB}$ is the effective gate voltage; $V_{FB} = \phi_{MG} - \phi_S$ is the flatband voltage; and ϕ_{MG} and ϕ_S are the work function of the metal gate and semiconductor in volts, respectively.

This chapter uses a pseudo-2D method to solve the 2D Poisson equation for subthreshold potential modeling. The pseudo-2D method was first proposed in Ref. [8]. After that, this method became the most widely used method to solve the 2D Poisson's equation to find the subthreshold potential of doped channel devices like MOSFET and TFET. In the pseudo-2D method, Poisson's equation, a 2D second-order partial differential equation, is transformed into a second-order linear differential equation by assuming a parabolic approximation to the potential along the y-direction. The parabolic approximation used to solve the 2D Poisson's equation is given in Equation (22.4).

$$\psi_I(x,y) = a_1(x) + b_1(x)y + c_1(x)y^2 \tag{22.4}$$

where $a_1(x)$, $b_1(x)$, and $c_1(x)$ are the x-dependent constants. By applying boundary conditions Equations (22.2) and (22.3), Equation (22.4) can be rewritten as

$$\psi_I(x,y) = \psi_{Is}(x)(1 + \theta_1 y - \theta_2 y^2) - \theta_3 y + \theta_4 y^2 \tag{22.4}$$

where $\theta_1 = \frac{\epsilon_{ox}}{\epsilon_s t_{ox}}$, $\theta_2 = \frac{\theta_1}{t_{si}}$, $\theta_3 = \theta_1 V_G'$, and $\theta_4 = \frac{\theta_2}{t_{si}}$.

One who needs only an analytical expression for surface potential ($\psi_{Is}(x)$) can quickly obtain it by substituting Equation (22.8) into Equation (22.2) and solving Equation (22.2) for $y=0$ or t_{si} [9]. In this chapter, such a surface potential model is derived for RTFET in Section 22.4. Here, in this section, the procedure to get an expression for potential at any point (x, y) in the channel is explained. Now, Equation (22.4a) can be modified as the expression for $\psi_{Is}(x)$ at an arbitrary point $\tilde{y}$ in the y-direction by replacing y with $\tilde{y}$.

$$\psi_{Is}(x) = \frac{\psi_{I\tilde{y}}(x) + \theta_3 \tilde{y} - \theta_4 \tilde{y}^2}{1 + \theta_1 \tilde{y} - \theta_2 \tilde{y}^2} \tag{22.5}$$

where $\psi_{I\tilde{y}}(x)$ is the potential at any arbitrary point $\tilde{y}$. Substituting Equation (22.5) into Equation (22.4a) and solving Equation (22.1) yields pseudo-2D Poisson's Equation (22.6), which is a second-order linear differential equation.

$$\frac{d^2\psi_{I\tilde{y}}(x)}{dx^2} - \frac{1}{\lambda_{\tilde{y}}^2}\psi_{I\tilde{y}}(x) = \pm\frac{qN_C}{\epsilon_s} + \frac{1}{\lambda_{\tilde{y}}^2}\left(\theta_3\tilde{y} - \theta_4\tilde{y}^2\right) - 2\theta_4 \tag{22.6}$$

A solution to Equation (22.6) is given by Equation (22.7).

$$\psi_{I\tilde{y}}(x) = C_1 e^{\frac{\lambda_{\tilde{y}}}{x}} + C_2 e^{-\frac{\lambda_{\tilde{y}}}{x}} + K_1 \tag{22.7}$$

where $\lambda_{\tilde{y}} = \sqrt{\frac{1+\theta_1\tilde{y}-\theta_2\tilde{y}^2}{2\theta_2}}$ is the root of the characteristic Equation (22.6), and it is called characteristic length or natural length of the device; $K_1 = 2\theta_4\left(\lambda_{\tilde{y}}^2 + \tilde{y}^2\right) - \theta_3\tilde{y} - (\pm)\frac{qN_C}{\epsilon_s}\lambda_{\tilde{y}}^2$ is the particular integral solution of Equation (22.6). The constants C_1 and C_2 can be identified using the left and right boundary conditions. The left and right boundary conditions of Region-I are referred to as potential at $x = 0$ and $x = L_G$, respectively, where the potential at $x = 0$ is a known and fixed value because the source terminal is connected to V_S. On the other hand, the potential at $x = L_G$ is unknown, and it varies with the gate voltage. The left and right boundary conditions are given in Equations (22.8) and (22.9), respectively.

$$\psi_{I\tilde{y}}(0) = \psi_{I0} = V_S + \phi_{MS} \tag{22.8}$$

$$\psi_{I\tilde{y}}(L_G) = \psi_{Ig} \tag{22.9}$$

where $\phi_{MS}(V)$ is the work-function difference between S/D metal and semiconductor, and ψ_{Ig} is unknown potential at the transmission point of Region-I and Region-II. Substituting back $\tilde{y} = y$ yields the expression for potential in Region-I as given in Equation (22.10)

$$\psi_I(x,y) = \frac{\psi_A \sinh\left(\frac{L_G - x}{\lambda_y}\right) + \psi_B \sinh\left(\frac{x}{\lambda_y}\right)}{\sinh\left(\frac{L_G}{\lambda_y}\right)} + K_1 \tag{22.10}$$

where $\psi_A = \psi_{I0} - K_1$ and $\psi_B = \psi_{Ig} - K_1$.

As discussed previously, potential in Region-II is assumed as a linear function. Thus, the equation for the line can be used to describe the potential at Region-II ($\psi_{II}(x,y)$) and is given in Equation (22.11).

$$\psi_{II}(x,y) = \frac{\psi_{Ic} - \psi_{Ig}}{L_C - L_G}\left(x - L_G\right) + \psi_{Ig} \tag{22.11}$$

where $\psi_{Ic} = V_D + \phi_{MS}$ is the potential at the drain end.

The unknown boundary potential ψ_{Ig} can be identified by verifying the conditions for continuity in the potential and electric field at $x = L_G$.

$$\psi_I(L_G,y) = \psi_{II}(L_G,y) = \psi_{Ig} \text{ and } \left.\frac{d\psi_I(x)}{dx}\right|_{x=L_G} = \left.\frac{d\psi_{II}(x)}{dx}\right|_{x=L_G} \tag{22.12}$$

Using Equation (22.12), ψ_{lg} is obtained as given in Equation (22.13).

$$\psi_{lg} = \frac{\psi_{lc}\lambda_y + \psi_C(L_C - L_G)}{\lambda_y + (L_C - L_G)\coth\left(\frac{L_G}{\lambda_y}\right)}$$ (22.13)

where $\psi_C = \psi_{I0}\operatorname{csch}\dfrac{L_G}{\lambda_y} + K_1\left(\coth\dfrac{L_G}{\lambda_y} - \operatorname{csch}\dfrac{L_G}{\lambda_y}\right)$ (22.14)

Now, the subthreshold potential in an RFET is described as follows

$$\psi(x,y) = \begin{cases} \psi_I(x,y) \ 0 \le x \le L_G, \ 0 \le y \le t_{si} \\ \psi_{II}(x,y) \ L_G \le x \le L_C, \ 0 \le y \le t_{si} \end{cases}$$ (22.15)

22.3.1.2 Super-Threshold Potential Model

To get a potential model that can find the potential even after the threshold voltage, one must consider the charge density due to electron/hole accumulation in the channel. The same strategy used in the subthreshold model, i.e., modeling potential in Region-I and Region-II separately, is adopted in the super-threshold model. Because the RFET works for both electron and hole accumulation, a general 2D Poisson's equation governing electron/hole accumulation in Region-I can be written as Equation (22.16).

$$\frac{\partial^2 \psi_I(x,y)}{\partial x^2} + \frac{\partial^2 \psi_I(x,y)}{\partial y^2} = \pm \frac{qn_i}{\epsilon_s} e^{\pm q\left(\frac{\psi_I(x,y)-(\pm)v}{kT}\right)}$$ (22.16)

where the right side of Equation (22.16) assumes a positive sign for electron accumulation and a negative sign for hole accumulation and n_i (cm^{-3}) is the intrinsic carrier concentration; $k \cong 8.62 \times 10^{-5}$ eVK^{-1} is the Boltzmann constant; T (K) is the ambient temperature; and v (V) is the quasi-Fermi potential, which is assumed to be constant and equal to V_D along the y-direction. According to the superposition principle [10], Equation (22.16) can be split into two equations, a one-dimensional (1D) non-homogeneous Poisson's equation and a 2D homogeneous Poisson's equation, and the sum of solutions of these equations gives the final solution as in Equation (22.17).

$$\psi_I(x,y) = \psi_1(y) + \tilde{\psi}_I(x,y)$$ (22.17)

where $\psi_1(y)$ and $\tilde{\psi}_I(x,y)$ are the solutions of the 1D non-homogeneous and 2D homogeneous Poisson's equations, respectively. The 1D non-homogeneous Poisson's equation is given by Equation (22.18), and its solution can be obtained by integrating Equation (22.18) twice. The solution of Equation (22.18) is given in Equation (22.19).

$$\frac{\partial^2 \psi_1(y)}{\partial y^2} = \pm \frac{qn_i}{\epsilon_s} e^{\pm q\left(\frac{\psi_I(y)-(\pm)v}{kT}\right)}$$ (22.18)

$$\psi_1(y) = v - (\pm)\frac{2kT}{q}\ln\left(\sqrt{\frac{t_{si}^2 q^2 n_i}{8\omega^2 kT\epsilon_s}}\cos\left(2\frac{\omega y}{t_{si}}\right)\right)$$ (22.19)

Using the top and bottom boundary conditions Equations (22.2) and (22.3), an expression to find ω at a given V_G is obtained as in Equation (22.20).

$$\frac{q(V_G' - v)}{2kT} - (\pm)\ln\left(\sqrt{\frac{8kT\epsilon_s}{t_{si}^2 q^2 n_i}}\right) = (\pm)\left(\ln\omega - \ln(\cos 2\omega) - \frac{2\epsilon_s t_{ox}}{\epsilon_{ox} t_{si}}\omega\tan 2\omega\right) \qquad (22.20)$$

The 2D homogeneous Poisson's equation is given by

$$\frac{\partial^2 \tilde{\psi}_I(x,y)}{\partial x^2} + \frac{\partial^2 \tilde{\psi}_I(x,y)}{\partial y^2} = 0 \qquad (22.21)$$

The method typically used to solve Equations (22.16) and (22.21) is the superposition method, along with the separation of the variable method, which yields an infinite series solution to the potential [10, 11]. This chapter explains an alternative method to solve Equations (22.16) and (22.21) by merging the pseudo-2D method into the superposition method. The pseudo-2D method is simple compared with the separation of the variable method. Thus, the technique used in this chapter can be beneficial for one who wants to solve the 2D Poisson's equation for modeling the super-threshold potential of a MOS transistor. Equation (22.21) is similar to Equation (22.1) when $N_C = 0$. Thus, the solution of Equation (22.21) becomes the same as in Equation (22.10). However, the constants and boundary conditions change according to the new scenario, because, in the solution of 1D Poisson's equation, channel potential corresponding to effective V_G is already accounted for in Region-I. Hence, to avoid the unwanted addition of channel potential in the final solution while superimposing 1D and 2D solutions, the 1% of effective voltage is assumed in the boundary condition Equation (22.23) to solve 2D homogeneous Poisson's equation, i.e., V_G' has to be replaced with 1% V_G'. Similarly, when superimposing the 1D and 2D solutions, the unnecessary addition of $\psi_1(y)$ to the left and right boundary values has to be avoided. Then $\psi_1(y)$ is reduced from ψ_{I0} and $\psi_{I\tilde{y}}(L_G)$ to avoid the unwanted addition of boundary potentials, and new left and right boundary conditions become Equations (22.22) and (22.23), respectively.

$$\tilde{\psi}_{I0} = \psi_{I0} - \psi_1(y) \qquad (22.22)$$

$$\tilde{\psi}_{I\tilde{y}}(L_G) = \tilde{\psi}_{Ig} = \psi_{Ig} - \psi_1(y) \qquad (22.23)$$

Then, the solution of Equation (22.21) is described by Equation (22.10) itself, i.e.,

$$\tilde{\psi}_I(x,y) = \psi_I(x,y) \qquad (22.24)$$

where $\psi_A = \tilde{\psi}_{I0} - K_1$ and $\psi_B = \tilde{\psi}_{Ig} - K_1$. The potential and electric field at $x = L_G$ is continuous. To match Region-I's potential and electric field with the Region-II's potential and electric field at $x = L_G$, $\psi_1(y)$ has to be reduced from Region-II's potential. However, this reduction is balanced in the final expression for potential at Region-II by adding back the term $\psi_1(y)$. A new definition for potential at Region-II ($\tilde{\psi}_{II}(x,y)$) is given in Equation (22.25).

$$\tilde{\psi}_{II}(x,y) = \psi_{II}(x,y) - \psi_1(y) \qquad (22.25)$$

Now, the super-threshold potential model of RFET is given in Equation (22.26).

$$\psi(x,y) = \begin{cases} \psi_1(y) + \tilde{\psi}_I(x,y) & 0 \le x \le L_G, \, 0 \le y \le t_{si} \\ \psi_1(y) + \tilde{\psi}_{II}(x,y) & L_G \le x \le L_C, \, 0 \le y \le t_{si} \end{cases} \qquad (22.26)$$

The subthreshold model discussed in Section 22.3.1.1 is compared with TCAD and is shown in Figure 22.4(a) where it can be seen that the analytical model exactly follows the TCAD simulations. In the subthreshold region, the surface potential varies linearly with the gate voltage, as shown in the inset of Figure 22.4(a), which means the model cannot predict the saturation of potential due to the pinning of the surface potential to S/D potential.

A comparison of the subthreshold and super-threshold model of RFET with TCAD simulation is shown in Figure 22.4(b). After a threshold voltage, the surface potential saturates in RFET. It can be seen from Figure 22.4(b) that the super-threshold model predicts the saturation behavior of potential compared with a subthreshold model, which shows a linear relation with V_G.

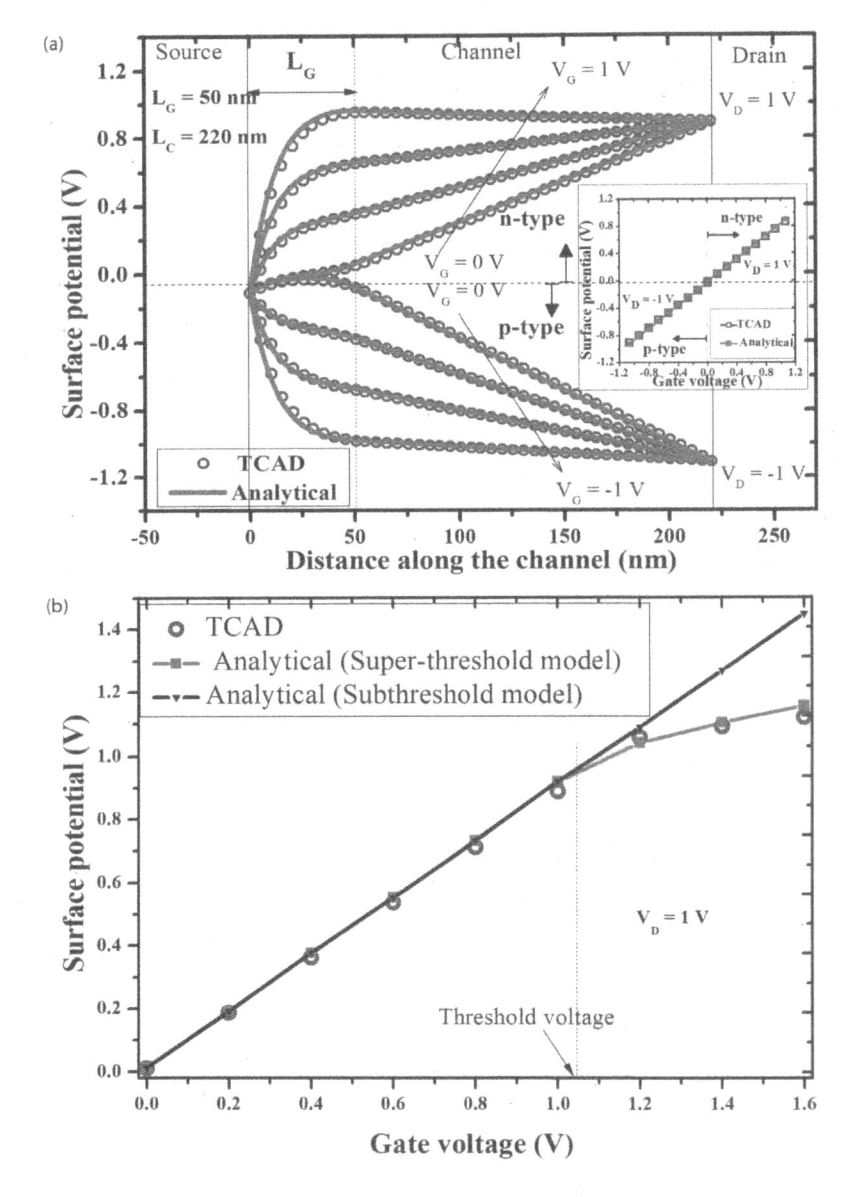

FIGURE 22.4 Comparison of the (a) subthreshold model of nRFET and pRFET with TCAD simulations for various gate voltages and the (b) subthreshold and super-threshold analytical potential model with TCAD simulation.

22.3.2 Current Model

The current conduction in RFET is due to the thermionic emission of carriers across the barrier and the Schottky barrier tunneling through the barrier. The thermionic emission is the cause of conduction when the device is off. Compared with the tunneling current, the thermionic current is negligible when the magnitude of $V_G > 0\ V$. Thus, the OFF current in the RFET is assumed to be a constant equal to the thermionic saturation current given in Equation (22.27) [12].

$$I_s = A_S A^* e^{-\left(\frac{\phi_{Be/h}}{kT}\right)} \tag{22.27}$$

where $\phi_{B_{e/h}}$ (eV) is the electron/hole tunneling barrier height, A_S is the area of the device, and $A^* = 120\ m^*/m_0$ is the Richardson constant. Tunneling current in an RFET is developed by integrating Landauer's formula [4]. For nRFET and pRFET operation, the tunneling current is obtained by Equations (22.28) and (22.29), respectively.

$$I_n = \frac{2q}{\hbar} \int_{\phi_{Be}-q\psi_I}^{\phi_{Be}} T_{WKB}\left(f_m(E) - f_s(E)\right) dE \tag{22.28}$$

$$I_p = \frac{2q}{\hbar} \int_{-\phi_{Bh}-q\psi_I}^{-\phi_{Bh}} T_{WKB}\left(f_m(E) - f_s(E)\right) dE \tag{22.29}$$

where $T_{WKB} = e^{-4\left(\frac{\sqrt{2m^*}}{3q\hbar\xi}\phi_{Be/h}^{1.5}\right)}$ is the tunneling probability function and f_m and f_s are the Fermi distribution functions of metal and semiconductor, respectively. Equation (22.28) gives the tunneling current of nRFET, which corresponds to all carriers tunneling from the metal's conduction band with energy $\phi_{B_e} - q\psi$ to the semiconductor's conduction band with energy equal to ϕ_{B_e}. Similarly, for the pRFET tunneling current, Equation (22.29), the energy range of tunneling carriers in the valence band of metal and semiconductor are $-\phi_{B_h} - q\psi$ and $-\phi_{B_h}$. By evaluating the integration Equations (22.28) and (22.29), the tunneling current in nRFET and pRFET is given by Equations (22.30) and (22.31), respectively.

$$I_n = \frac{2qkTT_{WKB}}{\hbar} \ln\left[\frac{\left(1 + e^{\frac{\phi_{Be}}{kT}}\right)\left(1 + e^{\frac{\phi_{Be}-2q\psi_I}{kT}}\right)}{\left(1 + e^{\frac{\phi_{Be}-q\psi_I}{kT}}\right)^2}\right] \tag{22.30}$$

$$I_p = \frac{2qkTT_{WKB}}{\hbar} \ln\left[\frac{\left(1 + e^{-\frac{\phi_{Be}}{kT}}\right)\left(1 + e^{-\frac{\left(\phi_{Be}+2q\psi_I\right)}{kT}}\right)}{\left(1 + e^{-\frac{\left(\phi_{Be}+2q\psi_I\right)}{kT}}\right)^2}\right] \tag{22.31}$$

The total current in nRFET/pRFET is given by

$$I_{total} = I_s + I_{n/p} \tag{22.32}$$

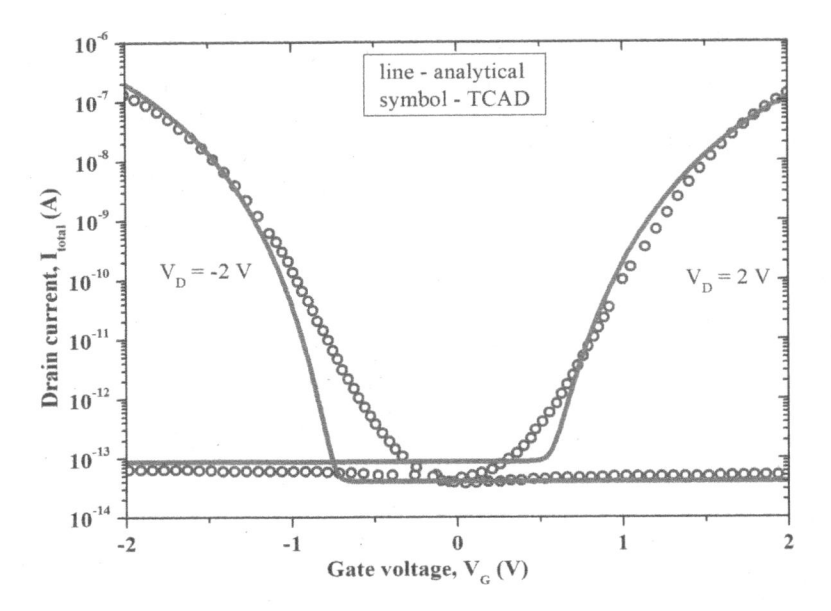

FIGURE 22.5   Current-voltage characteristics of an RFET.

A comparison of the current model of an RFET with TCAD simulation is shown in Figure 22.5. At lower gate voltages, a slight variation of current is seen in Figure 22.5. This deviation is attributed to the assumption of constant saturation current. Because the barrier height is chosen to be $\phi_{B_h} > \phi_{B_e}$ in the simulations, a pRFET shows a delayed response.

22.4   ANALYTICAL MODELING OF RTFET

The potential and current models of RTFET are explained in this section. A diagram illustrating various dimensions of RTFET is shown in Figure 22.6. From Figure 22.6, it can be seen that Region-I of an RTFET spans from 0 to L_1, Region-II spans from L_1 to L_2, and Region-III spans from L_2 to L_C.

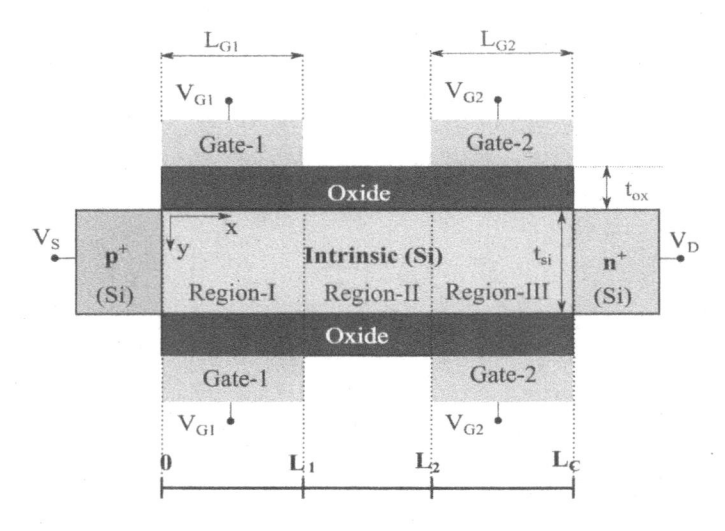

FIGURE 22.6   Structure and dimensions of RTFET.

22.4.1 POTENTIAL MODEL

RTFET has two gated regions and an ungated region, as shown in Figure 22.6. The strategy in the previous section is used to model potential in RTFET. The ungated region potential is assumed to be a linear function. The potential in the gated regions is solved using the pseudo-2D method, which assumes a parabolic approximation to the potential, where, in the case of RTFET, a subthreshold surface model is developed. However, one can extend the model to a super-threshold by adopting the method discussed in the previous section. The 2D Poisson's equation in Region-I and Region-III is given by Equation (22.33), where $i = I$ for Region I and $i = III$ for Region-III.

$$\frac{\partial^2 \psi_i(x,y)}{\partial x^2} + \frac{\partial^2 \psi_i(x,y)}{\partial y^2} = \pm \frac{qN_C}{\epsilon_s} \tag{22.33}$$

The same assumptions adopted for the RFET apply to the RTFET. Hence, the top and bottom boundary conditions Equations (22.2) and (22.3) used for the RFET apply to both Region-I and Region-III of the RTFET. Unlike the RFET, two gate voltages are used to program the RTFET. Thus, to find the potential in Region-I, V_G' in Equation (22.2) has to be replaced with V_{G1}'. Similarly, to obtain the potential model at Region-III, V_G' has to be replaced with V_{G2}'. By invoking parabolic approximation to solve Equation (22.33), at $y = 0$ the surface potential of the RTFET in Region-I and Region-III can be obtained as in Equations (22.34) and (22.35).

$$\psi_I(x) = C_1 e^{\frac{\lambda_s}{x}} + C_2 e^{-\frac{\lambda_s}{x}} + K_2 \tag{22.34}$$

$$\psi_{III}(x) = C_3 e^{\frac{\lambda_s}{x}} + C_4 e^{-\frac{\lambda_s}{x}} + K_3 \tag{22.35}$$

where $\lambda_s = \sqrt{\frac{\epsilon_s}{2\epsilon_{ox}} t_{si} t_{ox}}$ is the characteristic length of the device at $y = 0$; $K_2 = V_{G1}' \mp \frac{qN_C\lambda_s^2}{\epsilon_s}$ and $K_3 = V_{G1}' \mp \frac{qN_C\lambda_s^2}{\epsilon_s}$; and the constants C_1, C_2, C_3, and C_4 can be obtained by using the left and right boundary conditions. The left and right boundary conditions for Region-I are given in Equations (22.36) and Equation (22.37), and for Region III are given in Equations (22.38) and (22.39), respectively. Because Region-I and Region III are separated by an ungated region (Region-II), the boundary potentials at a distance $x = L_1$ and $x = L_2$ are unknown and floating with respect to respective gate voltages. The potential at $x = L_1$ and $x = L_2$ are represented by Equations (22.37) and (22.38), respectively. These potentials are determined using the conditions for continuity of potential and electric field later.

$$\psi_I(0) = \psi_{I0} = V_S - \frac{kT}{q} \ln \frac{N_A}{n_i} \tag{22.36}$$

$$\psi_I(L_1) = \psi_{lg1} \tag{22.37}$$

$$\psi_{III}(L_2) = \psi_{lg2} \tag{22.38}$$

$$\psi_{III}(L_C) = \psi_{lc} = V_D + \frac{kT}{q} \ln \frac{N_D}{n_i} \tag{22.39}$$

where N_A (cm^{-3}) and N_D (cm^{-3}) are acceptor and donor concentrations at source and drain regions, respectively. Note that the length of the gates L_{G1} and L_{G2} (refer to Figure 22.6) are equal for an

RTFET, i.e., $L_{G1} = L_{G2} = L_G$. The surface potential at Region-I and Region-III is expressed in Equations (22.40) and (22.41), respectively.

$$\psi_I(x,y) = \frac{\psi_{A1}\sinh\left(\frac{L_1-x}{\lambda_s}\right) + \psi_{B1}\sinh\left(\frac{x}{\lambda_s}\right)}{\sinh\left(\frac{L_G}{\lambda_s}\right)} + K_2 \tag{22.40}$$

$$\psi_{III}(x,y) = \frac{\psi_{A2}\sinh\left(\frac{L_C-x}{\lambda_s}\right) + \psi_{B2}\sinh\left(\frac{x-L_2}{\lambda_s}\right)}{\sinh\left(\frac{L_G}{\lambda_s}\right)} + K_3 \tag{22.41}$$

where $\psi_{A1} = \psi_{I0} - K_2$, $\psi_{B1} = \psi_{Ig1} - K_2$, $\psi_{A2} = \psi_{Ig2} - K_3$ and $\psi_{B2} = \psi_{Ic} - K_3$

The potential at Region-II is expressed using a linear approximation to potential as in Equation (22.42).

$$\psi_{II}(x,y) = \frac{\psi_{Ig2} - \psi_{Ig1}}{L_2 - L_1}(x - L_1) + \psi_{Ig1} \tag{22.42}$$

Using the continuity boundary conditions Equations (22.43) and (22.44), ψ_{Ig1} and ψ_{Ig2} are obtained as in Equations (22.45) and (22.46), respectively.

$$\psi_I(L_1) = \psi_{II}(L_1) = \psi_{Ig1} \text{ and } \psi_{II}(L_2) = \psi_{III}(L_2) = \psi_{Ig2} \tag{22.43}$$

$$\left.\frac{d\psi_I(x)}{dx}\right|_{x=L_1} = \left.\frac{d\psi_{II}(x)}{dx}\right|_{x=L_1} \text{ and } \left.\frac{d\psi_{II}(x)}{dx}\right|_{x=L_2} = \left.\frac{d\psi_{III}(x)}{dx}\right|_{x=2} \tag{22.44}$$

$$\psi_{Ig1} = \alpha_1 \operatorname{sech}\frac{L_G}{\lambda_s} + \gamma_1 K_2 + \gamma_2 K_3 \tag{22.45}$$

$$\psi_{Ig2} = \alpha_2 \operatorname{sech}\frac{L_G}{\lambda_s} + \gamma_1 K_3 + \gamma_2 K_1 \tag{22.46}$$

where $\alpha_1 = \psi_{A1}\gamma_1 + \psi_{B2}\gamma_2$, $\alpha_2 = \psi_{B2}\gamma_1 + \psi_{A1}\gamma_2$, $\gamma_1 = \frac{\beta+\coth\frac{L_G}{\lambda_s}}{2\beta+\coth\frac{L_G}{\lambda_s}}$, $\gamma_2 = \frac{\beta}{2\beta+\coth\frac{L_G}{\lambda_s}}$, and $\beta = \frac{\lambda_s}{L_2-L_1}$. Now, the surface potential $(\psi(x,y))$ of the RTFET can be expressed as

$$\psi(x,y) = \begin{cases} \psi_I(x) \ 0 \le x \le L_1 \\ \psi_{II}(x) \ L_1 \le x \le L_2 \\ \psi_{III}(x) \ L_2 \le x \le L_C \end{cases} \tag{22.47}$$

A comparison of the potential model with TCAD simulation for an nRTFET and pRTFET is shown in Figure 22.7(a and b), respectively. In the case of nRTFET, V_{G1} varied from 0 to 1.5 V. In pRTFET operation, V_{G2} is varied from 0 to −1.5 V. From Figure 22.7, it can be seen that the previously discussed potential model shows a good match with TCAD simulation for both nRTFET and pRTFET for various gate voltages.

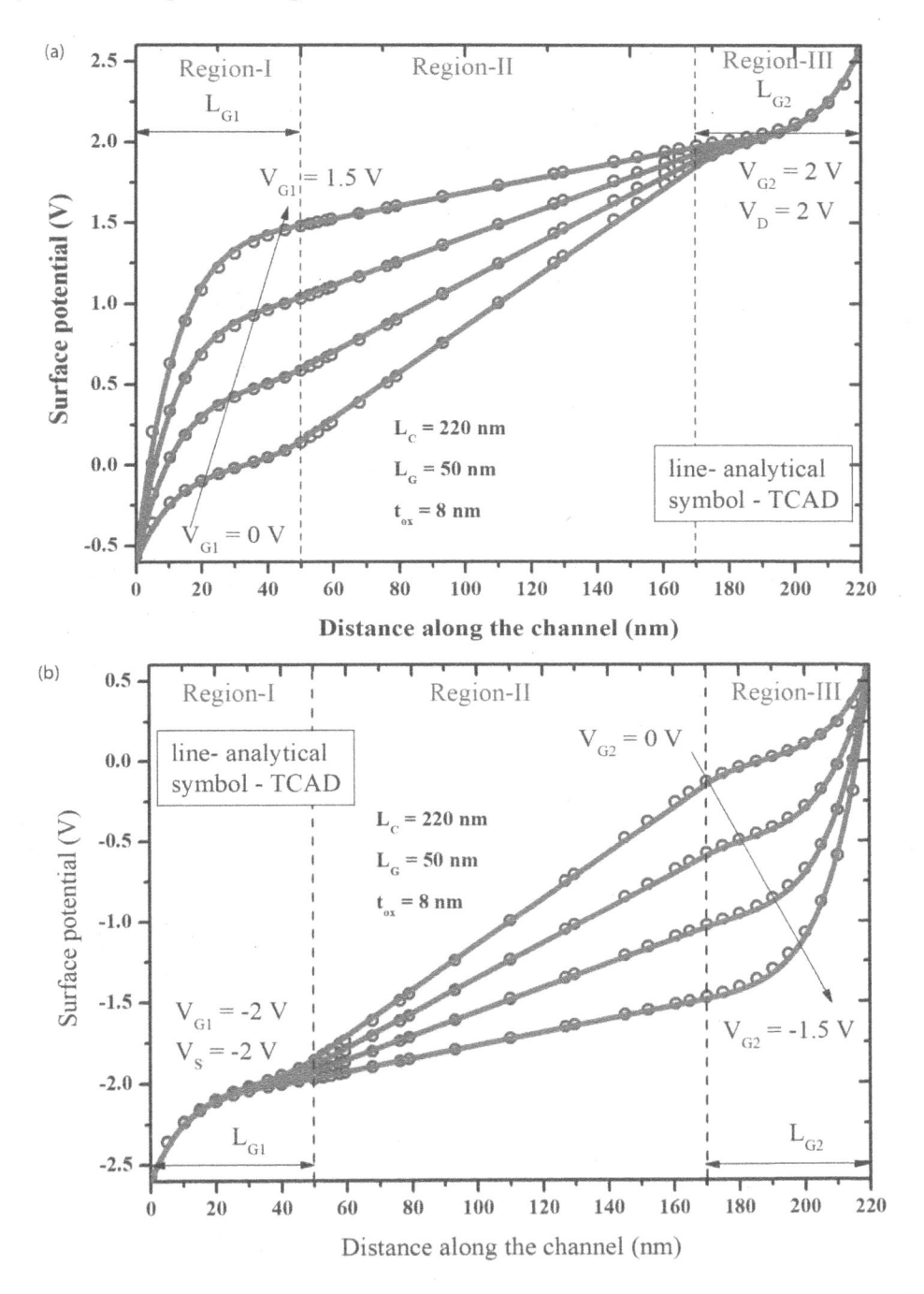

FIGURE 22.7 Surface potential variation in (a) an nRTFET and (b) a pRTFET for various gate voltages along the channel direction.

22.4.2 Tunneling Current Model

Tunneling current in an RTFET is derived by integrating a band-to-band generation rate (G_{BTB}) over the volume of the tunneling region as given in Equation (22.48). The expression of Kane's G_{BTB} model [13] given in Equation (22.49) contains an exponential and linear electric field component, which is quite complex to integrate. Numerical integration of Equation (22.49) is usually followed to find tunneling current in a TFET device [14]. In this chapter, an analytical expression for tunneling current in an RTFET is presented.

$$I = q \int G_{BTB} \, dV \tag{22.48}$$

$$G_{BTB} = A \xi \xi_{avg}^{D-1} e^{-B/\xi_{avg}} \tag{22.49}$$

where A, B, and D are Kane's tunneling parameters; $\xi_{avg} = E_g/ql_t$ is the average electric field; $\xi = \sqrt{\xi_x^2 + \xi_y^2}$ (Vcm^{-1}) is the local electric field; and l_t is the tunneling length. Figure 22.8 shows band alignment at a tunneling junction of RTFET. From Figure 22.8, it can be seen that l_t varies from l_1 to l_2.

The components of the local electric field ξ_x and ξ_y can be obtained from the potential model by differentiating potential $\psi_{I/III}(x,y)$ with respect to x and y.

$$\xi = \sqrt{\left[\lambda_s^{-1} \left(C_1 e^{\frac{x}{\lambda_s}} + C_2 e^{-\frac{x}{\lambda_s}} \right) \right]^2 + \left[b_1(x) + 2c_1(x) \right]^2} \tag{22.50}$$

Using the Taylor series expansion and ignoring higher-order terms Equation (22.50) is simplified to Equation (22.51) [15].

$$\xi = \xi_1 \left(1 + \frac{\Omega x + 2b(x)c(x)y}{\xi_1^2} \right) \tag{22.51}$$

where $\xi_1 = \sqrt{\lambda_s^{-2}(C_1 - C_2)^2 + b(x)^2}$ and $\Omega = \lambda_s^{-3}(C_1^2 - C_2^2)$.

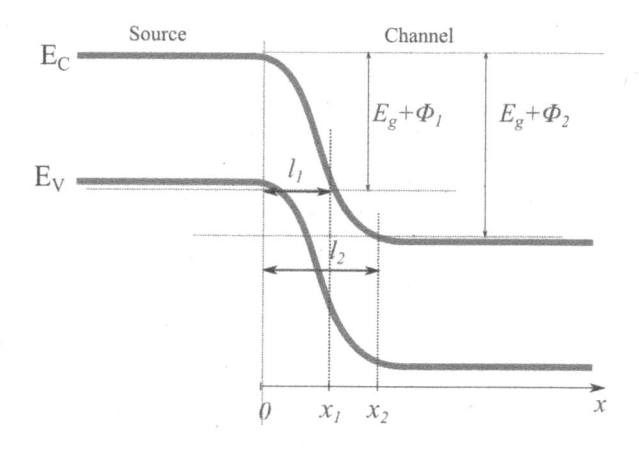

FIGURE 22.8 Band diagram of RTFET at the source-channel junction.

From Figure 22.8, it can be seen that tunneling of the carrier occurs from $x = x_1$ to $x = x_2$ along the x-direction. Assuming tunneling of carriers occurs from $y = 0$ to $y = t_{si}$ along the y-direction, and the width of the device is 1 μm along the z-direction, the integral Equation (22.48) can be expressed as

$$I = q \int_0^{t_{si}} \int_{x_1}^{x_2} A\xi_1 \left(1 + \frac{\Omega x + 2b(x)c(x)y}{\xi_1^2} \right) \left(\frac{E_g}{qx} \right)^{D-1} e^{-\frac{qB}{E_g}x} \, dx \, dy \tag{22.52}$$

where $x_1 = \lambda_s \frac{E_g + \Phi_1 - qK_2 - q(C_1 + C_2)}{q(C_1 - C_2)}$ and $x_2 = \lambda_s \frac{E_g + \Phi_2 - qK_2 - q(C_1 + C_2)}{q(C_1 - C_2)}$ are the lengths that correspond to the conduction band energy of the channel become equal to $E_g + \Phi_1$ and $E_g + \Phi_2$, respectively. By considering the variation of the polynomial term $(1/x^D)$ has less impact compared with the exponential term in Equation (22.45) and integrating Equation (22.52) yields the expression for the tunneling current in RTFET as in Equation (22.53) [15].

$$I = qA\xi_1 \left(\frac{E_g}{q} \right)^{D-1} \left[g_{x1x2}(x) \left(t_{si} + \frac{b(x)c(x)t_{si}^2}{\xi_1^2} \right) + h_{x1x2}(x) \left(\frac{\Omega t_{si}}{\xi_1^2} \right) \right] \tag{22.53}$$

Where $g_{x1x2}(x) = g(x1) - g(x_2)$ and $h_{x1x2}(x) = h(x1) - h(x_2)$.

$$g(x) = \frac{E_g}{qBx^D} e^{-\frac{qB}{E_g}x} \left(-x - \frac{E_g}{qB} \right) \tag{22.54}$$

$$h(x) = \frac{E_g}{qBx^D} e^{-\frac{qB}{E_g}x} \left(\frac{qB}{E_g} x^2 + 2x + 2\frac{E_g}{qB} \right) \tag{22.55}$$

A comparison of the analytical tunneling current model of RTFET with TCAD simulation is shown in Figure 22.9. The analytical current model shows a good match with TCAD simulation for lower gate voltages; however, there is a deviation for higher gate voltages.

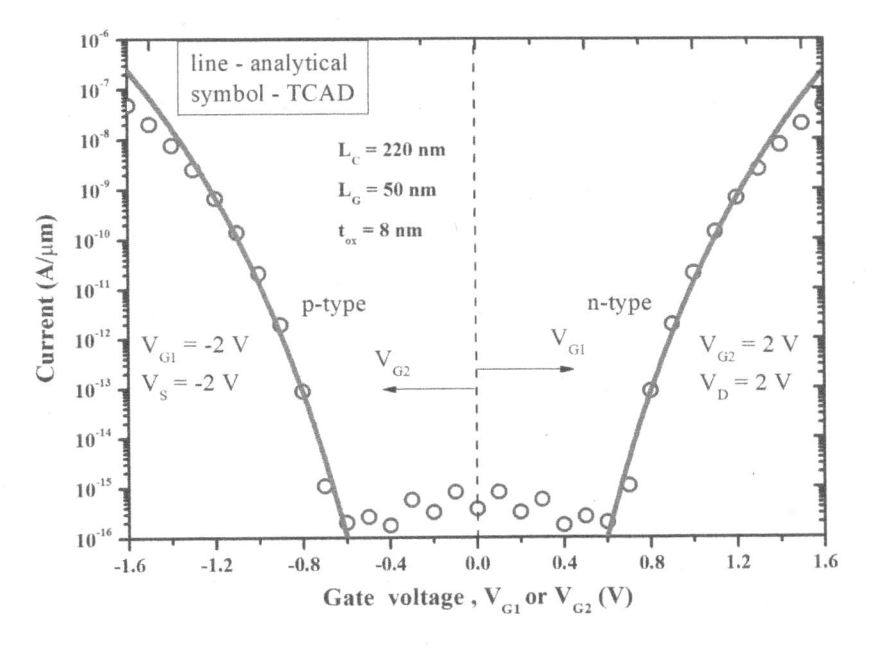

FIGURE 22.9 Current-voltage characteristics of an RTFET.

At the beginning of the derivation of the analytical model of the RTFET, it is stated that this model is a subthreshold model; hence, this model can only be used for analyzing subthreshold potential and current behavior of an RTFET. Thus, the deviation in current is due to the assumptions and approximations used in the derivation process.

22.5 SUMMARY

This chapter discussed the modeling techniques of two tunneling transistors, RFET and RTFET. The 2D Poisson's equation is solved in the gated region of RFET and RTFET. Parabolic and linear approximations are used in gated and ungated regions to model the entire channel potential of RFET and RTFET. This technique can be beneficial for one who models a MOS device with split-gate architecture. In Section 22.3.1.2, an alternative method to solve the 2D Poisson's equation containing exponentially varying charge density is reviewed. The discussed method in Section 22.3.1.2, i.e., merging of the pseudo-2D and superposition methods, simplifies the modeling effort. It helps to extend the subthreshold model of the RFET to an all-region or super-threshold model. The current model of an RFET and a RTFET is also discussed in this chapter. Tunneling current in an RFET is identified by integrating Launder's tunneling formula. On the other hand, to model current-voltage characteristics of an RTFET, Kane's band-to-band generation rate is analytically integrated over tunneling volume.

REFERENCES

1. Rai, S., Trommer, J., and Raitza, M., et al. 2018. Designing efficient circuits based on runtime-reconfigurable field-effect transistors. IEEE Trans. Very Large Scale Integr. VLSI Syst., 27 (3): 560–572.
2. Yao, Y., Sun, Y., Li, X., Shi, Y., and Liu, Z. 2020. Novel reconfigurable field-effect transistor with asymmetric spacer engineering at drain side. IEEE Trans. Electron Devices, 67 (2): 751–757.
3. Darbandy, G., Claus, M., and Schröter, M. 2016. High-performance reconfigurable Si nanowire field-effect transistor based on simplified device design. IEEE Trans. Nanotechnol., 15 (2): 289–294.
4. Larson, J. M., and Snyder, J. P. 2006. Overview and status of metal S/D schottky barrier MOSFET technology. IEEE Trans. Electron Devices, 53 (5): 1048–1058.
5. Ranjith, R., Komaragiri, R. S., and Suja, K. J. 2016. Reconfigurable tunnel field effect transistor exhibiting reduced ambipolar behavior. In Proc. INDICON, IEEE, 1–5.
6. Synopsys Inc, Sentaurus device user guide. CA, USA: Version L.2016.03, 2016.
7. Yan, R.-H., Ourmazd, A., and Lee, K. F. 1992. Scaling the Si MOSFET: From bulk to SOI to bulk. IEEE Trans. Electron Devices, 39 (7): 1704–1710.
8. Young, K. K. 1989. Short-channel effect in fully depleted SOI MOSFETs. IEEE Trans. Electron Devices, 36 (2): 399–402.
9. Ranjith, R., Jayachandran, R., Suja, K.J., and Komaragiri, R.S. 2018. Two dimensional analytical model for a reconfigurable field effect transistor. Superlatt. Microstruct., 114: 62–74.
10. Taur, Y., and Ning, T. H. 2013. Fundamentals of Modern VLSI Devices. Cambridge, UK: Cambridge University Press.
11. Keighobadi, D., Mohammadi, S., and Fathipour, M. 2019. An analytical drain current model for the cylindrical channel gate-all-around heterojunction tunnel FETs. IEEE Trans. Electron Devices, 66 (8): 3646–3651.
12. Kiziroglou, M., Li, X., Zhukov, A., De Groot, P., and De Groot, C. 2008. Thermionic field emission at electrodeposited Ni–Si Schottky barriers. Solid-State Electron., 52 (7): 1032–1038.
13. Gholizadeh, M., and Hosseini, S. E. 2014. A 2-D analytical model for double-gate tunnel FETs. IEEE Trans. Electron Devices, 61(5): 1494–1500.
14. Bardon, M. G., Neves, H. P., Puers, R., and Van Hoof, C. 2010. Pseudo-two-dimensional model for double-gate tunnel FETs considering the junctions depletion regions. IEEE Trans. Electron Devices, 57 (4): 827–834.
15. Ranjith, R., Suja, K. J., and Komaragiri, R. S. 2019. An analytical model for a reconfigurable tunnel field effect transistor. Superlatt. Microstr., 131: 40–52.

23 Flexi-Grid Technology
A Necessity for Spectral Resource Utilization

Divya Sharma, Shivam Singh, Anurag Upadhyay, and Sofyan A. Taya

CONTENTS

23.1 INTRODUCTION

A tremendous increase in an optical bandwidth requirement is observed because of extensive growth in Internet services, such as high-definition audio video streaming, broadcasting, cable TV (CATV), cloud computing, video conferencing, video on demand (VOD), social networking, etc. (Figure 23.1). The Cisco report states that worldwide mobile data traffic would expand sixfold from 2017 to 2022 at a 42% growth rate per year due to the high-performance connectivity over Wi-Fi, 4G, and 5G networks. Increased Wide Area Network (WAN) traffic is expected to reach 5.3 exabytes (EB) per month due to a high dependency on the cloud network. Internationally business Internet traffic will surpass 63.3 EB on a monthly basis in 2022, which is a threefold increase from 2017 [1]. As per the Hindustan Times Report of India, national Internet utilization is enhanced by 13% since the national lockdown was announced to fight against Covid-19. The telecom ministry stated that a consumption of 308 petabytes (PB) of data was observed on a daily basis during the week of March 22, 2020. During the COVID-19 pandemic situation, when most organizations (i.e., the whole education sector) were running on a digital plate-form, with the aid of student learning management systems and conferencing tools (e.g., Zoom, Google Meet, Microsoft Teams, etc.), there was a huge requirement for high-speed data traffic. This demands an efficient utilization of the available optical fiber capacity and evolution of next-generation optical technology.

ITU-T G.694.1 is a recommendation standard utilized for deploying dense wavelength-division multiplexing (DWDM) technology at a grid of 50 GHz in the C band (1530–1565 nm), which allows

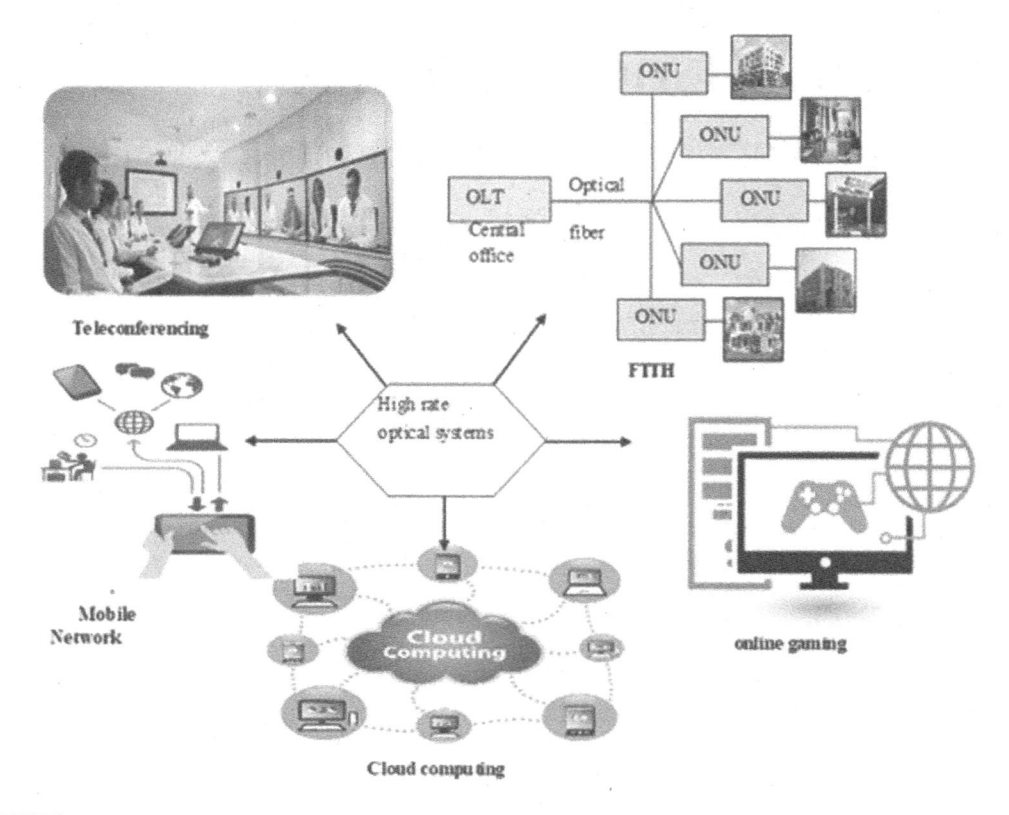

FIGURE 23.1   Application of high rate optical transmission (FTTH: Fiber to the Home, OLT: Optical Line Terminal, ONU: Optical Network Unit). (Adapted from [2].)

high throughput due to accommodating multiple simultaneous channels. Still, this 50-GHz ITU-T grid is not suitable in the current scenario because it causes wastage of optical spectrum at the low bit rate. This 50-GHz fixed grid is suitable for low data rate transmission per channel, say 10 Gb/s, 40 Gb/s, etc., along with using conventional modulation schemes in the presence of an intensity modulation direct detection (IM/DD) receiver [2].

23.1.1   Advance Hybrid Multiplexed Scheme

To achieve a high rate of transmission, i.e. greater than100 Gb/s, it is preferred to transmit information in multiple dimensions of an optical signal, which are frequency slots, intensity, polarization, phase, position, etc. (as shown in Figure 23.2). A very popular hybrid multiplexing scheme, polarization multiplexed (PM)-quadrature amplitude modulation (QAM)/quadrature phase shift keying (QPSK)/binary phase shift keying (BPSK), is based on transmitting information in the phase, intensity, and both polarizations of the light through modulation [3]. Among all these polarization multiplexed schemes, PM-QPSK is a preferential candidate because it doubles the channel transmission capacity along with supporting terahertz bandwidth. The PM-QPSK hybrid multiplexing scheme shows a compromise in energy and bandwidth efficiency and good receiver sensitivity along with delivering a large coverage distance with less complex hardware. This was possible because of small overhead in the transmitter [2]. A terahertz bandwidth capacity can be achieved through the PM-QPSK scheme at 100 Gb/s or greater. In Figure 23.3, an evolutionary growth is shown in transmission capacity utilizing various hybrid multiplexing schemes in the presence of a digital coherent receiver to satiate current traffic.

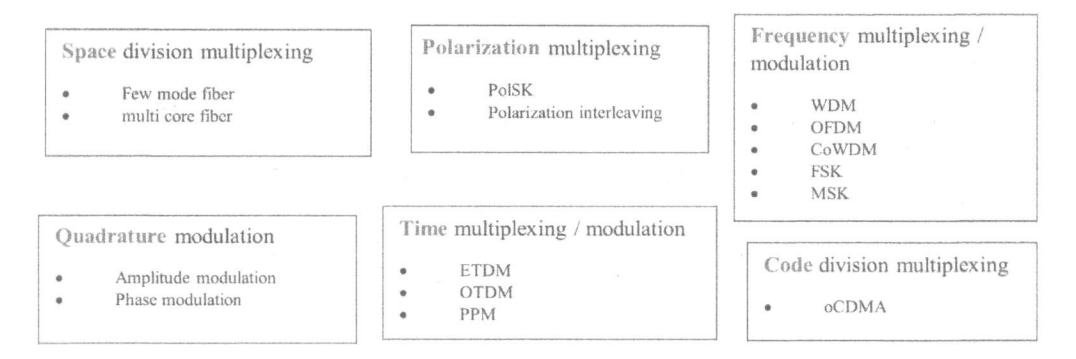

FIGURE 23.2 Hybrid multiplexed schemes (PolSK: Polarization Shift Keying, PPM: Pulse-position modulation, OTDM: Optical time division multiplexing, ETDM: Electrical time division multiplexing, WDM: Wavelength Division Multiplexing, OFDM: orthogonal frequency-division multiplexing, FSK: Frequency Shift Keying MSK: Minimum Shift Keying, oCDMA: Optical Code Division Multiple Access).

23.2 DIGITAL COHERENT RECEIVER

IM/DD technology was made possible in the 1970s for the sake of light transmission in the optical regime. In this method of the direct detection, receiver sensitivity was not dependent on the phase or polarization state of the laser light. In 2000, coherent technology was the center of attraction due to the deployment of real-time digital signal processing (DSP) technology and was capable of fulfilling exponentially grown traffic requirements. A coherent receiver along with the DSP is called a digital coherent receiver. It efficiently achieved two popular issues previously associated with coherent technology in the electronic domain; hence, it reduced complexity as well as processing time. DSP provides faster digital-to-analog conversion (DAC) and analog-to-digital conversion (ADC) [4].

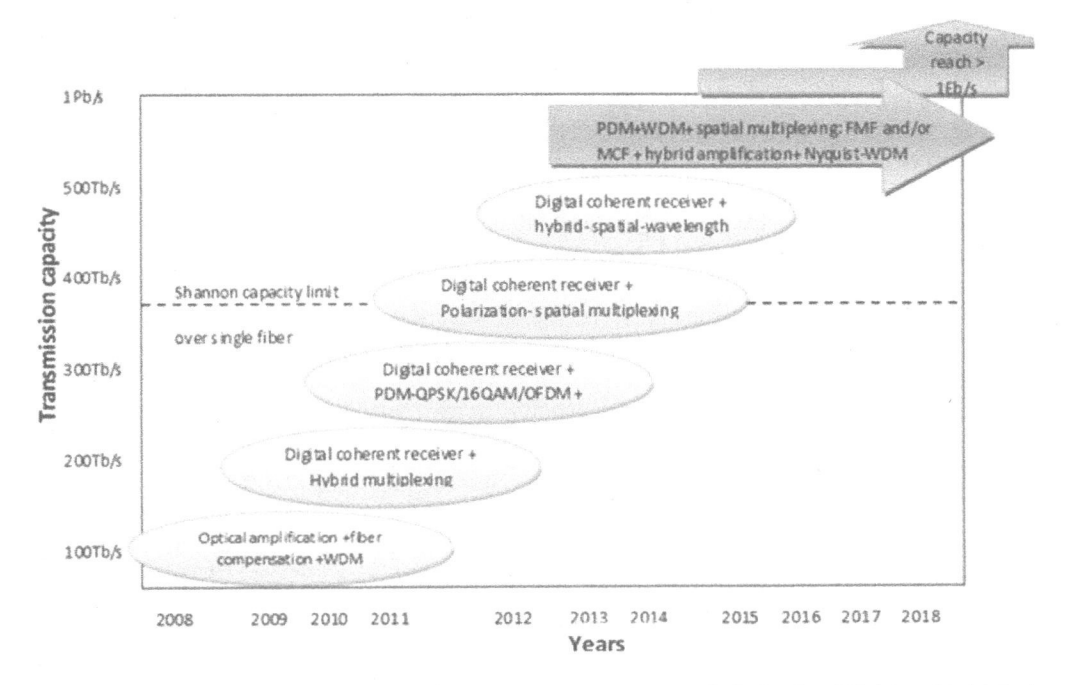

FIGURE 23.3 Advance trend in high-speed transmission (PDM: Polarization Division Multiplexing, MCF: Multi Core Fiber, FMF: Few Mode Fiber). (Adapted from [2].)

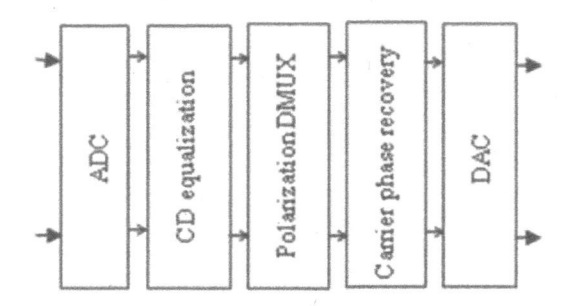

FIGURE 23.4   Algorithms involved in DSP.

Digital coherent receiver, a combination of coherent receiver along with DSP, permits a high data transmission rate, i.e. 100 Gb/s or greater [5], since transmitting information at 100 Gb/s results in linear and nonlinear channel impairments leading to performance degradation in the form of limited reach, poor eye opening, degraded bit error rate (BER) performance, poor quality factors, etc. Therefore, various algorithms are used at such a high data rate to reduce these nonlinear effects by employing real time DSP, as shown in Figure 23.4. Digital coherent technology supports various higher-order hybrid multiplexing schemes, such as PM-BPSK, PM-QPSK, and PM-QAM. By employing advance hybrid multiplexing schemes with higher-order the baud rate decreases; hence, lowering the sampling rate of ADC, which speeds up the DSP [6]. Hence, digital coherent technology supports greater than 100 Gb/s rate electronics, which is an advantage over employing regenerator and dispersion compensation fiber along with the standard single mode fiber (SSMF) to mitigate channel impairments.

23.2.1   SUPERCHANNEL TRANSMISSION

Data transmission capacity of an optical fiber can be increased with the help of enhancing the number of channels or by enhancing transmission capacity per channel. Due to an exhausted ITU-T C band, enhancing the number of channels seems impossible. In the L + C band, by employing optical amplifiers such as the Er^{3+} doped fiber amplifier (EDFA), transmission capacity can be enhanced. Hence, enhancing the per channel transmission capacity is more preferable. Assuming the availability of 100 GSamples/s ADCs, 1 Tb/s data transmission capacity over a single carrier can be attained with a greater higher-order modulation format, say PM-1024QAM. Utilizing such a huge constellation size format would result in multiple serious issues such as bad sensitivity, weak resistance to nonlinearity, high phase noise necessity, necessity of fast DACs at the transmitter, hardware bandwidth issues, and limited reach. Hence a terabit transmission superchannel, i.e. 1 Tb/s over a single channel, can be achieved by aggregating a suitable number of "subcarriers." A 1-Tb/s superchannel is achieved through 10 × 100 Gb/s PM-QPSK, which means 10 subcarriers, each with a carrying data rate of 100 Gb/s. To achieve good spectral efficiency (SE), each subcarriers must be tightly packed in frequency domain [7–9].

A superchannel contains multiple optical carriers dealt as a single transport entity. A superchannel is realized in terms of extreme DWDM wavelengths generated from the same optical line card. Hence, by employing adaptive filtering technologies (i.e., flexi-grid technology) a superchannel carries varying channels and yields different capacity and different spectral utilization. Figure 23.5 shows how a flexible grid technology yields flexible spectral allocation on the basis of bandwidth requirement and optical rate. Hence, a higher speed of 400 Gb/s and greater is achieved, along with enhanced SE, by tightly packing the channels instead of using a conventional 50-GHz fixed ITU-T grid [10, 11].

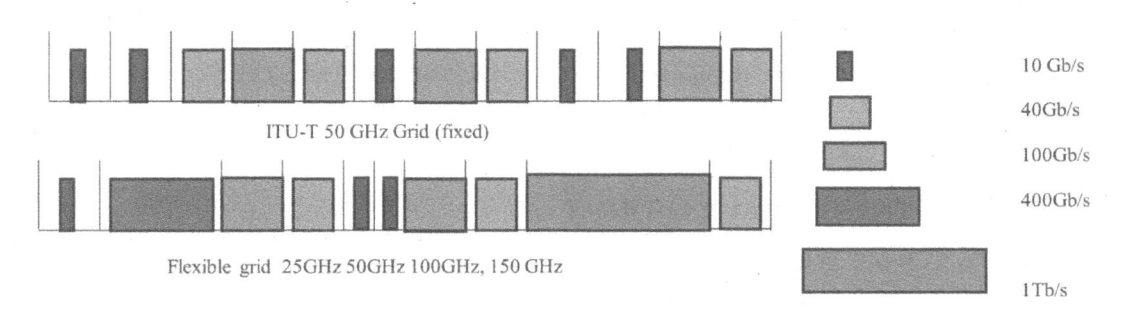

FIGURE 23.5 Fixed-grid (conventional DWDM) and flexi-grid technology.

Two wavelength selective switch (WSS) parameters responsible for flexi-grid systems design are attenuation or shaping granularity and switching granularity. At switching granularity, bandwidth is switched for adding or dropping superchannel, whereas at shaping granularity, the spectrum utilized by a single superchannel is shaped [12, 13].

Flexi-grid technology utilized in a superchannel overcomes the constraints of fixed ITU-T grid DWDM technology and enhances the channel capacity along with efficient optical spectrum resource utilization. A coherent superchannel supports next-generation optical infrastructure by enhancing the DWDM network capacity of a single channel (varying from 100 Gb/s to 1 Tb/s) with excellent transmission reach and high-speed data services. Such an enhancement in the SE is seen in the superchannel because of the higher-order advance modulation formats along with digital coherent technology [10].

In a conventional DWDM channel, it is necessary to guard band at both edges of the channel (Figure 23.5), which helps in optical switching, multiplexing, demultiplexing of the optical channel, etc. The presence of guard band leads to wastage of a huge spectrum, which is unutilized by the actual payload and reduces the fiber transmission capacity. Novel superchannel technology is a solution to the issue of conventional DWDM. Subcarrier aggregation is achieved either by coherent optical-orthogonal frequency division multiplexing (Co-OFDM) technique or by the Nyquist-WDM technique [14]. Table 23.1 shows that employing hybrid multiplexed schemes increases the spectrum utilization compared with the fixed ITU-T grid.

Limitations of the flexi-grid technology are cost and complexity of the associated hardware and software such as employing flexible grid transceivers, Flex-Reconfigurable optical add-drop multiplexer (ROADM), DAC, ADC and DSP module, etc. Control over this flexi-grid superchannel seems challenging by availing the variable spectrum on the ITU-T grid. This results in the evolution of Flex-ROADM, which assists in switching variable spectrum. To increase switching flexibility of the flexi-grid technology, tunable lasers, tunable filters, and a spectrum select switch are needed, which increases cost and complexity. Employing monolithic photonic integrated circuit (PIC) technology

TABLE 23.1

Comparison of Bandwidth Efficient Utilization of Fixed ITU-T Grid DWDM and Flexi-Grid Superchannel Technology [10]

Hybrid Scheme	Channel Bandwidth = Spectral Bandwidth + Guard Band	Fixed-Grid Solution	% Increase in Bandwidth Utilization
40Gb/s PM-QPSK	35 GHz = (25 + 10) GHz	One 50-GHz channel	43
100 Gb/s PM-QPSK	47.5 GHz = (37.5 +10) GHz	One 50-GHz channel	5
4 × 100 Gb/s PM-QPSK	85 GHz = (75 + 10) GHz	Four 50-GHz channel	135
10 × 100Gb/s PM-QPSK	200 GHz = (190 + 10) GHz	Ten 50-GHz channel	150

for the superchannel implementation is the best solution with low cost and complexity instead of implementing the superchannel through discrete optical components. Through PIC, all subcarriers are fused onto a single line to form a superchannel, which saves energy and lessens the complexity of the hardware; hence, advancement in semiconductor materials support the flexi-grid technology.

23.3 COHERENT OPTICAL (Co)-OFDM

The Co-OFDM superchannel concept is a parallel processing technique in which various optical subcarriers are individually modulated at some lower symbol rates, and then they are finally aggregated into a multicarrier system to yield a targeted data rate. By using the Co-OFDM technique, optical parallelization occurs in the frequency domain to attain a data rate beyond that supported by electronics. OFDM is employed to form a superchannel in two different ways: (1) OFDM-based modulation, where a box spectral-shaped signal is generated resulting in closely spaced modulated subcarriers and (2) Co-OFDM resulting in multiplexing of the modulated subcarriers [15]. To generate a superchannel using the OFDM technique, the following conditions must be satisfied [16]:

1. Spacing in between subcarriers must equal the symbol rate. It states that the modulated carriers must be frequency locked.
2. At the demultiplexing stage, to make the correct decision, modulated symbols on the carriers must be in time-sync to avoid the cross talk penalty and to maintain the orthogonality condition. Due to chromatic dispersion, the subcarriers traveling in the fiber are spread in the vicinity, which must be controlled at the time of demultiplexing through the dispersion compensation or by maintaining the orthogonality condition.
3. In general, modulated symbols in the frequency domain have a sinc shape. Hence, enough spectrum is required to modulate each subcarrier. At the receiver, a sufficient oversampling speed must be maintained to capture the sinc-shaped function by subcarriers.

Hence, Co-OFDM requires subcarrier spacing identical to the baud rate, which requires a large transmitter/receiver electric bandwidth. This seems hard to attain at current electronic equipment speed constraint, so Co-OFDM is preferred for lower baud rate transmission. For the sake of transmission capacity enhancement, employing multilevel modulations such as OFDM needs a high OSNR, sensitive to nonlinearities, and laser phase noise. Hence, employing OFDM for the superchannel generation yields limited reach along with low SE. Employing optical time division multiplexing (OTDM) could be a substitute of OFDM, but OTDM needs sharp timing synchronization and a costly optical dispersion mitigation technique [17]. Co-OFDM yields good SE, and robust dispersion and timing synchronization of both polarization branches.

23.3.1 MULTIBAND OFDM TECHNIQUE

In Figure 23.6(a), a single band transmitter (SB-Tx) is shown that consists of an MZI modulator driven by four analog drive signals (where I and Q components represent separate states of polarization) to generate a PM-OFDM/m-QAM modulated signal. Here, four DACs and OFDM-DSP are availed for the pulse shaping and pre-equalization process at the transmitter. In Figure 23.6(b), on combining parallelly multiple SB-Tx, a multiband transmitter (MB-Tx), i.e., multiband OFDM superchannel, is generated at the desired rate [15].

23.3.2 RECIRCULATING FREQUENCY SHIFTER (RFS) FOR OFDM

In 2009, Ma et al. transmitted a 1-Tb/s superchannel over 600 km SSMF with 27-dB back-to-back (B2B) optical signal-to-noise ratio (OSNR) at a BER of 10^{-3} in the presence of a frequency shifter. They reported 3.3 b/s/Hz SE without employing any optical compensation or Raman amplification [17].

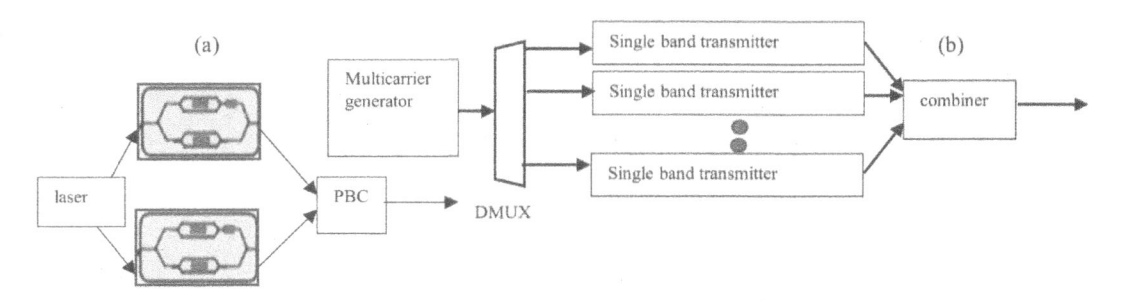

FIGURE 23.6 (a) Single band transmitter, i.e., PM-OFDM/m-QAM transmitter. (b) OFDM-based super-channel transmitter. DMUX, wavelength demultiplexing.

As shown in Figure 23.7, employing the recirculating frequency shifter (RFS) replicates numerous bands per step at each circulation on the initially applied optical OFDM signal. Hence, there is no need for high drive voltage for the external modulator. Also, a lag in the recirculating loop is kept equal to the integral multiple of the OFDM symbol periods, which leads to neighboring bands placed at the correct frequency grids and support synchronization in the OFDM frame [17].

23.3.3 OPTICAL COMB SOURCE FOR OFDM SUPERCHANNEL

In Figure 23.8 the optical comb source is shown, which reduces interchannel guard band by providing a fixed frequency spacing among subcarriers. It is done this way because independent lasers have failed to do so. A comb source exhibits free spectral range tunability, hence, it is easy for a single source to match the chosen symbol rate. An optical frequency comb source should deliver high-frequency stability, good spectral flatness, a tunable free spectral range, and low linewidth. A comb source is made up of electro-optic modulators composed of dual drive modulators. Gain switching the optical comb source is an alternate to the laser source [18, 19]. In Figure 23.8, for 32-QAM-OFDM generation, the external cavity laser (ECL), along with the MZI based on a five comb generator, is used at transmitter. WSS rejects undesired harmonics generated by five optical comb generator. When employing a four optical comb generator, there were quadruple the number of frequency locked carriers [20].

23.3.4 A BRIEF LITERATURE SURVEY OVER OFDM-BASED SUPERCHANNEL

In 2009 Yu et al. demonstrated an orthogonal superchannel of four 100-Gb/s orthogonal PM-RZ-QPSK superchannels on a 50-GHz ITU-T grid. They reported an SE of 4 b/s/Hz with 1040 km coverage using an SMF-28 fiber with EDFA and 22.8-dB span loss [21]. In 2010, Zhu et al. transmitted a 1.2-Tb/s no guard interleaved (NGI)-Co-OFDM superchannel over the span of a 100-km Ultra large

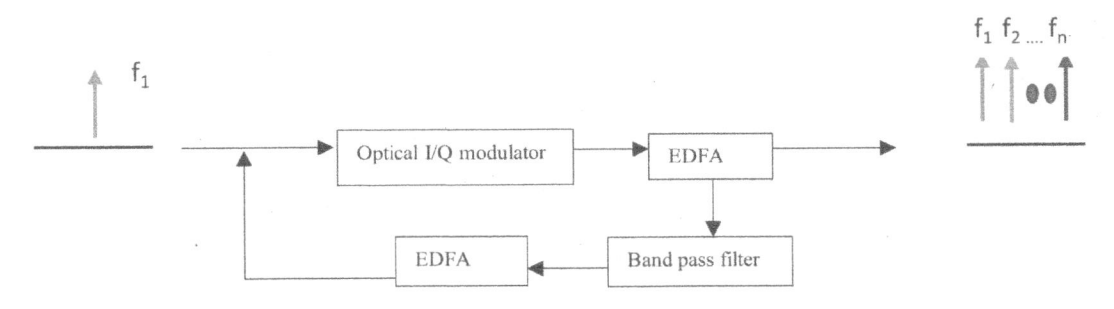

FIGURE 23.7 Recirculating frequency shifter (RFS) for superchannel generation.

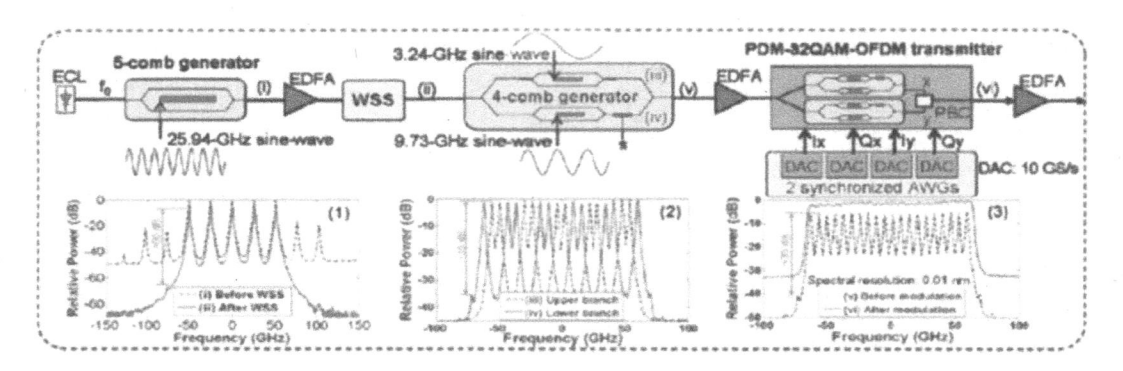

FIGURE 23.8 A 1.12-Tb/s 32-QAM-OFDM superchannel generation setup (Reprinted/adapted with permission from [20] © The Optical Society.)

area fiber (ULAF). In the presence of hybrid Raman and all-Raman amplifiers, transmitted distances were 5200 and 7200 km [18]. In 2011, Liu et al. generated a 1.12-Tb/s superchannel based on a Co-OFDM of 20×56 Gb/s PM-32-QAM frequency locked subcarriers, with a recorded SE of 8.6 b/s/Hz [20]. In 2012, Li et al. experimentally transmitted a 1-Tb/s PM-QPSK discrete Fourier transform spread (DFTS)-OFDM superchannel over a distance of 8000 km of SSMF in the presence of EDFA. Compared with conventional OFDM, they obtained a 0.6-dB enhancement in Q-factor and up to 1000 km reach [22]. In 2014, Tanimura et al. demonstrated a 4×28-Gbaud QPSK OFDM superchannel with a capacity of 224 Gb/s using fiber frequency conversion. They achieved transmission up to 1000 km [23]. In 2015, Song et al. transmitted a 400-Gb/s 16-QAM multiband OFDM superchannel (4 sub-bands spaced by a 2-GHz guard-band) over a distance of 10×100 km G.652 fiber with an SE of 5.13 b/s/Hz. Multiband OFDM eliminates ADC/DAC electronic constraints [24]. Later in 2016, M. Song et al. transmitted a 1-Tb/s multiband OFDM superchannel using 10×100 Gb/s 16QAM subcarrier over a distance of 1000 km of G.652 fiber [25]. In 2018, Geng et al. transmitted a 6.885-Tb/s superchannel using 180×12.75 Gbaud 8-QAM Co-OFDM with the aid of a Kerr soliton frequency comb (dissipative in nature). They achieved an SE of 2.625 b/s/Hz [26]. In 2020, Venkatasubramani et al. transmitted a 608-Gb/s Co-OFDM superchannel using gain switched comb through an optical carrier over a distance of 25 km and achieved excellent BER performance [27]. All literature survey work regarding OFDM-based superchannels is summarized in Table 23.2.

23.4 NYQUIST-WDM

Nyquist-WDM is a reliable solution for superchannel generation in the next-generation optical networks. Nyquist-WDM and Co-OFDM approaches offer similar complexity when putting a limit to time and frequency spacing. Nyquist-WDM provides compact spectral pre-filtering at the transmitter end, which results into intersymbol interreference (ISI)-free Nyquist criterion. Nyquist-WDM maintains channel spacing = symbol rate (Nyquist-spaced) or smaller than the symbol rate (super Nyquist-spaced) or more than the symbol rate (quasi Nyquist-spaced) as shown in Figure 23.9 [28–30]. Physically implementing a Nyquist filter yields a sharp spectral shaping, which is a challenging task. Yet Nyquist filtering is the preferential approach over deploying DAC with a huge bandwidth and a large sampling rate [31, 32].

23.4.1 A LITERATURE SURVEY OVER NYQUIST-WDM-BASED SUPERCHANNEL

In 2009, Salsi et al. successfully transmitted a Nyquist-WDM superchannel of 10×112 Gb/s PM-QPSK channels with very narrow spacing = $1.1 \times$ symbol rate. They observed coverage distances of 3300, 2300, 1120, and 800 km, respectively, by employing pure silica core fiber (PSCF),

TABLE 23.2

Research Work on OFDM-Based Superchannel

Superchannel	Strategy	Fiber Type	Capacity	Reach	OSNR/Spectral Efficiency (SE)	Reference
Co-OFDM	Frequency shifter	SSMF	1 Tb/s	600 km	27 dB/3.3 b/s/Hz	[17]
4 × 100 Gb/s orthogonal PM-RZ-QPSK	RZ pulse carver, optical filter	SMF-28	400 Gb/s	1040 km	22.8 dB	[21]
NGI-Co-OFDM	Hybrid Raman/ EDFA	ULAF	1.2 Tb/s	7200 km	-	[18]
20 × 56 Gb/s PM- 32-QAM Co-OFDM	Frequency locking, comb generator	-	1.12 Tb/s	-	8.6 b/s/Hz	[20]
DFTS-OFDM PM-QPSK	Discrete Fourier transform spread	SSMF + EDFA	1 Tb/s	8000 km	-	[22]
4 × 28 Gbaud QPSK	Frequency conversion	-	224 Gb/s	1000 km	-	[23]
16-QAM multiband OFDM	-	G.652 fiber	400 Gb/s	1000 km	5.13 b/s/Hz	[24]
10 × 100 Gb/s 16QAM OFDM	-	G.652 fiber	1 Tb/s	1000 km	-	[25]
180 × 12.75 Gbaud 8-QAM Co-OFDM	Dissipative Kerr soliton frequency comb	-	6.885 Tb/s	-	2.625 b/s/Hz	[26]
Co-OFDM	Gain switched comb		608 Gb/s	25 km		[27]

large effective area fiber (LEAF), SSMF, and non-zero dispersion shifted fiber (NZDSF) at a BER 3×10^{-3}. The experiment work is accomplished using EDFA amplification, a Nyquist optical filter, and RFS [31].

In 2010, G. Gavioli et al. transmitted a 1.12-Tb/s Nyquist-WDM superchannel composed of ten 112-Gb/s PM-QPSK subcarriers at a channel spacing 30.8 GHz, using narrow optical filtering [33].

Figure 23.10(a) shows an experimental setup of a 1.12-Tb/s Nyquist-WDM PM-QPSK-based superchannel generation. A laser source is modulated with an MZI modulator to give a 56-Gb/s QPSK signal, further narrow-filtered through a wave shaper filter with a bandwidth of 28 GHz (baud rate). The modulated channel is reached into an RFS and then through a recirculating loop, and a digital coherent receiver is present. Figure 23.10(b) shows how to achieve a BER of 3×10^{-3}. The OSNR required for the superchannel is 16.5 dB for B2B, which is 3 dB higher than single-channel transmission.

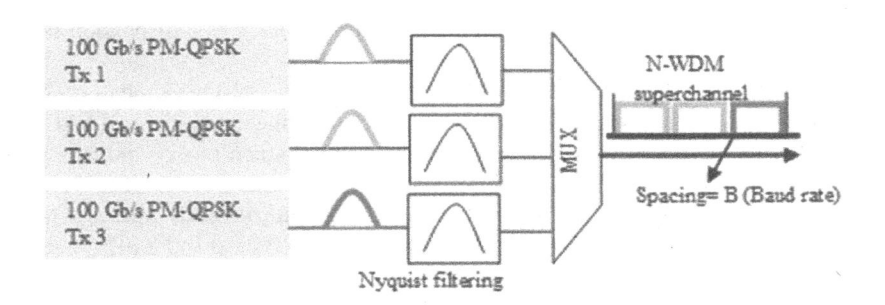

FIGURE 23.9 Nyquist-WDM superchannel.

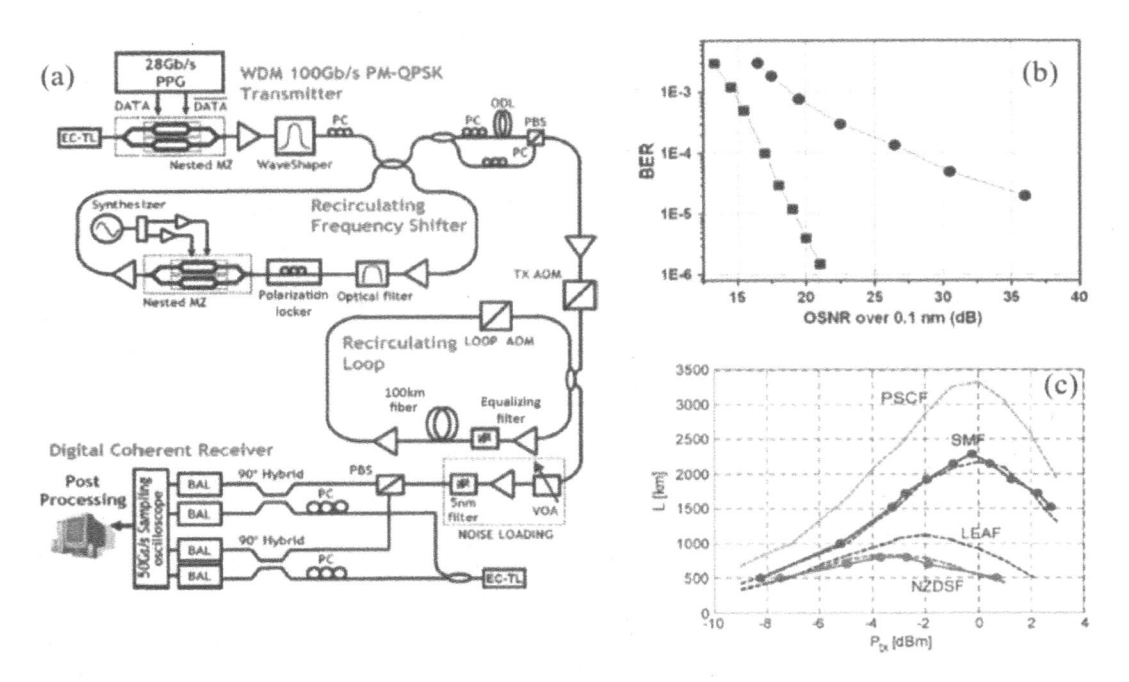

FIGURE 23.10 (a) Test bed for a 1.12-Tb/s Nyquist-WDM PM-QPSK-based superchannel. (b) B2B BER versus OSNR plot. Squares, single channel; circles, Nyquist-WDM conditions. (c) Transmission reach versus fiber launch power per channel. (Reprinted/adapted with permission from [33].)

In Figure 23.10(c) reach is plotted against the transmitted power. Coverage of 3300 km over SMF with an RFS, 1200 km over NZDSF, 1700km over LEAF, and 4900 km over PSCF were achieved [33].

In 2012, Chien et al. demonstrated a terabit superchannel with a noise-suppressed Nyquist-WDM approach along with aggregating a 100-Gb/s PM-QPSK subcarrier at Nyquist spacing ratios = 1.28. They realized that the coverage distance is up to 2800 km over SMF-28 fiber in the presence of EDFA amplification at BER = 2×10^{-3}. They accomplished the work by employing a 1-bit Maximum likelihood sequence estimation (MLSE) at the receiver section along with digital noise filtering [34].

In 2014, Igarashi et al. demonstrated a 140.7-Tb/s Nyquist-WDM superchannel by deploying 30 Gbaud Duobinary pulse-shaped seven 201-Gb/s PM-QPSK subcarriers. They reported a reach of 7326 km over seven-core fibers in the presence of seven-core EDFAs over a grid of 25 GHz along with reporting capacity-distance product = 1 Eb/s × km [35]. In 2014, Liu et al. transmitted 1024 Tb/s, i.e., 8 × 128 Gb/s Nyquist-WDM superchannel signal transmission, using PM-QPSK subcarriers. A polarization-insensitive fiberoptic parametric amplifier was utilized by them, which supported all optical signal processing and delivered doubled reach in a dispersion managed link with gain of 3.8 dB [36].

In 2015, Xu et al. performed transmission of a 5 × 230Gb/s PM-QPSK Nyquist-WDM superchannel with quasi Nyquist spacing of 60 GHz over an 820-km fiber link. The test bed is mentioned in Figure 23.11(a). In Figure 23.11(b), BER is plotted against launch power and it is realized that optimum launch power is 4 dBm for the single-channel transmission and 9 dBm for 5 × 230Gb/s superchannel transmission. In Figure 23.11(c) BER is plotted against OSNR values. To achieve min. BER = 4.5×10^{-3}, there exists a 1.2-dB OSNR penalty in the B2B case and 2 dB for 820 km. They achieved the transmission capacity of 1.15 Tb/s and SE of 3.5 b/s/Hz [37]. In 2016, Zhu et al. transmitted Nyquist-WDM superchannel and achieved a 1.5-dB quality factor enhancement after 3840 km for QPSK and a 1-dB quality factor gain with a 16 QAM modulated carrier after 800 km [38].

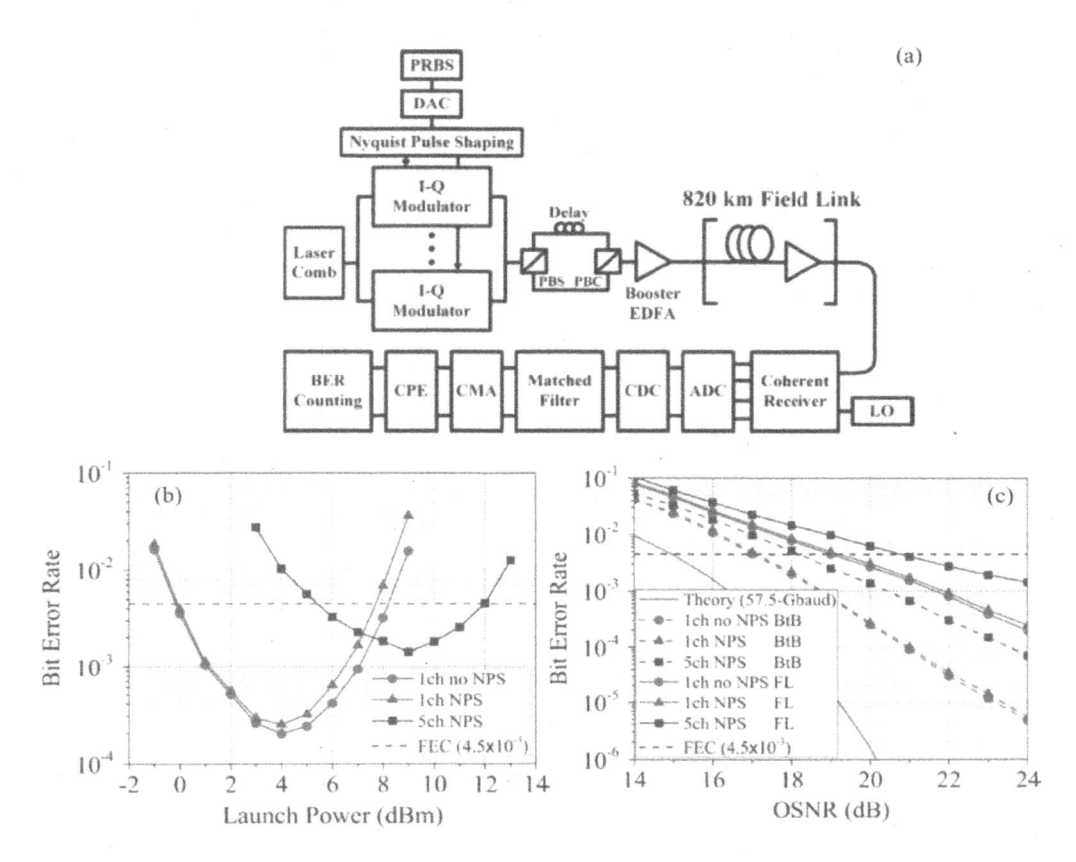

FIGURE 23.11 (a) Setup bed for quasi Nyquist-WDM superchannel using five 230-Gb/s PM-QPSK subcarriers. (b) BER is plotted against signal power per span. (c) BER is plotted against OSNR at the optimum signal power (Reprinted/adapted with permission from [37].)

In 2019, Sharma et al. transmitted a 550-Gb/s MLR-based Nyquist-WDM superchannel by simultaneously transmitting different 27.5-Gbaud polarization multiplexed subcarriers over PSCF together with the hybrid Raman amplifier (see Figure 23.12(a)). In Figure 23.12(b), min. acceptable BER 4×10^{-3} is plotted at OSNR values 10, 14, 17, and 24 dB. In Figure 23.12(c) coverage distances are 900, 2150, 1700, and 1060 km for PM-BPSK, PM-QPSK, PM-8 QAM, and PM-16 QAM subcarriers, respectively [39]. In 2020, Singh et al. transmitted Nyquist-WDM a 32×200 Gb/s PM-QPSK superchannel over a free space optics (FSO) link at a 60-GHz channel spacing and achieved the capacity of 6.4 Tb/s along with SE 3.33 bits/s/Hz [40]. All literature survey work regarding the Nyquist-WDM-based superchannel is summarized in Table 23.3.

23.4.2 OFDM VERSUS NYQUIST-WDM APPROACH OF SUPERCHANNEL GENERATION

In 2010, Bosco et al. applied two different techniques, i.e. OFDM and Nyquist-WD. For both techniques, channel spacing was kept equal to R_s (baud rate) to avoid cross talk and ISI. Nyquist-WDM has a rectangular spectrum with bandwidth = baud rate = R_s, and a sinc-shaped pulse in time domain [45]. As in Figure 23.13(a), channels are not overlapping, so cross talk and ISI are avoided. On the other hand, in Figure 23.13(b) there are rectangular pulses with duration = symbol time = $T_s = 1/R_s$, along with sinc function like the frequency spectrum. Here, cross talk is seen, which is avoided by proper filtering at the receiver on the condition that channels extracted from the carrier must be mutually coherent over T_s [46, 47].

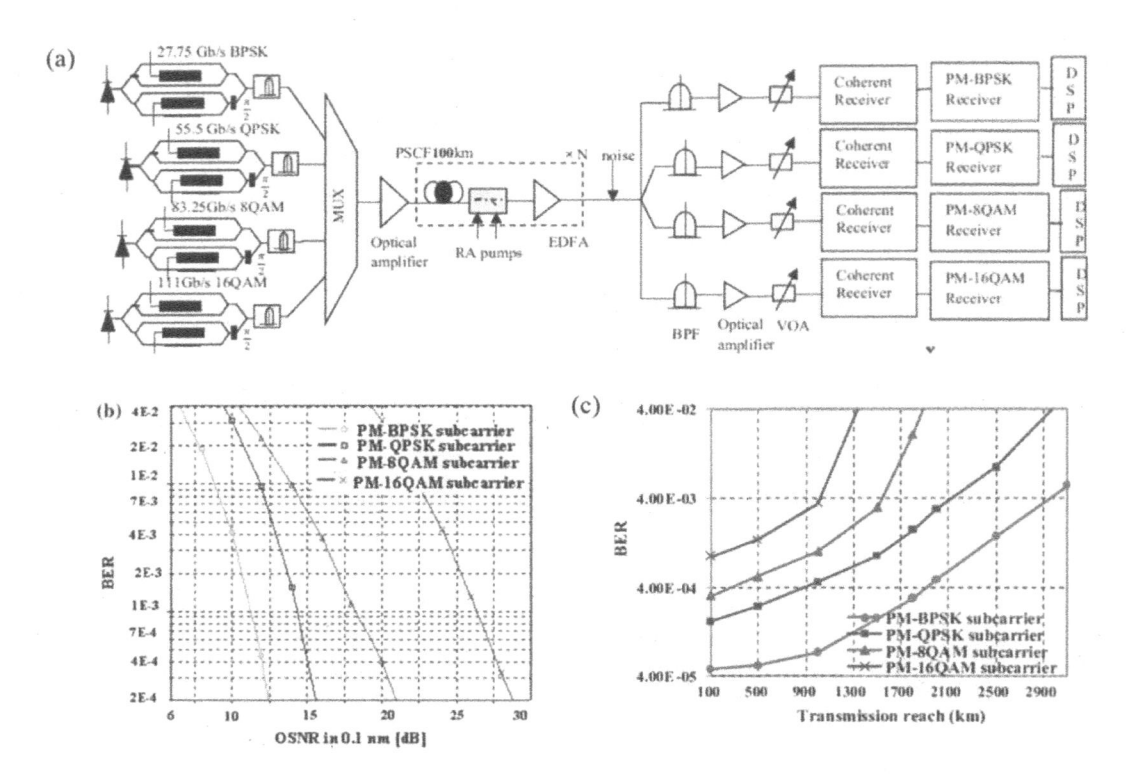

FIGURE 23.12 (a) Setup bed for mixed Nyquist-WDM superchannel through PM-BPSK, PM-QPSK, PM-8 QAM, and PM-16 QAM subcarriers. (b) BER is plotted against OSNR required. (c) BER versus maximum transmission reach is plotted (Reprinted/adapted with permission from [39].)

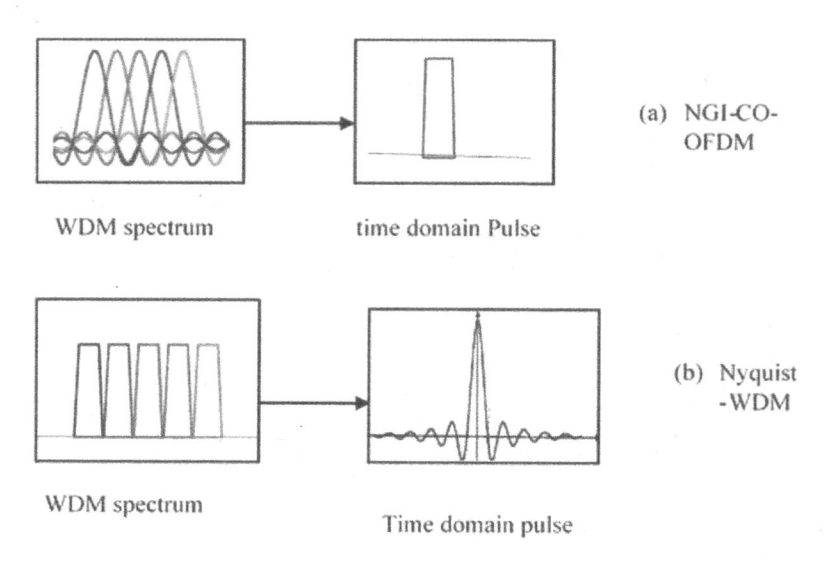

FIGURE 23.13 (a) WDM spectrum (left) and pulse in time domain (right) for NGI-Co-OFDM. (b) WDM spectrum (left) and pulse in time domain (right) for Nyquist-WDM.

TABLE 23.3

Research Work on the Nyquist WDM-Based PM-QPSK Superchannel [41]

Nyquist WDM	Grid	Type of Fiber	SE or Capacity	Reach (km)	OSNR/Q-factor, BER	Reference
10×112 Gb/s	$1.1 \times$ Symbol	SSMF		1120		[31]
PM-QPSK	rate	PSCF	1.12 Tb/s	3300	-	
		LEAF		2300		
		NZDSF		800		
96×100 Gb/s	33 and 25	SLEAF (150	3 b/s/Hz	10610	10 dB	[33]
PM-RZ-QPSK	GHz	μm^2), EDFA	4 b/s/Hz	4370		
100-Gb/s PM-QPSK	$1.28 \times$	SMF-28 and		2800	2×10^{-3}	[34]
noise suppressed	Symbol rate	EDFA				
Nyquist-WDM						
NGI-Co-OFDM				3200		
$7 \times 201 \times 100$ Gb/s	25-GHz grid	7 core (fibers	140.7 Tb/s	7326	-	[35]
PM-QPSK		and EDFAs)				
440-Gb/s PM-QPSK	100 GHz	ULAF	4 b/s/Hz	3600	-	[42]
5×230 Gb/s PM-QPSK	60 GHz	SSMF + EDFA	3.5 b/s/Hz,	820	3×10^{-4}	[37]
			1.15 Tb/s			
4×111 Gb/s PM-QPSK		PSCF	4.5 b/s/Hz	5500	4×10^{-3}	[43]
5×222 Gb/s PM-16QAM				2300		
55-Gb/s PM-BPSK		PSCF	4.16 b/s/Hz	2900	4×10^{-3}	[39]
111-Gb/s PM-QPSK				2150		
166.5-Gb/s PM-8QAM				1700		
222-Gb/s PM-16QAM				1060		
32×200 Gb/s PM-QPSK		FSO link	3.33 b/s/Hz,		-	[40]
			6.4 Tb/s			
166.5-Gb/s PM-8QAM		PSCF	5.25 b/s/Hz,	3600	4×10^{-3}	[44]
222-Gb/s PM-16QAM				2250		

During performance comparison, Bosco et al. observed that the Nyquist-WDM approach of superchannel generation is highly robust to practical implementation, e.g., constraint in the receiver analog bandwidth and number of SpS [45]. For B2B, Nyquist-WDM and Co-OFDM attain the identical highest sensitivity, but Co-OFDM, is more affected by nonlinear propagation than Nyquist-WDM [45]. Table 23.4 depicts a chart based on the performance comparison between Co-OFDM and Nyquist-WDM techniques.

TABLE 23.4

Comparison Between Co-OFDM and Nyquist-WDM Approach of Superchannel Generation [15]

Superchannel Generation Technique	Co-OFDM	Nyquist-WDM
Frequency locking needed	Yes	No
Transmission Bandwidth (BW)~B	Yes	Yes
ADC speed	1.5 B	2 B
DAC, DSP needed	Yes	Yes, with optical filters
Supporting data rate	Low	High

23.5 CONCLUSION

In this chapter, we have discussed one popular flexi-grid technology, the superchannel. In today's environment, when the whole world has shifted toward the digital era, it is mandatory to satiate huge bandwidth requirement demand from exponentially growing data traffic in limited available optical fiber bandwidth. We presented how, instead of using fixed ITU-T grid technology, flexi-grid technology, which is a next-generation technology, enhances the bandwidth utilization factor, increasing SE of the system. Also, we explained two superchannel generation strategies, i.e., Nyquist-WDM and Co-OFDM techniques. We presented a detailed literature survey as well as a comparison between both techniques.

REFERENCES

1. Cisco, "2020 global networking trend report," 1–95 (2020). www.https://daisyuk.tech/wp-content/uploads/2020/04/Cisco-2020-Global-Networking-Trends-Report.pdf
2. Divya Sharma, "Performance evaluation of advanced modulation techniques for lower and higher optical transmission rates," PhD thesis, Department of Electronics and Communication, Prayagraj, India: Motilal Nehru National Institute of Technology Allahabad (2018).
3. Sergejs Makovejs, "High-speed optical fiber transmission using advanced modulation formats," PhD thesis, Department of Electronic and Electrical Engineering, London, UK: University College London (2011).
4. D. van den Borne, V. A. Sleiffer, M. S. Alfiad, S. L. Jansen, and T. Wuth, "POLMUX-QPSK modulation and coherent detection: the challenge of long-haul 100G transmission," in 35th European Conference on Optical Communication, Conference (ECOC 2009), Vienna, Austria, 1–4 (September 2009).
5. Seb J. Savory, "Digital filters for coherent optical receivers," Optica Express, 16(2), 804–817 (2008).
6. K. C. Kao, and G. A. Hockham, "Dielectric-fiber surface waveguides for optical frequencies," Proceedings of IEEE, 113(7), 1151–1158 (1966).
7. G. Bosco, V. Curri, A. Carena, P. Poggiolini, and F. Forghieri, "On the performance of Nyquist-WDM terabit superchannels based on PM-BPSK, PM-QPSK, PM-8QAM or PM-16QAM subcarriers," Journal of Lightwave Technology, 29(1), 53–61 (2010).
8. M. Jana, L. Lampe, and J. Mitra, "Design of time-frequency packed WDM superchannel transmission systems," Journal of Lightwave Technology, 38(24), 6719–6731 (2020).
9. M. Jana, L. Lampe, J. Mitra, W. Jin, and K. Law, "Probabilistic shaping in time-frequency-packed terabit superchannel transmission," IEEE Photonics Technology Letters, 32(17), 1065–1068 (2020).
10. Y. Ujjwal, and J. Thangaraj, "Review and analysis of elastic optical network and sliceable bandwidth variable transponder architecture," Optical Engineering, 57(11), 110802 (2018).
11. Y. R. Zhou, and K. Smith, "Practical innovations enabling scalable optical transmission networks: real-world trials and experiences of advanced technologies in field deployed optical networks," Journal of Lightwave Technology, 38(12), 3106–3113 (2020).
12. S. Kumar, R. Egorov, K. Croussore, M. Allen, M. Mitchell1, and B. Basch, "Experimental study of intra- vs. inter-superchannel spectral equalization in flexible grid systems," in OFC/NFOEC Technical Digest 2013 OSA (2013).
13. N. Sharma, S. Agrawal, and V. Kapoor, "Performance optimization of OADM based DP-QPSK DWDM optical network with 37.5 GHz channel spacing," Optical Switching and Networking, 100606 (1-11) (2021).
14. Y. Chen, Y. Huang, Y. Han, J. Fu, K. Li, Y. Li, and J. Yu, "A novel phase noise suppression scheme utilizing gaussian wavelet basis expansion for PDM CO-OFDM superchannel," IEEE Photonics Journal, 12(2), 1–9 (2020).
15. S. Chandrasekhar, and X. Liu, "OFDM based superchannel transmission technology," Journal of Lightwave Technology, 30(24), 3816–3823 (2012).
16. S. Chandrasekhar, and X. Liu, "Experimental investigation on the performance of closely spaced multicarrier PDM-QPSK with digital coherent detection," Optica Express, 17, 12350–12361 (2009).
17. Y. Ma, Q. Yang, Y. Tang, S. Chen, and W. Shieh, "1-Tb/s per channel coherent optical OFDM transmission with subwavelength bandwidth access," in Optical Fiber Communication Conference and National Fiber Optic Engineers Conference (2009).
18. B. Zhu, X. Liu, S. Chandrasekhar, D. W. Peckham, and R. Lingle, "Ultra-long-haul transmission of 1.2-Tb/s multicarrier no-guard-interval CO-OFDM superchannel using ultra-large-area fiber," IEEE Photonics Technology Letters, 22(11), 826–828 (2010).

19. P. M. Anandarajah, R. Zhou, R. Maher, M. Deseada G. Pascual, F. Smyth, V. Vujicic, and Liam P. Barry, "Flexible optical comb source for super channel systems," in 2013 Optical Fiber Communication Conference and Exposition and the National Fiber Optic Engineers Conference, 8 (2013).

20. X. Liu, S. Chandrasekhar, X. Chen, P. J. Winzer, Y. Pan, T. F. Taunay, B. Zhu, M. Fishteyn, M. F. Yan, J. M. Fini, E. M. Monberg, and F.V. Dimarcello, "1.12-Tb/s 32-QAM-OFDM superchannel with 8.6-b/s/Hz intrachannel spectral efficiency and space-division multiplexed transmission with 60-b/s/Hz aggregate spectral efficiency," Optica Express, 19, B958–B964 (2011).

21. J. Yu, X. Zhou, M. F. Huang, D. Qian, P. N. Ji, T. Wang, and P. Magill, "400 Gb/s (4 × 100 Gb/s) orthogonal PDM-RZ-QPSK DWDM signal transmission over 1040 km SMF-28," Optica Express, 17(20), 17928–17933 (2009).

22. An Li, X. Chen, G. Gao, and W. Shieh, "Transmission of 1 Tb/s unique-word DFT-spread OFDM super-channel over 8000 km EDFA-only SSMF link," Journal of Lightwave Technology, 30(24), 3931–3937 (2012).

23. T. Tanimura, T. Kato, R. Okabe, S. Oda, T. Richter., R. Elschner, C. Schmidt-Langhorst, C. Schubert, J. C. Rasmussen, and S. Watanabe, "Coherent reception and 126 GHz bandwidth digital signal processing of CO-OFDM superchannel generated by fiber frequency conversion," in Optical Fiber Communication Conference (2014).

24. M. Song, E. Pincemin, V. Vgenopoulou, I. Roudas, E. M. Amhoud, and Y. Jaouën, "Transmission performances of 400 Gbps coherent 16-QAM multi-band OFDM adopting nonlinear mitigation techniques," in 2015 Tyrrhenian International Workshop on Digital Communications, 46–48 (2015).

25. M. Song, E. Pincemin, B. Bäuerle, A. Josten, D. Hillerkuss, J. Leuthold, and I. Tomkos, "Fibre nonlinearity limitations of 1 Tbps (10 × 100 Gbps) multi-band e-OFDM super-channel," Advanced Photonics 2016, OSA Technical Digest (online) (Optica Publishing Group, 2016), paper SpW1G.2.

26. Y. Geng, X. Huang, W. Cui, Y. Ling, B. Xu, J. Zhang, X. Yi, B. Wu, S.W. Huang, K. Qiu, and C. W. Wong, "Terabit optical OFDM superchannel transmission via coherent carriers of a hybrid chip-scale soliton frequency comb," Optics Letters, 43(10), 2406–2409 (2018).

27. L. N. Venkatasubramani, Y. Lin, C. Browning, A. Vijay, F. Smyth, R. D. Koilpillai, L. P. Barry, and D. Venkitesh, "Demonstration of 608 Gbps CO-OFDM transmission using gain switched comb," in CLEO: Science and Innovations (2020).

28. C. Liu, J. Pan, T. F. Detwiler, A. J. Stark, Y. T. Hsueh, G. K. Chang, and S. E. Ralph, "Joint ICI cancellation based on adaptive cross-channel linear equalizer for coherent optical superchannel systems," in Signal Processing in Photonic Communications (2012).

29. J. Yu, Z. Dong, H. C. Chien, Z. Jia, X. Li, D. Huo, M. Gunkel, P. Wagner, H. Mayer, and A. Schippel, "Transmission of 200 G PDM-CSRZ-QPSK and PDM-16 QAM with a SE of 4 b/s/Hz," Journal of Lightwave Technology, 31(4), 515–522 (2013).

30. D. Sharma, S. Devi, and Y. K. Prajapati, "832.5 Gb/s PM-8QAM superchannel with 5 b/s/Hz spectral efficiency," in David Harvey, Haranath Kar, Shekhar Verma, Vijaya Bhadauria (Eds)Advances in VLSI, Communication, and Signal Processing, 67–74, Singapore: Springer (2020).

31. M. Salsi, H. Mardoyan, P. Tran, C. Koebele, E. Dutisseuil, G. Charlet, and S. Bigo, "155 × 100 Gbit/s coherent PDM-QPSK transmission over 7,200 km," in 35th European Conference on Optical Communication (ECOC 2009), Vienna, Austria, 1–2 (2009).

32. D. Sharma, and Y. K. Prajapati, "Terabit Nyquist superchannel transmission using PM-QPSK subchannels," in 2018 3rd International Conference on Microwave and Photonics (ICMAP), 1–2 (2018).

33. G. Gavioli, E. Torrengo, G. Bosco, A. Carena, S. J. Savory, F. Forghieri, and P. Poggiolini, "Ultra-narrow-spacing 10-channel 1.12 Tb/s D-WDM long-haul transmission over uncompensated SMF and NZDSF," IEEE Photonics Technology Letters, 22(19), 1419–1421 (2010).

34. H. C. Chien, J. Yu, Z. Jia, Z. Dong, and X. Xiao, "Performance assessment of noise-suppressed Nyquist-WDM for Terabit superchannel transmission," IEEE Journal of Lightwave Technology, 30(24), 3965–3971 (2012).

35. K. Igarashi, T. Tsuritani, I. Morita, Y. Tsuchida, K. Maeda, M. Tadakuma, T. Saito, K. Watanabe, K. Imamura, R. Sugizaki, and M. Suzuki, "Super-Nyquist-WDM transmission over 7,326-km seven-core fiber with capacity-distance product of 1.03 Exabit/s·km," Optica Express, 22(2), 1220–1228 (2014).

36. X. Liu, H. Hu, S. Chandrasekhar, R. M. Jopson, A. H. Gnauck, M. Dinu, C. Xie, and P. J. Winzer, "Generation of 1.024-Tb/s Nyquist-WDM phase-conjugated twin vector waves by a polarization-insensitive optical parametric amplifier for fiber-nonlinearity-tolerant transmission," Optica Express, 22(6), 6478–6485 (2014).

37. T. Xu, J. Li, G. Jacobsen, S. Popov, A. Djupsjobacka, R. Schatz, Y. Zhang, and P. Bayvel, "Field trial over 820 km installed SSMF and its potential Terabit/s superchannel application with up to 57.5-Gbaud DP-QPSK transmission," Optics Communications, 353, 133–138 (2015).

38. C. Zhu, B. Song, L. Zhuang, B. Corcoran, and A. J. Lowery, "Subband pairwise coding for robust Nyquist-WDM superchannel transmission," Journal of Lightwave Technology, 34(8), 1746–1753 (2016).

39. D. Sharma, Y. K. Prajapati, and R. Tripathi, "0.55 Tb/s heterogeneous Nyquist-WDM superchannel using different polarization multiplexed subcarriers," Photonic Network Communications, 39, 120–128 (2020).

40. M. Singh, J. Malhotra, M. M. Rajan, V. Dhasarathan, and M. H. Aly, "Performance evaluation of 6.4 Tbps dual polarization quadrature phase shift keying Nyquist-WDM superchannel FSO transmission link: Impact of different weather conditions," Alexandria Engineering Journal, 59(2), 977–986 (2020).

41. D. Sharma, Y. K. Prajapati, and R. Tripathi, "Success journey of coherent PM-QPSK technique with its variants: a survey," IETE Technical Review, 37, 136–55 (2020).

42. J. Zhang, Z. Dong, H. C. Chien, Z. Jia, Y. Xia, and Y. Chen, "Transmission of 20×440-Gb/s super-Nyquist filtered signals over 3600 km based on single-carrier 110-GBaud PM QPSK with 100-GHz grid," in Optical Fiber Communication Conference (OFC-2014) (2014).

43. D. Sharma, Y. K. Prajapati, and R. Tripathi, "Spectrally efficient 1.55 Tb/s Nyquist-WDM superchannel with mixed line rate approach using 27.75 Gbaud PM-QPSK and PM-16QAM," Optical Engineering, 57(7), 1–12 (2018).

44. S. Devi, D. Sharma, Y. K. Prajapati, and R. Tripathi, "Independent and mixed transmission of 166.5Gb/s PM-8QAM and 222Gb/s PM-16QAM Nyquist-WDM superchannel for long haul metro network," International Journal of Communication Systems, 34(7), e4735 (2021). DOI: https://doi.org/10.1002/dac.4735

45. G. Bosco, A. Carena, V. Curri, P. Poggiolini, and F. Forghieri, "Performance limits of Nyquist-WDM and CO-OFDM in high-speed PM-QPSK systems," IEEE Photonics Technology Letters, 22(15), 1129–1131 (2010).

46. S. Benedetto, and E. Biglieri, Principles of Digital Transmission: With Wireless Applications, New York: Kluwer (1999).

47. Z. Jia, J. Yu, H. C. Chien, Z. Dong, and D. D. Huo, "Field transmission of 100 G and beyond: multiple baud rates and mixed line rates using Nyquist-WDM technology," IEEE Journal of Lightwave Technology, 30(24), 3793–3804 (2012).

Index